Geodynamics Series

Geodynamics Series

Earth's Core
Dynamics, Structure, Rotation

Véronique Dehant
Kenneth C. Creager
Shun-ichiro Karato
Stephen Zatman
Editors

Geodynamics Series   Volume 31

American Geophysical Union
Washington, D.C.

Library of Congress Cataloging-in-Publication Data
Earth's core : dynamics, structure, rotation / Veronique Dehant ... [et al.].
     p. cm. -- (Geodynamics series ; v. 31)
Papers from a Union Session organized at the fall 2000 AGU meeting in San Francisco.
Includes bibliographical references.
ISBN 0-87590-533-1
    1. Earth--Core. 2. Geodynamics. I. Dehant, Veronique, 1959- II. Series.

QE509.2 .E26 2003
551.1'12--dc21   2002038311

ISBN 0-87590-533-1
ISSN 0277-6669

CONTENTS

Geodesy

PREFACE

The Earth's core, which is made of liquid iron alloy solidified at its center, is the most dynamic part of our planet. Because the planet's magnetic field is generated within the core, the structure and temporal variation of the magnetic field carry important information on core dynamics. Seismology, along with experimental and theoretical mineral physics, has provided important perspectives on core structure and its mineralogy. Recently, however, these fields have seen important and exciting developments. We now know that the Earth's core plays a major role for Earth rotation at decadal time-scales as it does for diurnal timescales, as seen from the resonances it induces in nutation.

Seismology, high-pressure and high-temperature mineral physics, geochemistry, geomagnetism, geodynamics, Earth rotation, and geodesy are disciplines through which scientists can investigate core structure and dynamics. In an effort to promote this interdisciplinary approach, we have prepared the current volume for experts and students of the science to help each gain an overview of what we presently know about the core through review and research papers with latest data.

The origin of this work derives from a Union Session that we organized at the Fall 2000 AGU meeting in San Francisco: "Core Dynamics, Structure, and Rotation." The large attendance at the session and the general interest of the scientific community in the field initially encouraged us to edit this volume, which includes both contributions from session participants and invited complementary papers from noted scientists in their respective fields.

During the final phase of the preparation of the book, however, we suffered the loss of one of our editors, Stephen Zatman, who died suddenly in an automobile accident. Shocked and sadden by this tragedy, we have dedicated our book to Stephen (see "Dedication").

We thank David Bercovici, our AGU Oversight Editor, and the many manuscript reviewers, who helped us to sustain an outstanding level of accomplishment throughout this monograph. Finally, we thank AGU's Books Department staff—including our acquisitions editor, Allan Graubard, and production editor, Bethany Matsko—for their help in the preparation of this monograph.

Véronique Dehant
Kenneth C. Creager
Shun-ichiro Karato
Stephen Zatman

Stephen Zatman
Dedication

Stephen started graduate school at Harvard University in 1993, after doing his undergraduate work in physics at Cambridge University. With his enormous thirst to understand, he made an immediate impression on everyone with whom he came into contact. His keen sense of humor and ability to laugh at himself made endearing a somewhat "attack-dog" approach, which in any case mellowed with time. His quest for understanding never diminished.

Stephen's interest in his colleagues' work was deep and sincere. He could be counted on to help with others' problems. This carried over to his teaching. Some of us remember fondly spending one morning with a heavy top discussing various explanations of precession and nutation suitable for beginning geophysics students.

Graduate students vary greatly in the amount of supervision that they need, or which is pushed upon them. For Stephen, little was necessary, and Stephen was able to develop his research on his own, generating truly innovative work in the study of core dynamics (sub-decadal length of day variation, and dynamics of torsional oscillations within the core).

Subsequent to leaving Harvard, he developed several other research avenues, in particular his work on plate tectonics. Stephen developed an original and extremely useful theoretical formulation describing the force balance involved in intraplate deformation, leading to a straightforward explanation for why intraplate deformation usually involves a combination of extension and compression. A series of elegant papers on this subject by Stephen establishes a new line of inquiry about the evolution of plate boundaries. Such work in an area almost completely unconnected to his PhD thesis work is evidence of a scientist who was on the verge of "breaking out" into the broad community. It is a tragedy that we will now not be able to see this happen.

For the past year, Stephen had been an assistant professor in the Department of Earth and Planetary Sciences at Washington University. Stephen impressed everyone with his intelligence and his kindness. Stephen's research was heading in many interesting directions, including correlating high-frequency Oersted satellite geomagnetic data anomalies with large earthquakes. At a presentation to prospective graduate students, Stephen took thrice the time of any other faculty member. Even through he spoke at breakneck speed, he genuinely had a huge number of fascinating projects in mind for which he was seeking graduate students.

The areas of geophysics in which he was working will be weakened by his loss. Of course, the personal tragedy is much greater. He will be deeply missed, obviously by his colleagues, but more importantly by his family, in particular his wife and baby daughter.

Michael Bergman
Simons' Rock College

Jeremy Bloxham
Harvard University

Richard Holme
University of Liverpool

Mark Richards
University of California, Berkeley

Doug Wiens
Washington University of St. Louis

Michael Wysession
Washington University of St. Louis

Introduction

Véronique Dehant[1], Kenneth C. Creager[2], Shun Karato[3], Stephan Zatman[4]

After a general description of the core, we review the data, models, theories, hypotheses, and ideas about the core that are the subject of the book. We then delineate for the reader the different parts of the book and the papers presented within each part.

By the end of the 19th century, the existence of a dense core inside the Earth was deduced from its total mass and moment of inertia. The outer core of the Earth was discovered from seismology in 1906 by Richard Dixon Oldham. At long angular distance from an epicenter, he observed abnormally large travel times of seismic waves. His conclusion was that these waves, propagating deeply in the Earth, cross a central core, where the propagation speed is lower. In 1913, Gutenberg estimated the depth of the Core-Mantle Boundary (CMB) for the first time. In 1926, using Earth tides and seismic observations, Sir Harold Jeffreys showed that the core has a much smaller rigidity than the mantle and could be liquid. By then, it was already known from studies of waves associated with earthquakes, that there exists a zone of angular distances where no P wave arrives; this shadow zone is caused by the refraction of the P waves at the transition from the solid mantle to the fluid core. In 1936, Inge Lehmann interpreted waves that were nevertheless observed in the shadow zone as P waves that were refracted at a 5000 km depth discontinuity (PKIKP waves in modern terminol-ogy); there was thus inside the fluid core a region with different properties. This is the inner core, whose solidity was suggested in the 1940ies by Edward Bullen and proven in 1971 by A.M. Dziewonski and F. Gilbert using observations from the Earth's free oscillations.

The outer core is made of liquid metallic alloy: mainly Fe and Ni with some additional lighter elements such as oxygen and sulfur with a thickness of about 2900 km. The surface separating the fluid outer core from the mantle -- the core-mantle boundary (CMB) -- is characterized by a discontinuity such that its density jump is greater than between the "solid" Earth and the atmosphere, and its viscosity jump is the same as between the "solid" Earth and the oceans.

As Earth cools, the core solidifies, growing the solid inner core whose radius is currently about 1200 km. The inner core is made of a solid Fe-Ni alloy and may be partially molten. The surface separating the solid inner core from the liquid outer core is the inner core boundary (ICB).

Due to the temperature and pressure of the ICB, the liquid outer core is solidifying, producing latent heat and an increased concentration of light elements in the outer core. Due to the buoyancy associated with heating and chemical differentiation, the liquid outer core also undergoes convective motions, which enable it to generate a magnetic field.

This volume provides a synthesis of our understanding of the Earth's core from the viewpoint of several different fields (geomagnetism and dynamo theory, seismology, geodesy, and mineral physics), and the rapid progress of the last few years, which has largely resulted from their interrelation. The book contains four parts to provide the reader with a complementary view about the core from different disciplines.

[1] *Royal Observatory of Belgium, Brussels, Belgium*

[2] *Department of Earth and Space Sciences, University of Washington, USA*

[3] *Department of Geology, University of Minnesota, Minneapolis, USA*

[4] *Department of Earth and Planetary Sciences, Washington University in St Louis, USA*

Earth's Core: Dynamics, Structure, Rotation
Geodynamics Series 31
Copyright 2003 by the American Geophysical Union
10.1029/31GD01

The reader will find information about the core gathered from seismic data (normal mode observations, travel time, seismic velocities, refraction and diffraction of waves), from mineral rock physics (from laboratory experiment and quantum mechanical calculations), from geomagnetism (observation of the magnetic field, derived flow at the CMB, dynamo theory), and from deductions based on geodetic global observation.

Seismology provides observations to infer the three-dimensional variations in anisotropic elastic wavespeeds, density and attenuation factors, as well as the variation of wavespeeds over time during the past few decades. Three types of seismic data have been extensively analyzed to image the solid inner core: (1) *absolute travel times* of the phase PKP(DF), which propagates through the inner core as a compressional wave; (2) *differential travel times,* which are typically measured on a given seismogram by cross correlation between the phase PKP(DF) and a reference phase that has a similar path, but does not enter the inner core; and (3) *eigenfrequencies of Earth's normal modes* of vibration, which provide a more uniform sampling of the volume of the inner core.

There is broad agreement among seismologists analyzing all three wave types that the solid inner core is anisotropic – compressional waves propagating parallel to Earth's spin axis travel faster than those propagating in the plane perpendicular to this axis. Two papers by *Song* and by *Souriau and Poupinet* provide comprehensive reviews of constraints, primarily from differential travel times, on the 3-D variability of anisotropy within the inner core, as well as evidence that the isotropic average does not vary laterally. They each discuss evidence in favor of an isotropic uppermost inner core, with variable thickness ranging from about 400 km in the "eastern hemisphere" (40-180 deg E) to 100 km in the "western hemisphere", underlain by a strongly anisotropic lower inner core. This model is supported by new observations of *Li and Richards*. *Romanowicz et al.* combine new observations of high-quality absolute- and differential-time data to existing data sets to evaluate alternative explanations. Combining a simple model of inner core anisotropy with deep mantle heterogeneity, however, does not allow us to explain all the data. Nonetheless, by adding a model characterized by wave speeds that are 1% faster within the outer core "tangent cylinder" (axi-symmetric cylinder tangent to the inner core), scientists can explain the majority of observations without the need to invoke longitudinal variability in the inner core.

Four papers address constraints on the rate of relative rotation between the inner core and the mantle. *Laske and Masters* present a through analysis of constraints from normal mode "splitting functions" which are sensitive to large-scale lateral variations in inner core structure. Longitudinal shifts in their patterns measured from earthquakes from 1977 to 2001 suggest that the inner core is rotating slightly faster than the mantle at a rate of 0.13 to 0.11 deg/year. *Song* and *Souriau and Poupinet* each provide reviews of inner core rotation studies. *Song* analyzes sources of potential bias and concludes that the differential times provide strong evidence that the inner core is rotating faster than the mantle, but the rate remains uncertain (0.15 to 1.1 deg/year). In contrast, *Souriau and Poupinet,* by analyzing similar data and their potential biases, infer that existing seismological observations limit the relative rotation rate to be less than 0.2 deg/year, and that 0.0 cannot be ruled out. *Li and Richards* analyze a new set of differential times from nuclear explosions in Novaya Zemlya recorded at stations in Antarctica. Their observations, which have the important advantage of known source locations and sources close to one another, support the hemispherical model of inner core anisotropy and provide evidence in favor of inner core rotation.

Crossley examines the effect of a stably stratified layer in the core near the core-mantle boundary (CMB) and near the inner core on seismic normal modes as well as on the Chandler wobble (free oscillation of the Earth that appears in polar motion). This issue is relevant to Buffet and colleagues' recent proposal that a layer of sediments at the top of the core explains the conductivity needed to reconcile an electromagnetic coupling of the CMB with the observed nutation data (see *Dehant and Mathews*, in this volume). The possibility of a stably-stratified density profile has important implications for geodynamo models.

Mineral Physics observations provide key data to infer the compositions and geodynamic significance of geophysical observations. The composition of the core can be inferred through the comparison of seismological observations with laboratory and theoretical observations on properties of candidate materials. A phase diagram is the basis for such a discussion, which *Anderson* aptly summarizes. His analysis, which is based on a synthesis of experimental and theoretical observations, provides a refined phase diagram including new estimates of melting temperature and the density deficit of the outer core. For his part, *Saxena* discusses thermodynamic properties, which form the basis for phase diagram and other parameters such as the adiabatic temperature gradient. *Steinle-Neumann et al.* then present a detailed account of our current understanding of the physical properties of core materials with emphasis on the elasticity of e-iron. All of these studies support a notion that the inner core is likely to be made mostly of solid e-iron, which has highly anisotropic physical properties. Finally, *Bergman* provides an important review of current knowledge on the solidification of alloys with special reference to texture development through solidification at the inner-outer core boundary, an important process that may control the structure of the inner core.

Geomagnetism and Dynamo Theory elucidate the physical nature of the core through observation and theoretical consideration of the magnetic field on both long timescales (thousands and millions of years) and shorter timescales (such as decades).

Considering short timescales, one frequent topic of concern has been the process of physical coupling between the core and the mantle that causes decadal fluctuations in the length of the day and perhaps in polar motion. Here, a paper by *Kuang and Chao* addresses this topic in a theoretical manner, by examining the strength of different plausible coupling mechanisms in a numerical dynamo model.

A paper by *Zatman* considers the constraints implied by variations in the flow at the top of the core (estimated from observational geomagnetism) on the dynamics of the "tangent cylinder", the portion of the outer core most tightly coupled to the inner core.

Measurements of the field on longer timescales (the last 5 million years) inspire considerations of the behaviour of the dynamo on long timescales in a paper by *Gubbins*. Resemblances between different types of measurements (the time averaged magnetic field, polarity reversal pathways, seismology) suggests the possibility of a unified explanation for each of them, and here *Gubbins* proposes that this is due to thermal interactions between the core and mantle.

In a related vein, *Sumita and Yoshida* present a theoretical model for the evolution of the core due to thermal interactions between the various components (the mantle, inner core, and outer core), determining the implications for the structures of the inner and outer cores and a possible culprit for the anisotropy of the inner core.

Considering both long and short timescales, *Hollerbach* provides a framework for understanding the different timescales of the geodynamo, the processes that work on these timescales, and the mutual dependences.

Geodesy provides observations of length-of-day variations at decadal time scale. As *Ponsar et al.* show, however, it is possible to compare these measurements with computations performed using different hypotheses for the flow and the observed magnetic field.

Geodesy also provides observations of precession and nutations; that is, motion of the Earth rotation axis in Space. Precession is the secular motion of the rotation axis in space due to lunisolar attraction and, in more mi-nor ways, by the attraction of the other planets. Nutations are additional periodic motions induced by the variations of the relative position between the Earth and the Moon, the Sun, and eventually the other planets. Due to resonance in these motions with a normal (Free Core Nutation) mode, it is possible to obtain information about the core, and in particular on the viscosity of the inner core and on the electromagnetic coupling at the core-mantle boundary, as explained in *Dehant and Mathews'* paper.

Gravimetric observations employ super-conducting gravimeters (a small ball levitated by an electromagnetic field that responds to gravity variations, whether in the position of the ball or in the electric field required to maintain the ball at the same position) to measure gravity and its variations over time. The latter will be very accurate but contain local contributions such as pressure loading and attraction. A stacking of data from a network of superconducting gravimeters allows eliminating the local effects and provides us with useful information such as the period and damping of the Free Core Nutation (FCN). The results corroborate the FCN period deduced from nutation.

In conclusion: In view of the rapid progress encountered these last few years, it is important to present the reader with the current state of the art of particular questions or issues regarding the core from different viewpoints, and to illuminate the core from different disciplines. This book is a collection of up-to-date review and research papers in the different subjects in order to provide scientists and students with all information needed for understanding the core, its dynamics, structure, and rotation. By providing the reader with all aspects that should be treated for a better understanding of the core, we hope that this volume opens new doors onto the future.

V. Dehant, Royal Observatory of Belgium, 3 avenue Circulaire, B-1180 Brussels, Belgium. (e-mail: Dehant@oma.be)

C. Kreager, Geophysics, University of Washington, Seattle WA98195, USA. (e-mail: kcc@geophys.washington.edu)

S. Karato, Department of Geology, University of Minnesota, Minneapolis MN55455, USA.
(e-mail: shun-ichiro.karato@yale.edu)

S. Zatman, Department of Geology and Geophysics, University of California Berkeley, California 94720, USA

The Earth's Free Oscillations and the Differential Rotation of the Inner Core

Gabi Laske and Guy Masters

IGPP, Scripps Institution of Oceanography, University of California, San Diego, La Jolla, California

Differential rotation of the inner core has been inferred by several body-wave studies with most agreeing that a superrotation may exist with a rate between 0.2° and 3° per year. The wide range of inferred rotation rates is caused by the sensitivity of such studies to local complexities in structure which have been demonstrated to exist. Free-oscillation "splitting functions" are insensitive to local structure and are therefore better candidates for estimating differential IC rotation more accurately. We use a recently developed method for analyzing free oscillations which is insensitive to earthquake source, location and mechanism to constrain this differential rotation. In a prior study, we found that inner core differential rotation has been essentially zero over the last 20 years. We revisit this issue, including additional earthquakes and modes in our analysis. Our best estimate is a barely significant superrotation of 0.13±0.11°/yr, which is still consistent with the idea that the inner core is gravitationally locked to the mantle.

1. INTRODUCTION

The Earth's inner core enjoyed sudden public interest as "the planet within a planet" when seismologists found the first seismic evidence that it rotates faster than the mantle (Song and Richards, 1996; Su et al., 1996). The discovery was a timely one as some modern geodynamo simulations predicted this superrotation (Glatzmaier and Roberts, 1996), and earlier studies had already speculated that the inner core is likely rotating due to coupling through magnetic torques (e.g. Gubbins, 1981). Not all geodynamo calculations predict a superrotation though. For example, Kuang and Bloxham's (1997) dynamo calculations show that the inner core sometimes rotates faster and sometimes slower than the mantle which seems to be consistent with length–of–day observations (Buffett and Creager, 1999). The early seismic evidence came from the observation that differential body wave travel times between phases that turn deep in the outer core, PKP(BC), and phases that penetrate into the inner core, PKP(DF), change with time. Song and Richards

(1996) observed a 0.3s change over 30 years for differential times measured for paths from a source region in the South Sandwich Islands (SSI) in the Atlantic Ocean to global seismic network (GSN) station COL (College, Alaska). Using certain assumptions about the structure of the inner core, this change in time was converted into a 1°/yr superrotation of the inner core. The assumptions involved are actually quite strict and turned out later to be too simplistic. Though the physical cause of inner core anisotropy is not yet well understood (see e.g. Jeanloz and Wenk, 1988; Karato, 1993; Yoshida et al., 1997; Bergman, 1997), it was assumed that the inner core behaves roughly like a single anisotropic crystal with a fast symmetry axis closely but not exactly aligned with the rotation axis of the Earth. The slight tilt of the symmetry axis which has been inferred by several groups (Su and Dziewonski, 1995; Shearer and Toy, 1991; McSweeney et al., 1997; Song, 1997) allows any differential rotation of the inner core to manifest itself in temporal variations of the travel times of body waves emanating from a fixed source region and recorded by a fixed receiver. The idea of the inner core behaving like a single crystal was quickly questioned when temporal variations were found for some paths (e.g. Song and Richards, 1996 for the path from the SSI source region to station COL) but not for others (e.g. Souriau, 1998a for the path from the Novaja Zemlya nuclear test site to GEO-

Earth's Core: Dynamics, Structure, Rotation
Geodynamics Series 31
Copyright 2003 by the American Geophysical Union
10.1029/31GD02

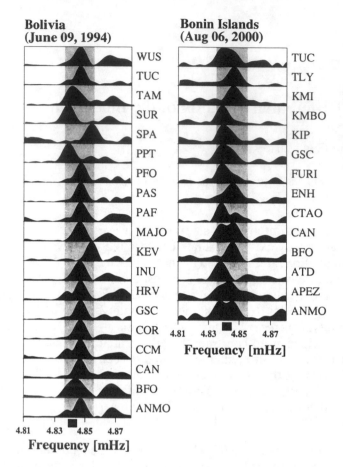

Bolivia (June 09, 1994)

Bonin Islands (Aug 06, 2000)

Figure 1. Hanning-tapered spectra of anomalously split inner-core sensitive mode $_{13}S_2$ from vertical component recordings of the Bolivian (June 09, 1994) and the Bonin Islands (August 06, 2000) earthquakes. The spectra are taken at stations of various global and regional seismic broad–band networks (GEOSCOPE, IRIS-IDA, IRIS-USGS, GEOFON and TERRAScope), starting 5h after the event and using 65h long records. Due to 3D structure of the Earth, the "spectral peak" exhibits fine–scale splitting, within a band defined as the splitting width (grey area). Some spectra even exhibit clearly split peaks (ANMO, CCM). Rotation and hydrostatic ellipticity of the Earth cause a splitting width of only $6.7\,\mu Hz$ (black bar at axis). The smaller peaks at 4.87mHz are the faster decaying mode $_9S_7$.

SCOPE station DRV, Dumont D'Urville in Antarctica). We now know that the path from SSI to COL is highly anomalous as it passes through regions of extremely heterogeneous structure having some of the largest gradients found so far (Creager, 1997). If unaccounted for, even simple large-scale heterogeneous structure can severely bias results for inner core rotation rates (Souriau, 1998b). Taking lateral heterogeneity into account Creager (1997) estimated the superrotation to be around 0.25°/yr which is significantly lower than initial estimates. Creager (2000) re-analyzed a large dataset of differential travel times and now sets the *lower* limit of the IC rotation rate to 0.15°/yr which is actually consistent

with his observations of 1997. This result would also be consistent with the small rate of 0.15°/yr found in a study that investigated the time–dependence of inner core scattered waves generated by sources at the Novaya Zemlya test site and observed at the LASA (Large Aperture Seismic Array) in Montana (Vidale et al., 2000). Song (2000) re-analyzed his dataset in a joint inversion for inner core structure and superrotation and now finds that the superrotation is 0.6°/yr which is less than 1°/yr but still the highest recently published value.

The process of measuring the time-dependence of the BC-DF differential times appears difficult for at least two reasons: 1) the data depend on the accurate location of earthquakes and it is certainly true that the locations of earlier events are less well known; 2) measuring the DF travel time for polar paths (e.g. the SSI-COL path) seems a difficult endeavor because the DF phase is often diffuse and quite small. A recent study suggests that the strategy of analyzing doublet events for which the waveforms are practically identical (Poupinet et al., 2000) instead of looking at the time-dependence of differential travel times is a more promising approach. Unfortunately, it seems that the definition of a doublet event is non-unique and the discovery of a time-dependent signal appears to be a matter of the analysis technique applied, the reference models used, and assumptions applied in regard of the polarity of the analyzed signals (Song, 2001; Poupinet and Souriau, 2001).

Another seismic dataset that is sensitive to inner core structure and hence has the potential of constraining inner core rotation is the normal mode dataset. It has long been known that compressional waves that travel parallel to the spin axis are significantly faster, arriving about 2 seconds earlier, than waves that travel in the equatorial plane (Poupinet et al., 1983). These observations eventually led to the discoveries described above. It has also been known, on the other hand, that free oscillations which sample the inner core are strongly split (Masters and Gilbert, 1981; see also Figure 1) by a structure which is dominantly axisymmetric and mimics the effect of an excess ellipticity of the Earth. Workers at Harvard (Woodhouse et al., 1986; Morelli et al., 1986) inferred that anisotropy of the inner core was the main reason for the anomalous observations in both mode splitting as well as body wave travel times. Normal modes provide a powerful tool to constrain differential inner core rotation that is, in certain ways, superior to the body wave method. Free oscillations are natural low-pass filters of 3D structure so long-wavelength phenomena, such as IC rotation, are prime study targets. Free oscillations "see" the Earth as a whole, so the observation of how a free oscillation splitting pattern changes with time and any inference on IC rotation is not biased by effects from localized structures. It is also not necessary to know the physical cause of these patterns (anisotropy or heterogeneity). All that needs to be observed is if they change with time.

The credit of being the first to study IC rotation with normal modes goes to Sharrock and Woodhouse (1998). They investigated the fit of "splitting functions" (see below) to the data of 5 inner-core sensitive modes under the assumption of a rotating inner core and inferred a *westward* rotation rate of 1 to more than 2°/yr, which is obviously inconsistent with the body wave observations. We recently developed a new technique to analyze free oscillation splitting (Masters et al., 2000a) and re-examined inner–core sensitive modes for differential rotation. A convenient feature of this technique is that it is insensitive to errors in source location and mechanism. The analysis of nine modes resulted in a best–fitting IC rotation rate of $0.01 \pm 0.21°$/yr eastward and we concluded that the inner core is most likely gravitationally locked to the mantle (Laske and Masters, 1999). The brevity of that paper did not allow us to discuss some of the details of the method or extend our discussion to subjects that are of obvious concern. For example, inner–core sensitive modes are also quite sensitive to mantle structure and it was suggested that our results depend on the "mantle correction" we are using to study the inner core (Creager, 2000). Here, we demonstrate in a synthetic test that our forward modelling strategy is capable of recovering an inner core rotation, if it exists. We also show that there is indeed some variability in the results when using different global mantle tomographic models but that the results are actually remarkably consistent. It turns out that inconsistencies in inferred rotation rates seem to be mode–specific rather than model–specific and we discuss possible causes for this. We also discuss cases for which our method cannot be applied. Foremost among these are strongly coupled modes and uncoupled modes that overlap in frequency with other modes of high harmonic degree.

2. SPECTRA, RECEIVER STRIPS AND SPLITTING FUNCTIONS

In this section, we briefly introduce the data and the essentials of mode seismology and summarize the autoregressive method we use to analyze mode splitting. For details about the AR method the reader is referred to Masters et al. (2000a). The 1990ies have seen a renaissance in free oscillation seismology, not least because there were more than 25 extremely large earthquakes (Table 1) that were recorded on typically 100 observatory–quality broadband seismic instruments. All of these earthquakes excited numerous free oscillation overtones that are sensitive to inner core structure. Figure 1 shows typical examples of spectra for inner-core sensitive mode $_{13}S_2$. Non-spherical structure splits the $2\ell + 1$ singlets of a mode $_nS_\ell$. The effects of rotation and hydrostatic ellipticity of the Earth split the set of 5 singlets of $_{13}S_2$ by 6.7μHz. When comparing the spectral lines at stations PPT (Papete, Tahiti) and SPA (South Pole) it becomes immediately clear that the actual splitting of this mode is much larger (more that 15μHz), i.e. this mode is anomalously split due to strong heterogeneity within the Earth. Because the

shape of the spectra depends on the source–receiver geometry as well as the source mechanism, the peaks vary for different earthquakes at the same station (e.g. compare the spectra for stations ANMO or BFO). For each earthquake and each mode, we apply a trick to collapse the information contained in the roughly 100 time series (or spectra) into a set of only $2\ell + 1$ "receiver strips", without losing any information about 3D structure.

Our starting point is the representation of the time series of an isolated split multiplet at station j, first given by Woodhouse and Girnius (1982):

$$u_j(t) = \sum_{k=1}^{2\ell+1} R_{jk} a_k(t) e^{i\bar{\omega} t} \quad \text{or} \quad \mathbf{u}(t) = \mathbf{R} \cdot \mathbf{a}(t) e^{i\bar{\omega} t} \quad (1)$$

where the real part is understood. The j'th row of $\mathbf{R}$ is a $2\ell+1$ vector of spherical harmonics which describe the motion of the spherical-earth singlets at the j'th receiver and is readily calculated. $\bar{\omega}$ is the multiplet degenerate frequency and $\mathbf{a}(t)$ is a slowly varying function of time given by

$$\mathbf{a}(t) = \exp(i\mathbf{H}t) \cdot \mathbf{a}(0) \quad (2)$$

where $\mathbf{a}(0)$ is a $2\ell + 1$ vector of spherical-earth singlet excitation coefficients which can be computed if the source mechanism of the event is known. $\mathbf{H}$ is the "splitting matrix" of the multiplet and incorporates all the information about 3D structure, i.e.

$$H_{mm'} = (a + mb + m^2 c)\delta_{mm'} + \sum \gamma_s^{mm'} c_s^t \quad (3)$$

where $-\ell \leq m \leq \ell$; $-\ell \leq m' \leq \ell$ and $t = m - m'$. a, b and c describe the effects of rotation and hydrostatic ellipticity (Dahlen, 1968), $\gamma_s^{mm'}$ are integrals over three spherical harmonics which are easy to compute (e.g. Dahlen and Tromp, 1998) and the "structure coefficients", c_s^t, are given by

$$c_s^t = \int_0^a \mathbf{M}_s(r) \cdot \delta\mathbf{m}_s^t(r) r^2 \, dr. \quad (4)$$

$\delta\mathbf{m}_s^t$ are the expansion coefficients of the 3D aspherical Earth structure: $\delta\mathbf{m}(r, \theta, \phi) = \sum \delta\mathbf{m}_s^t(r) Y_s^t(\theta, \phi)$ and $\mathbf{M}_s$ are integral kernels which can be computed (Woodhouse and Dahlen, 1978; Woodhouse, 1980; Henson, 1989). Strictly speaking, equation 1 is not quite correct since both $\mathbf{R}$ and $\mathbf{a}(0)$ should include small renormalization terms (see Dahlen and Tromp, 1998, equations 14.87 and 14.88). Technically, we would need to know the splitting matrix before we can apply the renormalization which would require an iterative approach. However, for the (isolated) modes considered here, ignoring the renormalization terms leads to errors in $\mathbf{R}$ and $\mathbf{a}(0)$ on the order of a part in 10^3 and does not affect our results in any significant way. The $Y_s^t = X_s^t(\theta)e^{it\phi}$ is a spherical harmonic of harmonic degree s and azimuthal order number t. An isolated mode of harmonic degree ℓ is sensitive

Table 1. Earthquakes used in this Study

Event Name	Year.Day	Depth [km]	Moment [10^{20}Nm]	No. of records	Day since 1 Jan. 1977
Arequipa/South. Peru	2001.174	26	49	85	8940
Bonin Islands Region	2000.219	394	1.2	120	8619
South Indian Ocean	2000.170	10	7.9	133	8570
Southern Sumatera	2000.156	33	7.5	130	8556
Santiago del Estero, Argentina	2000.114	608	0.3	83	8514
USSR/China Border	1999.098	566	0.5	118	8133
Molucca Sea/Ceram Sea	1998.333	33	4.5	71	8003
Balleny Islands Region	1998.084	33	18.2	81	7754
Kamchatka	1997.339	33	5.3	110	7644
Fiji Islands	1997.287	166	4.6	88	7592
Santa Cruz Islands	1997.111	33	4.4	93	7416
Peru	1996.317	33	4.6	97	7256
Fiji Islands	1996.218	550	1.4	96	7157
Flores	1996.169	587	7.3	90	7108
Andreanof Isl., Aleutians	1996.162	33	8.1	97	7101
Irian Jaya	1996.048	33	24.1	103	6987
Minahassa Penins., Celebes	1996.001	24	7.8	91	6940
Kuril Islands	1995.337	33	8.2	83	6911
Jalisco, Mexico	1995.282	33	11.5	96	6856
Chile	1995.211	46	12.2	111	6785
Loyalty Islands	1995.136	20	3.9	80	6710
Honshu	1994.362	26	4.9	87	6571
Kuril Islands	1994.277	54	30.0	101	6486
Hokkaido	1994.202	471	1.1	76	6411
Bolivia	1994.160	631	26.3	88	6369
Java	1994.153	18	5.3	85	6362
Fiji Islands	1994.068	562	3.1	83	6277
South of Mariana Isl.	1993.220	59	5.2	72	6064
Hokkaido	1993.193	16	4.7	72	6037
Fiji Islands	1992.193	377	0.8	65	5671
Bolivia	1991.174	558	0.9	56	5287
New Britain	1990.364	178	1.8	44	5112
Philippines	1990.197	25	4.1	43	4945
Sakhalin Island	1990.132	605	0.8	41	4880
Macquarie Islands	1989.143	10	13.6	52	4526
Alaska	1987.334	10	7.3	41	3986
Andreanof Isl.,Aleutians	1986.127	33	10.4	42	3414
Chile	1985.062	33	10.3	29	2984
New Ireland	1983.077	70	4.6	32	2268
Banda Sea	1982.173	450	1.8	21	1999
New Hebrides	1980.199	33	4.8	21	1294
Colombia	1979.346	24	16.9	26	1076
Kuril Islands	1978.340	91	6.4	24	705
Sumbawa	1977.231	33	35.9	18	231
Tonga Islands	1977.173	65	13.9	16	173

to even-order structure only, up to harmonic degree $s = 2\ell$. If the structure within the Earth is axisymmetric (e.g. rotation/ellipticity only), then the splitting matrix is diagonal, the individual singlets can be identified by the index m and the only singlet visible at a station at the Earth's poles is the $m = 0$ singlet.

Using equations (1) and (2) we now form the "receiver strips" for each event:

$$\mathbf{b}(t) = \mathbf{R}^{-1} \cdot \mathbf{u}(t) = \exp[i(\mathbf{H} + \mathbf{I}\bar{\omega})t] \cdot \mathbf{a}(0). \qquad (5)$$

We actually work in the frequency domain using spectra of Hanning-tapered records in a small frequency band about a multiplet of interest. Examples for the events of Figure 1 are given in Figure 2. The spectral lines in this diagram are proportional to the spectra of individual singlets, if axisymmetric structure dominates the splitting matrix. This is almost the case for modes which sample the inner core. Note that not every set of receiver strips exhibits a high signal-to-noise ratio. Depending on the source depth and mechanism, some of the largest earthquakes may very well not excite a specific mode particularly well and hence cannot be included in the analysis for this particular mode. This is the case for the Indian Ocean 2000 event for mode $_{13}S_2$ (Figure 2), while on the other hand this earthquake produced high–quality strips for $_{13}S_3$. For each event, we assign error bars to the receiver strips by performing a standard linear error propagation (e.g. Jackson, 1972) using the residual variance as a measure of data error.

We now use the autoregressive nature of the receiver strips to make our analysis technique independent of earthquake location and source mechanism. The receiver strips satisfy a recurrence in time (using equation 5):

$$\mathbf{b}(t + \delta t) = \mathbf{R}^{-1} \cdot \mathbf{u}(t + \delta t)$$

$$= \exp[i(\mathbf{H} + \mathbf{I}\bar{\omega})(t + \delta t)] \cdot \mathbf{a}(0) = \mathbf{P}(\delta t)\mathbf{b}(t)$$

so

$$\mathbf{b}(t + \delta t) = \mathbf{P}(\delta t)\mathbf{b}(t)$$

where

$$\mathbf{P}(\delta t) = \exp[i\,\delta t\,(\mathbf{H} + \mathbf{I}\bar{\omega})] \qquad (6)$$

and equation (6) has no term that depends on the seismic source. We can now set up an inverse problem for the propagator matrix $\mathbf{P}$, using the strips of all events simultaneously and then determine the splitting matrix $\mathbf{H}$ from $\mathbf{P}$ using the eigenvalue decomposition of $\mathbf{P}$ (Masters et al., 2000a). The matrix $\mathbf{H}$ we retrieve in this process is in general non-Hermitian. If we think of structure as having a real (elastic) and imaginary (anelastic) part, we can use the unique representation

$$\mathbf{H} = \mathbf{E} + i\mathbf{A} \qquad (7)$$

where $\mathbf{E} = \frac{1}{2}(\mathbf{H} + \mathbf{H}^H)$ and $i\mathbf{A} = \frac{1}{2}(\mathbf{H} - \mathbf{H}^H)$ and superscript H indicates Hermitian transpose. Both $\mathbf{E}$ and $\mathbf{A}$ are Hermitian and can be written:

$$E_{mm'} = (a + mb + m^2)\delta_{mm'} + \sum_s \gamma_s^{mm'} c_s^t$$
$$A_{mm'} = \sum_s \gamma_s^{mm'} d_s^t. \qquad (8)$$

The γs are the geometrical factors of equation (3), the c_s^t are the elastic structure coefficients, and the d_s^t are the anelastic structure coefficients. After removing the effects of the Earth's rotation and hydrostatic ellipticity, equations (8) can be regarded as a pair of linear inverse problems for $\mathbf{c}$ and $\mathbf{d}$

and we can explicitly include penalties for rough structure (i.e., high s) and so remove structure which is not required to fit the data. It is convenient to visualize the geographic distribution of structure as sensed by a mode by forming the elastic splitting function (Woodhouse and Giardini, 1985):

$$f_E(\theta, \phi) = \sum_{s,t} c_s^t Y_s^t(\theta, \phi) \qquad (9)$$

and an equivalent function for anelastic structure where the c_s^t are replaced by d_s^t.

Figure 3 shows an example of the procedure for mode $_{13}S_2$. We use the receiver strips of 13 earthquakes that occurred between the 94 Fiji Islands and the 98 Molucca Sea events to determine the splitting matrix for this mode. After subtracting the effects of the Earth's rotation and hydrostatic ellipticity, each of the Hermitian matrices $\mathbf{E}$ and $\mathbf{A}$ are inverted for structure coefficients c_s^t and d_s^t which are then used to compute the splitting functions. The splitting functions are largely dominated by zonal structure ($t = 0$) but the smaller non-zonal components are very robustly determined, i.e. the patterns do not change significantly when varying

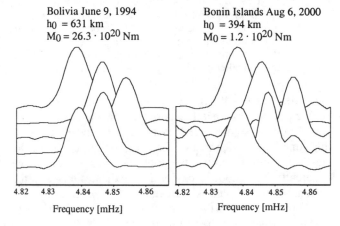

Bolivia June 9, 1994
$h_0 = 631$ km
$M_0 = 26.3 \cdot 10^{20}$ Nm

Bonin Islands Aug 6, 2000
$h_0 = 394$ km
$M_0 = 1.2 \cdot 10^{20}$ Nm

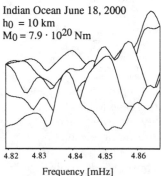

Indian Ocean June 18, 2000
$h_0 = 10$ km
$M_0 = 7.9 \cdot 10^{20}$ Nm

Figure 2. Receiver strips for mode $_{13}S_2$ for the Bolivia and the Bonin Islands events. Even though the latter was 20 times smaller, it excited the mode well enough to produce high signal-to-noise strips. On the other hand, the greater Indian Ocean event on June 19, 2000 did not excite this particular mode sufficiently well to be considered in the analysis of inner core rotation.

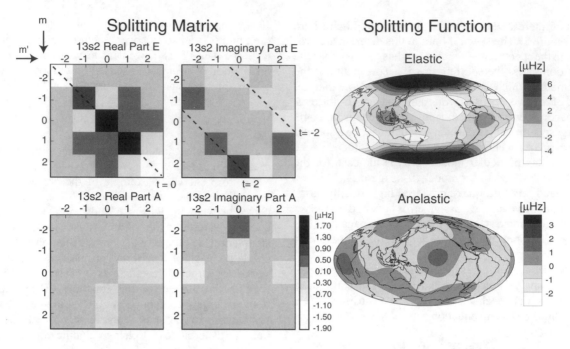

Figure 3. Left: Observed complete splitting matrix for mode $_{13}S_2$, decomposed into its elastic (E) and anelastic (A) parts. Receiver strips of 13 earthquakes between the 94 Fiji Islands and the 98 Molucca Sea events entered the inversion. The signal down the diagonal of the matrices is caused by zonal structure (t=0), as constrained by the selection rules for a mode. Also indicated are the contributions from $t = \pm2$–structure (which is sectoral for $s = 2$). Right: Splitting functions obtained from the splitting matrix on the left. The signal from anelastic structure is typically much smaller than that from elastic structure.

the set of events or the frequency–band chosen to determine **H**. We also observe that the anelastic signal is much smaller than the elastic signal (note that the scale for the elastic splitting function is twice that of the anelastic one). Unlike the elastic splitting functions, the anelastic ones do not change coherently from mode to mode. This indicates that with the current set of earthquakes, we cannot yet determine anelastic structure reliably. We therefore will ignore the effects due to anelastic structure in the following sections.

3. LOOKING FOR A TIME–DEPENDENT SIGNAL

3.1. The Obvious Approach

A straightforward approach to search for a time-dependent signal in the splitting functions is to determined the splitting matrices using only old events and only recent events and then compare the resulting elastic splitting functions. An example is shown in Figure 4 for mode $_{13}S_2$. Using the 13 events described above, we determine the "recent" splitting function. The temporal distribution of the events gives us the splitting function at the time of the Flores Sea event (June 17, 1996). All 11 events between the 77 Tonga Islands and the 89 Macquarie Islands events are used to determine the "past" splitting function. The temporal distribution of these events gives the splitting function for August 18, 1982. An assumed

differential inner core rotation rate of 1°/yr should let some of the patterns in the splitting function be out of phase by 14°. However, the splitting functions are remarkably similar and a phase shift of any pattern is not obvious (compare left panels in Figure 4). A complicating factor in this comparison is the fact that inner–core sensitive modes are quite sensitive to structure of the Earth's mantle (e.g. Figure 7). To study the signal solely generated by inner core structure, the observed splitting functions should therefore be corrected for mantle signal which we assume does not change with time. To predict the mantle corrections we use our model SB10L18 (Masters et al., 2000b). We prefer this model over others because, as opposed to other models, mode data were included in the construction of this model and, perhaps equally important, bulk sound speed was determined independently of shear velocity. Bulk sound speed is negatively correlated with shear velocity at the base of the mantle, a region were anomalies are quite large. The negative correlation implies that perturbations in shear velocity change differently from those of compressional velocity. Using a shear–velocity model and standard scaling relationships between V_s and V_p to calculate the mantle predictions can potentially cause a bias in the splitting functions. The mantle correction for SB10L18 is shown in Figure 4 together with the "residual" splitting functions that now display only the contribution from the inner core (we assume a spherically symmetric outer core). The

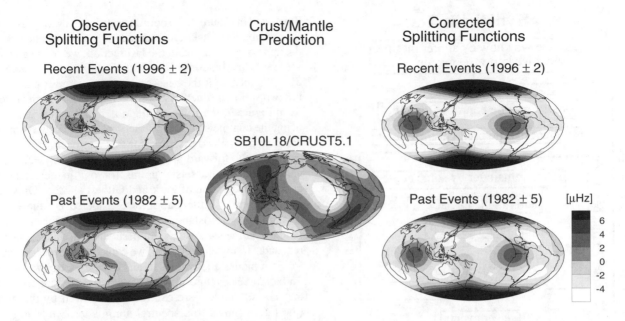

Observed
Splitting Functions

Crust/Mantle
Prediction

Corrected
Splitting Functions

Recent Events (1996 ± 2)

Recent Events (1996 ± 2)

SB10L18/CRUST5.1

Past Events (1982 ± 5)

Past Events (1982 ± 5)

[μHz]

6
4
2
0
-2
-4

Figure 4. Left: Observed splitting functions for mode $_{13}S_2$, using 13 recent events (94 Fiji Islands up to 98 Molucca Sea) and using 11 past events (77 Tonga Islands up to 89 Macquarie Islands). Middle: The prediction for the contribution to the splitting function from crustal (CRUST 5.1 of Mooney et al. 1998) and mantle structure (SB10L18 of Masters et al. 2000b). Right: Splitting function after subtracting the crustal and mantle signal. The remaining signal must come from structure in the core, most likely the inner core. The two splitting functions are extremely similar and no obvious shift between them is visible.

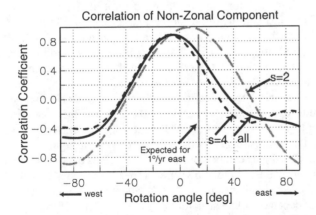

Correlation of Non-Zonal Component

Correlation Coefficient

0.8
0.4
0.0
-0.4
-0.8

-80   -40   0   40   80
◄— west   Rotation angle [deg]   east —►

Expected for
1°/yr east

s=2
s=4 all

Figure 5. Correlation between the two "core" splitting functions of Figure 4, as function of rotation angle for the "past" splitting function. Only the non-zonal component is considered (the dominant zonal component forces the correlation to be above 0.8, independent of the rotation angle). Harmonic degrees 2 and 4 require different angles for the highest correlation which is inconsistent with the inner core rotating as a rigid body.

excellent agreement of the residual splitting functions (right panels in Figure 4) does not suggest a relative rotation. To quantify our comparison, we determine the "best–fitting" rotation angle that gives the highest correlation between the two maps (Figure 5). Only the non-zonal parts are considered in the calculations for the fit (including the zonal part always give a correlation well above 0.8). The best rotation to map

the "past" splitting function into the "recent" one is a westward (!) rotation by about 7°, which is in concordance with the Sharrock and Woodhouse (1998) results but inconsistent with the body wave studies. A disturbing fact is that the components of different harmonic degree s require different best–fitting rotation angles (i.e. degree 2 requires an eastward rotation while degree 4 requires a westward one). This is physically implausible if the inner core rotates as a rigid body. From this comparison one is left to conclude that either inner core rotation does not exist or that the splitting functions are not determined precisely enough to allow such a comparison. The latter is likely, especially for the "past" splitting functions for which far less seismograms are available than for the recent ones. Furthermore, harmonic degree 4 structure is probably less well determined than degree 2 structure because its spectral amplitudes are smaller. We therefore seek alternative ways to search for inner core rotation and a forward approach is described in the next section.

3.2. A Forward Approach and a Synthetic Test

A better way of testing for inner core rotation is by a hypothesis test that the inner core is differentially rotating about the rotation axis (Figure 6). We assume that the modern data accurately constrain the current splitting function of the mode and that we can properly correct for structure in the mantle. The corrected splitting function by assumption reflects only inner core structure which can now be rotated about the rotation axis using an assumed rotation rate. We then add the

Hypothesis Test

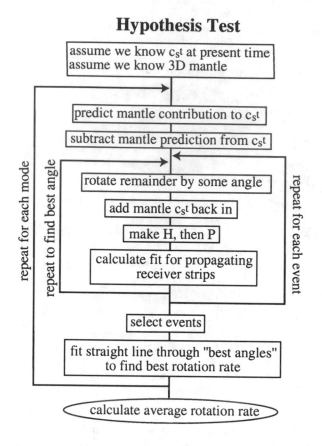

Figure 6. Flow chart for the hypothesis test for differential inner core rotation.

mantle contribution back in, construct a synthetic **H** from which we compute a synthetic **P** (note that we ignore the anelastic part in this test). This **P** is used in equation (6) to test if the assumed rotation rate provides a good fit to the receiver strips, **b**, for a given event. For a given mode, all events have to give the same rotation rate. And for the hypothesis to be acceptable, ALL the modal splitting functions should appear to be rotating at the same rate. The data have to meet certain criteria to be considered in the hypothesis test which will be outlined in the next section. In this section we perform a synthetic test to convince ourselves that this forward approach is indeed capable of recovering an assumed differential rotation of the inner core. We perform this test for two inner–core sensitive modes, $_{13}S_2$ and $_{15}S_3$.

The sensitivity of these modes to mantle structure is quite different, especially near the core–mantle boundary where the question of using different scaling relationships between V_s and V_p is important (Figure 7). We construct a model (Model 1) that is a 9-layer simplified version of our shear velocity model S16B30 (Masters et al., 1996) and a simple model of inner core structure that has a large c_2^0 component and smaller c_2^2 and c_4^3 components. The contributions of synthetic mantle and inner core to the splitting functions are

illustrated in Figure 8. To demonstrate how severely the mantle contribution masks an assumed inner–core rotation rate in the approach discussed in the last section, we create a second set of splitting functions for which the inner core is rotated by 20°. We then seek the best–fitting rotation angle that brings the two splitting functions for $_{13}S_2$ in phase. As can be seem from Figure 9, the rotation angle is largely underestimated using the complete splitting functions, so an accurate mantle correction is indeed essential for that approach.

To test the forward approach laid out in this section, we calculate synthetic seismograms for our model using the coupled–mode code of Park and Gilbert (1986). Only self-coupling and 1D attenuation is considered but the Earth's rotation and hydrostatic ellipticity is included in the calculations and the same steps are performed in this test as with real data. Using equation (5), we construct the receiver strips for both modes and two earthquakes (the 94 Bolivia and the 96 Flores Sea events) from the synthetics (Figure 10). Note that not all singlets are excited equally well by the events (the figure shows true spectral amplitudes, while Figure 2 shows normalized ones). The results from the hypothesis test are shown in Figure 11. Using equation (6) we determine the misfit for the receiver strips for both events and modes, assuming inner core rotation angles between -90 and 90°. The minimum in the misfit curves are at the expected rotation angle (+20°), to within 1°. Slight changes in the mantle model for the correction do not alter the outcome of this test significantly. For example, model 2 is S16B30 used in the original parameterization (30 natural B-splines radially) and CRUST5.1 (Mooney et al., 1998) added near the surface (though the crustal contribution to these modes is quite small). We find that the rotation angle is not recovered reliably only in cases when the mantle model is changed quite significantly. Model 3 is SB10L18 and the prediction for this model can be quite different from that of S16B30 for certain modes (e.g. compare those for $_{13}S_2$ of Figures 4 and 8). The angles can be both under– as well as over–estimated though the example shown underpredicts the angle by 4 to 10°. We should point out that the success of our hypothesis test depends on the structure of the inner core. If the signal were 5 times smaller, or the structure was a pure c_2^2 term then our method could not reliably recover a given rotation angle. The observations in the next section will show that the signal caused by structure in the inner core is not as small and not as simple as this, so the results of our synthetic test are more pessimistic than we should expect for the real Earth.

4. SPLITTING FUNCTIONS FOR REAL DATA

We have analyzed a suite of 15 inner–core sensitive modes and determined the "recent" splitting functions which are shown in Figure 12. The sensitivity of these modes to structure within the Earth can be estimates from inspecting the energy densities in Figure 7. Note that most splitting functions

Energy Densities for the Modes used in this Study

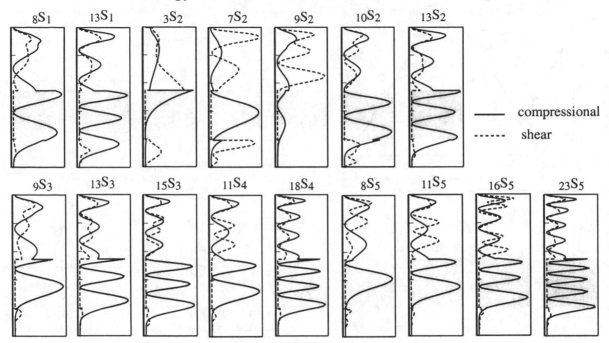

Figure 7. Energy densities for a 1D Earth model for compression and shear as function of radius for the modes in this study. The sensitivity to structure in a 3D Earth slightly varies laterally which is taken into account in the calculation of the splitting functions. The grey area marks the outer core where the shear energy density is zero.

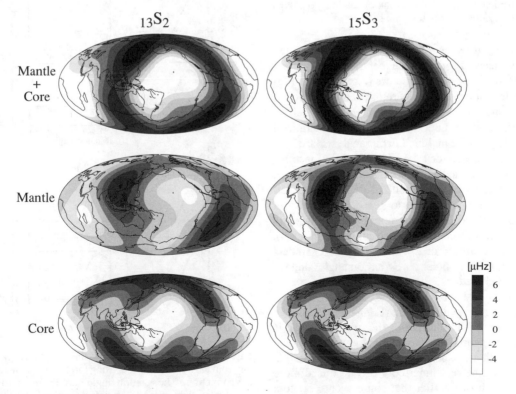

Figure 8. Splitting functions for the synthetic test for modes $_{13}S_2$ and $_{15}S_3$. The "mantle" is a simplified version of S16B30 (Masters et al., 1996). The "core" signal comes from a 200km thick layer at the top of the inner core (contributions from c_2^0, c_2^2, and c_4^3).

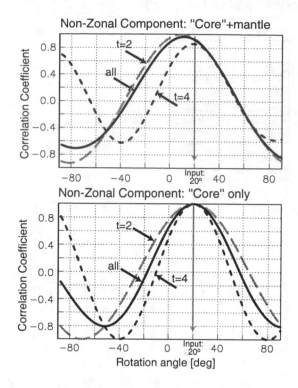

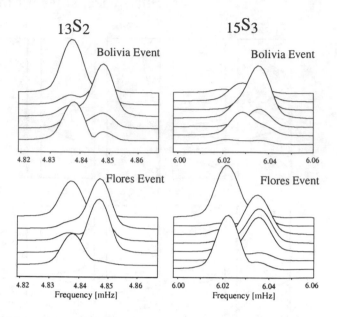

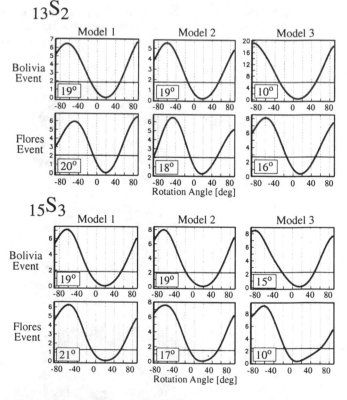

Figure 9. Correlation between two synthetic splitting functions for which the "core" part was rotated by 20°. Only the non-zonal component is considered. Note that the rotation angle is recovered only after the mantle contribution has been subtracted.

are largely dominated by degree 2 zonal structure, though this is somewhat masked for high–frequency modes that are increasingly sensitive to mantle structure (e.g. compare Figures 12 and 13 for mode $_{23}S_5$). There are some exceptions to this (e.g. $_9S_3$ and, to a much lesser extent, $_3S_2$) and we suspect that these splitting functions are less well determined and should probably not be included in our analysis (see later section for possible causes for this). The splitting functions for modes of low harmonic degree are quite simple because, as mentioned above, an isolated mode of degree ℓ is sensitive only to structure of degree up to $s = 2\ell$. It turns out that such modes are dominated even more by zonal structure after they have been corrected for mantle signal (Figure 14). This dominance is so strong for $\ell = 1$ modes that it hampers a reliable determination of inner core rotation angles and we therefore do not include these in our analysis.

Typical misfit curves for the hypothesis test for mode $_{13}S_2$ as those obtained for the synthetic test (Figure 11) are shown in Figure 15. Note that the trough of the misfit function is narrower than in the synthetic experiment suggesting that the structure in the synthetic experiment was too simple and that rotating structure in the real inner core can be traced more reliably. We have shown earlier that some events excite a specific mode better than others so the receiver strips with a poor signal-to-noise ratio have to be identified and discarded. We devise a series of tests that the receiver strips have to pass.

Figure 10. Receiver strips for modes $_{13}S_2$ and $_{15}S_3$ made from coupled-mode synthetic seismograms (self–coupling only) for the 94 Bolivia and 96 Flores Sea events.

Figure 11. Misfit functions for the two modes and events resulting from the hypothesis tests using three different mantle models for the correction. The rotation angle corresponding to the minimum misfit is marked in the lower left corner. Models 1 and 2 usually give the right angle to within a degree, while model 3 can give a significantly smaller angle. The horizontal grey line marks the fit for the zonal component.

Observed Splitting Functions

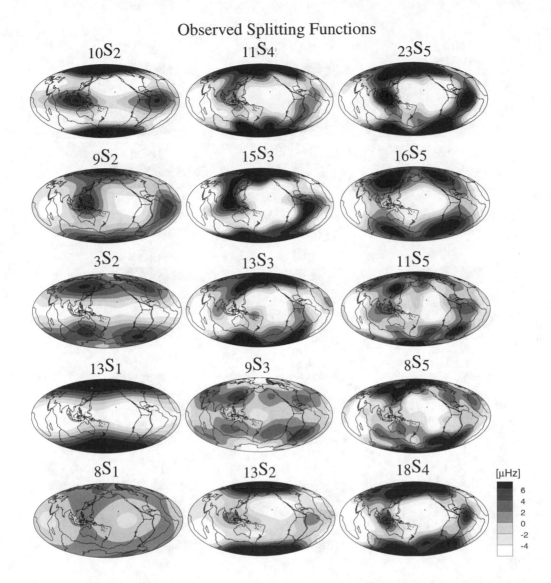

Figure 12. Splitting functions for the 15 inner–core sensitive modes analyzed in this study. The result for mode $_9S_3$ is somewhat uncertain. (The effects due to rotation and hydrostatic ellipticity have been removed).

The most obvious criterion is that the fit to the receiver strips with any non-zero inner core rotation must be better than that using no rotation. Receiver strips with small initial misfit indicate that the event either did not excite this mode particularly well or that the mode couples to other modes and are so discarded. The reconstructed splitting matrix has to give a variance reduction greater than 70% and the receiver strips should be fit close to their error bars (the variance reduction is typically more than 90%). A large residual misfit or a small variance reduction may indicate either a large anelastic contribution to the splitting matrix or strong coupling with another mode, both of which is ignored here. A non-zero rotation angle also has to yield a better fit to the data than the zonal structure alone (horizontal grey lines in Figure 15). In a few cases, different mantle corrections yield significantly different rotation angles. We regard this as an indicator of

serious noise contamination and discard these events, for a particular mode. Figure 15 shows three examples for events that fail the tests for mode $_{13}S_2$. All three events fail the test because the strips are noisy and the initial misfit is too small. In addition to this, the variance reduction of two of them is also too small (8514, 7101), one has no obvious minimum for a reasonable rotation angle (7101) and for one the improvement of fit to the strips by the non-zonal component is insignificant (7256).

5. DETERMINING THE FINAL ROTATION RATES

The final two steps to determine the best–fitting differential inner core rotation rate are quite straight forward. In the

Crust+Mantle Predictions

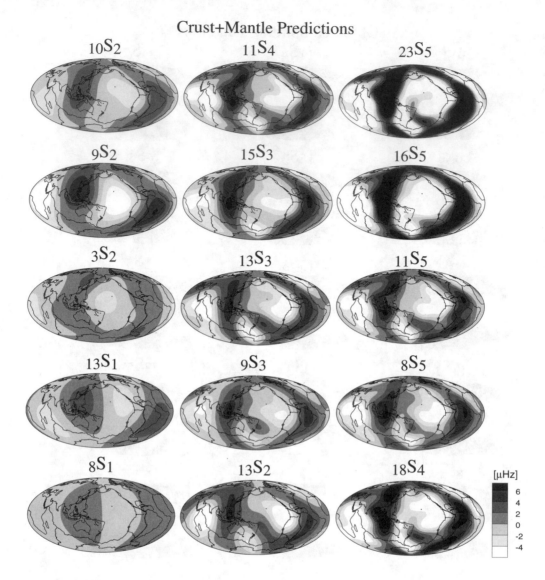

Figure 13. Contributions to the splitting functions of Figure 12 from crustal (Crust 5.1, Mooney et al., 1998) and mantle (SB10L18, Masters et al., 2000b) structure. The signal is usually moderate (and smaller than the observed signal) but is significant for high–frequency modes $_{16}S_5$ (6.83 mHz), $_{18}S_4$ (7.24 mHz) and $_{23}S_5$ (9.29 mHz) which sense the shallow mantle.

first step, we determine the rotation rate for each mode. This is done by fitting a straight line to the rotation angles for all events. In the second step, we average the results over all modes. The first step is the crucial one because the assignment of error bars effectively weighs the results of different events. This process is somewhat subjective as the forward modelling process has no formal error propagation and we examine 3 different ways. High signal–to–noise events typically yield small residual misfit as well as a large relative difference between the fit given by the total splitting function vs. the zonal part only (i.e. grey line vs. bottom of the trough in Figure 15). We take the residual misfit and the relative trough depth as two possible error bars, σ_{RM} and σ_{TD}.

The typical halfwidth of a trough in the misfit function (the half width of the trough below the grey line) is about 25° so we scale both errors by 12.5°, which we regard as conservative choice. For the third type of error, we simply assume uniform errors for all events, i.e. average scaled σ_{TD} over all events and re-assign the average to all events. All three sets of errors yield about the same rotation rates, where the residual misfit error, σ_{RM}, exhibits a slightly larger scatter in the rotation rates for different modes. We take this as indicative that the exact choice of error bars is less significant than we had originally anticipated. Our preferred error bars are the σ_{TD} because they yield the most consistent results. Figure 16 shows the rotation angles as function of event date for the nine

Corrected Splitting Functions

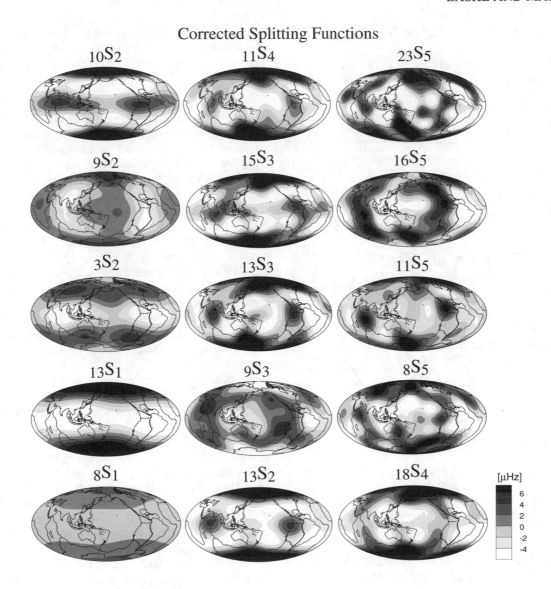

modes for which more than 20 events passed the tests mentioned above. Some modes are marginally consistent with an eastward inner core rotation rate of 1°/yr (e.g. $_{23}S_5$) but others are clearly not (e.g. $_{10}S_2$, $_{18}S_4$), and all modes give rotation rates smaller than 0.67°/yr. The rotation rates for all modes are summarized in Figure 17 and the average over all modes, the final differential inner core rotation rate is 0.13±0.11°/yr. This rate is somewhat higher than the 0.01 ± 0.21°/yr given in (Laske and Masters, 1999) but is within their error bars. Our result is also consistent with recent body wave studies that give eastward rotation rates between 0.15 and 0.25°/yr (e.g. Creager, 2000; Vidale et al., 2000).

6. DOES THE ROTATION RATE DEPEND ON THE MANTLE CORRECTION?

An inspection of Figure 7 and the comparison of the splitting functions for the mantle corrections in Figures 4 and 8

suggest that the final inner core rotation could depend on the mantle correction we apply. We therefore repeated our analysis for a variety of mantle models that have recently been published. These include our older shear–velocity model S16B30 (Masters et al., 1996), our recent high–resolution model SB4L18 (Masters et al., 2000b), the recent Harvard model S362D1 (Gu and Dziewonski, 1999) and the recent Berkeley V_{SH} model SAW24B16 (Megnin and Romanowicz, 2000). We also calculate mantle corrections for the recent Caltech model using both their shear and compressional velocity models S20RTS and P20RTS (Ritsema and van Heist, 2000). Figure 18 shows the misfit curves for mode $_{13}S_3$ for the 94 Bolivia event. The misfit functions are strikingly similar for the different mantle models and yield the same rotation angles to within 3° of our estimate using model SB10L18. The Bolivia event has a high signal–to–noise ratio but we want to stress that the great similarity between misfit func-

Misfit as Function of Rotation Angle for 13S2

Events which Pass the Selection Process

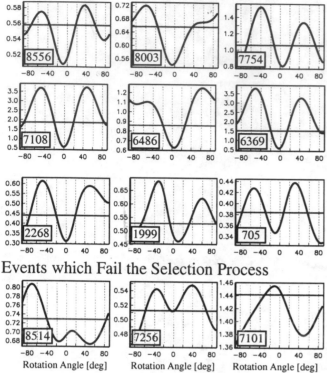

Events which Fail the Selection Process

Figure 15. Examples for misfit functions for mode $_{13}S_2$ of events which pass the selection process and which fail. The horizontal grey line marks the fit for the zonal component and the number in the lower left corner in each diagram is the day since Jan. 01, 1977, e.g. day 6369 is the Bolivia event (see Table 1).

tions is rather typical and not an exception. We take this as indicative that current different mantle models give the same inner core rotation rates. That this is indeed the case is further stressed by the results shown in Figure 17. Almost all rotation rates obtained for different mantle models lie within the error bars of the rates obtained with SB10L18, our preferred mantle model. In fact, variations in rotation rate seem to depend on the mode, not on the mantle correction. For example, almost all models give westward rotation rates for modes $_3S_2$, $_{10}S_2$, $_{13}S_2$ and $_{16}S_5$. With one exception the final average rotation rates using the different models lie within the error bars of the SB10L18-value ($0.13 \pm 0.11°$/yr). The rates are: $0.07°$/yr (SB4L18), $0.08°$/yr (S/P20RTS), $0.22°$/yr (SAW24B16), $-0.05°$/yr (S362D1) and $0.01°$/yr (S16B30). The variation in these values indicates that the mantle correction does influence the final rotation rate somewhat. It is rather obvious however that rotation rates outside of the range of roughly -0.05 to $0.25°$/yr are clearly inconsistent with our mode data. Our preferred values are those obtained with joint S-/P- models (SB10L18 and S/P20RTS) as these take into account the anomalous relative behavior of shear and compressional velocity anomalies at the base of the mantle. The two values (0.13 and 0.08) are in striking agreement.

7. DISCUSSION

Early on in our analysis, we found that $\ell = 1$ modes can constrain inner core rotation only poorly because the mantle-corrected splitting functions are dominated by a large zonal component. We can identify two more types of modes that, at this point, cannot be used in our analysis. The first type of modes are not coupled to other modes (or only weakly coupled) but coincide in frequency with other modes of relatively high ℓ. One such mode is $_2S_3$ which has considerable shear-energy in the inner core so would be ideal to study inner core rotation. Its degenerate frequency in the PREM model is 1.24219 mHz (Dziewonski and Anderson, 1981). Immediately adjacent to it are $_0T_7$(1.22070mHz), $_0S_7$ (1.23179mHz) and $_1T_1$ (1.23611mHz) and these modes can potentially couple (see also Deuss and Woodhouse, 2001). According to the selection rules for coupling modes, none of these modes couple through rotation or ellipticity but limited coupling can occur through 3D structure. In this mode group and for realistic Earth models, the mode pairs that couple strongest are $_0S_7$ with $_0T_7$ and $_0S_7$ with $_1T_1$, while the coupling between the $l = 7$ modes and $_2S_3$ exists but is extremely weak. Even without significant coupling, $_0S_7$ is close enough in frequency to $_2S_3$ that the receiver strips of $_0S_7$ and $_2S_3$ need to be computed simultaneously. If the coupling between $_2S_3$ and $_0S_7$ were negligible, we could then consider only the strips for $_2S_3$ and perform our hypothesis test. For a given mode pair, we need at least $2 \times (l'+l+1)$ records per event to do this, which gives 22 records for the mode pair $_0S_7-_2S_3$. Some records for the events listed in Table 1 are not suitable for a particular mode because of insufficient data volume (e.g. when an instrument stopped recording significantly earlier than Q-cycles of the mode), so that the hypothesis test cannot be performed for $_2S_3$ for events prior to 1986. For obvious reasons, this is not desirable and we therefore exclude modes like $_2S_3$ from our analysis.

Another group of modes that we exclude from this study are strongly coupled modes. As with weakly coupled modes, we would include both modes to compute the receiver strips. But in this case, the strips of a mode can no longer be treated as those for an isolated mode and the complete splitting matrix, including the cross-coupling blocks have to be determined. This fact complicates the hypothesis test immensely as all four blocks (the two self-coupling blocks of the two modes and the two cross-coupling blocks) need to be decomposed and re-assembled using assumed rotation rates. We therefore discard such modes. This is particularly unfortunate, since this affects some of the $\ell = 2$ modes which are particularly sensitive to inner core structure but sometimes strongly coupled to radial modes. An example is the pair $_7S_2-_2S_0$ (Masters et al., 2000c). $_7S_2$ has a considerable

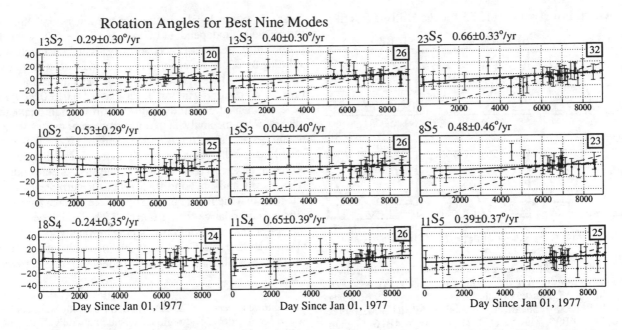

Figure 16. Rotation angles and inferred differential rotation rate for the "best nine modes" (i.e. high number of events). The solid line marks the best fitting straight line that has zero angle at the median time for which the "recent" splitting function was determined. Grey dashed lines mark assumed rotation rates of 1° and 3° per year. For most modes, these high rates are inconsistent with the measured angles. The number of events for each mode is given in the upper right corner.

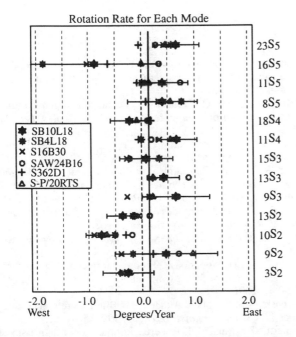

Figure 17. Inner core rotation rates obtained for 13 inner core–sensitive modes, using our preferred mantle model SB10L18. Also shown are the results obtained using other mantle models (for details see Figure 18). The results using different models are remarkably consistent and variations seem to be mode specific, not model specific. The least squares fitting rotation rate, using all modes, is 0.13±0.11°/yr (light grey area). A rotation rate of 0.25°/yr is therefore marginally consistent with our data.

amount of shear–energy at the top of the inner core (Figure 7) and so carries invaluable information about the inner core. However, the coupling to $_2S_0$ is strong enough to inhibit a clear identification of the $m = 0$–line of $_7S_2$. Even if we treat $_7S_2$-$_2S_0$ as coupled modes, the splitting function we retrieve from the self–coupling block of $_7S_2$ does not have the expected shape of a dominant c_2^0 component with large positive local frequency shifts being at the poles. We rather observe negative frequency shifts at the poles (similar to the splitting function of $_9S_3$ in Figure 12). Synthetic tests with coupled–mode seismograms using simple realistic structures for the inner core cannot reproduce this observation so this phenomenon is currently not understood. Deuss and Woodhouse (2001) find significant distortions of individual synthetic spectra of mode $_{13}S_2$ when coupling this mode with $_5S_0$. The frequency shifts due to coupling are predicted to be rather small for these two modes, possibly because the modes are separated in frequency by almost 40μHz. We notice though that the cross–coupling blocks can get quite large (e.g. coupling through the Earth's ellipticity). We do not observe a similar distortion of the receiver strips of $_{13}S_2$ as we see for $_7S_2$ so we expect the bias in the splitting functions introduced by ignoring coupling to be rather small. We notice however that both $_{13}S_2$ and $_{10}S_2$ (which is equally weakly coupled to $_4S_0$) give westward rotation for the inner core. A westward rotation is also seen for $_3S_2$ the coupling properties of which are quite complex (see also Zürn et al., 2000). It is not obvious why inner core rotation rates for these modes

Misfit Function for 13S3 for the Bolivia Event

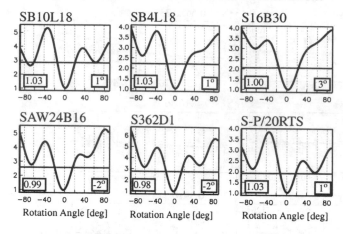

Figure 18. Misfit function for mode $_{13}S_3$ for the 94 Bolivia Event obtained for different mantle models. Joint Vs/Vc or Vs/Vp models: SB10L18: Masters et al. (2000b); S-P/20RTS: Ritsema and van Heijst (2000). Shear velocity models: SB4L18: Masters et al. (2000b); S16B30: Masters et al. (1996); SAW24B16: Megnin and Romanowicz (2000); S362D1: Gu and Dziewonski (1999). The minimum misfit as well as the optimal rotation angle are given.

should systematically be biased westward, but for the sake of argument we could assume that this is the case and eliminate their results. Even when not taking into account the rates for the $\ell = 2$–modes, the final inner core rotation rate is as small as 0.34±0.13°/yr eastward which is inconsistent with the rotation rate of 0.6°/yr, a recent result from Song's re-iterated body wave data (Song, 2000) and we can rule out any greater rate with confidence.

A caveat when analyzing modes using the isolated–mode assumption is that only even degree structure can be determined. It is known from body wave studies that the heterogeneity at the top of the inner core has a strong $s = 1$ signal that is roughly divided into a western and an eastern hemisphere (Tanaka and Hamaguchi, 1997; Creager, 2000). The fact that isolated modes are insensitive to such structure does not affect the ability of a mode analysis to track down the differential rotation, provided the inner core rotates as a rigid body. Structure of uneven harmonic degree can potentially be determined by analyzing coupled modes. For a coupling mode pair (same type only, i.e. S-S, T-T) to be sensitive to $s = 1$ structure the selection rules predict that the harmonic degree of the modes need to be different by an odd number. The analysis of the coupling blocks for modes $_{13}S_2$ and $_9S_7$ can therefore potentially recover $s = 1$ structure of the inner core, though the coupling between these modes by our Model 1 (mantle and simple inner core) appears to be relatively weak. Further treatment of this will be the subject of a future contribution.

In summary, we believe a modal analysis is the best way to determine inner core rotation. First, we are dealing with large-scale vibrations which are insensitive to errors in event locations and to local structure in the inner core. Second, the method is independent of the earthquake source mechanism. Third, we do not need to worry about how much of the splitting functions we observe are caused by heterogeneity or anisotropy – all we care about is whether they change with time. After obtaining an initial estimate of 0.01±0.21°/yr we have augmented our study with data from additional earthquakes. With the analysis of a larger suite of modes we now obtain a small superrotation of 0.13±0.11°/yr. This result is still marginally consistent with the small rotation rates reported in most recent body wave analyses but rules out any rates significantly beyond 0.35°/yr. Our current best value indicates that the inner core is super–rotating at a barely significant rate relative to the mantle. This is in accord with the notion that the inner core is gravitationally locked to the mantle (Buffett and Creager, 1999) in which case the inner core could exhibit small time–dependent differential rotation rates.

Acknowledgments. The data used in this study were collected at a variety of global and regional seismic networks (IRIS-USGS, IRIS-IDA, GEOSCOPE, TERRAscope, BDSN, GEOFON) and obtained from the IRIS-DMC, GEOSCOPE, NCEDC, GEOFON and BFO. We thank Ken Creager, Ruedi Widmer-Schnidrig and Jeroen Tromp for discussion and helpful reviews. This research has been supported by National Science Foundation grants EAR98-09706 and EAR00-00920.

REFERENCES

Bergman, M.I., Measurements of elastic anisotropy due to solidification texturing and the implications for the Earth's inner core, *Nature, 389*, 60–63, 1997.

Buffett, B.A., and K.C. Creager, A comparison of geodetic and seismic estimates of inner–core rotation, *Geophys. Res. Let., 26*, 1509–1512, 1999.

Creager, K.C., Inner core rotation rate from small-scale heterogeneity and time-varying travel times, *Science, 278*, 1248–1288, 1997.

Creager, K.C., Inner Core Anisotropy and Rotation, In: *Earth's Deep Interior: Mineral Physics and Tomography, AGU Monograph, 117*, eds Karato et al, Washington DC, pp.89–114, 2000.

Dahlen, F.A., The normal modes of a rotating, elliptical earth, *Geophys. J. R. Astron. Soc., 16*, 329–367, 1968.

Dahlen, F.A., and J. Tromp, *Theoretical Global Seismology*, Princeton University Press, 1998.

Deuss, A., and J.H. Woodhouse, Theoretical free–oscillation spectra: the importance of wide band coupling, *Geophys. J. Int., 146*, 833–842, 2001.

Dziewonski, A.M., and D.L. Anderson, Preliminary reference Earth model, *Phys. Earth Planet. Inter., 25*, 297–356, 1981.

Glatzmaier, G.A., and P.H. Roberts, Rotation and Magnetism of Earth's Inner Core, *Science, 274*, 1887–1891, 1996.

Gu, Y.J., and A.M. Dziewonski, Mantle Discontinuities and 3-D Tomographic Models, *EOS Trans. AGU, 80*, F717, 1999.

Gubbins, D., Rotation of the Inner Core, *J. Geophys. Res., 85*, 11,695–11,699, 1981.

Henson, I.H., Multiplet coupling of the normal modes of an elliptical, transversely isotropic Earth, *Geophys. J. Int., 98*, 457–459, 1989.

Jackson, D.D., Interpretation of inaccurate, insufficient and inconsistent data, *Geophys. J. R. Astron. Soc.*, 28, 97–109, 1972.

Jeanloz, R., and H.-R. Wenk, Convection and anisotropy of the inner core, *Geophys. Res. Lett.*, 15, 72–75, 1988.

Karato, S., Inner Core Anisotropy Due to the Magnetic Field–Induced Preferred Orientation of Iron, *Science*, 262, 1708–1711, 1993.

Kuang, W., and J. Bloxham. An Earth–like Numerical dynamo model, *Nature*, 389, 371–374, 1997.

Laske, G., and G. Masters, Limits on the Rotation of the Inner Core from a new Analysis of Free Oscillations, *Nature*, 402, 66-68, 1999.

Masters, G., and F. Gilbert, Structure of the inner core inferred from observations of its spheroidal shear modes, *Geophys. Res. Lett.*, 8, 569–571, 1981.

Masters, G., S. Johnson, G. Laske, and H. Bolton, A shear velocity model of the mantle, *Phil. Trans. R. Soc. Lond.*, 354A, 1385–1411, 1996.

Masters, G., G. Laske, and F. Gilbert, Autoregressive estimation of the splitting matrix of free oscillation multiplets, *Geophys. J. Int*, 141, 25–42, 2000a.

Masters, G., G. Laske, and H. Bolton. A. Dziewonski, The relative behavior of shear velocity, bulk sound speed, and compressional velocity in the mantle: implications for chemical and thermal structure, In: *Earth's Deep Interior: Mineral Physics and Tomography, AGU Monograph, 117, eds Karato et al, Washington DC*, pp.63–87, 2000b.

Masters, G., G. Laske, and F. Gilbert, Matrix autoregressive analysis of free-oscillation coupling and splitting, *Geophys. J. Int*, 143, 478–489, 2000c.

McSweeney, T.J., K.C. Creager, and R.T. Merrill, Depth extent of inner core seismic anisotropy and implications for geomagnetism, *Phys, Earth. Planet. Int.*, 101, 131–156, 1997.

Megnin, C., and B. Romanowicz, The 3D shear velocity of the mantle from the inversion of body, surface, and higher mode waveforms, *Goephys. J. Int.*, 143, 709–728, 2000.

Mooney, W.D., G. Laske, and G. Masters, CRUST 5.1: A global crustal model at 5° X 5°, *J. Geophys. Res.*, 103, 727–747, 1998.

Morelli, A., A.M. Dziewonski, and J.H. Woodhouse, Anisotropy of the inner core inferred from $PKIKP$ travel times, *Geophys. Res. Lett.*, 13, 1545–1548, 1986.

Park, J., and F. Gilbert, Coupled free oscillations of an aspherical dissipative rotating earth: Galerkin theory, *J. Geophys. Res.*, 91, 7241–7260, 1986.

Poupinet, G., R. Pillet, and A. Souriau, Possible heterogeneity of the earth's core deduced from $PKIKP$ travel times, *Nature*, 305, 204–206, 1983.

Poupinet, G., A. Souriau, and O. Coutant, The existence of an inner core super–rotation questioned by teleseismic doublets, *Phys. Earth Planet. Int.*, 118, 77–88, 2000.

Poupinet, G., and A. Souriau, Reply to Xiadong Song's comment on "The existence of an inner core super–rotation questioned by teleseismic doublet", *Phys. Earth Planet. Int.*, 124, 275–279, 2001.

Ritsema, J., and H. Van Heijst, Seismic imaging of structural heterogeneity in Earth's mantle: evidence for large-scale mantle flow, *Science Progress*, 83, 243–259, 2000.

Sharrock, D.S., and J.H. Woodhouse, Investigation of time dependent inner core structure by the analysis of free oscillation spectra, *Earth, Planets, and Space*, 50, 1013–1018, 1998.

Shearer, P.M., and K.M. Toy, $PKP(BC)$ veresus $PKP(DF)$ differential travel times and aspherical structure in Earth's inner core, *J. Geophys.Res.*, 96, 2233–2247, 1991.

Song, X., and P.G. Richards, Seismological evidence for differential rotation of the Earth's inner core, *Nature*, 382, 221–224, 1996.

Song, X., Anisotropy of the Earth's inner core, *Reviews of Geophys.*, 35, 297–313, 1997.

Song, X., Joint inversion for inner core rotation, inner core anisotropy, and mantle heterogeneity, *J. Geophys. Res.*, 105, 7931–7943, 2000.

Song, X., Comment on "The existence of an inner core super–rotation questioned by teleseismic doublets" by Georges Poupinet, Annie Souriau, and Oliver Coutant, *Phys. Earth Planet. Int.*, 124, 269–273, 2001.

Souriau, A., New seismological constraints on differential rotation rates of the inner core from Novaya Zemlya events recorded at DRV, Antarctica, *Geophys. J. Int.*, 134, F1–5, 1998a.

Souriau, A., Earth's inner core – is the rotation real?, *Science*, 281, 55–56, 1998b.

Su, W-J., and A.M. Dziewonski, Inner core anisotropy in three dimensions, *J. Geophys. Res.*, 100, 9831–9852, 1995.

Su, W., A.M. Dziewonski, and R. Jeanloz, Planet within a planet: rotation of the inner core of the Earth, *Science*, 274, 1883–1887, 1996.

Tanaka, S., and H. Hamaguchi, Degree one heterogeneity and hemispherical variation of anisotropy in the inner core from $PKP(BC) - PKP(DF)$ times, *J. Geophys. Res*, 102, 2925–2938, 1997.

Vidale, J.E., D.A. Dodge, and P.S. Earle, Slow differential rotation of the Earth's inner core indicated by temporal changes in scattering, *Nature*, 405, 445–448, 2000.

Woodhouse, J.H., The coupling and attenuation of nearly resonant multiplets in the earth's free oscillation spectrum, *Geophys. J. R. Astron. Soc.*, 61, 261–283, 1980.

Woodhouse, J.H., and F.A. Dahlen, The effect of a general aspherical perturbation on the free oscillations of the earth, *Geophys. J. R. Astron. Soc.*, 53, 335–354, 1978.

Woodhouse, J.H., and D. Giardini, Inversion for the splitting function of isolated low order normal mode multiplets, *EOS Trans. AGU*, 66, 300, 1985.

Woodhouse, J.H., D. Giardini, and X.-D. Li, Evidence for inner core anisotropy from free oscillations, *Geophys. Res. Lett.*, 13, 1549–1552, 1986.

Woodhouse, J.H., and T.P. Girnius, Surface waves and free oscillations in a regionalized earth model, *Geophys. J. R. Astron. Soc.*, 68, 653–673, 1982.

Yoshida, S., I. Sumita, and M. Kumazagawa, Growth model of the inner core coupled with the outer core dynamics and the resulting elastic anisotropy, *J. Geophys. Res.*, 101, 28085–28103, 1997.

Zürn, W., G. Laske, R. Widmer-Schnidrig, and F. Gilbert, Observation of Coriolis coupled modes below 1 mHz, *Geophys. J. Int.*, 143, 113–118, 2000.

G. Laske, IGPP 0225, U.C. San Diego, La Jolla, CA 92093-0225. (e-mail: glaske@ucsd.edu)

G. Masters, IGPP 0225, U.C. San Diego, La Jolla, CA 92093-0225. (e-mail: gmasters@ucsd.edu)

Study of Inner Core Structure and Rotation Using Seismic Records from Novaya Zemlya Underground Nuclear Tests

Anyi Li and Paul G. Richards

Lamont-Doherty Earth Observatory, and Department of Earth and Environmental Sciences of Columbia University, Palisades, New York

Data of differential travel times between core phases PKP(BC), PKP(AB) and PKP(DF) recorded at four Antarctica stations (SNA, NVL, DRV and SBA) from Novaya Zemlya underground nuclear tests at the North test site have been analyzed to investigate the structure and rotation of the Earth's inner core. The differential travel time residuals (referenced to the *iasp91* Earth model) spanning 24 years show that: The inner core anisotropy is strong (about 3%) in the uppermost 200 km beneath eastern Africa, and the strong anisotropy exists up close to the top of the inner core; but the anisotropy is weak (about 0.5–1%) beneath the eastern Indian Ocean region, and possibly the weak anisotropy there extends down to about 600 km below the inner core boundary. For the ray paths to NVL, DRV and SBA, the differential travel time residuals tend to decrease with calendar year, providing evidence in support of inner core differential rotation relative to the mantle and the crust.

INTRODUCTION

Our knowledge of inner core seismic velocity structure has improved greatly over the last 20 years, but is still incomplete in important ways. To explain the anomalous splitting of the Earth's free oscillation modes first reported by *Masters and Gilbert* [1981] and the travel time anomaly of inner core phase PKP(DF) first reported by *Poupinet et al.* [1983], *Morelli et al.* [1986] and *Woodhouse et al.* [1986] proposed that the inner core velocity structure is cylindrically anisotropic with symmetry axis aligned with the Earth's rotation axis. Several subsequent studies assumed the anisotropy is homogeneous and sought to estimate the direction of the fast axis. *Su and Dziewonski* [1995] processed 313,422 PKP(DF) arrival times of International Seismological Centre Bulletins for years 1964–

1990, and confirmed the cylindrical elastic anisotropy, but with symmetry axis tilted 10.5° from the Earth's spin axis at 160°E. *McSweeney et al.* [1997] gave the axis of the cylindrically symmetric inner core anisotropy as 80°N, 120°E by analyzing 879 differential PKP travel times. *Song and Richards* [1996] gave the axis as 80°N, 175°E using global PKP differential time data. But later studies showed that the inner core anisotropy is inhomogeneous. It changes not only radially, but also laterally. *Song and Helmberger* [1995] suggested that the top 150 km of the inner core is only weakly anisotropic (less than 1%). *Tanaka and Hamaguchi* [1997] reported a degree one heterogeneity pattern of the inner core anisotropy in the uppermost 500 km. Further studies by *Creager* [1999] with nearly 2000 PKP differential times indicated that the inner core anisotropy is weak in the uppermost 400–700 km of a quasi-eastern hemisphere (40°–160°E), but strong throughout all other parts of the inner core.

Studies of PKP differential travel times utilize two of the three seismic core phases: PKP(DF), PKP(BC) and PKP(AB). In seismic records, PKP(DF) appears at distances from about 114° till 180°, while PKP(BC) and

Earth's Core: Dynamics, Structure, Rotation
Geodynamics Series 31
10.1029/31GD03

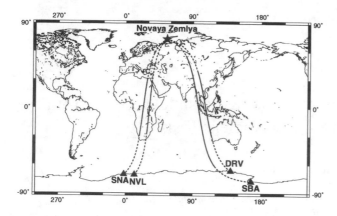

Figure 1. World map showing Novaya Zemlya and PKP ray paths from Novaya Zemlya to four Antarctica stations: SNA, NVL, DRV and SBA. The solid lines indicate in map view the PKP(DF) paths within the inner core.

PKP(AB) arrive after PKP(DF) beginning at about 146°, and disappear from records at about 155° and 177° respectively. The ray paths of PKP(DF) and PKP(BC) are very close to each other in the crust, the whole mantle and most of the outer core. PKP(BC) turns at the bottom of the outer core, whereas PKP(DF) turns in the inner core. Therefore, using PKP(BC) as the reference phase for PKP(DF), the differential time BC–DF can eliminate most of the uncertainty due to Earth structure above the core. When PKP(BC) is not available at larger distance, PKP(AB), which also turns in the outer core, can be used as the reference phase for PKP(DF) to form differential time AB–DF, which may be affected more than BC–DF by inhomogeneous deep mantle structure.

Based on the cylindrically anisotropic inner core model with symmetry axis tilted by a few degrees from the spin axis of the Earth, *Song and Richards* [1996] first reported seismological evidence of BC–DF differential time change of about 0.1 s per decade for inner core differential rotation relative to the mantle. They mainly used seismic data from South Sandwich Islands (SSI) earthquakes recorded at College, Alaska (COL). And they estimated a differential rotation rate of about 1° per year eastward. *Creager*'s [1997] estimate of 0.2–0.3° (with lower limit 0.05°) per year eastward was the first to recognize the effect of lateral heterogeneity of inner core compressional wave speed, in the interpretation of travel time changes through the inner core. He also used records from South Sandwich Islands events to Alaska stations. *Song and Li* [2000] analyzed travel time data from Alaska earthquakes to the South Pole station (SPA) and found travel time changes, giving additional support for inner core differential rotation. Their estimated rotation rate is about 0.6° per year eastward. *Song* [2000] collected South Sandwich Islands earthquake rec-

ords at College, Alaska (COL) for 45 years and at the Alaska Seismic Network (ASN) for 22 years. His inferred inner core rotation rate ranges from 0.3° to 1.1° per year eastward. *Vidale et al.* [2000] studied the temporal changes of scattering waves from inner core fine-scale heterogeneity by stacking, and estimated that the inner core rotates about 0.15° per year faster than the mantle.

Several studies have found no sign of inner core rotation. *Souriau* [1998a] studied Novaya Zemlya nuclear test records at Dumont d'Urville, Antarctica (DRV), and found no obvious trend of PKP(BC–DF) residuals over about 24 years. Her results did not rule out the possibility of a moderate eastward rotation of up to about 1° per year. *Laske and Masters* [1999] applied a new method for analyzing free oscillation splitting functions that are insensitive to earthquake source, location and mechanism, and concluded that the mean value of inner core differential rotation rate lies between ±0.2° per year over the past 20 years.

All these studies tend to constrain the rate of inner core differential rotation to be small or nearly zero. And the zero rate, the special case of no differential rotation, is tolerated by some studies [*Souriau*, 1998a; *Laske and Masters*, 1999]. Additional evidence is still desired to answer whether or not the Earth's inner core rotates relative to the mantle and the crust, to augment the studies of *Creager* [1997], *Song* [2000], and *Vidale et al.* [2000], etc.

A basic problem of studying inner core structure and rotation using travel time data is the possibility of hypocenter mislocation, especially when using historic events. And signals from different events located tens of kilometers apart may be contaminated quite differently by heterogeneous crust and mantle structures. Our solution to these problems is to use underground nuclear tests with highly accurate hypocenter positions within one small area. In this paper, we report our measurements and analyses of differential travel time residuals of core phases from Novaya Zemlya underground nuclear tests at two Antarctica stations: Sanae, Antarctica (SNA) and Scott Base, Antarctica (SBA). We also analyze the published data for two additional stations: Novolazarevskaya, Antarctica (NVL) and Dumont d'Urville, Antarctica (DRV) (Figure 1 and Table 1). From the high quality data of 24 years period, we have obtained gross characteristics of inner core anisotropy along these four ray paths, and supporting evidence for inner core differential rotation relative to the mantle and the crust.

DATA ANALYSIS AND RESULTS

From September 1964 to October 1990, the Former Soviet Union carried out 39 underground nuclear tests at Novaya Zemlya test sites (Figure 1), with 4 tests at the South site (around 70.78°N, 53.88°E), and all others at the North

Table 1. Data Information and Results: N is number of data for each station; Δ is distance from Novaya Zemlya to the stations; H is the PKP(DF) turning point depth beneath inner core boundary; TA is the percentage travel time anomaly for the inner core ray path; σ, a, σ_a, b and σ_b are described in the text. NVL† row shows the analyzed results of NVL data after the Threshold Test Ban Treaty (TTBT) took effect on March 31, 1976; and NVL‡ row shows the analyzed results of NVL data without the 6 events in years 1975–1978.

Station	Time Span year.month	N	Δ °	H km	TA %	σ s	a s	σ_a s	b s/year	σ_b s/year
SNA	75.10–88.12	9	148.0	173	3.0	0.11	10.33	16.58	-0.0035	0.0084
NVL	66.10–90.10	21	146.5	147	3.1	0.13	31.63	8.53	-0.014	0.0043
NVL†	TTBT–90.10	12			3.1	0.11	51.26	13.47	-0.024	0.0068
NVL‡	no 75–78	15			3.1	0.080	29.98	5.14	-0.014	0.0026
DRV	66.10–88.12	13	150.3	222	0.53	0.055	5.24	4.24	-0.0023	0.0021
SBA	66.10–84.10	19	163.5	616	1.1	0.077	12.58	6.77	-0.0053	0.0034

site (around 73.33°N, 54.81°E). The North test site, which supplied all the events studied in this paper, is at the northern end of the south island of Novaya Zemlya. A valley to the south separates this site from other mountains in the south island. And a straight valley going from southeast to northwest divides the test site into two small mountain areas. This rugged topography with two mountainous terrains and low margins sets a strong limit on the absolute event locations, because all of these nuclear tests are underground and exploded within the body of the mountains.

Novaya Zemlya nuclear tests at the North test site are advantageous for study of inner core structure and rotation for several reasons: First, all these tests were carried out within 1 km from the ground surface, so the focal depth in practice can be taken as 0 km for all events used in this study. Second, accurate epicenter positions are available from *Marshall et al.* [1994] as discussed by *Richards* [2000]. The event locations match well the high terrain of the rugged mountain topography. Third, all the events we have used are located in a small rectangular area with east-west length less than 16.5 km, and north-south length less than 12.3 km. So the maximum separation between ray turning points in the core of these events is less than 10 km. The small scatter of these event locations leads the travel time data of core phases PKP(DF), PKP(BC) and PKP(AB) of different events to have almost the same contamination from inhomogeneous structure of the source and receiver region crust and the mantle. Fourth, although most of these tests were composed of double or multiple explosions, the explosions composing each test were located sufficiently close in space and time to constitute effectively a simple point source. So the PKP waveforms of these events are simple and similar to each other. The measured travel time data are more consistent than those from naturally occurring earthquakes. And the last point is that, these events are at high latitude, so the angles between the Earth's spin axis and ray paths to the four Antarctica stations used in this study are small,

between 6° and 14°. So the PKP(DF) travel time anomaly resulting from inner core anisotropy is expected to be large or at least noticeable based on the cylindrically anisotropic inner core model [*Morelli et al.*, 1986; *Woodhouse et al.*, 1986] and later corrections to it by *Tanaka and Hamaguchi* [1997] and *Creager* [1999]. The travel time data are thus favorable to study inner core structure, and there would be more chance to detect the differential rotation of the Earth's inner core.

Seismic records (short period, vertical component) of Novaya Zemlya nuclear tests were collected at two Antarctica stations SBA and SNA (Figure 1). 20 SBA records were scanned from original paper seismograms in New Zealand, where the paper archive for that station is maintained now by the Institute of Geological and Nuclear Sciences. 5 SNA records were scanned from original paper records in South Africa from the SNA archive maintained by the Geological Survey of South Africa. Another 1 SBA record and 4 SNA records were magnified and printed from the World-Wide Standardized Seismograph Network film chip library at the Lamont-Doherty Earth Observatory, and then scanned from the printed paper records. All these records were then digitized (see Figure 2 for a sample), and differential travel times between core phases PKP(DF) and PKP(AB) or PKP(BC) could be measured by cross-correlation.

Nine BC–DF differential travel times from 1975 to 1988 were measured by waveform cross-correlation between the PKP(DF) and PKP(BC) phases on each of the digitized SNA records (Table 1). The differential time residuals were then computed using the *iasp91* Earth model [*Kennett and Engdahl*, 1991] for convenience of data analysis, and plotted vs. calendar time in Figure 3a as circles.

We have been able to use the published results of *Ovchinnikov et al.* [1998] who scanned the original records with 60 millimeter per minute time scale, picked arrivals by eye, averaged five measurements by different people for

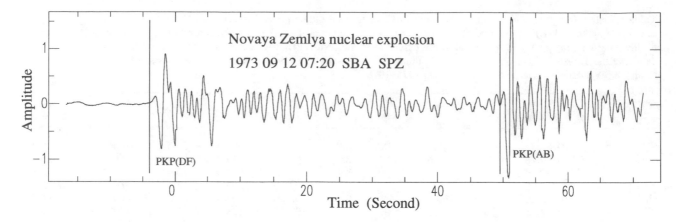

Figure 2. A sample of digitized record: Novaya Zemlya underground nuclear test on Sept 12, 1973 at station SBA, short period vertical (SPZ) component. Zero time corresponds to 12:20.

each arrival, and reported 21 BC–DF differential times for NVL from 1966 to 1990 (Table 1). Their NVL residual time data are plotted in Figure 3b.

We have also obtained the published results of *Souriau* [1998a] who reported BC–DF differential times for DRV. For the four sets of DRV data as described by *Souriau* [1998a], the BC–DF residuals by cross-correlation have the smallest scatter using locations from *Marshall et al.* [1994]. So we use this data set in our own analysis, excluding the residual of the 1981 event because of its low quality [*Souriau*, 1998a]. We also deleted the residual of the Oct 27, 1973 event for the reason discussed below. In this way we obtain 13 DRV residual time data from 1966 to 1988 as shown in Table 1 and Figure 3c.

For SBA records, we measured 21 AB–DF differential travel times by waveform cross-correlation between the PKP(DF) and PKP(AB) phases of each record, with a Hilbert transform on the PKP(DF) waveform. The two SBA residuals of event Nov 2, 1974 and Oct 18, 1975 and the one DRV residual of event Oct 27, 1973 are the only data we have from the South test site of Novaya Zemlya. These three residuals are consistently more than 0.33 s larger after distance normalization than the average of the other residuals which are all from the North test site. Possible reasons include structural heterogeneity in the source region, the mantle and the inner core. Because the South test site is about 290 km away from the North site, information from inner core may be easily obscured by perturbations from the crust and mantle. Therefore these three data were deleted from our data sets in later analysis. So we have 19 SBA residual data from 1966 to 1984 which are shown in Table 1 and Figure 3d. Because the event sources are so near each other, distance normalization is not necessary for all our data. In total, 32 nuclear test events at the North test site are used in this study.

Least square linear regression [*Press et al.*, 1992] was applied to the residual data of the four Antarctica stations to obtain the coefficients of best-fit lines in the form:

$$\Delta t = a + b \cdot T$$

where Δt is the differential time residual (BC–DF for SNA, NVL and DRV, AB–DF for SBA), and T is calendar time in years. Since the measurement uncertainties were not known in advance, we assume they all have the same standard deviation σ, and that the data for each station could be fitted by a straight line. The σ value of the line, the intercept a, slope b and their variances σ_a and σ_b can then be assigned (Table 1). The best-fit lines of the four stations are plotted in Figure 3 as solid lines.

DISCUSSION

Because PKP(BC) and PKP(AB) diverge from PKP(DF) in the deep mantle, BC–DF and AB–DF differential times may be contaminated by heterogeneous deep mantle structures. By carefully checking the ray paths from Novaya Zemlya to SNA, NVL, DRV and SBA in the background of the mantle tomography model of P velocity anomaly described by *Breger et al.* [1999] and *Breger et al.* [2000], we see that all these four ray paths avoid the known P velocity anomaly regions and Ultra Low Velocity Zones. Therefore, our differential time residual data are probably not affected by these mantle anomalies, and it is appropriate to seek explanations in terms of inner core properties. We next discuss the inner core anisotropy characteristics derived from our data, and then show evidence supporting inner core rotation. We also discuss two mechanisms that could explain the systematic PKP(DF) travel time change.

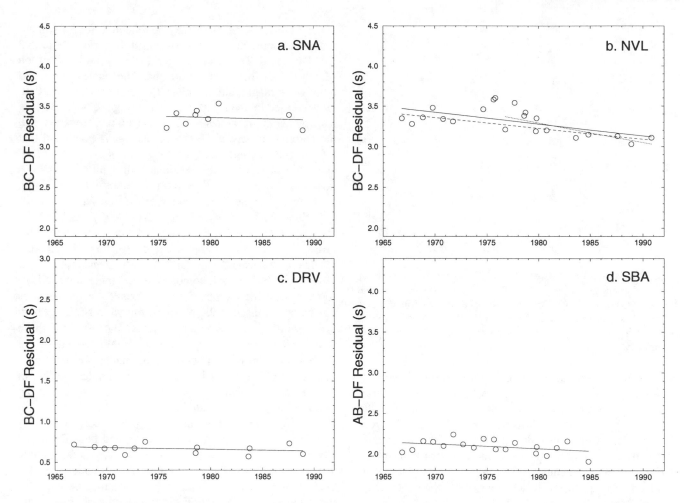

Figure 3. Differential travel time residuals vs. calendar time of SNA, NVL, DRV (BC–DF) and SBA (AB–DF). They are shown as circles. The solid lines best fit the data. In Figure 3b, the dotted line best fits NVL data after TTBT, and the dashed line best fits NVL data without the 6 points in years 1975–1987.

From Figure 3 and Table 1, we can see that the average travel time anomalies along inner core ray paths from Novaya Zemlya to SNA and NVL are 3.0% and 3.1% respectively, indicating strong anisotropy. But along the ray paths to DRV and SBA, the anomalies are only 0.53% and 1.1% (weak anisotropy). The depth of PKP(DF) turning points from the inner core boundary are 173 km, 147 km, 222 km and 616 km for ray paths to SNA, NVL, DRV and SBA respectively. As shown in Figure 1, the ray paths to SNA and NVL are very close to each other. The lateral separation of turning points of the two rays is about 170 km when projected on the inner core boundary. The ray paths to DRV and SBA are close, too. The lateral separation at the inner core boundary is about 165 km. Based on these observations, we conclude that the anisotropy is strong (about 3%) for the inner core region beneath eastern Africa, which is traversed by SNA and NVL ray paths. And that strong anisotropy is present in the uppermost 200 km of this inner

core region. Since the PKP(DF) turning point depth is only 147 km into the inner core for the NVL ray path, very possibly here the anisotropy is strong even from the very top of the inner core. On the other hand, from our data for paths to DRV and SBA, we know that the inner core anisotropy is weak (about 0.5–1%) beneath the eastern Indian Ocean region. Since the travel time anomaly is an average along the whole PKP(DF) ray path within the inner core, the 1.1% anomaly of SBA path and the 0.53% anomaly of DRV path suggest that the deeper inner core sampled by the path to SBA (turning point depth 616 km in inner core) has somewhat stronger anisotropy than the shallower part sampled by the path to DRV (turning point depth 222 km in inner core), and the anisotropy increase with depth is limited in this region of weak anisotropy. The general weak anisotropy here may extend from the inner core boundary down to about 600 km in the inner core. These results indicate that the heterogeneity of inner core anisotropy may exist

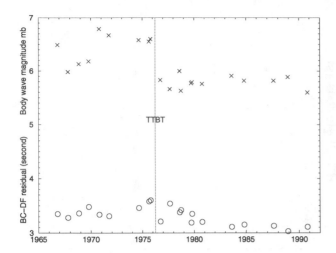

Figure 4. Comparison between NVL BC–DF residuals and body wave magnitudes m_b vs. calendar time. Circles show residual times, and crosses show m_b magnitudes. The vertical dotted line is on March 31, 1976, when the Threshold Test Ban Treaty (TTBT) was to take effect, and a big drop in explosion yields did in fact occur after this date.

right from very near the inner core boundary and down to at least half of the inner core radius, which indicates a complex inner core structure. Our results are consistent with the conclusions of *Tanaka and Hamaguchi* [1997] and *Creager* [1999], and extend the radial thickness of the inner core part with weak anisotropy at the quasi-eastern hemisphere [*Creager*, 1999] to about 600 km, thus confining the transition from weak to strong anisotropy to be in the depth range from 600 to 700 km.

In Table 1, all the four (SNA, NVL, DRV and SBA) best-fitting slope b values are negative. This indicates a systematic change of PKP(DF) travel times with respect to calendar year along all the four ray paths. However, the confidences of the four slope b values are different. The slope of SNA includes a horizontal straight line ($b = 0$) within the $\pm 1\sigma_b$ range (-0.0035 ± 0.0084). Possible reasons for this are the small time span (13 years), too few data points (only 9) (Table 1), and/or small slope value. The PKP(BC) phase at SNA was off scale for the larger tests, but PKP(DF) was too weak to pick arrivals for the smaller ones, because the distance from Novaya Zemlya to SNA is near caustic. Zero slope is outside the $\pm 3\sigma_b$ range for NVL (-0.014 ± 0.0043), is just outside the $\pm 1\sigma_b$ range for DRV (-0.0023 ± 0.0021), and is outside the $\pm 1.5\sigma_b$ range for SBA (-0.0053 ± 0.0034).

The NVL residual data in Figure 3b depart substantially from a straight line between 1975 and 1980. *Souriau* [1998b] suggested this was caused by instrument modification, but we have been assured (by *V. M. Ovchinnikov*, personal communication) that this is not the case. Another pos-

sibility is that the residuals may be related to the large magnitudes of some nuclear tests during this period. Note that the Threshold Test Ban Treaty (TTBT) took effect on March 31, 1976. Right before that, the nuclear tests had body wave magnitude m_b about 6.5, and shortly after that, m_b decreased to about 5.8. But comparing the NVL BC–DF residuals with m_b values as shown in Figure 4, we see that the correlation between residuals and m_b is poor, and that for the time period 1976 to 1991, the m_b values are all about the same while the residuals still exhibit an obvious downward trend. Least square linear regression analysis on NVL data after the TTBT taking effect gives a more negative slope $b = -0.024$ and $\sigma_b = 0.0068$ (Table 1, NVL† row). Zero slope is again outside of the $\pm 3\sigma_b$ range. This best-fit line is shown dotted in Figure 3b. As to the BC–DF residual bulge around 1977 in Figure 3b, *Ovchinnikov et al.* [1998] ascribed it to inhomogeneity of inner core anisotropy. A possible explanation is a strongly heterogeneous object passed through the Novaya Zemlya to NVL ray path in years 1975–1978. We then omit the 6 data points in these years, and do the linear regression again. We have the same slope value $b = -0.014$ with much smaller variance $\sigma_b = 0.0026$ (the NVL‡ row in Table 1, and the dashed line in Figure 3b). All these analyses lead us to conclude that the negative slope of NVL data is robust.

The DRV path through the weakest inner core anisotropy area has a slope value most near zero, and both DRV and SBA paths passing through weak anisotropy areas include zero slope values within $\pm 2\sigma_b$ range. This is understandable even in the presence of inner core rotation, because for weak anisotropy, the PKP(DF) travel time anomalies are generally small, so the travel time change would not be large in general, and thus difficult to detect. Therefore, to study inner core rotation, strong anisotropy paths are preferred.

Summarizing all above discussions of the NVL, DRV and SBA slopes and σ_b values, we conclude that the PKP(DF) travel times increase with calendar year. This means that during the 24 years period something did happen to change the travel times for these three PKP(DF) ray paths. And because of the very small scatter of nuclear test event locations at the North test site, and the short time span of the data (24 years) in geological time scale, the PKP(DF) travel time change is most likely to result from the inner core, not from the mantle. That is to say, at different calendar time the PKP(DF) ray sampled different parts of the inner core with different compressional wave velocity structure. At this point, the differential rotation of the inner core relative to the mantle is the natural conclusion.

To explain the systematic PKP(DF) travel time change with calendar year in term of inner core rotation, two possible causes have been proposed: (a) change in fast direction of inner core anisotropy [*Song and Richards*, 1996], especially for the local areas sampled by PKP(DF) rays, and (b) lateral heterogeneity of inner core P wave velocity [*Creager*, 1997; *Song and Li*, 2000; *Song*, 2000]. Possibly, both mechanisms operate to affect PKP(DF) travel times. At present, we cannot distinguish the different reasons to cause PKP(DF) travel time change, although we note that cause (a) above, in conjunction with a tilted fast axis as proposed by *McSweeney et al.* [1997] or *Song and Richards* [1996] would lead to interpretation of our travel time changes in terms of an inner core rotation in the eastward direction. In any case, detailed knowledge of inner core velocity structure is the key to precise rate estimation of the inner core differential rotation relative to the mantle and the crust.

CONCLUSION

Differential PKP travel time data for a period of 24 years from Novaya Zemlya underground nuclear tests at the North test site to four Antarctica stations SNA, NVL, DRV and SBA have been collected and analyzed. Results show that the inner core anisotropy is strong (about 3%) beneath the eastern Africa (for SNA and NVL paths), and possibly the strong anisotropy begins from very near the inner core boundary. The inner core anisotropy is weak (about 0.5–1%) beneath the eastern Indian Ocean region (for DRV and SBA paths), and this weak anisotropy part of inner core may occupy the whole outer half of the inner core radius. Anisotropy increases little with depth in this generally weak anisotropy region. The PKP differential time residual data of NVL, DRV and SBA indicate that travel times through the inner core increased slightly with calendar time, which support the differential rotation of the Earth's inner core relative to the mantle. More knowledge of inner core structure is needed to refine estimates of the actual rotation rate.

Acknowledgments. We thank Martin Reyners of the Institute of Geological and Nuclear Sciences, New Zealand for arranging access to original paper records of station SBA; and Gerhard Graham of the Geological Survey of South Africa for scanning original paper records of station SNA. We also thank V. M. Ovchinnikov for NVL data, A. Souriau for DRV data, and X. D. Song, W. Y. Kim, and J. Shi for their kind help in this work. This research was supported by NSF grant EAR-98-05245. This is Lamont-Doherty Earth Observatory contribution No. 6316.

REFERENCES

Breger, L., B. Romanowicz, and H. Tkalcic, PKP(BC–DF) travel time residuals and short scale heterogeneity in the deep earth, *Geophys. Res. Lett.*, *26*, 3169–3172, 1999.

Breger, L., H. Tkalcic, and B. Romanowicz, The effect of D'' on PKP(AB–DF) travel time residuals and possible implications for inner core structure, *Earth Planet. Sci. Lett.*, *175*, 133–143, 2000.

Creager, K. C., Inner core rotation rate from small-scale heterogeneity and time-varying travel times, *Science*, *278*, 1284–1288, 1997.

Creager, K. C., Large-scale variations in inner core anisotropy, *J. Geophys. Res.*, *104*, 23127–23139, 1999.

Kennett, B. L. N., and E. R. Engdahl, Traveltimes for global earthquake location and phase identification, *Geophys. J. Int.*, *105*, 429–465, 1991.

Laske, G., and G. Masters, Limits on differential rotation of the inner core from an analysis of the Earth's free oscillations, *Nature*, *402*, 66–69, 1999.

Marshall, P. D., D. Porter, J. B. Young, and P. A. Peachell, Analysis of short-period seismograms from explosions at the Novaya Zemlya test site in Russia, *Atomic Weapons Establishment Report O 2/94*, H.M. Stationery Office, London, 1994.

Masters, G., and F. Gilbert, Structure of the inner core inferred from observations of its spheroidal shear modes, *Geophys. Res. Lett.*, *8*, 569–571, 1981.

McSweeney, T. J., K. C. Creager, and R. T. Merrill, Depth extent of inner-core seismic anisotropy and implications for geomagnetism, *Phys. Earth Planet. Inter.*, *101*, 131–156, 1997.

Morelli, A., A. M. Dziewonski, and J. H. Woodhouse, Anisotropy of the inner core inferred from PKIKP travel times, *Geophys. Res. Lett.*, *13*, 1545–1548, 1986.

Ovchinnikov, V. M., V. V. Adushkin, V. A. An, The relative rotational velocity of the Earth's inner core, *Doklady Earth Sciences*, *363*, 1117–1119, 1998.

Poupinet, G., R. Pillet, and A. Souriau, Possible heterogeneity of the Earth's core deduced from PKIKP travel times, *Nature*, *305*, 204–206, 1983.

Press, W. H., S. A. Teukolsky, W. T. Vetterling, and B. P. Flannery, *Numerical recipes in C*, 2nd edition, pp. 661–665, Cambridge University Press, New York, NY, 1992.

Richards, P. G., Accurate estimates of the absolute location of underground nuclear tests at the northern Novaya Zemlya Test Site, Proceedings, 2nd Workshop on IMS Location Calibration, Oslo, Norway, 20–24, March 2000.

Song, X. D., Joint inversion for inner core rotation, inner core anisotropy, and mantle heterogeneity, *J. Geophys. Res.*, *105*, 7931–7943, 2000.

Song, X. D., and D. V. Helmberger, Depth dependence of anisotropy of Earth's inner core, *J. Geophys. Res.*, *100*, 9805–9816, 1995.

Song, X. D., and A. Li, Support for differential inner core superrotation from earthquakes in Alaska recorded at South Pole station, *J. Geophys. Res.*, *105*, 623–630, 2000.

Song, X. D., and P. G. Richards, Seismological evidence for differential rotation of the Earth's inner core, *Nature*, *382*, 221–224, 1996.

Souriau, A., New seismological constraints on differential rotation of the inner core from Novaya Zemlya events recorded at DRV, Antarctica, *Geophys. J. Int.*, *134*, F1–5, 1998a.

Souriau, A., Detecting possible rotation of Earth's inner core, response to comments by P. G. Richards, X. D. Song, and A. Li, *Science*, *282*, 1227a, 1998b.

Su, W. J., and A. M. Dziewonski, Inner core anisotropy in three dimensions, *J. Geophys. Res.*, *100*, 9831–9852, 1995.

Takahashi, T., and W. A. Bassett, High-pressure polymorph of iron, *Science*, *145*, 483–486, 1964.

Tanaka, S., and H. Hamaguchi, Degree one heterogeneity and hemispherical variation of anisotropy in the inner core from PKP(BC)–PKP(DF) times, *J. Geophys. Res.*, *102*, 2925–2938, 1997.

Vidale, J. E., D. A. Dodge, and P. S. Earle, Slow differential rotation of the Earth's inner core indicated by temporal changes in scattering, *Nature*, *405*, 445–448, 2000.

Woodhouse, J. H., D. Giardini, and X.-D. Li, Evidence for inner core anisotropy from free oscillations, *Geophys. Res. Lett.*, *13*, 1549–1552, 1986.

A. Li and P. G. Richards, Lamont-Doherty Earth Observatory of Columbia University, Palisades, NY 10964.
(e-mail: anyili@ldeo.columbia.edu; richards@ldeo.columbia.edu)

On the Origin of Complexity in PKP Travel Time Data.

B. Romanowicz, H. Tkalčić[1], L. Bréger[2]

Seismological Laboratory, University of California, Berkeley.

In order to investigate the origin of short spatial scale features in PKP travel time data and to determine whether a complex inner core anisotropy model is required, we have assembled a new global dataset of handpicked absolute PKP(DF) travel times, and completed existing datasets of hand-picked relative PKP(AB-DF) and PKP(BC-DF) travel times. We discuss in detail the trends of relative and absolute PKP travel time residuals at the global scale, as well as for a well sampled set of paths between the south Atlantic and Alaska.

We discuss the relative merits of several types of models: a) a model of hemispherical anisotropy in the inner core previously proposed to explain PKP(BC-DF) travel time residuals on the global scale; b) a model combining weak constant anisotropy in the inner core with strong heterogeneity in the deep mantle; c) a model involving structure in the outer core associated with the tangent cylinder to the inner core, with axis parallel to the rotation axis, a feature described in magnetohydrodynamical models of the outer core.

Because absolute PKP(DF) travel time residuals exhibit the same hemispherical pattern as relative PKP(BC-DF) and PKP(AB-DF) data, when plotted at the location of the bottoming point of DF in the inner core, we infer that the causative structure must at least partly originate in the core. However, the transition between anomalous and normal structure is quite abrupt, and hemispherical inner core anisotropy models fail to reproduce the characteristic "L shape" of PKP(BC-DF) travel time residuals, when plotted as a function of the angle of the ray in the inner core with the rotation axis (ξ). Models involving mantle heterogeneity compatible with other mantle sensitive data can explain PKP(AB-DF) travel times, but fail to explain 3 sec of average PKP(BC-DF) anomaly observed for paths bottoming in the western hemisphere, for $\xi \sim 20 - 30^o$, even when a model of constant anisotropy in the inner core, compatible with mode splitting data, is also included. On the other hand, models with $\sim 1\%$ faster velocity inside an outer core region roughly delimited by the inner core tangent cylinder allow for rapid transitions, are compatible with rends in absolute PKP(DF) and PKP(BC) times observed in Alaska, and can reproduce the L-shaped feature

---

[1]Now at: IGPP, University of California, San Diego
[2]Now at: BARRA, Berkeley, CA

Earth's Core: Dynamics, Structure, Rotation
Geodynamics Series 31
Copyright 2003 by the American Geophysical Union
10.1029/31GD04

of the PKP(BC-DF) travel time data. Sustained heterogeneity in the outer core could arise within polar vorteces in and around the tangent cylinder, as suggest by recent dynamical and magnetic investigations. Such models are also compatible with most normal mode splitting data and present less departure from axial symmetry than the hemispherical inner core anisotropy models. When trying to physically explain them, both types of models present challenges, and should be pursued further.

1. INTRODUCTION

The first observation that PKP(DF) waves travel faster through the earth's inner core along polar paths (paths quasi-parallel to the earth's rotation axis) than on equatorial paths, was made almost 20 years ago [Poupinet et al., 1983]. Subsequently, it was proposed that this could be due to inner core anisotropy, which would explain the PKP observations [Morelli et al., 1986] as well as observations of anomalous splitting of core sensitive free oscillations [Masters and Gilbert, 1981; Woodhouse et al., 1986]. These observations were later confirmed in many studies, both for PKP travel times (e.g. [Shearer et al., 1988; Shearer, 1991; Creager, 1992; Vinnik et al., 1994; Su and Dziewonski, 1995; Song, 1996]) and for core modes [Ritzwoller et al., 1988; Li et al., 1991].

The early inner core anisotropy models were cast in terms of constant transverse isotropy with fast axis parallel to the earth's axis of rotation, as would be expected if the anisotropy were due to the alignment of hcp-iron crystals with the axis of rotation (e.g. [Stixrude and Cohen, 1995]). Proposed physical mechanisms for anisotropy have involved convection in the inner core [Jeanloz and Wenk, 1988], magnetic effects [Karato, 1995, 1999], gravitational interaction with the mantle (e.g. [Buffett and Creager, 1999]) or texturing of iron during inner core solidification [Bergman, 1997]. Most of these mechanisms, except perhaps gravitational interaction in the mantle, imply axisymmetry of the anisotropic structure.

As data have accumulated and revealed more details, inner core anisotropy models have become more complex. Depth dependence of the strength of anisotropy was proposed [Su and Dziewonski, 1995; Tromp, 1993] and helped explain a long standing discrepancy between the travel time and mode observations [Tromp, 1995], even better so when departures from a simple radial model are also considered [Romanowicz et al., 1996]. To explain strong anomalies for polar paths in northeastern Eurasia and Alaska, it was proposed that the axis of symmetry of the anisotropy could be tilted with respect to the earth's axis of rotation [Su and Dziewonski, 1995; McSweeney et al., 1997], but Souriau et al. [1997] demonstrated that this result was not statistically robust, due to the uneven sampling of the globe by the PKP data. The most intriguing observation to date, in our opinion, was made by Tanaka and Hamaguchi [1997], who observed that only one hemisphere, extending roughly from longitude 177°W to 43°E ("Quasi-western" hemisphere) was anisotropic, a fact later confirmed by Creager [1999], who noted that the strength of anisotropy was different in the two hemispheres, but both supported the same Voigt average velocity. On the other hand, Song and Helmberger [1998] proposed that the top of the inner core is isotropic and separated from the central anisotropic part by a discontinuity of varying depth. However, the isotropic part cannot be, on average, thicker than 100-200 km, to account for constraints from anomalous splitting of core sensitive modes [Durek and Romanowicz, 1999]. In order to account for the difference in the two hemispheres, as well as the existence of an isotropic region at the top of the inner core, Creager [2000] and Garcia and Souriau [2000] recently proposed very similar models which comprise a discontinuity within the inner core, separating an isotropic outer core from an anisotropic inner core. The ellipsoidal shape of this discontinuity is shifted with respect to the center of the inner core, so that the isotropic part is thicker in the eastern (400km) than in the western ($< 100km$) quasi-hemisphere.

Other complexities in the PKP(BC-DF) and PKP(AB-DF) travel time data have recently been documented by Bréger et al.[1999, 2000a,b], who pointed out how important it is to account accurately for the influence of strong heterogeneity at the base of the mantle, before making inferences on inner core anisotropy from the observation of core sensitive phases. It is difficult to find physical mechanisms to explain the increasingly complex structure of the inner core anisotropy required by recently accumulated high quality broadband data, and in particular, the hemispherical differences, given that the inner core is thought to be close to the melting point of its constituents. In view of the mounting evidence for strong heterogeneity in the deep mantle (e.g. [Garnero and Helmberger, 1996; Bréger and Romanowicz, 1998; Ritsema et al., 1998]) and the uneven distribution of PKP observations on polar paths around the globe [Bréger et al., 2000a], it is important to consider whether the complexity originates in the inner core or

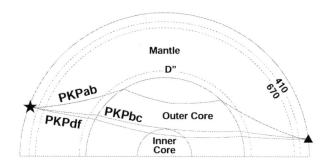

Figure 1. Vertical cross-section through the earth showing the paths of the three PKP phases

elsewhere, and whether it might all be accounted for by mantle structure.

In what follows, we discuss possible origins of the most significant, first order features of PKP travel time data, measured on the rapidly growing collection of short period and broadband records, at the global scale, and also, more specifically, for a set of intriguing paths between the south-Atlantic and Alaska.

2. DESCRIPTION OF THE DATASETS AND SOME SPECIFIC TRENDS

We have assembled a comprehensive dataset comprising PKP(AB-DF), PKP(BC-DF) differential travel times, and PKP(DF) absolute travel times, which we measured on vertical component records from broadband and short period stations worldwide for the time period 1990-1998 [*Tkalčić et al.*, 2002], and complemented by datasets collected by several other authors. The geometry of the various PKP phases is shown in Figure 1.

We measured differential travel times by cross-correlation of the two phases involved. The details of the measurement technique are given in *Tkalčić et al.* [2002]. The complete dataset combines our data with those of *McSweeney et al.* [1997], *Creager* [1999], *Tanaka and Hamaguchi* [1997], *Souriau* (personal communication) and *Wysession* (personal communication). These data have been carefully inspected for inconsistencies between authors, duplications, and errors. In particular, we made systematic plots of variations as a function of back-azimuth for groups of neighboring stations, as well as variations as a function of azimuth for groups of neighboring events. This allowed us to eliminate clear outliers, but it was possible only for equatorial and quasi-equatorial paths ($\xi > 35^o$), for which data are numerous. The corresponding differential travel times consistently show variations around the mean not exceeding $\pm 2.5 sec$ for PKP(AB-DF), and $\pm 1.5 sec$ for PKP(BC-DF). The error in measurement

for these equatorial paths is estimated to be $\leq 0.5 sec$ and we were able to eliminate practically all residuals exceeding respectively $\pm 2.5 sec$ (AB-DF) and $\pm 1.0 sec$ (BC-DF) as outliers. For polar paths, data are fewer, so that this type of verification is not possible. We note however that, for quasi-polar paths ($\xi < 35^o$) for which numerous measurements are available, such as at stations of the dense Alaska network, there are indications of consistent variations over short distances, as we will discuss further below.

We also measured absolute PKP(DF) travel times, whenever possible, and present this new dataset here for the first time. The distance range spanned by the data is 145^0 to 175^0. For these measurements, we cannot take advantage of the accuracy of waveform comparison, and we must rely on direct picks of the onset of the DF phase, which is often emergent, especially for polar paths. Therefore, the measurement error is larger in general, on the order of $1 sec$ for equatorial paths, and up to $2 sec$ in some cases, for polar paths. We thus expect a larger scatter in the data. However, absolute measurements are of great interest for the study of inner core anisotropy, and are the basis of most inferences made using data collected from ISC bulletins (e.g.[*Poupinet et al.*, 1983; *Morelli et al.*, 1986; *Shearer*, 1988; *Su and Dziewonski*, 1995]). Moreover, global variations in absolute DF (in particular differences between polar and equatorial paths) are largely in excess of the measurement error.

Figure 2 shows the variations, as a function of angle ξ, of PKP(BC-DF), PKP(AB-DF) and PKP(DF) travel time residuals, referenced to model AK135 [*Kennett and Engdahl*, 1991], and corrected for ellipticity. The epicentral data used in the computation of residuals are those from the EHB catalog [*Engdahl et al.*, 1998]. We note the larger scatter of the DF data, as expected, and the large spread of values for $\xi \leq 35^o$. On average, residuals are several seconds larger for polar paths than for equatorial paths, consistent with all previous studies. The raw datasets, however, do not exhibit a smooth variation with ξ as would be expected for simple models of inner core anisotropy. Rather, the curves are L-shaped and there is a sharp break around $\xi = 30^o$, with many residuals larger by 2-4 sec for the more polar paths. Several events in the south Atlantic (south Sandwich Islands in the BC-DF distance range and Bouvet Islands in the AB-DF distance range) observed at stations of the Alaska network contribute to the large concentration of data points for $20^o \leq \xi \leq 30^o$ and exhibit a large scatter, which has been attributed to heterogeneity in the inner core (e.g.[*Creager*, 1997; *Song*, 2000]). We will discuss these data in detail. The distinct "line" of negative anomalies between 0 and -1 sec,

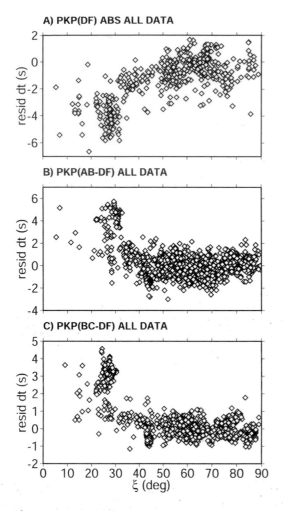

A) PKP(DF) ABS ALL DATA

B) PKP(AB-DF) ALL DATA

C) PKP(BC-DF) ALL DATA

Figure 2. Variations of travel time residuals for PKP (DF) (absolute measurements, top), PKP(BC-DF) (middle) and PKP(AB-DF) (bottom) as a function of the angle ξ made by the inner core leg of the path with the earth's rotation axis. Residuals are referred to model AK135 and have been corrected for ellipticity.

around $\xi = 43°$, in the BC-DF dataset, (also present in the AB-DF dataset) corresponds to the 03/29/1993 South Atlantic earthquake observed on the dense California short period networks.

In Figure 3a, we show the variations with ξ of the PKP(BC-DF) travel time residuals, after replacing the two clusters mentioned above by summary rays, and distinguishing the quasi-eastern and quasi-western hemispheres, according to the definition of *Tanaka and Hamaguchi* [1997]. Indeed, we confirm the differences in trends for both hemispheres, with practically no dependence with ξ in the quasi-eastern hemisphere. As noted previously [*Tanaka and Hamaguchi,* 1997; *Creager,* 1999], there is also a difference of 1 sec on average, for non polar angles ξ, between BC-DF residu-

als in the quasi-eastern and quasi-western hemispheres, with the former being faster. For comparison, we plot, in Figure 3 (b)(c), the predictions of two inner core anisotropy models that provide good fits to the average observed trends. In Figure 3 (b), the predictions for each hemisphere are calculated separately for two constant anisotropy models proposed in the literature. The quasi-eastern hemisphere could support anisotropy in the inner core of strength less than 1%, whereas the anisotropy required to explain data in the quasi-western hemisphere is close to 3%. In Figure 3 (c), we show the predictions of *Creager's* [2000] 2-layer model of the inner core. Some slight discrepancies in the location of the boundary between the eastern and western parts of the model are apparent, but on average this model fits

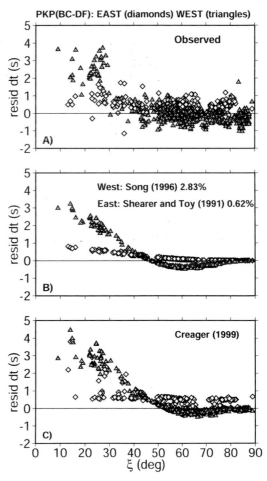

PKP(BC-DF): EAST (diamonds) WEST (triangles)

Observed

West: Song (1996) 2.83%

East: Shearer and Toy (1991) 0.62%

Creager (1999)

Figure 3. PKP(BC-DF) travel time residuals as a function of ξ, distinguishing quasi-eastern (diamonds) and quasi-western (triangles) hemispheres. A) Observed; B) predictions of two different models of constrant transverse isotropy in the inner core (strength indicated); C) Predictions of Creager's [2000] off-centered hemispherical inner core anisotropy model. In this plot, Alaska network data have been replaced by summary rays.

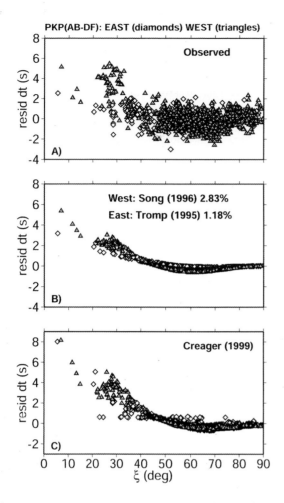

Figure 4. Same as Figure 3 for the PKP(AB-DF) dataset. Note that a slightly stronger anisotropy model is plotted in B) for the eastern hemisphere.

the trends in the data well to first order. However, in the western hemisphere, both models (3b, 3c) overestimate the residuals observed in the range $\xi = 30 - 35^o$, and underestimate them in the range $\xi = 20 - 30^o$, indicating that significant features in the data, namely the L shape rather than smooth increase of residuals as a function of ξ, are not explained by hemispherical anisotropy.

In Figure 4, we present the results of the same analysis for the PKP(AB-DF) dataset. In this case, the constant anisotropy required in the quasi-eastern hemisphere (Figure 4 (b)) is stronger than for PKP(BC-DF). On the other hand, the western hemisphere model with 2.83% anisotropy underestimates the dispersion of the residuals in the $\xi = 20 - 30^o$ range by a factor of two. A model with 3.5% anisotropy would fit the data in this range better, but would overestimate the residuals at smaller angles. Creager's hemispherical model (Figure 4 (c)) underestimates residuals in the range $\xi = 20 - 30^o$

by 1-2 sec and overestimates the residuals at $\xi < 20^o$ by up to 2.5 sec. Inspection of the data shows that, unlike for BC-DF, there is no clear distinction in the data between eastern and western hemispheres at $\xi > 35^o$ (non polar paths).

3. COMPLEX INNER CORE ANISOTROPY OR STRONG HETEROGENEITY AT THE BASE OF THE MANTLE?

In the previous section, we discussed how hemispherical models of inner core anisotropy can reproduce some of the trends in the PKP travel time data, but fail to reproduce the characteristic "L-shape" of the variations of residuals with angle ξ. Here we consider the possible contribution of heterogeneity at the base of the mantle. As is now well established, lateral heterogeneity increases and changes style in the last few hundred kilometers above the core mantle boundary (CMB), reaching rms variations in S velocity in excess of 2% in D". Although recent S tomographic models differ from each other in their details, they all agree that the spectrum of heterogeneity changes from white to red at the bottom of the mantle, where degree 2 predominates (e.g. [*Masters et al.*, 1996; *Grand*, 1997; *Liu et al.*, 1998; *Mégnin and Romanowicz*, 2000; *Ritsema et al.*, 2000]), with a distinctive spatial pattern showing two large low velocity regions under Africa and in the central Pacific, surrounded by a "ring" of fast velocities, as first shown in *Dziewonski et al.* [1977]. While S tomographic models successfully retrieve the large scale patterns of heterogeneity, they underestimate the strength of lateral variations, at least in some regions, by a factor of 2 or 3, as has been shown by comparison of observed and predicted differential travel time anomalies of S-SKS and Sdiff-SKS waves, in the well sampled "corridor" across the Pacific Ocean [*Bréger et al.*, 1998], as well as from measurements of diffracted P and S waves on the global scale (e.g. [*Wysession*, 1996]).

Forward modeling studies of S, ScS, and Sdiff waves in regions sampling the Pacific Plume [*Bréger and Romanowicz*, 1998; *Bréger et al.*, 2001] and the African Plume [*Ritsema et al.*, 1998; *Ni and Helmberger*, 1999] have documented strong gradients in the regions bordering these plumes in D", with lateral variations in excess of ±5% over distances of 200-400km. While not necessarily completely correlated, similar characteristics are expected in the P velocity distribution at the base of the mantle. Indeed, PcP-P data at large distances confirm the presence of short wavelength variations of at least ±2% in some well sampled regions [*Tkalčić et al.*, 2002]. In addition, there is evidence for regions of localized ultra low velocities (ulvz's), with P velocity anomalies in excess of 10% (e.g. [*Garnero*

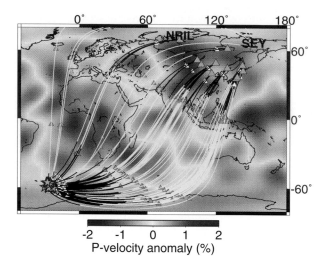

Figure 5. Surface projections of PKP wavepaths from sources in south Sandwich Islands to stations in Eurasia. Yellow thick lines correspond to the inner core legs of PKPdf. We also plotted the points where PKPdf and PKPab enter and exit the outer core (white and green triangles). Model PAW24B16, obtained by converting S-velocity model SAW24B16 [*Mégnin and Romanowicz*, 2000] into a P-velocity model using the scaling relation $dlnVs/dlnVp = 2$, is used as background.

and Helmberger, 1996]). Recently, *Bréger et al.* [2001] showed that, by considering an existing tomographic model of the mantle [*Grand*, 1997] increasing the amplitude of lateral variations in D" and including ulvz's, a significant portion of the trend with ξ of PKP(AB-DF) travel time residuals could be explained without even accounting for anisotropy. Indeed, as shown in Figure 1, the PKP(AB) wavepath grazes the core-mantle boundary and thus interacts with structure in D" much more than the corresponding PKP(DF) path. We thus expect, as first pointed out by *Sacks et al.* [1979] and further considered by *Sylvander and Souriau* [1996], that PKP(AB-DF) differential travel times may be strongly affected by heterogeneity in D". A major concern is that the distribution of PKP paths in the distance range appropriate for PKP(AB) observations and for angles ξ smaller than 40° is very non-uniform, with a majority of paths originating in the south Atlantic, specifically in the seismically active region of the south Sandwich Islands (e.g.[*Bréger et al.*, 2000a]), located near the border of the African superplume.

Figure 5 shows the geometry of paths from south Sandwich Islands to stations in Eurasia. Plotted in the background is a P velocity model (PAW24D) obtained by scaling the tomographic S velocity model SAW24B16 [*Mégnin and Romanowicz*, 2000] using a ratio $dlnVs/dlnVp = 2$. Indicated are entry points of DF

and AB into the core. We note that for north to northeast trending paths, the AB phase interacts with the low velocity "African superplume" structure, whereas the DF phase stays largely outside of it. In Figure 6, we present various attempts at modeling trends in the PKP(AB-DF) travel time residuals, as a function of azimuth or ξ, for events originating in the south-Sandwich Island region, to stations in Eurasia and Alaska. In Figure 6 (a)(b), we compare the PKP(AB-DF) observations with the predictions of tomographic model PAW24D, shifted upward by 1sec (to account for an obvious baseline shift on these paths) and of the hemispherical model of inner core anisotropy, presented in Figure 4. The tomographic model fails to predict the large spread of residuals at azimuths greater than 270°, which correspond to South-Sandwich to Alaska paths. The hemispherical anisotropy model improves the average fit in these azimuths (as well as in the azimuth range $0 - 40^o$), but still fails to explain the large scatter in the Alaska data. Finally, in Figure 6 (c), we show the predictions of a D" model (TRH_{KC}) constructed by *Tkalčić et al.*[2002] using a combination of globally distributed PKP(AB-DF) and PcP-P data, corrected for mantle structure using the *Karason and vander-Hilst*[2000] mantle P model to a depth of 300 km above the CMB. Model TRH_{KC} predicts the scatter in the Alaska data better, as well as the longer wavelength trends with azimuth. In fact, *Tkalčić et al.* [2002] have shown that over 80% of the variance in the PKP(AB-DF) data can be explained by such a model, without requiring hemispherical anisotropy in the inner core.

While it is not too surprising that PKP(AB-DF) travel time residuals can be explained largely by mantle heterogeneity, especially since model TRH_{KC} was constructed to fit such data, additional insight can be gained from the analysis of PKP(BC-DF) data, noting that PKP(BC) and PKP(DF) travel on very close paths throughout the mantle. Figure 7 shows the comparison of observed and predicted PKP(BC-DF) travel time residuals for paths originating in the south-Sandwich Islands, for model TRH_{KC} (Figure 7 (a)) and the hemispherical inner core model (Figure 7 (b)). We note that the mantle model reproduces a large fraction of the local scatter in the data, but fails to predict the 3 sec average BC-DF travel time anomaly for Alaska paths. On the other hand, the hemispherical inner core anisotropy model does predict 3 sec of BC-DF anomaly in Alaska, but fails to produce the full observed scatter in these data. In Figure 7 (c), we show the predictions of a model which combines the D" model of *Tkalčić et al.*[2002] with a constant inner core anisotropy model of about 1.5%. The maximum strength of anisotropy is constrained by the necessity to fit small residuals for

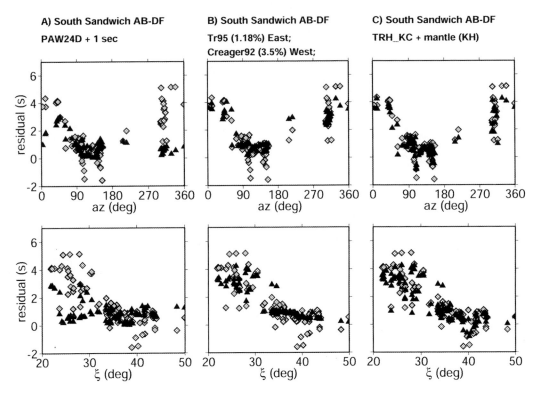

Figure 6. Variations as a function of ξ (bottom) and Azimuth (top) of PKP(AB-DF) travel time residuals for south Sandwich Island events. Comparisons of observations (diamonds) with predictions (triangles) are shown for A) mantle model PAW12B16 shifted upward by 1sec; B) hemispherical inner core anisotropy model: Tromp (1995) in the quasi eastern hemisphere and Creager (1992) in the quasi-western hemisphere; C) mantle model combining a tomographic model [*Karason and van der Hilst*, 2001] down to 300km above the CMB and the D" model TRH_KC of *Tkalčić et al.*[2002].

azimuths between 70° and 160°. On the other hand, the strength of lateral heterogeneity in D" derived in the models of *Tkalčić et al.*[2002] is also constrained by the scatter observed in PKP(BC-DF) data on non-polar paths. We see that a model such as shown in Figure 7 (c) fails to predict the average 2.5-3 sec of PKP(BC-DF) residuals on paths from south Sandwich Islands to Alaska.

We infer from Figures 6 and 7 that we cannot completely explain both the local scatter and the large scale variations in the south Sandwich events subset of PKP(BC-DF) by a model of heterogeneity in D" combined with a simple model of constant weak anisotropy in the inner core as might be compatible with normal mode splitting data. For this subset of data, it is necessary to combine D" heterogeneity with inner core anisotropy of at least 3.5%.

4. DIFFERENT GLOBAL PROJECTIONS OF THE PKP TRAVEL TIME RESIDUALS

In Figure 8a,b,c, we compare the global distribution of PKP(DF), PKP(DF-AB) and PKP(DF-BC) travel time residuals plotted at the location of the bottoming point of DF in the inner core on the one hand, and on the other, at the entry point of DF into the outer core, in the northern hemisphere. When plotted at the DF bottoming point, (only polar paths, for $\xi < 40^0$ are shown for clarity), all three datasets show the same, well documented quasi-hemispherical pattern. It is important to note that the absolute PKP(DF) residuals also show the hemispherical trend, which, on the other hand, is not clearly present in absolute PKP(BC) times, implying that it likely originates at least partly in the core. There are however notable outliers, in particular for paths to stations in Europe, bottoming at longitudes near 310^oE (both in PKP(DF) and PKP(DF-AB)). Also, the cluster of well sampled paths from the south Atlantic to Alaska "hides" many points with small residuals in the western hemisphere (those that account for the spread in residuals for $20^o < \xi < 30^o$ in Figure 1). The transition from fast to slow in the middle of the Pacific occurs very rapidly, although somewhat further east (by over 40° in longitude) in the PKP(DF-AB) dataset than in the absolute PKP(DF) one.

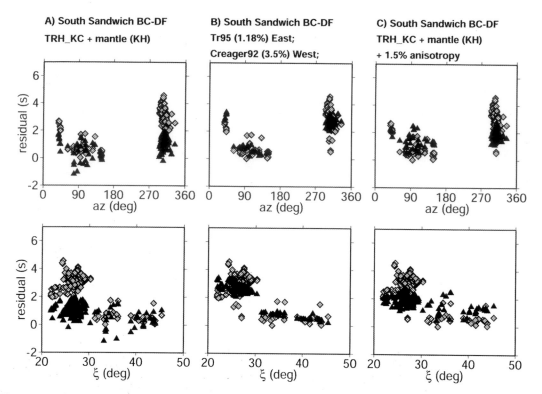

Figure 7. Same as Figure 6 for PKP(BC-DF) observations (diamonds) and predictions (triangles): A) TRH_KC model [*Tkalčić et al., 2002*]; B) Hemispherical inner core anisotropy model; C) TRH_KC plus constant inner core anisotropy model of strength 1.5% [*Romanowicz and Bréger, 2000*].

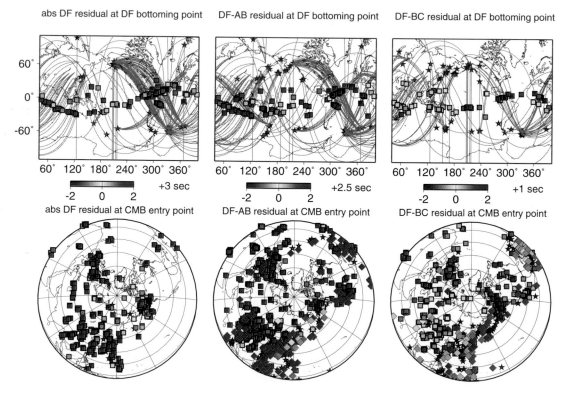

Figure 8. Travel time residuals for quasi-polar paths ($\xi < 40°$) plotted at the position of the bottoming point of the path in the inner core (top) and at the entry point into the core in the northern hemisphere (bottom). Left: absolute PKP(DF) residuals; The color code is centered at $\delta t = -3.0 sec$; Middle: PKP(DF-AB) residuals. The color code is centered at $\delta t = -2.5 sec$; Right: PKP(DF-BC) residuals. The color code is centered at $\delta t = -1.0 sec$.

On the other hand, the projections at the DF entry point into the core for these two subsets show a very similar pattern: except for a few isolated outliers, all the very anomalously fast paths concentrate in the polar region, which is unfortunately not well sampled in its center. In particular, in this projection, the cluster of mild residuals (blue) from an event in the central Pacific to stations in Europe is now compatible with other less polar and not anomalous paths to stations in Europe. Not visible at this scale, the cluster of points from south Atlantic to Alaska indicates that the more anomalous paths are on the north pole side, delineating a transition from "normal" to anomalous structure. We will return to this in more detail in Figure 9. The PKP(DF-BC) polar plot (Figure 8 (c)) is compatible with the two previous ones: if heterogeneity located in a polar region is responsible for the observed patterns, one would expect to observe anomalous PKP(DF-BC) only at the border of this region: in its center, both PKP(DF) and PKP(BC) would sense the anomaly, resulting in a small differential residual. Unfortunately, the current sampling of the polar regions is insufficient, due in particular to very noisy data combined with highly attenuated PKP(DF) on polar paths (e.g. [*Souriau and Romanowicz, 1996*]). Whether or not the anomalous structure involves the entire "polar cap" is not clear at this point, but we note that, in any case, it requires much less departure from axial symmetry in the core, than the hemispherical inner core anisotropy model.

In Figure 9, we show in more detail the distribution of absolute DF and absolute BC travel time anomalies, plotted at the entry point of the rays into the core, for 5 south-Sandwich events, and two events located further east in the south Atlantic (90/04/30 Bouvet Island event: latitude = -54.34^o; longitude = 1.341^o, depth = 7.7 km and 96/09/20 event south of Africa: latitude = -53.01^o; longitude = 9.855^0, depth = 6.7 km), recorded on the Alaska network, from which we were able to obtain waveforms with clear onsets of PKP(DF) and PKP(BC) or PKP(AB) (courtesy of R. Hansen). Figure 9 illustrates the rapid transition from normal to strongly anomalous paths from south-east to northwest under northwestern Canada and Alaska. Except for one point around $lat = 49^o, lon = -143^o$ (where the DF measurement is for an Alaska event observed at SPA and the BC measurement for a south Sandwich Island event observed in Alaska), both DF and BC (and AB) absolute times are compatible with a structure trending SW-NE, located near the CMB, with a strong gradient from fast to slow in the NW to SE direction. This structure could be a quasi-vertical "slab" of high velocity in the deep mantle, which would need to be very thin to be as yet undetected by standard mantle tomographic approaches. However, it could also be on

the core side, which would in particular make the DF and BC observations even more compatible. Thus, the non-incompatibility of the BC data (crosses) and the DF data (circles) suggests that an origin outside of the inner core, for these anomalies, is not inconceivable, in contrast to inferences made by Creager (1997) and Song (2000) based on the analysis of differential BC-DF and AB-DF travel times from these events.

Figure 10 shows a closer view of the trends of absolute DF (diamonds) and absolute BC (triangles) as a function of ξ, azimuth and epicentral distance, for the 5 south Sandwich Island events which we measured. The patterns seen in Figure 10 confirm that the BC residuals track the DF residuals, when plotted on the station side, although spatial variations have smaller amplitudes, and the trend is clearest in the plots as a function of epicentral distance. This cannot be explained by structure on the source side, where the azimuthal spans of these events partially overlap, as shown in Figure 11. On the other hand, because the variation with epicentral distance is smaller for BC than for DF, a residual trend is observed in the DF-BC data (Figure 12). When considering differential travel times, uncertainties in source location or depth, as well as near source and near station effects are eliminated. Because a similar trend is observed in absolute BC, the structure re-

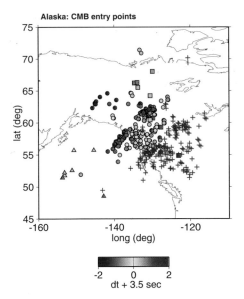

Figure 9. Absolute PKP(DF) residuals (full symbols) and PKP(BC) and PKP(AB) residuals (crosses) as a function of position of the DF entry point into the core on the Alaska side, for different events in the south Sandwich Islands and south Atlantic. 90/04/30 Bouvet Island event DF data are indicated by squares. Also shown are DF entry points for events in Alaska observed at south pole station SPA. The colors indicate relative values of the residual around the mean for each event and the color code is centered at $\delta t = -3.5$ sec.

sponsible for the variations with epicentral distance in PKP(DF-BC) should be outside of the inner core.

Figures 8-12 thus indicate that the transition from normal to anomalous paths happens over very short spatial scales. If the anomalous region is in the inner core, then a hemispherical model is necessary, and a physical explanation needs to be found for such an improbable structure. On the other hand, heterogeneity outside of the inner-core remains a possibility. We cannot rule out the possible contribution to observations in Alaska from a thin quasi-vertical slab in the lower mantle.

5. POSSIBLE ALTERNATIVE MODELS TO THE HEMISPHERICAL INNER CORE ANISOTROPY

In previous sections we have seen that it is difficult to explain PKP(BC-DF) travel time residuals with a realistic global mantle model that would not violate constraints from splitting of modes sensitive to mantle structure (e.g. [*Romanowicz and Bréger*, 2000]), that a hemispherical model of inner core anisotropy is the simplest explanation for strongly anomalous PKP data on polar paths, but shows some inconsistencies, and that there are indications from the data, in particular from the dense recordings of south-Atlantic events in Alaska that at least part of the anomaly could originate in the vicinity of the core-mantle boundary. We have previously argued that models that allow outer core heterogeneity, as first proposed by *Ritzwoller et al.*[1986] and *Widmer et al.*[1992] could provide an alternative explanation for the strongly anomalous PKP travel time data, as well as splitting data for most normal modes sensitive to core structure. Here we further argue that a hypothetical structure bounded approximately by the cylinder tangent to the inner core, with axis parallel to the earth's rotation axis, a region singled out in models of core dynamics (e.g. [*Hollerbach and Jones*, 1995; *Olson et al.*, 1999]) could create the types of trends observed in the data (e.g. [*Romanowicz and Bréger*, 2001]). Within the anomalous region, bounded by the tangent cylinder, P velocity would be about 1% faster than outside.

Figure 13 shows a comparison of the global PKP(BC-DF) dataset, plotted as a function of ξ (Figure 13 (a)), with predictions from two simple models of outer core heterogeneity of the type described above (Figure 13 (b)(c)). Both models are able to reproduce the characteristic L shape of the BC-DF trend as a function of angle ξ. The fit to individual data points depends on the details of the model, which we do not attempt to quantify any further here. Fits appear slightly better if the cylinder is tilted about 15° with respect to

the earth's rotation axis (Figure 13 (c)). However, this may be an artefact due to uneven sampling, and to the fact that the real structure may be more complex than can be accounted for by such a simple model. Indeed, some models of the dynamics of the outercore indicate the presence of irregular vorteces around the periphery of the tangent cylinder (e.g. [*J. Arnou*, personal communication; *Hulot et al.*, 2002]), so that the detailed shape of the borders of the region of fast velocity may not be exactly cylindrical. Because the sampling of polar paths largely misses the central part of the tangent cylinder, it is not possible to determine if the whole volume of the latter would contain faster than average P velocity. The main point here is to illustrate that such a class of models is geometrically plausible. This is further emphasized in Figure 14, where we only show polar paths, separated according to whether the station or an event located in southern polar regions. One feature of the data is that both subsets thus obtained show an L shaped trend (as a function of ξ), but the vertical portion of the L occurs in different ξ ranges (possibly due to uneven sampling). With slightly different cylindrical models, the trends in each of the subsets can be well reproduced. We note, in particular, that a simple model appears to also provide an explanation for the negative PKP(BC-DF) travel time anomalies around $\xi \sim 45°$ that correspond to a south Atlantic earthquake (93/03/29, lat = $-52.96°$; lon = $27.37°$; depth 24.1km) observed on the dense short period California networks, as measured by *McSweeney et al.* ([1997].

Such a heterogeneous model of the outer core, related to structure in and/or around the tangent cylinder, is compatible with free oscillation splitting data [*Romanowicz and Bréger,* 2000]. An associated negative density anomaly of the order of -0.5% inside the tangent cylinder has been suggested from normal mode data analysis [*Widmer et al.*, 1992; *Romanowicz and Bréger*, 2000], although it may be possible to fit normal mode data without any density anomaly in the outer core *Widmer et al.*, 1992]. The physical plausibility of sustained lateral heterogeneity in the outer core is generally rejected on the basis of simple dynamical arguments in the vigorously convecting outer core (e.g. [*Stevenson*, 1987]). However, because the circulation within the tangent cylinder appears to be largely isolated from that outside of it [*Hollerbach and Jones*, 1995; *Olson et al.*, 1999], one could imagine that light elements expelled from the inner core during crystallization might concentrate inside vortices in and around the tangent cylinder, giving rise to higher velocities. Whether or not the actual balance of forces allows a higher concentration of light elements in a region of the outercore, remains to be determined. At this point, both models proposed to explain first order fea-

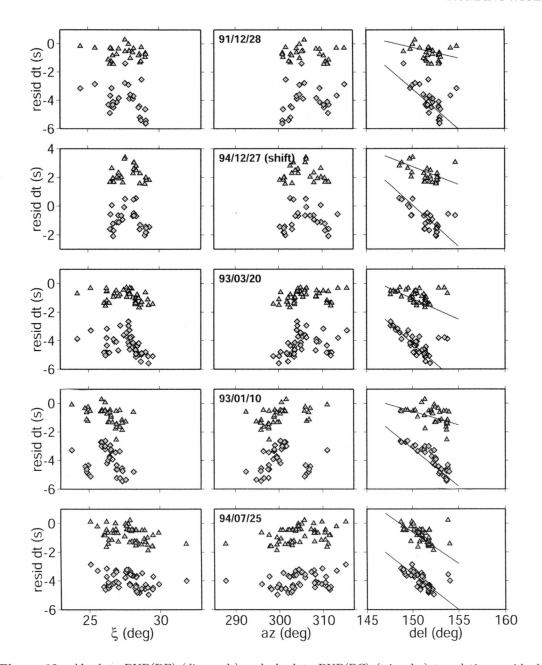

Figure 10. Absolute PKP(DF) (diamonds) and absolute PKP(BC) (triangles) travel time residuals measured across the Alaska network for five of the south Sandwich Islands events discussed in the text, plotted as a function of ξ (left), azimuth (middle) and epicentral distance (right). Note that for the 91/12/27 event, all absolute measurements have to be shifted by +3 sec, probably due to an error in the relocated epicentral parameters.

tures in the PKP travel time data: complex inner core anisotropy or outer core structure, present challenges for interpretation in terms of physical processes.

6. CONCLUSIONS

Our analysis of absolute and relative PKP travel time residuals on the global scale indicates that the hemi-

spherical pattern previously documented in PKP(BC-DF) data is also present in PKP(AB-DF), and more significantly, in PKP(DF), a priori favoring an interpretation in terms of hemispherical anisotropy in the inner core, as has previously been proposed.

It is not possible to explain this hemispherical pattern and its amplitude by a combination of realistic heterogeneity in the deep mantle and constant inner

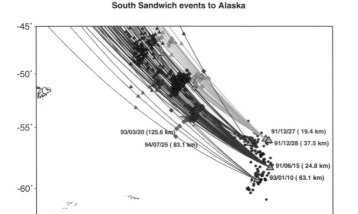

Figure 11. Location of the 6 south Sandwich Island earthquakes discussed in the text. Ray paths to the Alaska network are plotted, as well as the location of core entry points of PKP(DF) (diamonds), and PKP(BC) (triangles). Different colors are used to distinguish paths from different source locations. Only differential travel times were available for the event of 91/06/15.

core anisotropy. The hemispherical anisotropy model is however difficult to explain physically, and also fails to explain the L-shaped pattern of PKP(BC-DF) residuals as a function of ξ, as well as the details of the distribution of residuals on south Sandwich to Alaska paths. The latter could indicate the presence of a thin, quasi-vertical fast velocity slab in the deep mantle. An alternative interpretation of the trends observed, in particular the L-shape in the trend of PKP(BC-DF) travel times as a function of ξ, could involve outer core structure in the vicinity of the inner core "tangent cylinder", an important feature in outer core dynamical models, which, in particular, exhibit separate circulation within and outside the tangent cylinder. Faster than average P velocity (by 0.8 to 1%) could arise inside the tangent cylinder and/or in vortices surrounding it (e.g. [*Hulot et al.*, 2002]), and could be related to stronger concentration of light elements, as they are expelled from the inner core during crystallization. This interpretation also does not require the major departure from axial symmetry implied by the hemispherical inner core model. However, it is generally assumed that the outer core is well mixed, which does not allow any detectable heterogeneity in the outer core. Yet, such models do not account for effects of turbulence. As long as a valid physical explanation for strong non axial symmetry in the inner core, as implied by the hemispherical models, has not been found, such an alternative model may be of interest.

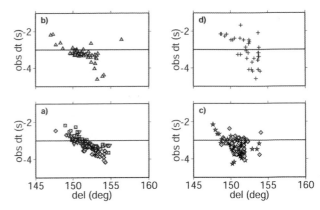

Figure 12. Variations of PKP(BC-DF) relative travel time residuals as a function of distance for different south Sandwich Island events observed at stations of the Alaska network. a) 91/12/27, 91/06/15 and 91/12/28 events. ; b) data at stations COL and INK of the global seismic network; c) 04/07/25 and 93/03/20 events; d) 93/01/10 event.

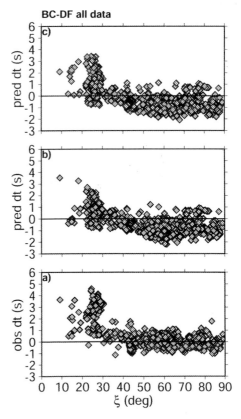

Figure 13. Observed (Bottom) and predicted (middle,top) PKP(BC-DF) travel time anomalies as a function of ξ, for two models involving cylindrical heterogeneity in the outer core. The P velocity is higher by 1% inside a cylinder of radius 1400 km surrounding the inner core, with, in b), axis parallel to the rotation axis, and in c) axis inclined towards lat $=75^{o}N$, lon $= -110^{o}E$. Both models explain over 50% of the variance in the data.

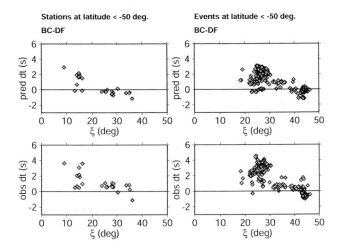

Figure 14. PKP(BC-DF) travel time residuals observed for polar paths. Left panels: paths corresponding to stations at latitudes $< -50°S$; right panels: paths corresponding to events at latitudes $< -50°S$. Stations include SPA, SYO , PMSA , SNA and SBA. Events include earthquakes in the south Atlantic, southern Indian and south Pacific ocean. Bottom: observations; top: predictions for outercore models with 1% higher velocity inside a cylinder surrounding the inner core, of radius 1250 km, of axis pointing towards lat $=75°N$, lon $=-110°E$ (top left) and lat $= 75°N$, lon $= -170°E$ (top right). The inclination of the cylinder may be an artifact due to the simplicity of the model. There are also no strong constraints on the structure in the central part of the cylinder.

Acknowledgments. This paper was completed while BR was on sabbatical leave at the CNRS in Paris, France. We acknowledge partial support from NSF grant EAR#9902777. It is BSL contribution #0209.

REFERENCES

Bergman,M. I. Measurements of elastic anisotropy due to solidification texturing and the implications for the Earth's inner core, *Nature, 389,* 60-63, 1997.

Bréger, L. and B. Romanowicz, Thermal and chemical 3D heterogeneity in D", *Science, 282,* 718-720, 1998.

Bréger, L., B. Romanowicz, and H. Tkalčić, PKP(BC-DF) travel times: New constraints on short scale heterogeneity in the deep earth? *Geophys. Res. Lett., 26,* 3169-3172, 1999.

Bréger, L., H. Tkalčić and B. Romanowicz, The effect of D" on PKP(AB-DF) travel time residuals and possible implications for inner core structure, *Earth Planet. Sci. Lett., 175,* 133-143, 2000a.

Bréger, L., B. Romanowicz and S. Rousset, New constraints on the structure of the inner core from P'P', *Geophys. Res. Lett., 27,* 2781-2784, 2000b.

Bréger, L., B. Romanowicz and C. Ng, The Pacific plume as seen by S, ScS, and SKS, *Geophys. Res. Lett., 28,*1859-1862, 2001.

Buffett, B. A. and K. C. Creager, A comparison of geode-tic and seismic estimates of inner-core rotation, *Geophys. Res. Lett., 26,* 1509-1512, 1999.

Creager, K.C., Anisotropy of the inner core from differential travel times of the phases PKP and PKIKP, *Nature, 356,* 309-314, 1992.

Creager, K.C., Inner core rotation rate from small-scale heterogeneity and time-varying travel times, *Science, 278,* 1284-1288, 1997.

Creager, K.C., Large-scale variations in inner core anisotropy, *J. Geophys. Res., 104,* 23,127-23,139, 1999.

Creager, K. C., Inner core anisotropy and rotation, in *Earth's deep interior: Mineral Physics and Seismic Tomography from the atomic to the global scale,* Geophysical Monograph 117, AGU, 2000.

Durek, J. and B. Romanowicz, Inner core anisotropy inferred by direct inversion of normal mode spectra, *Geophys. J. Int., 139,* 599-622. 1999.

Dziewonski, A. M., B. Hager and R. J. O'Connell, Large scale heterogeneities in the lower mantle, *J. Geophys. Res., 82,* 239-255, 1977.

Engdahl, E. R., van der Hilst, R. D. and Buland, R. P., Global teleseismic earthquake relocation with improved travel times and procedures for depth determination, *Bull. Seismol. Soc. Am., 88,* No 3, 722-743, 1998.

Garcia, R. and A. Souriau, Inner core anisotropy and heterogeneity level, *Geophys. Res. Lett., 27,* 3121-3124, 2000.

Garnero, E. J. and D.V. Helmberger, Seismic detection of a thin laterally varying boundary layer at the base of the mantle beneath the Central-Pacific, *Geophys. Res. Lett., 23,* 977-980, 1996.

Grand, S.P., R. D. van der Hilst, and S. Widiyantoro, Global seismic tomography: a snapshot of convection in the earth, *GSA Today, 7,* 1-7, 1997.

Hollerbach, R. and C. A. Jones, On the magnetically stabilizing role of the Earth's inner core, *Phys. Earth Planet. Inter., 87,* 171-181, 1995.

Hulot, G., C. Eymin, B. Langlais, M. Madea and N. Olsen, Small-scale structure of the geodynamo inferred from Oersted and Magsat satellite data, *Nature, 416,* 620-623, 2002. Jeanloz, R. and Wenk, H.R., Convection and anisotropy of the inner core, *Geophys. Res. Lett. 15,* 72-75, 1988.

Karason, H and R. D. vander Hilst, Tomographic imaging of the lowermost mantle with differential times of refracted and diffracted core phases (PKP, Pdiff), *J. Geophys. Res., 106,* 6569-6587, 2001.

Karato, S., Seismic anisotropy of the earth's inner core resulting from flow induced by Maxwell stresses, *Nature, 402,* 871-873, 1999.

Karato, S., Inner core anisotropy due to magnetic field-induced preferred orientation of iron, *Science,262,* 1708-1711, 1993.

Li, X-D., D. Giardini, and J. H. Woodhouse, Large-scale three-dimensional even-degree structure of the Earth from splitting of long-period normal modes, *J. Geophys. Res., 96,* 551-557, 1991.

Liu, X. F. and A.M. Dziewonski, Global analysis of shear wave velocity anomalies in the lowermost mantle, in *The core-mantle boundary region,* Geodyn. ser. vol. 28, edited by M. Gurnis et al., pp 21-36, AGU, Washington, D.C., 1998.

Masters, G. and F. Gilbert, Structure of the inner core inferred from observations of its spheroidal shear modes, *Geophys. Res. Lett., 8*, 569-571, 1981.

Masters G., Johnson, S., Laske, G. and Bolton, B., A shear-velocity model of the mantle, *Philos. Trans. R. Soc. Lond. A, 354*, 1,385-1,411, 1996.

McSweeney, T. J., K. C. Creager and R. T. Merrill, Depth extent of inner-core seismic anisotropy and implications for geomagentism, *Phys. Earth Planet. Inter., 101*, 131-156, 1997.

Mégnin, C. and Romanowicz, B., 2000. The 3D shear velocity structure of the mantle from the inversion of body, surface, and higher mode waveforms, *Geophys. J. Int., 143*, 709-728, 2000.

Morelli, A., A.M. Dziewonski, and J. H. Woodhouse, Anisotropy of the core inferred from PKIKP travel times, *Geophys. Res. Lett., 13*, 1545-1548, 1986.

Ni, S. and D. V. Helmberger, Low-velocity structure beneath Africa from forward modeling, *Earth Planet. Sci. Lett., 170*, 497-507, 1999.

Olson, P., U. Christensen and G. Glatzmeier, Numerical modeling of the geodynamo: mechanisms of field generation and equilibration, *J. Geophys. Res, 104*, 10,383-10,404, 1999.

Poupinet, G., R. Pillet and A. Souriau, Possible heterogeneity of the Earth's core deduced from PKIKP travel times, *Nature, 305*, 204-206, 1983.

Ritsema, J., S. Ni , D. V. Helmberger, and H.P. Crotwell, Evidence for strong shear velocity reductions and velocity gradients in the lower mantle beneath Africa, *Geophys. Res. Lett., 25*, 4245-4248, 1998.

Ritsema, J. , H. van Heijst and J. Woodhouse, Complex shear wavev velocity structure imaged beneath Africa and Iceland, *Science, 286*, 1925-1928, 1999.

Ritzwoller, M., G. Masters, and F. Gilbert, Observations of anomalous splitting and their interpretation in terms of aspherical structure, *J. Geophys. Res., 91*, 10203-10228, 1988.

Romanowicz, B. and L. Bréger, Anomalous splitting of free osccilations: a reevaluation of possible interpretations, *J. Geophys. Res., 105*, 21,559-21,578, 2000.

Romanowicz, B., X.-D. Li, and J. Durek, Anisotropy in the inner core; could it be due to low-order convection? *Science, 274*, 963-966, 1996.

Sacks, I. S., Snoke, J. A. and Beach, L., Lateral heterogeneity at the base of the mantle revealed by observations of amplitudes of PKP phases, *Geophys. J. R. astr. Soc., 59*, 379-387, 1979.

Shearer, P.M., K.M. Toy, and J.A. Orcutt, Axi-symmetric Earth models and inner core anisotropy, *Nature, 333*, 228-232, 1988.

Shearer, P.M., PKP(BC) versus PKP(DF) differential travel times and aspherical structure in the Earth's inner core, *J. Geophys. Res., 96*, 2233-2247, 1991.

Song, X.-D., Anisotropy in central part of inner core, *J. Geophys. Res., 101*, 16,089-16,097, 1996.

Song, X.-D., Joint inversion for inner core rotation, inner core anisotropy and mantle heterogeneity, *J. Geophys. Res., 105*, 7931-7943, 2000.

Song, X.-D. and D.V. Helmberger, Seismic evidence for an inner core transition zone, *Science, 282*, 924-927, 1998.

Souriau, A., P. Roudil and B. Moynot, Inner core differential rotation, facts and artefacts,*Geophys. Res. Lett., 24*, 2103-2106, 1997.

Stevenson, D. J., Limits on lateral density and velocity variations in the Earth's outer core, *Geophys. J. R. Astr. Soc., 88*, 311-319, 1987.

Stixrude, L. and R. Cohen, High-pressure elasticity of iron and anisotropy of Earth's inner core, *Science, 275*, 1972-1975, 1995.

Sylvander, M. and A. Souriau, P-velocity structure of the core-mantle boundary region inferred from PKP(AB)-PKP(BC) differential travel times", *Geophys. Res. Lett., 23*, 853-856, 1996.

Su, W. and A.M. Dziewonski, Inner core anisotropy in three dimensions, *J. Geophys. Res., 100*, 9831-9852, 1995.

Tanaka, S. and H. Hamaguchi, Degree one heterogeneity and hemispherical variation of anisotropy in the inner core from PKP(BC)-PKP(DF) times, *J. Geophys. Res., 102*, 2925-2938, 1997.

Tkalčić, H., B. Romanowicz and N. Houy, Constraints on D" structure using PKP(AB-DF), PKP(BC-DF) and PcP-P travel time data from broadband records, *Geophys. J. Int., 149*, 599-616, 2002.

Tromp, J., Support for anisotropy of the Earth's inner core from splitting in free oscillation data, *Nature, 366*, 678-681, 1993.

Tromp, J., Normal-mode splitting observations from the great 1994 Bolivia and Kuril Islands earthquakes: constraints on the structure of the mantle and inner core, *GSA Today, 5*, 137-151, 1995.

Vinnik, L., B. Romanowicz and L. Bréger, Anisotropy in the center of the Inner Core, *Geophys. Res. Lett., 21*, 1671-1674, 1994.

Vinnik, L., Farra, V. and Romanowicz, B. Observational evidence for diffracted SV in the shadow of the Earth's core. *Geophys. Res. Lett. 16*, 519-522, 1989.

Widmer, R., G. Masters, and F. Gilbert, Observably split multiplets-data analysis and interpretation in terms of large-scale aspherical structure, *Geophys. J. Int., 111*, 559-576, 1992.

Woodhouse, J.H., D. Giardini, and X.-D. Li Evidence for inner core anisotropy from splitting in free oscillation data,*Geophys. Res. Lett., 13*, 1549-1552, 1986.

Wysession, M. E., Large-scale structure at the core-mantle boundary from diffracted waves, *Nature, 382*, 244-248, 1996.

Yoshida, S., Sumita,I. & Kumazawa, M. Growth model of the inner core coupled with outer core dynamics and the resultant elastic anisotropy.*J. Geophys. Res., 101*, 28,085-28,103, 1996.

B.Romanowicz, H. Tkalcic, L. Bréger, Berkeley Seismological Laboratory, Berkeley, CA, 94720, USA. (e-mail: barbara@seismo.berkeley.edu

Three-Dimensional Structure and Differential Rotation of the Inner Core

Xiaodong Song

University of Illinois at Urbana-Champaign, Illinois

In this paper, I review the 3-D structure and differential rotation of the inner core based on seismological observations made during the past 5 years. Inner core structure appears to vary strongly on all scales (from 1000 kms to a few kms). The uppermost 100-400 km of the inner core is nearly isotropic but significant anisotropy (~3% on average) persists in the remainder of the inner core to the Earth's center. A new inner-core seismic triplication has been observed, providing direct evidence for a transition from isotropy to anisotropy in the inner core. The western hemisphere is much more anisotropic, averaged over the upper few hundred kms of the inner core, than the eastern hemisphere. These observations lead to a proposal of an inner-core transition zone model, with an irregular boundary separating the isotropic upper inner core (UIC) and the anisotropic lower inner core (LIC). I suggest that such a model may provide a unified explanation of both the observed depth variation and the observed lateral variation of the inner-core anisotropy, as well as of the well-known anomalous PKP(DF) waveforms along polar paths at distances of about 150°. Our waveform modeling of one sample of the inner core from earthquakes in the South Sandwich Islands to Alaska stations indicates a huge amplitude of anisotropy, 7-8%, in the lower inner core, which is two to three times the previous estimates of the anisotropy assuming a uniform model. Significant topography (over 100 km of relief across 30° distance) of the UIC/LIC boundary may explain the steep lateral gradient in the inner core previously observed along the South Sandwich Islands – Alaska path. These fine 3-D structures of the inner core provide new insights into the physical properties of the inner core as well as essential markers for the detection of the inner-core rotation. A growing body of evidence now supports the eastward rotation of the inner core (super-rotation) relative to the mantle. A westward rotation can be ruled out. The current estimates of the rotation rate range from 0.15 to 1.1°/yr. Inner core super-rotation has been suggested to be an artifact resulting from mantle heterogeneity, heterogeneous event magnitudes, and, in particular, event mislocations. We conclude that none of these biases could explain the

Earth's Core: Dynamics, Structure, Rotation
Geodynamics Series 31
Copyright 2003 by the American Geophysical Union
10.1029/31GD05

observations that travel times through the inner core along certain pathways have changed, which implies an inner-core motion.

1. INTRODUCTION

The Earth's solid inner core, separated from the mantle by the fluid outer core, will rotate relative to the mantle if subjected to an external torque. Gubbins [1981] first suggested that electromagnetic forces between the electrically conducting inner core and the magnetic field generated in the fluid outer core would cause a differential rotation of the inner core. In a computer simulation of a three-dimensional, self-consistent geodynamo, Glatzmaier and Roberts [1995] predicted that the inner core is driven to rotate by electromagnetic coupling, generally at a few degrees per year faster than the mantle. Motivated in part by the work of Glatzmaier and Roberts [1995], we searched for evidence for differential inner-core rotation by seismic waves that pass through the core [Song and Richards, 1996]. In particular, we found that the differential travel times between the BC and DF branches of PKP waves (Figure 1) along the pathway from earthquakes in South Sandwich Islands (SSI) to station COL (College, Alaska) increased systematically by 0.3 s over a 28-year time interval. We estimated that the inner core rotates at about 1°/year faster than the mantle. There is now growing support for such inner-core rotation, with a rotation rate of a fraction of a degree per year (Table 1), but this proposal remains under debate.

In this paper, I review seismological observations that demonstrate inner-core rotation and discuss possible artifacts and uncertainties in determining the rotation. I first focus my discussion on recent work concerning 3-D structures and anisotropy of the inner core. Three-dimensional fine structures of the inner core provide not only new insights into the physical properties and the evolution and dynamics of the inner core, but also provide essential markers for the detection of the inner-core rotation. The studies discussed in the paper involve various branches of seismic PKP body-wave phases, which penetrate the core (Figure 1). They include waves that pass through the inner core, the DF branch (also called PKIKP); waves reflected off the inner-core boundary (ICB), the CD branch (also called PKiKP); and waves that turn in the middle and the bottom of the outer core, the AB and BC branches, respectively. Differential travel times between different branches, such as differential BC-DF times between DF and BC branches and differential AB-DF times between DF and AB branches, as well as total waveforms involving various branches, are often used to reduce earthquake location errors and biases from upper-mantle heterogeneities.

2. 3-D STRUCTURE OF THE INNER CORE

Evidence for the inner-core anisotropy, first proposed by Morelli et al. [1986] and Woodhouse et al. [1986], was well established from studies of travel times and normal-mode data collected in the 1990s [e.g., Creager, 1992; Song and Helmberger, 1993; Tromp, 1993, 1995; Shearer, 1994; Vinnik et al., 1994; Su and Dziewonski, 1995; Song, 1996; McSweeney et al., 1997]. The anisotropy appears to be dominantly cylindrical, with the axis of symmetry aligned approximately with the north-south (NS) spin axis of the Earth. Compressional P waves traversing the inner core along the NS direction are about 3% faster than those propagating parallel to the equatorial plane (see recent reviews by Song [1997], Creager [2000], Tromp [2001], and Song [2002]). More recent studies, however, suggest that the inner core is much more complex than a simple model of uniform axisymmetric anisotropy.

2.1. Inner-core Transition Zone: An Isotropic Upper Inner Core Overlaying An Anisotropic Lower Inner Core

Inner-core anisotropy varies strongly with depth and significant anisotropy appears to extend to the center of the Earth. Differential PKP AB-DF travel times are consistent with an anisotropy of about 3.0% averaged over the whole in the inner core [Vinnik et al., 1994; Song, 1996; McSweeney et al., 1997; Sun and Song, 2001].

Though the inner core as a whole is anisotropic, the outermost part of the inner core is nearly isotropic. The idea was first suggested by Shearer [1994], based on DF arrival times from the International Seismological Centre (ISC) bulletins, and by Song and Helmberger [1995a], based on detailed modeling of PKP waveforms.

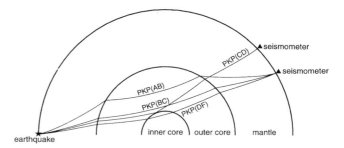

Figure 1. Ray paths of various branches of PKP waves, which penetrate the Earth's core. PKP(CD) is also called PKiKP and PKP(DF) is also called PKIKP. PKP waves are commonly used in body-wave studies of the Earth's core.

Table 1. Estimates of inner core rotation

Model	Reference	Rate, °/yr	Data Used
1	Song and Richards [1996]	1.1	BC-DF times from SSI to COL, Alaska
2	Su and Dziewonski [1996]	3.0	ISC PKP(DF) absolute times
3	Creager [1997]	0.2-0.3	BC-DF from SSI to Alaska
4	Souriau [1998]	< 1	BC-DF times from Novaya Zemlya to DRV
5	Dziewonski and Su [1998]	not detectable	ISC PKP(DF) absolute times
6	Ovchinnikov et al. [1998]	≥0.8	BC-DF times from Noyaya Zelmya to NVL
7	Laske and Masters [1999]	$0\pm0.20(1\sigma)$	Normal modes
8	Souriau and Poupinet [2000]	< 1	Worldwide absolute and differential times
9	Poupinet et al. [2000]	< 0.2	12 SSI event pairs at COL
10	Song [2000a]	0.3-1.1	BC-DF times from SSI to Alaska
11	Song [2000b]	same as Models 1 and 10	65 SSI event pairs at COL
12	Song and Li [2000]	0.6	BC-DF from Alaska to South Pole
13	Vidale et al. [2000]	0.15	Inner core scattering waves
14	Song [2001]	same as Models 1 and 10	Same 12 SSI event pairs as Model 9
15a	Li and Richards [2001]	weak time shift	Novaya Zelmya to SNA, DRV
15b	Li and Richards [2001]	robust time shift	Novaya Zelmya to NVL, SBA
16	Xu and Song [2002]	0.29-0.50	BC-DF times from SSI to Beijing stations
17	Li and Richards, in preparation	0.35-0.79	1 SSI doublet event pair at COL

The upper 150 km of the inner core is best resolved from PKP waveforms at distances of 130° to 146° [e.g., Choy and Cormier, 1983; Cummins and Johnson, 1988; Song and Helmberger, 1992]. Song and Helmberger [1995a] explored the depth dependence of inner-core anisotropy using PKP waveforms of near-polar paths at distances 120° to 173°. The results suggest that the upper 150 km is only weakly anisotropic (less than 1%) and the upper 60 km appears to be isotropic. More recent studies confirmed that the upper 100-200 km of the inner core is nearly isotropic [Song and Helmberger, 1998; Creager, 2000; Niu and Wen, 2001; Garcia and Souriau, 2000; Ouzounis and Creager, 2001a]. Figure 2 shows an example of the dramatic differences between waves that sample the uppermost inner core (at distances from 120° to 140°) and waves that travel in a similar direction, but sample a greater depth (at about 150°). Synthetics from PREM model [Dziewonski and Anderson, 1981] predict the data at small distances very well without any requirement of anisotropy. But the DF arrival at 150° (bottom panel) is 2.9 s earlier than the PREM prediction, consistent with an anisotropy of over 3% averaged over the whole sampling depth.

A new seismic triplication was identified from PKP(DF) waves traveling along polar paths, providing direct evidence for a sharp inner-core transition from the uppermost inner core to its depth [Song and Helmberger, 1998]. Song and Helmberger [1998] proposed an inner-core transition-zone model with an isotropic upper inner core (UIC) overlying an anisotropic lower inner core (LIC) at a depth of about 200 km below the ICB (Figure 3).

Figure 4 shows one best example of evidence for an inner-core triplication. Normally seismograms from earth-west (EW) paths, like those in the second column from the left, show DF waveforms quite similar to the corresponding BC waveforms with the exception of a slight broadening and less high-frequency content in the DF waveforms due to inner-core attenuation [Doornbos, 1974]. In this example, the time differences between DF and BC arrivals and their waveforms are well predicted by the smooth reference model PREM2 [Song and Helmberger, 1995b]. In contrast, the DF phases from the NS paths arrive earlier than predicted by the reference model, as attributed to inner-core anisotropy, and have much broader waveforms than the corresponding BC or AB phases (Figure 4; 3rd column from the left). A model with a P velocity jump of 4.3% at 250 km below the ICB reproduces the broad DF waveforms of the NS paths reasonably well (the rightmost column). The broad "DF" waveforms in these synthetics are the results of overlapping energy of the waves that are refracted in the lower inner core (LIC) (the DFcd branch) and the waves that are refracted at the bottom of the upper inner core (UIC) (the DFab branch) or diffracted along the boundary (diffraction beyond the b-cusp) (Figure 3). To produce broad enough waveforms, the boundary needs to be relatively sharp (~50 km in thickness). However, there is some trade-off in the model between the transition depth and the velocity jump [Song and Helmberger, 1998].

Inner-core triplication can also been seen in short-period waveforms. Song and Helmlberger [1998] reported direct reflections from the inner-core transition

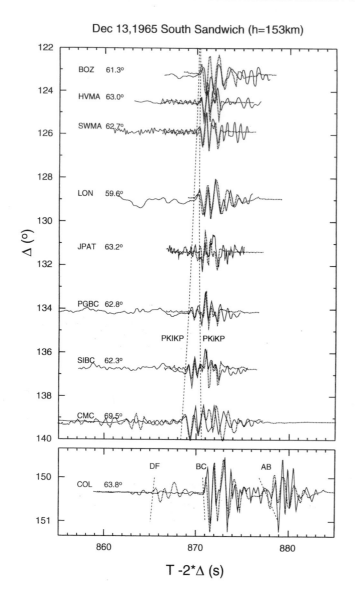

Dec 13,1965 South Sandwich (h=153km)

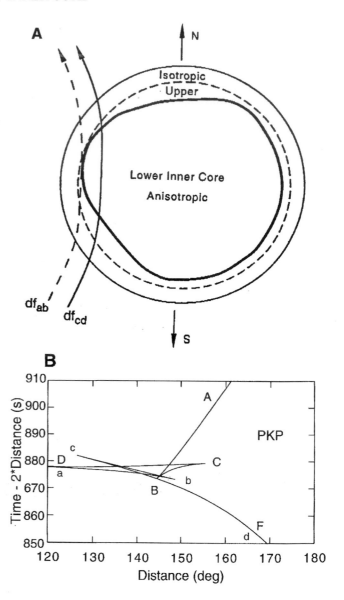

Figure 2. Distinct differences of PKP waves sampling the very top of the inner core and those sampling the deeper part [from Song and Helmberger, 1995a]. The data (solid) are from December 13, 1965 SSI earthquake recorded at stations in North America. The synthetics (dotted) and travel-time predictions (dashed lines) are calculated for PREM (dotted) [Dziewonski and Anderson, 1981]. The synthetics match the data at distances 123° to 139° very well but the DF prediction is much delayed around 150°, indicating isotropy at the very top of the inner core and strong anisotropy at depth.

boundary (the DFbc branch, Figure 3) from an earthquake in Drake Passage to the German Regional Seismic Network at distances between 120 to 130°. Figure 5 (left) shows a short-period record section from a SSI earthquake to the Alaska Seismic Network (ASN) at

Figure 3. Inner-core transition-zone model proposed by Song and Helmberger [1998]. **(A)** A schematic illustration of isotropic upper inner core (UIC) and anisotropic lower inner core (LIC) structure. The UIC/LIC boundary (solid line) would give rise to multiple paths at certain distances for seismic waves traveling nearly NS through the inner core, producing distorted waveforms in long-period seismograms and multiple arrivals in short-period seismograms. The boundary is speculated to be irregular, which may explain recent reports of large scatter in inner-core travel times. **(B)** Travel-time curves of PKP for an Earth model that include a two-layered inner core with a velocity discontinuity at the boundary. Because of the discontinuity, waves that go through the inner core produce three branches (triplication) of arrivals, instead of one branch (DF): waves that turn in the upper inner core (DFab), waves that are reflected at the boundary (DFbc), and waves that turn in the lower inner core (DFcd).

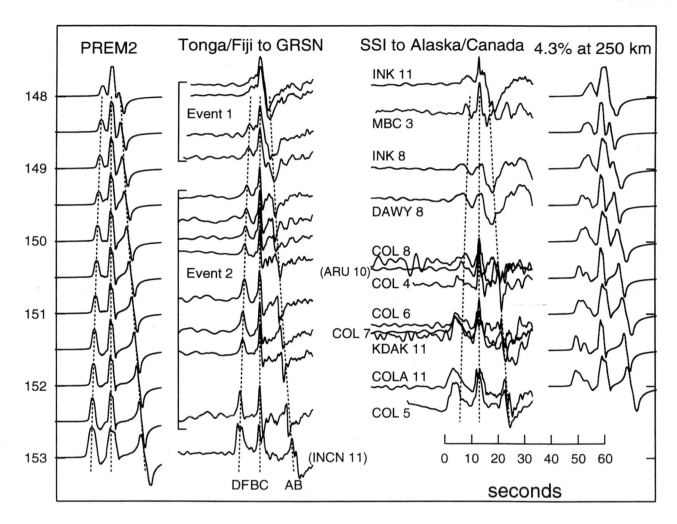

Figure 4. Evidence for a triplication from broadband displacement waveforms caused by an inner-core transition zone [from Song and Helmberger, 1998]. From left to right columns, synthetics for PREM2 [Song and Helmberger, 1995b], data from EW paths, data from NS paths, and synthetics for a model with a 4.3% P velocity jump at 250 km below the ICB, respectively. The waveforms of the DF phases from the EW paths are similar to the corresponding BC waveforms. In contrast, the DF phases from the NS paths arrive much earlier and have much broader waveforms than the corresponding BC and AB phases. The timing as well as broad DF waveforms from the NS paths are reasonably reproduced in the synthetics for the model with a velocity jump inside the inner core.

larger distances [Song and Xu, 2002]. Two distinct arrivals are clearly visible in the DF time window, which are not in the BC time window. This can be explained by an inner-core transition structure. Synthetics for a model with 250-km thick isotropic UIC overlying an anisotropic LIC with 8% anisotropy (or a velocity jump of about 5% at 250 km below the ICB for the SSI-ASN paths) reproduces the multiple DF arrivals reasonably well (Figure 5, right), although the details of the model may vary.

The details of an UIC/LIC transition zone have yet to be resolved. But because of the strong evidence for lack of anisotropy at the uppermost inner core, while there is strong anisotropy in the lower inner core, such a transition seems inevitable. The amplitude of anisotropy in the lower inner core, newly inferred from the modeling of the transition zone, is much greater than that found in any of the previous studies of the inner-core anisotropy. A velocity jump of 4.3% or 5% at 250 km below the ICB, as discussed above for the SSI-Alaska path, is equivalent to an anisotropy amplitude of about 7-8% along the fastest NS direction.

Ouzounis and Creager [2001a] inverted a set of differential PKP times sampling the same region and sug-

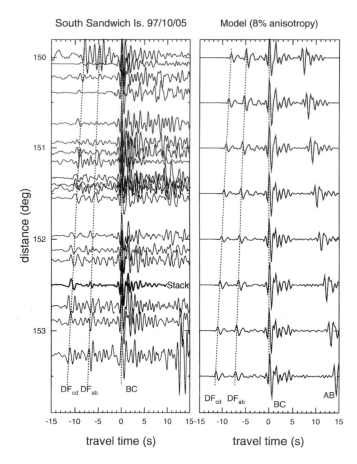

Figure 5. Evidence for a triplication from short-period waveforms caused by an inner-core transition zone (modified from Song and Xu [2002]). **(Left)** Short-period vertical seismograms from October 5, 1997 SSI earthquake to stations in Alaska. The traces are aligned and normalized with respect to BC. The DF segments have been amplified by a factor of three for visualization. The highlighted trace is a stack of the data. The stacked trace was obtained by aligning the segments before and after BC arrivals relative to the first signals of DF and relative to BC, respectively. Two distinct arrivals are clearly visible in the DF time window in almost all the traces but not in the BC coda. **(Right)** Generalized ray synthetics calculated for a model with a 250-km thick isotropic UIC and 8% anisotropy in the LIC (or 5.1% velocity jump for this particular path). We use the stacked BC waveform as the source-time function and the inner-core attenuation $t^*=0.35$ s of PREM2 (or Q about 385 at 152°). The DF segments have also been amplified by a factor of three to be compared with the data. The two DF arrivals come from refractions turning above (DFab) and below (DFcd) the discontinuity, respectively. The synthetics generally match relative times between the two DF arrivals and the BC phase of the data. In addition, the synthetics generally match the anomalously small DF amplitudes of the data.

gested that thickness of the isotropic layer is 50-150 km and that the anisotropy in the lower inner core is as high as 5%. They used a linear inversion, which assumes the same ray paths as those for an isotropic inner core. This appears to be the main cause for the differences in their estimates of the depth of the boundary and the amplitude of the anisotropy at depth and the estimates provided by Song and Helmberger [1998] and Song and Xu [2002] as discussed above. A non-linear inversion by Ouzounis and Creager [2001b] is compatible with the model of a 250-km isotropic layer overlying a lower inner core with 8% anisotropy. The boundary between these two layers is likely not sharp enough to produce reflections because no such reflections from the boundary were observed by Ouzounis and Creager [2001b].

Finally, it has been well recognized that short-period DF waveforms are anomalous near 150° along NS paths [Creager, 1992; Song and Helmberger, 1993], but the cause of this anomaly has been a mystery [e.g., Masters, 1993; Shearer, 1994; Song, 1997]. The DF amplitudes are small and waveforms are complex and sometimes show multiple arrivals. Examples are shown by Figure 5 (the DF waveforms have been amplified by a factor of 3), although distinct multiple DF arrivals are not always as easy to see as in this figure. We suggest that a UIC/LIC transition zone may be the cause of the anomaly [Song and Xu, 2002]. The interference between multipathed rays from such a structure would produce broadened long-period waveforms [Song and Helmberger, 1998] and, at short periods, multiple arrivals or complex waveforms (e.g. when affected by a complex source-time function or by a short-wavelength irregular boundary). Note the model in Figure 5 predicts the small short-period DF amplitudes very well, assuming a reasonable value for inner-core attenuation with $t^*=0.35$ s (or about Q=385 at 152°) as in PREM2 [Song and Helmberger, 1995b]. The small DF amplitudes are essentially the results of defocusing, in that the total DF energy is partitioned to multiple paths through such a transition zone.

2.2. Lateral Variations

There is now growing evidence for strong lateral variation in inner-core structure. The inner core appears to vary on all scales, from the scale of half a hemisphere [Tanaka and Hamaguchi, 1997; Creager, 1999; Niu and Wen, 2001], to the scale of a few hundred km [Creager, 1997; Song, 2000a], to the scale of a few km [Cormier et al., 1998, Vidale and Earle, 2000]. If inner-core attenu-

ation is due to scattering, Cormier *et al.* [1998] suggests that the required inner-core fabric may consist of crystals with a P velocity contrast of 5-12% at scale lengths of 0.5-2 km.

Recently Niu and Wen [2001] hand-picked 200 globally distributed CD-DF times. They confirmed that the upper 100 km of the inner core is isotropic. More importantly, the results suggest a difference of velocity between two hemispheres at the uppermost inner core, with the quasi-eastern hemisphere (40-180°E) being 0.8% isotropically faster than the quasi-western hemisphere (180°W-40°E). At intermediate depth, 100 to 400 km below the ICB, differential BC-DF times suggest that the inner core possesses a hemispherical pattern of a different form. The quasi-western hemisphere is strongly anisotropic (with the average of the anisotropy of about 3%) but the quasi-eastern hemisphere is nearly isotropic [Tanaka and Hamaguchi, 1997; Creager, 1999]. However, in the central inner core (600 km or so below the ICB), significant anisotropy (~3%) seems to exist in both hemispheres [Song, 1996; Creager, 2000; Sun and Song, 2001]. At a smaller scale, Creager [1997] and Song [2000a] observed a sharp lateral-velocity gradient in the inner core from SSI earthquakes to ASN stations (Figure 6). The velocity perturbation increases by about 1% (or about 1.3 s in differential BC-DF times) from east to west over a distance of about 30°.

In proposing an inner-core transition zone model with an irregular boundary (Figure 3), Song and Helmberger [1998] speculated that the irregular boundary may also provide an explanation for the observed lateral variation of anisotropy, in additional to providing an explanation for the depth variation. If the boundary is deeper in the eastern hemisphere than in the western hemisphere, the model would explain the smaller BC-DF residuals in the eastern hemisphere as observed by Tanaka and Hamaguchi [1997] and Creager [1999]. Indeed, using a globally distributed data set of BC-DF and AB-DF times, Creager [2000] inverted for a model containing a discontinuity separating the isotropic and the anisotropic regions with smooth topography. He found that the trends of the data versus ray direction, distance, and longitude are generally consistent with a model with the discontinuity at depth of 100 km in the western hemisphere and at depth of 100 to 400 km in the eastern hemisphere.

At a smaller scale, local topography of the discontinuity may provide an explanation to the sharp lateral gradient sampled by the SSI-ASN paths. A simple calculation using 1-D ray tracing suggests that if the

depth of the discontinuity deepens 90 km, from 190 to 280 km, from west to east over about 30° distance (A to A′) with a velocity jump of 5% across the boundary along this ray direction (or equivalent to 8% anisotropy at the LIC), the lateral trend of the BC-DF times can be explained (Figure 6). This is speculative, however. The lateral gradient could be the result of different degree of alignment of iron crystals. In addition, there appears to be evidence for a strong lateral gradient at the isotropic uppermost inner core in the same region [Niu and Wen, 2001].

The symmetry axis of the inner-core anisotropy may be tilted from the spin axis but the inference is uncertain [Shearer and Toy, 1991; Creager, 1992; Su and Dziewonski, 1995; Song and Richards, 1996; McSweeney *et al.*, 1997; Souriau *et al.*, 1997; Dziewonski and Su, 1998]. Su and Dziewonski [1995] inferred, from DF

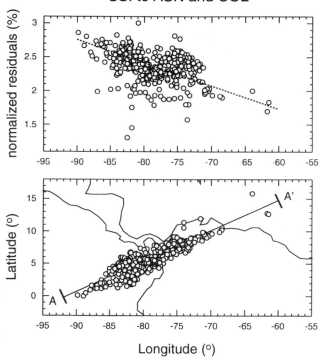

Figure 6. **(Top)** Residuals of BC-DF times from SSI earthquakes in 1990s to stations in Alaska, normalized by the travel times through the inner core. **(Bottom)** Turning points of the DF rays in the inner core. Note the residuals decrease sharply from west to east, which can be predicted (top, dashed line) by an inner-core transition zone model with an isotropic layer at the top and 5% of anisotropy at depth and the boundary deepening from 190 to 280 km from west to east (A to A′).

r
b
i
s
1
c
v
a
a
t
d
1
o
n
a

ς
I

ii
o
w
r
[ς
2
sɛ
o
fɛ
tl
ir
tl
D
si
tı
aι
n
sι
E
re
m

re
eɛ
19
(E
Sε
eɛ
aı
th

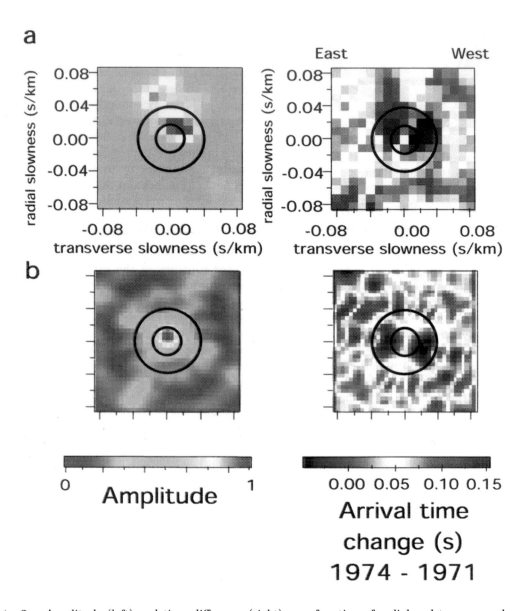

Plate 3.   Amplitude (left) and time difference (right) as a function of radial and transverse slowness [modified from Vidale *et al.*, 2000]. **(a)** Observed ICS from two Novaya Zelmya nuclear explosions recorded at LASA. **(b)** Simulated ICS with 0.45 rotation calculation. Seismic waves that are vertically incident, for example, appear in the middle of such a plot, at 0.0 km/s transverse and radial slowness. The ring of rays that would graze the inner-core boundary and core-mantle boundary are shown by the black circles in each frame. Right side shows the distribution of change in arrival time between 1971 and 1974 as a function of radial and transverse slowness. The observed ICS arrivals **(a)** from the east are later (red) in 1974 than in 1971, and the ICS arrivals from the west are earlier (blue). The pattern is reproduced with the simulated ICS with 0.45° eastward rotation calculation **(b)**, suggesting an inner-core rotation of 0.15° per year.

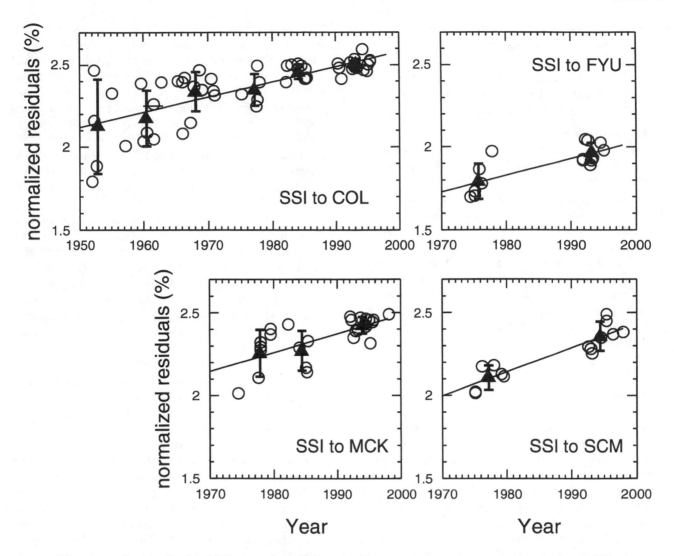

Figure 8. Residuals of BC-DF times from SSI earthquakes recorded at COL and three ASN stations as a function of earthquake time of occurrence [from Song, 2000a]. The residuals are calculated relative to PREM [Dziewonski and Anderson, 1981], and are normalized by the times (about 120 to 130 s) that the rays travel through the inner core. The observed residuals are indicated by open circles. The mean (solid triangles) and plus or minus one standard deviation (error bar) are shown for the residuals over about 5-year periods. The lines are the linear regressions of the residuals on event occurrence times. The BC-DF residuals increase consistently with time at all four stations. The time-dependence is interpreted as evidence for differential inner-core rotation.

Su [1998] of the temporal variation of the total pattern of the inner-core anomalies suggests the rotation is not detectable from the ISC data.

Souriau [1998a] examined BC-DF times from Novaya Zemlya nuclear explosions recorded at DRV, Antarctica over 24 years and did not observed robust temporal change. She concluded that an eastward rotation of 3°/yr (as in Su et al. [1996]) is not possible, but a rate of 1°/yr or lower could not be ruled out. A test at the worldwide scale using PKP absolute and differential times was also consistent with no rotation but

could not rule out a rotation of less than 1°/yr [Souriau and Poupinet, 2000]. Li and Richards [2001] studied BC-DF or AB-DF times from Novaya Zemlya nuclear explosions recorded at four stations (SNA, NVL, DRV, and SBA) in Antarctica. They observed robust time-dependence at NVL and SBA, weak time-dependence at DRV, but little time-dependence at SNA.

Results from recent normal-mode studies on inner-core rotation are uncertain. Sharrock and Woodhouse [1998] appear to favor a westward rotation component, but Laske and Masters [1999] found that inner-core rotation is less than $\pm 0.2^{\circ}$ per year (to within one standard deviation) over the last 20 years.

4. POTENTIAL BIASES IN DETERMINATION OF INNER-CORE ROTATION

The claim that the inner core is rotating relative to the mantle faces challenges and is still under debate. First, some attempts have not resolved rotation [Souriau, 1998a; Souriau and Poupinet, 2000; Laske and Masters, 1999], as pointed out above. Second, some authors [Souriau et al., 1997; Souriau, 1998a,b,c; Poupinet et al., 2000] argue that the detection of the inner-core rotation from the BC-DF times of SSI-COL path [Song and Richards, 1996] could be the artifacts of potential biases from event mislocations, mantle heterogeneity, and heterogeneous event magnitudes. These are important questions, which have been examined by Song [2000b, 2001] and Song and Li [2000]. I address briefly the main criticisms below.

4.1. Systematic Event Mislocations

The chief problem with the detection of inner-core rotation using differential BC-DF times is possible systematic event mislocations. The observed temporal variations could potentially be an artifact of the use of different global networks used to locate the earthquakes in different years. This problem is particular important for SSI earthquakes, since stations within 20° of epicentral distances are rare. However, at present, there are six independent lines of arguments against the proposal that event mislocation is the cause for the observed temporal changes along the SSI-COL path. The first three points, discussed in detail by Song [2000a], are: (1) The temporal trend at COL is extremely robust, no matter what earthquake locations (Earthquake Data Report, EDR; relocations by Engdahl et al., EHB; and the Joint Hypercenter Determination, JHD) are used. (2) Station distributions used to locate the SSI events do not show obvious shifts over time. (3) Data recorded at the Alaska Seismic Network over a period of 8 years (in the 1990s) show time-dependence. There is no reason to believe that any systematic change in station distributions and location procedures occurred during this short period. (4) Li and Richards [2000] found a doublet which shows a time shift of BC-DF times that is consistent with the temporal changes along the SSI-Alaska path observed by Song and Richards [1996] and Song [2000a]. (5) The standard errors in horizontal locations from the EDR or the ISC seems to be less than 10 km based on study of SSI doublets [Li and Richards, 2000].

The errors are not enough to explain the observed temporal change, which would require a systematic mislocation of 50 km towards the azimuthal direction over 30 years. (6) The pattern of increasing BC-DF times over time at COL does not exist in the AB-BC times [Song and Richards, 1996], confirmed by a recent "doublets" study [Song, 2000b]. Because the difference in ray parameters between AB and BC is more than twice the difference between BC and DF, systematic mislocation would have resulted in greater changes with time in differential AB-BC times than in differential BC-DF times.

Poupinet et al. [2000] examined the mislocation issue with a novel "doublet" technique. They measured the difference d(BC-DF) between the differential BC-DF times of each pair and the difference d(AB-BC) between the differential AB-BC times of the same pair and compared the {d(BC-DF), d(AB-BC)} measurements with the effect of event mislocations. They examined 12 pairs of PKP seismograms from the SSI-COL path and conclude that the apparent variations with time along the SSI-COL path observed by Song and Richards [1996] are the artifacts due to poor earthquake locations. However, in a reanalysis of the same event pairs [Song, 2001], I find that their results are systematically biased by the choice of radial earth models and by incorrectly assuming that one of the phases had reversed polarity. Upon correcting for these biases, I conclude that these event pairs are consistent with an inner-core rotation.

One of the critical problems in Poupinet et al.'s [2000] analyses, as discussed in Song [2000b], is that the "doublets" they selected have large d(AB-BC) values, indicating the event pairs are, in fact, far apart. In this case, the interpretation of {d(BC-DF), d(AB-BC)} measurements is sensitive to biases from the 1D reference Earth model used and heterogeneities in the mantle and the inner core [Song, 2000b]. Using a procedure very similar to Poupinet et al.'s [2000] "doublet" technique, I re-examined the SSI-COL path [Song, 2000b]. The results re-affirm the time-dependence of BC-DF times, and rule out the possibility that the observed time-dependence of the BC-DF times is caused by earthquake mislocation.

Specifically, I chose a total of 65 event pairs that best resemble "doublets" in terms of location and waveform similarities. For example, I only used events with d(AB-BC)<0.5 s. The {d(BC-DF), d(AB-BC)} measurements are compared with "mislocation bands" determined for mislocations up to 50 km (i.e. within a cylinder of 100 km in diameter in horizontal directions and 100 km in height in the radial direction) (Figure 9). If the DF travel times change with time, the {d(BC-

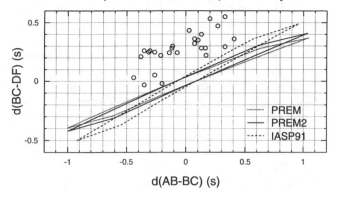

Time lapse between event pairs > 15 years

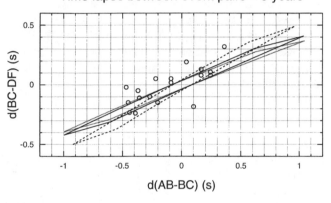

Time lapse between event pairs < 5 years

Figure 9.   Study of inner-core rotation using event pairs from the SSI-COL path [from Song, 2000b]. Each data point (circle) represents the difference d(BC-DF) between the differential BC-DF times of each pair and the difference d(AB-BC) between the differential AB-BC times of the same pair. The "mislocation bands" (lines) are constructed by calculating the predicted values of {d(BC-DF), d(AB-BC)} for the 1D reference models for event pairs between the reference source ($-56.2°$, $-27.4°$, 100 km) and all sources located within a cylinder of 50 km in horizontal and radial directions surrounding the reference source. **(Top)** The measurements between pairs with time lapses greater than 15 years are all above the "mislocation bands", consistent with the time-dependence of the BC-DF times. **(Bottom)** The measurements between pairs with time lapses less than 5 years are scattered around "mislocation bands", indicating no noticeable time dependence in such short durations.

DF), d(AB-BC)} measurements should fall outside the mislocation band. Figure 9 shows that the measurements between pairs with time lapses of greater than 15 years (a total of 29 pairs) are all above the "mislocation bands", consistent with the proposal that the BC-DF times changed with time. The measurements between pairs with time lapses less than 5 years (a total of 18 pairs) are scattered around "mislocation bands", indicating no noticeable time-dependence during the short

durations of time, within the resolution limit of the "mislocation bands". Note the "mislocation band" for IASP91 (Kennett and Engdahl, 1991) is significantly different from those for PREM [Dziewonski and Anderson, 1981] and PREM2 [Song and Helmberger, 1995b] at $|d(AB - BC)| > 0.4$ s. This is one of the reasons that it is important to use event pairs with small d(AB-BC) values; it is also one of the reasons that Poupinet *et al.* [2000] erred in their conclusion, as discussed in Song [2000b].

Temporal variation in the BC-DF times is directly visible in certain seismograms (Figure 10). Figure 10A shows overlays of pairs of seismograms with similar AB-BC times, but with increased BC-DF times for the later events. The similarity of the AB-BC times re-affirms the similarity of the event locations. Thus, the noticeable changes in the BC-DF times cannot be explained by differences in location. Figure 10B shows another event pair. The later event of the pair has a smaller AB-BC time but a larger BC-DF time. Clearly, it is difficult to prove that the shift of the BC-DF time is due to mislocation, because changes in BC-DF times caused by location errors (either in epicentral distances or in focal depths) would be amplified in AB-BC times with the same sign.

4.2. Biases From Mantle Heterogeneity

Mantle heterogeneity is unlikely to explain the BC-DF temporal change along the SSI-ASN path or the Alaska-SPA path for the following reasons. First, the distributions of earthquakes at different time periods are uniform [Song, 2000a; Song and Li, 2000]. Second, we examined the mantle biases from joint inversions with and without mantle correction terms and found that time-dependent α terms are statistically robust (with confidence levels of over 99% for both pathways) regardless whether mantle corrections are considered. In fact, the inclusion of mantle corrections increases the estimate of the rotation rate from the SSI-ASN data set [Song, 2000a].

4.3. Uncertainty In Tilt of Anisotropy Axis

In a series of papers [Souriau *et al.*, 1997; Souriau 1998a,b,c], Souriau and co-workers stated that the detection of the inner-core rotation by Song and Richards [1996] relies on the assumption that the anisotropy axis is tilted with respect to the rotation axis. They questioned the detection of the inner-core rotation because Souriau *et al.* [1997] found that the tilt of the anisotropy axis cannot be resolved from ISC arrival times. Indeed the determination of the tilt of the anisotropy axis

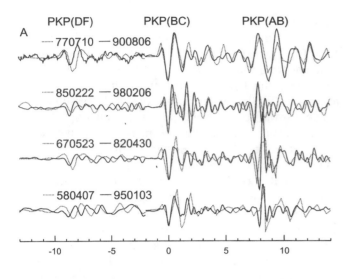

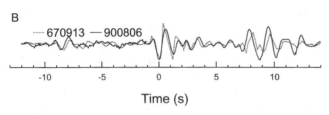

Figure 10. Pairs of SSI earthquakes recorded at COL [modified from Song, 2000b]. All the DF wave trains have been amplified by five times for visualization. **(a)** Event pairs with similar AB-BC times but increased BC-DF times for the later events. **(b)** A SSI event pair with a smaller AB-BC time but a larger BC-DF time for the later event. Earthquake mislocation cannot explain the increases in the BC-DF times, because mislocation would have caused more than twice as much increases in the AB-BC times.

affects the determination of the rotation rate as discussed above. This, however, does not change our first-order observation that the data change with time, which calls for some type of inner-core motion. The time-dependences of BC-DF residuals along certain pathways that we have sought to detect do not rely upon whether the averaged anisotropy axis of the inner core is tilted. Viable markers for the detection of the inner-core rotation can be provided by a local lateral-velocity gradient (first noticed by Creager, 1997) or a tilt of a local anisotropy axis at the part of the inner core sampled by the fixed path.

4.4. Biases From Event Magnitudes

Souriau *et al.* [1997] observed an apparent correlation between the differential BC-DF travel-time residuals at COL used by Song and Richards [1996] and the magnitudes of the events. They suggested that the observed time-dependence of the differential travel times could

be an artifact related to the heterogeneity of the event magnitudes during the period considered, because Song and Richards [1996] used more events with smaller magnitudes during later time period.

To examine this, Song and Li [2000] selected events of a specific magnitude range so that the magnitudes of the events are distributed uniformly in time. We examined BC-DF times from the SSI-COL path as well as from the Alaska-SPA path. The BC-DF residuals at both COL and SPA show clear increases with time. But the linear regressions of the residual on event magnitude have slopes of close to zero at both COL and SPA; the slight negative slope at COL and the slight positive slope at SPA are well within the corresponding one standard error. Thus, there is no evidence for a correlation between travel-time residuals and event magnitudes as suggested by Souriau *et al.* [1997]. It is more likely that the apparent correlation observed by Souriau *et al.* [1997] resulted from the fact that the average magnitudes of the events considered change with time.

5. DISCUSSION AND CONCLUSION

A 3-D picture of inner-core structure and anisotropy is starting to emerge, which demonstrates that the inner core is one of the most anisotropic and heterogeneous regions of the earth. The inner-core structure appears to vary strongly on all scales, from the hemispherical scale to the scale of a few km. The uppermost inner core appears to be isotropic but significant anisotropy (~3% on average) appears to persist in the rest of the inner core right to the center of the Earth. The details of the transition from isotropy at the upper inner core to anisotropy at the lower inner core are still being mapped out. At some places, the transition appears to be sharp enough to produce an inner-core seismic triplication.

Although still speculative, I argue that an inner-core transition zone (from an isotropic upper inner core to an anisotropic lower inner core), with an irregular boundary, as proposed first by Song and Helmberger [1998], may provide a unified explanation to observed depth variations and lateral variations of anisotropy, and to observed anomalous DF waveforms along polar paths. Specifically, if the boundary is deeper in the eastern hemisphere than in the western hemisphere, the model would explain why BC-DF anomalies are much smaller in the eastern hemisphere than in the western hemisphere, as observed by Tanaka and Hamaguchi [1997] and Creager [1999]. A significant relief of the transition boundary (perhaps as much as 100 km over 30° distance) may explain the steep lateral gradients of PKP times in some areas of the inner core, such as beneath central America, as sampled by the SSI-ASN data [Creager, 1997; Song, 2000a]. Such a transition-zone model also provides an explanation to the enigmatic observation that the DF waveforms along polar paths are complex and their short-period amplitudes are very small.

If we accept a transition-zone model, our waveform modeling of the SSI-Alaska path indicates that the anisotropy in the lower inner core is 7-8%, two to three times the previous estimates of anisotropy, which were calculated assuming a uniform model. This proposal raises a new question on the cause of the inner-core anisotropy. The source of the anisotropy is believed to be preferred orientation of hexagonal close-packed (hcp) iron [e.g., Brown and McQueen, 1986; Anderson, 1986; Jephcoat and Olson, 1987; Sayers, 1989; Stixrude and Cohen, 1995]. The theoretical calculation of Stixrude and Cohen [1995] showed that a perfectly aligned aggregate of hcp-iron crystals (or a gigantic single crystal) calculated at the room temperature but at inner-core pressure would explain the P anisotropy of 3% previously determined. However, recent laboratory measurements at inner-core pressure [Mao et al., 1998] and theoretical calculation of elasticity of hcp-iron at inner-core temperature and pressure [Steinle-Neumann et al., 2001] suggest a much higher degree of anisotropy, which may account for the increased estimate of the seismic anisotropy.

A growing body of evidence now supports the initial claim of inner-core super-rotation [Song and Richards, 1996]. The current estimates of the rotation rate is uncertain, ranging from 0.15 to 1.1°/yr. All the estimates suggest that the inner core is rotating eastwards; westward rotation can be ruled out. Some attempts to detect inner-core rotation have failed, but one should expect that not all searches for inner-core rotation will succeed. For example, if there is no significant lateral variation of the inner-core structure at the sampling region, observations of a fixed path won't "see" the inner-core motion. Such a situation may explain why data from Novaya Zelmya explosions recorded at DRV, Antarctica did not see rotation [Souriau, 1998a]; these data sample the inner core beneath the Southeast Asia (in the eastern hemisphere), where inner-core anisotropy appears to be small [Tanaka and Hamaguchi, 1997]. On the other hand, the combination of a steep lateral spatial gradient sampled by the SSI-Alaska path and the resolvability of inner-core rotation by the joint-inversion method using even a very short-duration (less than 10 years) of data offers an exciting opportunity to confirm and constrain the rotation in the very near future. The ASN also offers a rich data set for determining inner-core rotation using doublets [Li and Richards, 2000]; once found, a doublet can be examined at potentially over 100 stations.

Souriau (and co-workers) [Souriau et al., 1997; Souriau 1998a,b,c] and Poupinet et al. [2000] raise important questions about the existence of inner-core rotation described by Song and Richards [1996]. They argue that the time-dependence along the SSI-COL path could be artifacts of biases from mantle heterogeneity, heterogeneous event magnitudes, and, in particular, event mislocations. We have examined each one of the arguments and conclude that none of them could explain the first-order observation that travel times through the inner core along certain pathways have changed through time, which implies an inner-core motion.

Acknowledgments. I thank Steve Marshak for detailed editing, which improves greatly the readability of the paper. I thank two anonymous reviewers for helpful reviews, Xiaoxia Xu for assistance in figure preparation, and John Vidale for sending me their figures in digital forms. Supported by NSF grant EAR 01-06544.

REFERENCES

Anderson, O.L., Properties of iron at the Earth's core conditions, *Geophys. J. R. Astron. Soc., 84*, 561-579, 1986.

Brown, J.M., and R.G. McQueen, Phase-transitions, gruneisen-parameter, and elasticity for shocked iron between 77-GPA and 400-GPA, *Geophys. J. R. Astron. Soc., 91*, 7485-7494, 1986.

Choy, G.L., and V.F. Cormier, The structure of the inner core inferred from short-period and broadband GDSN data, *Geophys. J. R. Astron. Soc., 72*, 1-21, 1983.

Creager, K.C., Anisotropy of the inner core from differential travel times of the phases PKP and PKIKP, *Nature, 356*, 309-314, 1992.

Creager, K.C., Inner core rotation rate from small-scale heterogeneity and time-varying travel times, *Science, 278*, 1284-1288, 1997.

Creager, K.C., Large-scale variations in inner core anisotropy, *J. Geophys. Res., 104*, 23127-23139, 1999.

Creager, K.C., Inner Core Anisotropy and Rotation, in *Earth's Deep Interior: Mineral Physics and Tomography From the Atomic to the Global Scale*, edited by S. Karato, A.M. Forte, R.C. Liebermann, G. Masters, L. Stixrude, AGU monograph, 2000.

Cummins, P., and L. Johnson, Short-period body wave constraints on properties of the Earth's inner core boundary, *J. Geophys. Res., 93*, 9058-9074, 1988.

Doornbos, D.J., The anelasticity of the inner core, *Geophys. J. R. Astron. Soc., 38*, 397-415, 1974.

Dziewonski, A.M., and W.-J. Su, A local anomaly in the inner core, *Eos Trans. AGU, 79*(17), Spring Meet. Suppl., S218, 1998.

Garcia, R. and A. Souriau, Inner core anisotropy and heterogeneity level, *Geophys. Res. Lett., 27*, 3121-3124, 2000.

Gillard, D., A.M. Rubin, and P. Okubo, Highly concentrated seismicity caused by deformation of Kilauea's deep magma system, *Nature, 384*, 343-346, 1996.

Glatzmaier, G.A., and P.H. Roberts, A three-dimensional convective dynamo solution with rotating and finitely conducting inner core and mantle, *Phys. Earth Planet. Inter.*, *91*, 63-75, 1995.

Gubbins, D., Rotation of the inner core, *J. Geophys. Res.*, *86*, 11695-11699, 1981.

Jephcoat, A., and P. Olson, Is the inner core of the Earth pure iron, *Nature*, *325*, 332-335, 1987.

Laske, G., and G. Masters, Limits on differential rotation of the inner core from an analysis of the Earth's free oscillations, *Nature*, *402*, 66-69, 1999.

Li, A.Y., and P.G. Richards, Using doublets to study inner core rotation and catalog precision, *Eos, Trans. AGU*, *81(48)*, Fall Meet. Suppl., F885, 2000.

Li, A.Y., and P.G. Richards, Study of inner core travel structure and rotation using Novaya Zelmya nuclear explosions, *AGU monograph: Core Structure, Dynamics and Rotation*, submitted, 2001.

Li, Y.G, J.E. Vidale, K. Aki, F. Xu, and T. Burdette, Evidence of shallow fault zone strengthening after the 1992 M7.5 Landers, California, earthquake *Science*, *279*, 217-219, 1998.

Mao, H.K., J.F. Shu, G.Y. Shen, *et al.*, Elasticity and rheology of iron above 220 GPa and the nature of the Earth's inner core *Nature*, *396*, 741-743, 1998.

Masters, G., Core models ring true, *Nature*, *366*, 629-630, 1993.

McSweeney, T.J., K.C. Creager, and R.T. Merrill, Depth extent of inner-core seismic anisotropy and implications for geomagnetism, *Phys. Earth Planet. Inter.*, *101*, 131-156, 1997.

Morelli, A., A.M. Dziewonski, and J.H. Woodhouse, Anisotropy of the inner core inferred from PKIKP travel times, *Geophys. Res. Lett.*, *13*, 1545-1548, 1986.

Niu, F.L., and L.X. Wen, Hemispherical variations in seismic velocity at the top of the Earth's inner core, *Nature*, 410, 1081-1084, 2001.

Ouzounis, A., and K.C. Creager, Isotropy overlying anisotropy at the top of the inner core, *Geophys. Res. Lett.*, *28*, 4331-4334, 2001a.

Ouzounis, A., and K.C. Creager, Radial transition from isotropy to strong Anisotropy in the upper inner core, *Eos Trans. AGU, 82(47)*, Fall Meet. Suppl., F943, 2001b.

Ovchinnikov, V.M., V.V. Adushkin, and V.A. An, On the velocity of differential rotation of the Earth's inner core, *Dokl. Akad. Nauk.*, *362*, 683-686, 1998.

Poupinet, G., W.L. Ellsworth, and J. Frechét, Monitoring velocity variations in the crust using earthquake doublets: an application to the Calaveras Fault, California, *J. Geophys. Res.*, *89*, 5719-5731, 1984.

Poupinet, G., A. Souriau, and O. Coutant, The existence of an inner core super-rotation questioned by teleseismic doublets, *Phys. Earth. Planet. Inter.*, *118*, 77-88, 2000.

Sayers, C.M., Seismic anisotropy of the inner core, *Geophys. Res. Lett.*, *16*, 267-270, 1989.

Sharrock, D.S., and J.H. Woodhouse, Investigation of time dependent inner core structure by the analysis of free oscillation spectra, *Earth Planets Space*, *50*, 1013-1018, 1998.

Shearer, P.M., and K.M. Toy, PKP(BC) versus PKP(DF) differential travel times and aspherical structure in the Earth's inner core, *J. Geophys. Res.*, *96*, 2233-2247, 1991.

Song, X.D., Anisotropy in central part of inner core, *J. Geophys. Res.*, *101*, 16,089-16097, 1996.

Song, X.D., Anisotropy of the earth's inner core, *Rev. Geophys.*, *35*, 297-313, 1997.

Song, X.D., Joint inversion for inner core rotation, inner core anisotropy, and mantle heterogeneity, *J. Geophys. Res.*, *105*, 7931-7943, 2000a.

Song, X.D., Time dependence of PKP(BC)-PKP(DF) times: Could it be an artifact of potential systematic earthquake mislocations? *Phys. Earth. Planet. Inter.*, *122*, 221-228, 2000b.

Song, X.D., The Earth's core, in *International Handbook of Earthquake and Engineering Seismology*, edited by W.H.K. Lee, H. Kanamori, P.C. Jennings, and C. Kisslinger, Volume 1, Chapter 56, Academic Press, San Diego, in press, 2002.

Song, X.D., Comment on "The existence of an inner core super-rotation questioned by teleseismic doublets" by Georges Poupinet, Annie Souriau, and Olivier Coutant, *Phys. Earth Planet. Inter.*, 124, 269-273, 2001.

Song, X.D., and D.V. Helmberger, Velocity structure near the inner core boundary from waveform modeling, *J. Geophys. Res.*, *97*, 6573-6586, 1992.

Song, X.D., and D.V. Helmberger, Anisotropy of Earth's inner core, *Geophys. Res. Lett.*, *20*, 2591-2594, 1993.

Song, X.D., and D.V. Helmberger, Depth dependence of anisotropy of Earth's inner core, *J. Geophys. Res.*, *100*, 9805-9816, 1995a.

Song, X.D., and D.V. Helmberger, A P-wave velocity model of the Earth's core, *J. Geophys. Res.*, *100*, 9817-9830, 1995b.

Song, X.D., and D.V. Helmberger, Seismic evidence for an inner core transition zone, *Science*, *282*, 924-927, 1998.

Song, X.D., and A.Y. Li, Support for differential inner core superrotation from earthquakes in Alaska recorded at South Pole station, *J. Geophys. Res.*, *105*, 623-630, 2000.

Song, X.D., and P.G. Richards, Observational evidence for differential rotation of the Earth's inner core, *Nature*, *382*, 221-224, 1996.

Song, X.D., and X.X. Xu, Inner core transition zone and anomalous PKP(DF) waveforms from polar paths, *Geophys. Res. Lett.*, *29(4)*, 10.1029/2001GL013822, 2002.

Souriau, A., New seismological constraints on differential rotation of the inner core from Novaya Zemlya events recorded at DRV, Antarctica, *Geophys. J. Int.*, *134*, F1-5, 1998a.

Souriau, A., Earth's inner core - Is the rotation real? *Science*, 281, 55-56, 1998b.

Souriau, A., Detecting possible rotation of Earth's inner core, response to comments by P. G. Richards, X. D. Song, and A. Li, *Science*, 282, 1227a, 1998c.

Souriau, A., and G. Poupinet, Inner core rotation: a test at the worldwide scale *Phys. Earth Planet. Inter.*, *118*, 13-27, 2000.

Souriau, A., P. Roudil, and B. Moynot, Inner core differential rotation: facts and artefacts, *Geophys. Res. Lett.*, *24*, 2103-2106, 1997.

Steinle-Neumann, G., L. Stixrude, R.E. Cohen, *et al.*, Elasticity of iron at the temperature of the Earth's inner core, *Nature*, *413*, 57-60, 2001.

Stixrude, L., and R.E. Cohen, High-pressure elasticity of

iron and anisotropy of Earth's inner core, *Science, 267,* 1972-1975, 1995.

Su, W.J., and A.M. Dziewonski, Inner core anisotropy in three dimensions, *J. Geophys. Res., 100,* 9831-9852, 1995.

Su, W.J., A.M. Dziewonski, and R. Jeanloz, Planet within a planet: Rotation of inner core of Earth, *Science, 274,* 1883-1887, 1996.

Sun, X.L., and X.D. Song, PKP travel times at near antipodal distances: Implications for inner core anisotropy and lowermost mantle structure, *Earth Planet. Sci. Lett.,* in press, 2002.

Tanaka, S., and H. Hamaguchi, Degree one heterogeneity and hemispherical variation of anisotropy in the inner core from PKP(BC)-PKP(DF) times *J. Geophys. Res., 102,* 2925-2938, 1997.

Tromp, J., Inner-core anisotropy and rotation, *Annu. Rev. Eearth Planet. Sci., 29,* 47-69, 2001.

Tromp, J., Support for anisotropy of the Earth's inner core from free oscillations, *Nature, 366,* 678-681, 1993.

Tromp, J., Normal-mode splitting observations from the Great 1994 Bolivia and Kuril Islands Earthquakes: con-straints on the structure of the mantle and inner core, *GSA Today,* Vol.5, No.7, July, 1995.

Vidale, J.E., D.A. Dodge, and P.S. Earle, Slow differential rotation of the Earth's inner core indicated by temporal changes in scattering, *Nature, 405,* 445-448, 2000.

Vidale, J.E., and P.S. Earle, Fine-scale heterogeneity in the Earth's inner core *Nature, 404,* 273-275, 2000.

Vinnik, L., B. Romanowicz, and L. Breger, Anisotropy in the center of the inner core, *Geophys. Res. Lett., 21,* 1671-1674, 1994.

Woodhouse, J.H., D. Giardini, and X.-D. Li, Evidence for inner core anisotropy from free oscillations, *Geophys. Res. Lett., 13,* 1549-1552, 1986.

Xu, X.X., and X.D. Song, Evidence for inner core super-rotation from time-dependent differential PKP travel times observed at Beijing Seismic Network, *Geophys. J. Int.,* submitted, 2002.

X.D. Song, Department of Geology, University of Illinois, Urbana, IL 61822. (email: xsong@uiuc.edu)

Inner Core Rotation: A Critical Appraisal

Annie Souriau

Observatoire Midi-Pyrénées, CNRS/GRGS, 31400 Toulouse, France

Georges Poupinet

LGIT-CNRS, Observatoire, BP 53X, 38041 Grenoble, France

The detection of a possible inner core rotation relies on the existence of a departure from cylindrical symmetry inside the inner core. We first investigate the general properties of the inner core anisotropy and isotropic velocity. The major feature is an apparent hemispherical pattern in the anisotropy level, which is mostly due to an hemispherical variation in the thickness of the isotropic layer surrounding the central anisotropic inner core, with about 3.5% anisotropy and with fast axis parallel to the Earth rotation axis. The thickness of this layer is about 400 km beneath the quasi-eastern hemisphere (40°E to 180°), but only 100 km beneath the western hemisphere. A sharp transition seems to be present between the two hemispheres near longitude 175°W. There is no strong evidence of a tilt of the anisotropy axis, even at regional scale. On the other hand, the level of heterogeneity in isotropic velocity appears to be low, except at very short wavelength, where scatterers seem to be present. An attempt to detect an inner core differential rotation with respect to the mantle is performed at worldwide scale in analysing the coherence, over increasing time intervals, of the propagation anomalies of rays which sample the inner core, compared to rays with nearly similar paths which turn in the liquid core. No rotation can be detected (within ± 1° per year). A temporal variation of the anomaly at longitude 175°W can also not be detected, bounding the rotation rate at 0.1° per year, if this anomaly effectively originates inside the inner core. We also investigate the path from South Sandwich Island (SSI) to College, Alaska, which was the strongest evidence for an inner core super-rotation. Using a doublet technique, which does not require accurate event locations, we demonstrate that the travel time variations observed along this path are strongly affected by event mislocation biases which have evolved with time. This is confirmed from the PKP mean residuals observed in northern and southern azimuths with respect to SSI. All together, the different methods, based on

Earth's Core: Dynamics, Structure, Rotation
Geodynamics Series 31
Copyright 2003 by the American Geophysical Union
10.1029/31GD06

differential travel times, scattered waves or eigenmodes, lead to rotation rates in the range $0 \pm 0.2°$/yr (Table 2), a precision which is limited by the perturbing effect of mantle heterogeneities, and by the resolution of the seismic methods with the presently available data.

INTRODUCTION

The possibility that the inner core exhibits a differential rotation with respect to the mantle has been proposed for about two decades from the modelling of the magnetic field [*Gubbins*, 1981]. Recent models of the geodynamo [*Glatzmaier and Roberts*, 1996] have proposed a fast, eastward super-rotation, with angular velocities of the order of 2 to 3 deg/yr. This very large value, which may correspond to a full rotation in 120 years, has motivated seismologists to attempt to observe it. The issues of a differential rotation are quite important, not only for constraining the models of the magnetic field, but also because it has important implications on the geodynamic processes in the Earth [*Buffett*, 1997], on the mechanisms of dissipation inside the Earth, and on inner core properties, in particular its viscosity, its heterogeneity level and the way its anisotropy is generated.

The various attempts to measure the inner core differential rotation have led so far to conflicting results. *Song and Richards* [1996] analysed the differential travel times of the two seismic phases PKP(DF) and PKP(BC) for the polar path from South Sandwich Island (SSI) to station College (COL) in Alaska. PKP(DF) samples the inner core, whereas PKP(BC) has its turning point in the liquid core and is thus used as a reference phase. Song and Richards observed a 0.3s increase in the travel time anomaly over 30 years, that they ascribed to a variation of the apparent anisotropy along this path, related to a tilt of the inner core anisotropy symmetry axis. Assuming a tilt value of 11°, they obtained an inner core super-rotation of 1.1°/yr. A still larger rotation rate of about 3°/yr, first proposed by *Su et al.* [1996] from a worldwide analysis of the absolute PKP(DF) travel times reported by the International Seismological Centre (ISC), was later dismissed: *Dziewonski and Su* [1998] concluded that the inner core tilt and rotation are not detectable with confidence from ISC data.

A tilted anisotropy symmetry axis has been contradicted by more recent studies [*Souriau et al.*, 1997; *Creager*, 1999], so methods based on the drift of heterogeneity beneath a particular seismic path have also been applied. *Creager* [1997] re-analysed the South Sandwich Island to COL data, together with the spatial pattern of PKP(BC)-PKP(DF) anomalies at an Alaska array, which he interpreted in terms of inner core structure. He found a rotation rate between 0.05 and 0.3°/yr, depending on the contribution of mantle heterogeneities beneath Alaska (possible heterogeneities beneath South Sandwich Island were not taken into account). A simultaneous inversion for source structure, receiver structure, inner core heterogeneities and inner core rotation for SSI events recorded at four Alaska stations led to values of 0.3 to 1.1 °/yr [*Song*, 2000a], but it is difficult to resolve all these parameters with paths that are nearly parallel, so that the results are questionable. The same criticisms apply still drastically to a path from Alaska to the single station SPA at South pole [*Song and Li*, 2000], which led with a similar analysis to a super-rotation of 0.6°/yr. Moreover, as will be discussed later in this paper, the crucial problem of event mislocations is not fully resolved in these studies, although the use of differential travel times lessens the effect of location errors. The use of doublets provides probably the best way to discriminate between differential rotation and event mislocations. This technique, applied to the path from South Sandwich Island to COL, reveals mislocation biases that have evolved with time [*Poupinet et al.*, 2000].

The effects of event mislocations and source side structure are minimized with the use of nuclear tests. For the polar path connecting Novaya Zemlya events to station DRV in Antarctica, no travel time variation could be observed over 26 years [*Souriau*, 1998a]. As noted by *Song and Li* [2000], however, the presence of heterogeneities beneath this path is not well established. Observations at other Antarctic stations (SPA, SBA) also led to small or insignificant variations. A differential travel time increase of 0.34 ± 0.66s, insignificant due to the large error bar, is observed at SPA over 37 years, whereas a very small decrease of 0.09 ± 0.07s over 19 years is observed at SBA [*Li et al.*, 1998]. For this last path, a slightly larger value with a smaller error bar (-0.0053 ± 0.0034s per year) is also reported [*Richards et al.*, 1998]. Only at NVL [*Ovtchinnikov et al.*, 1998] have large and sharp variations been observed, about 0.2 s in 3 years (see *Souriau* [1998b, 1998c] for a discussion of these results).

Another method relies on the use of waves scattered on small scale inner core heterogeneities. Such heterogeneities have been detected in the uppermost 300 km of the inner core, thanks to the energy they generate in the coda of PKiKP, which reflects at the inner core boundary [*Vidale and Earle*, 2000]. The comparison of the diffraction pattern observed at the LASA array in Montana (USA) for two Novaya Zemlya nuclear tests separated by 3 years, led to a 0.15°/yr rotation rate [*Vidale et al.*, 2000]. Although this method is promising, its application is limited by the difficulty finding sources that are identical. In *Vidale et al.'s* [2000] study, the two events were separated by 500 to 1000 m, which is not negligible with respect to the size of the identified scatterers, about 2 km. Moreover, the change in the pattern of the scattered energy may reflect more complex phenomena than just scatterer advance or recede in time.

Because of difficulties in unambiguously identifying the presence of inner core heterogeneities or anisotropy variations beneath a particular path, studies involving longer wavelengths are of interest. Eigenmodes allow such a worldwide approach. A first analysis of the time variations of the splitting functions of inner core sensitive modes led to a westward rotation [Sharrock and Woodhouse, 1998], but the poor data coverage led to large error bars. A second study based on a larger data set analysed the variations of the splitting functions of 9 core sensitive eigenmodes during the last 20 years, leading to a rotation rate which is indistinguishable from zero (0 ± 0.2 °/yr) [Laske and Masters, 1999]. Body waves with worldwide distributed paths also provide a global approach. Either polar PKP(DF) residuals or differential PKP(BC)-PKP(DF) residuals have been considered, in order to decrease the mantle contribution compared to the inner core contribution in the residuals. In each case, a null rotation rate leads to the best coherence of the residuals at different epochs, but the resolution is poor, of the order of 1°/yr [Souriau and Poupinet, 2000].

In this article, we will attempt to clarify what seismological data are telling us about inner core rotation. We first review the conditions that are required for being able to detect a differential rotation from seismological methods. The most important one is the presence inside the inner core of structures that depart from cylindrical symmetry with respect to the Earth rotation axis. We then present various studies that have attempted to detect an inner core rotation, with a critical analysis of the hypotheses, results and possible biases related to each method.

THE STRUCTURE OF THE INNER CORE AND THE DEPARTURE FROM CYLINDRICAL SYMMETRY

The inner core is assumed to have the same spin axis as the mantle. Thus, the detection of a differential rotation is possible only if the inner core exhibits a departure from cylindrical symmetry with respect to the Earth rotation axis. This may occur in several ways: a- a tilt of the anisotropy symmetry axis with respect to the Earth rotation axis, b- heterogeneities in the distribution or in the nature of the anisotropy; c- heterogeneities in bulk and shear modulus, i.e., in isotropic velocity average.

Anisotropy has been recognized for about two decades as one of the most important properties of the inner core. This feature has been detected from the travel time anomalies of the PKP(DF) waves [Poupinet et al.; 1983, Morelli et al., 1986] and from the splitting of core sensitive eigenmodes [Woodhouse et al., 1986]. The early models proposed an axisymmetrical anisotropy with cylindrical symmetry, with P-wave velocities in the direction of the Earth rotation axis

about 3% to 3.5% higher than in the perpendicular direction in the central part of the inner core [e.g. Creager, 1992; Shearer, 1994; Vinnik et al., 1994; Song, 1996)]. There are now many evidences that the anisotropy departs from this simple structure [e.g., Tanaka and Hamaguchi, 1997], but there is not yet a clear 3D image of the inner core anisotropy distribution. In this section, we review briefly the possible departures from uniform axisymmetrical anisotropy that could be used to detect an inner core rotation.

Tilt of the symmetry axis

A tilt of the anisotropy symmetry axis of about 10° with a pole close to 80°N, 120°E has been proposed from the analysis of the differential travel times of core phases for the outermost part of the inner core [Creager, 1992; McSweeny et al., 1997]. A worldwide analysis of PKP(DF) data from ISC bulletins [Su and Dziewonski, 1995], as well as a joint analysis of normal modes and body waves [Romanowicz et al., 1996] also favor a slightly tilted symmetry axis. Although this tilt is compatible with most of the observations, it is not required by the data: several analyses have demonstrated that it reflects in a large part the uneven sampling of the Earth by the seismic paths [Souriau et al., 1997] and the lateral variations of the anisotropy [Creager, 1999]. The cylindrical symmetry of the anisotropy at the worldwide scale now is effectively dismissed [Tanaka and Hamaguchi, 1997; Creager, 1999; Garcia and Souriau, 2000].

That a particular region of the inner core exhibits a tilted fast axis is quite possible, however. For example, a 10° tilt has been proposed beneath Australia from PKP(BC)-PKP(DF) anomalies [Isse and Nakanishi, 2000]. Determining such a tilt requires one to have many paths with various orientations sampling a particular region in the same depth range. Fig.1 shows PKP(DF) turning points and ray orientations beneath two well sampled regions, Africa and Central America. High quality BC-DF differential travel times with various orientations are available for these two regions. The BC-DF travel times have been extracted from a file of about 1300 carefully picked arrival times collected by ourselves and several other authors [Creager, 1992; Song and Richards, 1996; McSweeny et al., 1997; Souriau and Romanowicz, 1997; Tanaka and Hamaguchi, 1997; Bréger et al., 1999; Tkalcic et al., 2001]. The residuals with respect to the reference Earth model ak135 [Kennett et al., 1995] have been normalized to their path length inside the anisotropic part of the inner core, assuming that the uppermost 100 km of the inner core is isotropic [Song, 1996]. Beneath Africa, the data sample many directions and the residuals are well fitted with an hexagonal symmetry model with fast axis parallel to Earth rotation axis:

$$\delta v_P / v_P = \alpha + \varepsilon \cos^2\theta + \gamma \sin^2 2\theta \qquad (1)$$

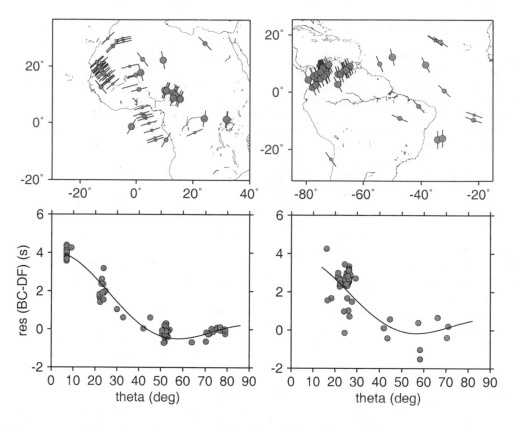

Figure 1. Inner core anisotropy beneath two equatorial regions: Africa and central America. Top figures: location of the PKP(DF) turning points inside the inner core, and azimuths of the ray paths. Symbol size increases with (BC-DF) residuals. Bottom figures: (BC-DF) differential travel time anomalies plotted as a function of the angle θ of the PKP(DF) ray inside the inner core, with respect to the Earth rotation axis. Also drawn are the best fits for a model of hexagonal anisotropy with the fast axis parallel to the Earth rotation axis (see Table 1). Note the large scale coherence of the anisotropy for these two regions.

coefficients, is nearly independent of the longitude. This important result has been extended to shorter wavelengths (5° to 20°) using a stochastic analysis of the PKP(DF) residuals of equatorial paths, which are poorly sensitive to anisotropy [*Garcia and Souriau*, 2000, 2001]. This analysis shows that the coherent part of the signal included in the PKP(DF) residuals is nearly the same as the coherent signal in PKP(BC), thus inner core heterogeneities, if any, are not strong enough to generate a significant coherent signal in the PKP(DF) travel times. This suggests that, for wavelengths larger than 5°, anisotropy must be the prevailing mechanism to explain PKP(DF) residual variations.

The large scatter of the polar residuals for a same turning point longitude, in particular beneath the western hemisphere (Fig. 2b), suggests however that some short wavelength heterogeneities are present inside the inner core, or that an important mantle contamination is still present in spite of the use of differential travel times. Another way to analyse the inner core structure is to measure the ray parameter q for rays turning inside the inner core, as q is

directly related to the P-velocity at the turning point: $q=r/v_P$, where r is the radius at the turning point. Assuming that the liquid core has no lateral heterogeneity and that it is well modelled by ak135, the (BC-DF) differential ray parameter may be obtained from pairs of nearby events recorded at the same station. The ray parameter is directly obtained from $dT/d\Delta$, where dT and dΔ are the differences in residuals and in epicentral distance for the two nearby paths. It may also be obtained from the slope of the travel time anomaly as a function of distance. Fig. 5 shows the residual variations as a function of distance for a few specific paths sampling either the western or the eastern hemisphere, and corresponding to different angle θ with respect to the Earth rotation axis. Also shown are the slopes expected for a model similar to the model of Fig. 4, with the anisotropy coefficients obtained by *Garcia and Souriau* [2000], a thickness of the isotropic layer of 400 km beneath the quasi-eastern hemisphere and 100 km beneath the western hemisphere.

Fig. 5 shows that, if we except NRIL, which exhibits very scattered data, the observed slopes are similar to those

where $\delta v_P/v_P$ is the relative velocity perturbation with respect to the mean velocity v_P, and θ the angle between the Earth rotation axis and the ray path inside the inner core. The data are binned in domains of 5° in θ. Model (1) explains 96% of the data variance (Fig.1). The good fit obtained with this model shows in addition that a large scale coherence is present in the anisotropy pattern in that part of the Earth. For the paths with the turning point beneath Central America, a similar model, without tilted axis, explains 86% of the variance of the data binned in the same way. Moreover, the ε and γ coefficients are very close to those of Africa, and to those obtained by *Creager* [1999] for the whole western hemisphere (Table 1), a result which shows the large spatial coherence of the anisotropy. Note however that for Central America the error bars are rather large, and a slightly tilted axis would explain the data equally well.

Long wavelength anomalies: the hemispherical pattern of the anisotropy

Heterogeneity in the distribution of the anisotropy has been detected at hemispherical scale from the differential travel times of PKP(BC) and PKP(DF): the quasi-eastern hemisphere (from 40°E to 180°E) appears less anisotropic than the western hemisphere [*Tanaka and Hamaguchi*, 1997], with only 1% anisotropy against about 3.5%. This feature appears clearly in Fig. 2, which shows the residuals for polar and equatorial paths extracted from our (BC-DF) data file, as a function of turning point longitude. We denote by "polar paths" those for which the ray path inside the inner core is nearly parallel to the polar axis (angle θ with respect to the Earth rotation axis $\leq 30°$), by "equatorial paths" those which are nearly parallel to equatorial plane ($\theta > 60°$). The reference model is ak135 [*Kennett et al.*, 1995]. As noted by *Creager* [1999], this model is presently the best isotropic reference model for core phases. Residuals for the paths in the quasi-eastern hemisphere are about 2s larger than those in the western hemisphere. This figure is very similar to that obtained by *Creager* [1999, his Fig. 3], although the overlap between our two datasets is small, less than 20% of our (BC-DF) data. Polar and equatorial variations are proportional to each other and opposite, as expected for an hexagonally symmetric cylindrical anisotropy with fast axis parallel to the Earth rotation axis. The (BC-DF) variations may be approximately fitted by a sine function dependent on longitude. However, a discontinuous model with sharp velocity jumps near longitudes 40° and 180° is suggested by the polar data (Fig. 2b). The discontinuity near 180° is still more evident if the residuals at station SPA (South Pole) are considered, as they allow a direct scanning in longitude with a single station.

The hemispherical pattern is also revealed by the PKP(DF) residuals extracted from the EHB file [*Engdahl et al.*, 1998] of teleseismically relocated, high quality events.

Fig. 3a shows the residuals with respect to model ak135 for nearly polar paths corresponding to θ-values less than 30°. The data are divided in two classes according to the longitude of the ray turning points: the quasi eastern hemisphere corresponds to longitudes 40°E to 180°E, the western hemisphere to longitudes 180°W to 40°E. At large epicentral distance ($\Delta>150°$, thus for turning point deeper than 200 km below inner core boundary), the PKP(DF) residuals are smaller for the western hemisphere (fast) than for the quasi-eastern hemisphere (slow), denoting a clear difference in the anisotropy level (Fig. 3b). As most of the paths sampling the western hemisphere originate at South Sandwich Island, the mantle structure beneath South Atlantic could bias this pattern [*Bréger et al.*, 2000]. We have checked that the hemispherical pattern is still present when South Sandwich Island events are removed.

The structure responsible for this hemispherical pattern has been inferred from the inversion of PKP(DF) and (BC-DF) data [*Creager*, 2000; *Garcia and Souriau*, 2000]. An isotropic layer of variable thickness at the top of the inner core is weakly required by the data. The presence of such a layer has been proposed in previous studies [e.g. *Song*, 1996, *Ouzounis and Creager*, 2001], but eigenmodes suggest it must be thin [*Durek and Romanowicz*, 1999]. A maximum thickness of about 100 km is obtained beneath the western hemisphere, 400 km beneath the eastern hemisphere [*Garcia and Souriau*, 2000]. This discrepancy explains the absence of significant (BC-DF) travel time variation as a function of θ beneath the eastern hemisphere, as the turning point depth remains in the uppermost 360 km below inner core boundary for distances less than 155°. Beneath the isotropic layer, the inversion reveals an uniform 3 to 3.5% anisotropy. The structure of the inner core thus appears as very simple (Fig. 4): a central body with uniform, hexagonal anisotropy is asymmetrically surrounded by an isotropic layer. The transition between the two structures seems to be smooth [*Ouzounis and Creager*, 2001], but the nature (sharp or smooth) of the transition between eastern and western hemispheres is not fully elucidated. A hemispherical velocity heterogeneity has also been observed in the uppermost 50 km of the inner core [*Niu and Wen*, 2001; *Garcia*, 2002]. The hemispherical structure of the inner core thus appears as a robust feature, which will provide a powerful tool to infer inner core rotation, in particular with the use of polar paths, which are strongly sensitive to anisotropy.

Heterogeneities or anisotropy?

Assuming an hexagonally symmetric anisotropy with fast axis parallel to the Earth rotation axis, *Creager* [1999] demonstrated the absence of variation in isotropic average velocity at hemispherical scale for depths larger than about 100 km below inner core boundary: the mean velocity, obtained from a linear combination of the anisotropy

Table 1: Anisotropy parameters (in %) for the depth range 100-450 km, assuming an hexagonal anisotropy structure with fast axis parallel to Earth rotation axis: $\delta v_P/v_P = \alpha + \varepsilon \cos^2\theta + \gamma \sin^2 2\theta$, and mean velocity perturbation δv_{iso} (in km/s) inferred from the PKP(BC)-PKP(DF) residuals for two regions of the inner core. The reference model is *ak135 [Kennett et al.*, 1995]. The values obtained by *Creager* [1999] for the whole western hemisphere (his model "*PKP_BC* western") are given for comparison. Note the great coherency of the anisotropy through the whole hemisphere.

	Turning point beneath Africa	Turning point beneath Central America	Western hemisphere [*Creager*, 1999]
Nb of data	81	101	
α ($\pm 1\sigma$)	0.14 (0.28)	0.51 (0.76)	-0.09 (0.09)
ε ($\pm 1\sigma$)	3.16 (0.33)	2.97 (0.69)	3.83 (0.78)
γ ($\pm 1\sigma$)	-1.77 (0.33)	-1.83 (0.99)	-1.61 (0.09)
δv_{iso} (km/s)	0.03 (0.006)	0.06 (0.02)	0.04 (0.004)

predicted by the anisotropic model for most of the paths considered, thus there is no need for large scale heterogeneity. The only anomalous slope is that for South Sandwich Island events recorded in Alaska: if all the data are considered together, the observed slope is about 0.3 s/deg, almost three times larger than the predicted one. This cannot reflect a very large anisotropy, as it is incompatible with the mean value of the BC-DF residuals. Interpreted in terms of inner core heterogeneity, it leads to a 6% heterogeneity with a 10° to 15° geographic lateral extension. However, if the data are separated in two classes according to the path location (with mostly COL to the West, DAWY, WHY and INK to the East), there is a good agreement between the observed slopes and those predicted by the anisotropy model: the ray parameter does not require the presence of inner core heterogenity. Thus the jump in the residuals near 150° likely reflects the presence of strong heterogeneities at the source or at the station side, rather than an inner core heterogeneity.

Inner core heterogeneities at short or intermediate wavelengths have been proposed in many studies along various paths [e.g. *Creager*, 1997; *Souriau*, 1998; *Bréger et al.*, 1999; *Song*, 2000a; *Song and Li*, 2000]. However in none of these cases can a possible explanation in terms of mantle heterogeneity be dismissed. On the other hand, heterogeneities at very short wavelengths with size of the order of 2 km may be present in the uppermost 300 km of the inner core [*Vidale and Earle*, 2000]. These heterogeneities with velocity anomalies of about 1.2% provide the simplest way to explain the energy present in the coda of the PKiKP wave observed at LASA Array (Montana) for Novaya Zemlya events. If their existence is general in the uppermost inner core, they may be an interesting tool to track the inner core rotation [*Vidale et al.*, 2000].

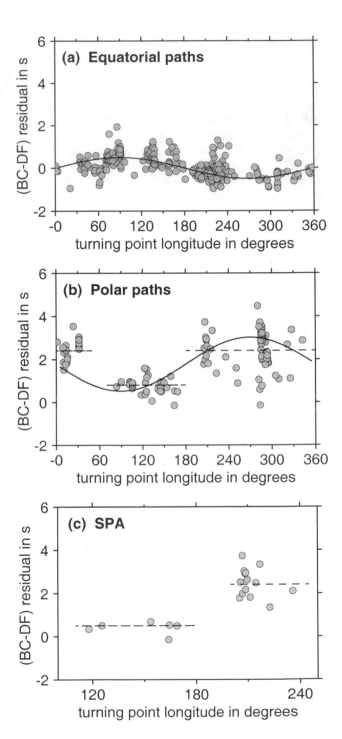

Figure 2. Hemispherical pattern of the residuals. PKP(BC)-PKP(DF) travel time residuals as a function of the longitude of the ray turning point, for: a- equatorial paths (ray propagating in the inner core with an angle $\theta > 60°$ with respect to the Earth rotation axis), b- polar paths ($\theta < 30°$). Also shown is a fit of the residuals with homothetical sine functions for the two data sets. c-Residuals for station SPA (South pole), as a function of longitude, showing a sharp residual variation near longitude 180°-200°.

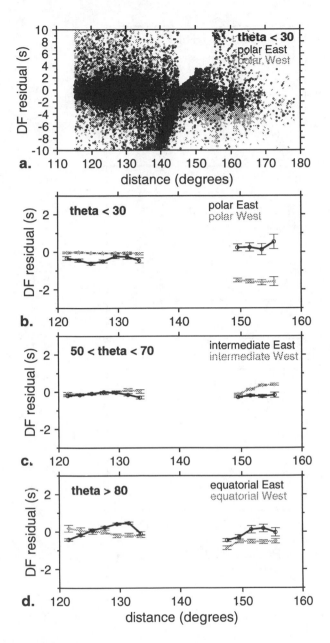

To summarize, the anisotropy has the characteristics of a medium with hexagonal symmetry and a fast axis parallel to rotation axis. Its hemispherical pattern, with an apparent strongest anisotropy in the western hemisphere, is due in a large part to the presence of an isotropic layer of varying thickness (from 100 to 400 km) in the uppermost inner core, which overlays an anisotropic central body. Small scale heterogeneities in anisotropy (variations in fast axis direction or in amplitude) are possible, but are very difficult to detect. On the other hand, heterogeneities in average isotropic velocity seem to be absent, except small scatterers in the uppermost inner core, and perhaps an hemispherical structure in the uppermost 50 km of the inner core .

INNER CORE ROTATION: INVESTIGATIONS AT THE WORLDWIDE SCALE

An investigation of inner core rotation at worldwide scale may take advantage of the hemispherical variation of the anisotropy, which is a well identified pattern, and of other possible long wavelength structures, which are sampled by a large number of paths.

A first approach consists in analysing the dispersion of the travel time residuals for data sets separated by increasing time intervals. For PKP(BC), which does not sample the inner core, the dispersion of the residuals will reflect mislocation errors, clock errors and reading errors. For PKP(DF), it includes in addition the variations with time of the structures sampled by the rays inside the inner core, when the inner core rotates with respect to the mantle. For short time intervals, these structures will not have

Figure 3. Travel time anomalies of PKP(DF) for: polar paths, with raw residuals (a), and mean value (b); mean value for intermediate paths (c) and for equatorial paths (d). The angle θ between the Earth rotation axis and the ray path at its turning point is θ<30° for polar paths, 50°<θ<70° for intermediate paths, θ>80° for equatorial paths, with the additional condition that the turning point latitude is in the range [-30°, 30°]. Data from the EHB file of *Engdahl et al.* [1998]. The data are sorted in two regions according to the longitude λ of the turning point of the PKP(DF) ray: blue: western hemisphere, 180°W<λ<40°E; red: eastern hemisphere, 40°E<λ<180°E. The residual variations at large distance reflect the properties of hexagonal anisotropy with fast axis parallel to Earth rotation axis (see Fig. 1).

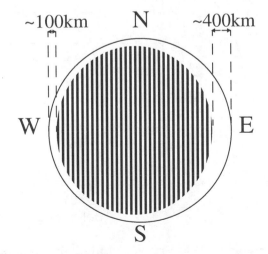

Figure 4. Inner core structure deduced from the inversion of PKP(DF) and differential PKP(BC)-PKP(DF) travel time residuals. Scheme of a meridian cross-section showing a central part with an uniform, 3% hexagonal anisotropy surrounded by an isotropic layer of variable thickness [*Garcia and Souriau*, 2000].

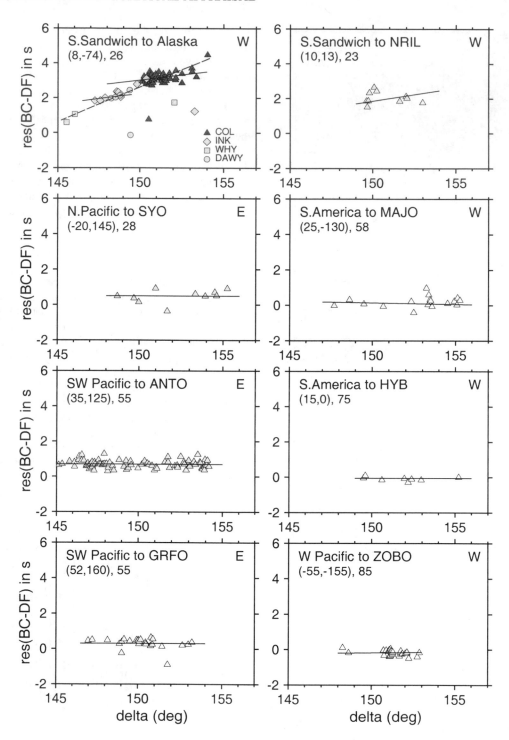

Figure 5. Determination of the PKP(BC-DF) ray parameter (slope of the travel time as a function of distance) for various paths corresponding to different turning point locations and to different values of the angle θ of the ray orientation with respect to Earth rotation axis. The three numbers at the top left of each figure correspond to mean latitude and longitude at the turning point, and to the θ-value. E or W at the top right denotes the hemisphere of the turning point (E: longitude 40°E-180°E, W: 180°W- 40°E). The dashed line in the top left figure is the mean trend of the data. The full lines correspond to the slopes expected for the anisotropy model of *Garcia and Souriau* [2000] (Fig. 4).

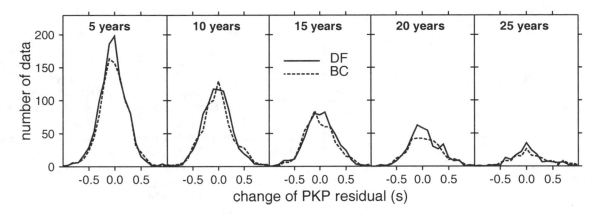

Figure 6. Variation of PKP travel time residuals with time, for various time lapses from 5 to 25 years, compared to the variation observed for PKP(BC). Each datum is the mean residual for sectors in azimuth and distance around seismic zones. No significant change is observed when the time lapse increases.

varied so much, and we have checked that PKP(DF) residuals have nearly the same dispersion as PKP(BC) residuals: the errors are globally nearly the same for the two phases. If long time intervals are considered, the inner core structures sampled by PKP(DF) will have changed, and the PKP(DF) residuals will have a dispersion larger than for short time intervals; but this must not be observed for PKP(BC).

Fig. 6 shows the histograms of the PKP(DF) and PKP(BC) residuals corresponding to different time windows, from 5 years to 25 years. Residuals have been extracted from the ISC file for the period 1964-1995. In order to decrease the influence of the variable distribution of hypocenters and stations with time, a binning procedure has been applied. Classes in azimuths j and distances k are defined around each seismic region i. A mean residual is computed for the whole period for each station inside the class (ijk), then this mean value is substracted from the residuals of the station in order to keep only the time variable part of the signal. A mean residual r_{ijk} is then computed for each class in azimuth and distance for each seismic region. Fig. 6 is the histogram of the differences of the residuals $r_{ijk}(t_2)-r_{ijk}(t_1)$, for time intervals t_2-t_1 of 5, 10, 15, 20 and 25 years. It has been computed for PKP(DF) and for PKP(BC). This figure shows that the histograms of BC and DF are identical within ±0.1 s. A rough simulation has been performed in considering a hemispherical anisotropy pattern with sinusoidal residual variations with longitude, as shown in Fig. 2 but with 4s amplitude, and a worldwide distribution of data with all possible orientations, as for the binned data. A rotation of 20° leads to a standard deviation increase of the order of 0.1 s. This allows us to discard rotation rates larger than about 1°/yr.

Another approach at worldwide scale has been considered previously [*Souriau and Poupinet*, 2000]. It is also based on the decrease of coherence of the residuals of core phases

with time, if the inner core is rotating: the region of the inner core sampled by a ray connecting one seismic region to one station will vary with time if the inner core is rotating. If the contribution of the inner core to residuals is prevailing over that of the mantle, computating the residuals in the inner core frame, and not in the mantle frame as usually done, must increase the coherence of the residuals. The use of PKP(DF) with polar paths, or (BC-DF), allows to decrease the mantle contribution to the residuals. A rotation rate of 0 ± 1°/yr is obtained. Unfortunately, as previously, the resolution is rather poor, not better than 1°/yr; it is mostly limited by the number of old polar paths.

INNER CORE ROTATION: EVIDENCE FROM RESIDUAL VARIATIONS ALONG PARTICULAR PATHS

The first attempts to detect an inner core rotation focussed on the variations over several decades of the differential travel time residuals of two core phases along particular paths connecting a focal region to a station, e.g. *Souriau* [1989, and GRL paper 9L7552R, 1989, unpublished] for Tuamotu events to Warramunga array in Australia, *Roudil* [1996] for Tonga to France, *Song and Richards* [1996] for Tonga to Germany, Kermadec to Norway and South Sandwich Island to Alaska, *Collier and Helffrich* [2001] for Tonga to UK. One of the phases samples the inner core (PKiKP or PKP(DF)), the other one is taken as reference phase. The main technical limitations of such studies are first the difficulty to find old, well documented data, with unchanged instrument locations, and second the quality of the records, in particular the time definition (marks on the records). However, even if these limitations may be overcome, the two most severe problems are the presence of a detectable inner core feature (heterogeneity or anisotropy variation), and possible biases due to mantle heterogeneities

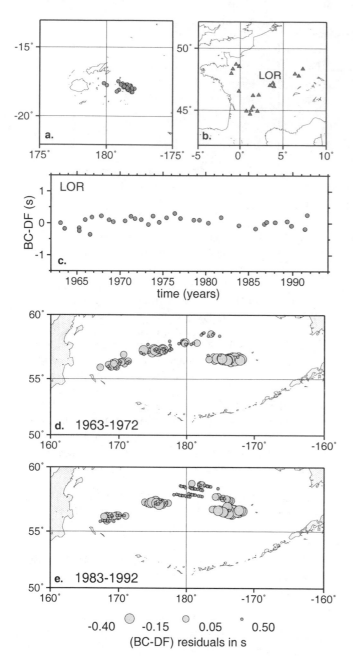

Figure 7. Travel time anomalies for the paths from deep Tonga events to France. Locations of epicenters (a) and stations (b); (c): Example of (BC-DF) residual variation as a function of time at station Lormes (LOR); (d-e): Values of the (BC-DF) residuals reported at the ray turning point for two epochs distant by 20 years (1963-72 and 1983-92). Note the similarity of the observed pattern for the two epochs [*Souriau and Poupinet*, 2000].

and source locations. Some of the paths which have been investigated led to the absence of significant temporal variation, as for example the paths from Tonga to Europe [*Roudil*, 1996; *Song and Richards*, 1996; *Souriau and Poupinet*, 2000; *Collier and Helffrich*, 2001], most of the

paths from Novaya Zemlya to Antarctica (see introduction). Only a few paths led to well identified variations, in particular the path from South Sandwich Island to station COL in Alaska, which has been investigated by several authors [*Song and Richards*, 1996; *Creager*, 1997; *Song*, 2000a, 2000b; *Poupinet et al.*, 2000], and which will be reconsidered later in this paper. In what follows, we will investigate the path from Tonga to France, as it provides a good example of spatial variations that may mimic temporal variations. We will also consider the paths at 180° longitude, which possibly sample a sharp variation of structure inside the inner core. Finally, we will discuss the use of teleseismic doublets for discriminating between temporal variations and event mislocations.

The path from Tonga to France

The differential travel times BC-DF reported in Fig. 7 correspond to 64 deep Tonga events spanning from 1964 to 1992, and recorded at 14 stations of the Laboratoire de Détection Géophysique of the French Atomic Energy Commission. The events (Fig. 7a) correspond to a small cluster of hypocenters in latitude, longitude and depth, whereas the stations (Fig. 7b) are distributed over 10 degrees in longitude. This geographic configuration gives turning points that stretch East-West over 24 degrees beneath the Bering Sea: it is thus well suited to detect the drift of an inner core structure. If we consider the residuals at particular stations (for example LOR in Central France), some temporal variations seem to be present (fig. 7c). However, the same fluctuations are not apparent (with a possible time shift) at stations which are located eastward or westward (see *Souriau and Poupinet* [2000], Fig. 4). Moreover, Fig. 7d-e shows that the residuals at the ray turning point for two epochs separated by twenty years have the same geographic pattern. This may denote either the absence of rotation (if the residual variations are due to inner core structure), or uncorrected mantle heterogeneities. The presence of residuals that differ by 0.3s for two families of rays with the same turning points (e.g. at longitude 175°) and with similar θ-values argues strongly in favour of a mantle origin. Thus even in particularly good conditions (very deep events located nearly at the same place), it is impossible to get rid of the effects of mantle heterogeneities, which may induce residuals as large as the precision required to detect inner core rotation, of the order of a few tenths of seconds. This holds a fortiori for other paths for which such a drastic data selection cannot be applied, in particular for the path from South Sandwich Island to COL.

The drift of the heterogeneity near longitude 180°

As noted previously (Fig. 2), a strong change in BC-DF residual is observed near longitude 175°W (see also *Collier and Helffrich*, 2001). It is also apparent in absolute

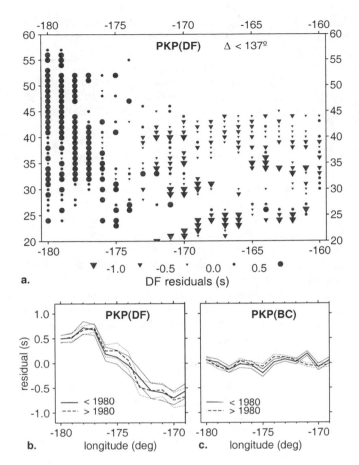

Figure 8. a- Detection of a sharp heterogeneity at longitude – 175°. Each value, plotted at the ray turning point, is the mean of the PKP(DF) residuals for paths with turning point in the same 1°×1° domain, and with epicentral distance Δ ≤137°. b- Mean residual as a function of longitude for the same data, for two epochs: 1964-1980 (full line) and 1981-1996 (dashed line). No drift of the heterogeneity with time may be observed. c- Same for BC data, indicating that the heterogeneity may be inside the inner core.

PKP(DF) residuals. Fig. 8a shows the mean residuals for summary rays binned according to the turning point latitude and longitude, for distances less than 137°. Means obtained from less than 30 values have been discarded. A clear jump is observed at 175°W, with lower residuals (higher velocities) in the western hemisphere, i.e. east of the meridian 175°W. The opposite pattern is observed at large distance (Δ>155°), with a jump at longitude ~180°. A similar pattern is not observed for PKP(BC), suggesting that this anomaly is located in the inner core. Note however that PKP(BC) does not sample exactly the same mantle regions as the two PKP(DF) data sets, thus it is impossible to fully discard a mantle contribution to the observed anomaly. The advantage of considering the distance Δ<137° for PKP(DF) is that the DF and BC paths, with same turning point and same orientation, sample the highly heterogeneous D" layer

nearly at the same place. This is not the case if DF and BC correspond to the same epicentral distance: For Δ~150°, the D" patches sampled by the two rays are distant by more than 4° (240 km).

If the inner core is differentially rotating, the detected inner core heterogeneity must drift with time. Fig. 8b shows the variations of the PKP(DF) residuals at distances <137° as a function of longitude at two epochs, 1964-1980 and 1981-1995. Taking into account the error bars, a mean shift of 1.5° between these two epochs distant of 16 years would be easily detected. Such a shift is not observed. If the anomaly originates effectively inside the inner core, this limits the rotation rate to values lower than 0.1°/yr. It is however impossible to completely discard a mantle contribution to this anomaly.

The use of doublets

An important source of error is event mislocations. Contrary to mantle heterogeneities, they vary in time as a consequence of the evolution of global seismic networks, thus they may mimic an inner core rotation. The most efficient way to discriminate between mislocation errors and inner core rotation is to use doublets. Doublets are pairs of events located nearly at the same place and generating signals with similar waveforms at a same station [*Poupinet et al.*, 1984]. The use of teleseismic doublets for testing inner core rotation is based on the existence of three core phases at distances between 146 and 156° (Fig. 9): PKP(DF), which turns inside the inner core, and PKP(BC) and PKP(AB), which turn at different depths inside the liquid core. The differential travel times (BC-DF) and (AB-BC) are both sensitive to distance variations (thus to event mislocations), but only (BC-DF) is sensitive to perturbations induced by inner core rotation. The main advantage of using doublets is that neither accurate locations nor travel time residual computations are needed to investigate the inner core rotation.

Fig. 9 illustrates how to use doublets to infer inner core rotation. One event (noted 0) is taken as reference. All the differential travel time will be measured with respect to the times observed for this event. In the diagram dt(BC)-dt(DF) as a function of dt(AB)-dt(BC), the corresponding point is thus at (0,0). If the distance Δ increases (event 1), the differential times (BC-DF) and (AB-BC) both increase. In the diagram dt(BC)-dt(DF) versus dt(AB)-dt(BC), the corresponding point will move along a line whose slope is governed by the values of the ray parameters dt/dΔ of the three phases. If in addition the depth increases, point (1) will move to the right and down with respect to this line. The opposite occurs if the depth decreases. This allows to define a "mislocation band" (shaded area in Fig. 9), which corresponds to the differential travel time perturbations for all the events whose epicentral distance or depth differ by less than 50 km with respect to the reference event: this value of 50 km is basically the maximum difference in

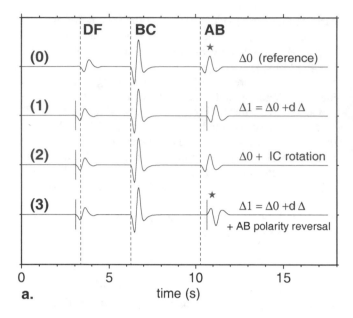

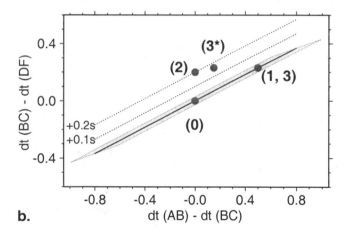

location which is considered (or the mislocation error), it is probably underestimated for depth. The variations in latitude and longitude are responsible for the length of this band, the variations in depth are responsible for its width. If the event is located at the same place as the reference event, but detects an inner core rotation (event 2), only DF will be perturbed. The point will be located above or below the 0-point. A rotation will generate points outside the "mislocation band". For example, a DF residual decrease by 0.1s or 0.2s will give points along the two dotted lines of Fig. 9b.

A serious difficulty may occur if one of the phase exhibits a polarity reversal compared to the reference event (event 3): this may occur if the different phases do not take off on the same side of the nodal plane. If the polarity reversal is taken into account, the point will lie inside the mislocation band. If it is not corrected, an erroneous phase correlation, as noted with stars in Fig. 9a, will lead to a point outside the mislocation band (point 3*), which mimics a rotation. This problem is crucial, because this situation is not rare, and because it is often difficult to identify phase reversals on events which are not perfect doublets. The simplest way to detect the polarity reversals is to use the coherence between events of the same epoch: If a point outside the mislocation band (as 3*) is found for pairs of recent (or old) events, then a phase reversal has been detected and must be corrected. Only after these corrections have been applied to recent events and old events separately is it possible to compare the old events with the recent events [*Poupinet et al.*, 2001]. If this is not done, the results may be erroneous [*Song*, 2000b].

Fig. 10 shows an example of a doublet for South Sandwich Island events recorded at College (COL), Alaska. These two events have been selected in the IRIS database according to their distance difference (less than 20 km), depth difference (less than 40 km), and their focal solutions, which give strong DF and BC phases with the same polarity. The differences (DF_1-DF_0), (BC_1-BC_0) and (AB_1-AB_0) are measured from the cross-correlation of the two events (Fig. 10) in order to estimate very accurately the differential travel times $dt(BC)-dt(DF) = \{(BC_1-BC_0) - (DF_1-DF_0)\}$ and $dt(AB)-dt(BC) = \{(AB_1-AB_0) - (BC_1-BC_0)\}$. The measured values are directly reported on the "mislocation band" diagram, without need of any additional computation. This makes this method particularly robust.

The results for the South Sandwich events recorded at COL are presented in Fig. 11a, for pairs of recent events and pairs of old events (open symbols). The pairs of events with short time lapse (open symbols) allow us to define an experimental "mislocation band". Superimposed is the theoretical mislocation band computed for the anisotropic inner core model of *Garcia and Souriau* [2000], derived from ak135 with a 3.5% anisotropy below 100 km depth. The mislocation band is computed for a mean distance

Figure 9. Scheme explaining how to use doublets to detect a possible inner core rotation. We report the theoretical computations of differential PKP travel times for two nearby events. a- waveforms with the three core phases; b- differential delays $[\{dt(BC) - dt(DF)\}_i - \{dt(BC) - dt(DF)\}_0]$ as a function of $[\{dt(AB) - dt(BC)\}_i - \{dt(AB) - dt(BC)\}_0]$. (0): perfect doublet: event similar to the reference earthquake, (1) same as (0) with increase in epicentral distance, (2): as (0) with inner core rotation (perturbation of DF only), (3): as (1), but with AB polarity reversal. The star indicates a wrong phase correlation which mimics an inner core rotation (point 3* in (b)). The shaded area in (b) is the "mislocation band" corresponding to variations in epicentral distance and depth of 50 km with respect to the reference event. The two dashed lines correspond to DF perturbations of 0.1 and 0.2 s (inner core rotation). In the absence of rotation, the points must remain close to the "mislocation band".

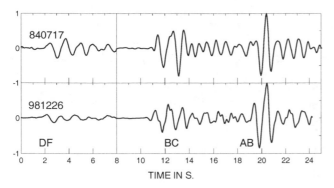

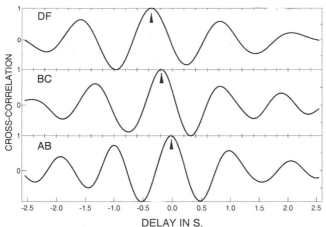

Figure 10. Example of doublet for two South Sandwich Island events recorded at COL, Alaska (1984 07 17 and 1998 12 26). Top: Seismograms showing the three core phases, band-pass filtered at 0.2-2.0 Hz. The first eight seconds are amplified by a factor of 5. Bottom: Estimation of dt(DF), dt(BC), dtAB) from the cross-correlations between the two PKP(DF) phases, the two PKP(BC) phases and the two PKP(AB) phases, used to determine the point position with respect to the "mislocation band" of Fig. 9.

150.6°, a mean depth 100 km, and a θ-value of 26°, which correspond to the geometry of the paths considered. The band width corresponds to focal depth variations of ±70 km, which corresponds to the depth range of the events considered, but it does not include possible errors on depth determinations (of the order of ± 50 km). We observe a very good agreement between this theoretical curve and the observations. The open circles outside this band reflect both the depth mislocations and the reading errors, of the order of 0.05s.

Full symbols (Fig. 11a) correspond to the doublets associating one old event (taken as reference event) and one recent event, for the events previously used by *Song and Richards* [1996] and *Poupinet et al.* [2001] to infer inner core rotation, and for two new doublets extracted from the IRIS database (squares). The time lapse between events range from 13 to 31 years. Note that these points lie inside

or slightly above the mislocation band, far below the 0.3s expected from *Song and Richard's* (1996) results. Moreover, their scatter is not significantly larger than the scatter observed for short time lapses. This scatter increases when the distance between the two events increases (points on the right of the figure), because the doublet technique becomes sensitive to structure variations between the two nearby paths. We note that most of the points are on the positive (right) side of the mislocation band, indicating that most of the recent events are more distant to COL than the old events: this is fortuitous in the data selection.

Most of the points are also above the mislocation band. This could indicate either that the recent events are shallower than the old events, or that inner core is rotating slowly. In order to test this hypothesis, we have plotted the {dt(BC)-dt(DF)} delay of these points from the median of the mislocation band, as a function of the time lapse between the two events (Fig. 11b). An increase in time is expected if it is due to inner core rotation. This is not observed. This result argues against significant rotation. The mean residual increase with time deduced from Fig. 11b is −0.0025 ± 0.0030 s/yr (error bar 1σ), which is basically 0, and in any case much smaller than the variation of about + 0.01 s/yr reported by *Song and Richards* [1996] and *Song* [2000a]. The conversion of this drift into a rotation rate requires the knowledge of the structure beneath the relevant path. If we adopt the same heterogeneity model as *Song* [2000], this leads to a rotation rate of −0.17 ± 0.21 °/yr (error bar 1σ).

The previous results suggest that the time variation of the BC-DF residuals observed by *Song and Richards* [1996] for this path, which was the strongest evidence for a fast inner core rotation, is in fact for a large part the result of mislocation bias, old events being systematically shifted to the north with respect to their actual position. A simple way to confirm this bias is to compute, for each event, the PKP mean residual differences for stations in northern azimuths (COL, IMA, INK, MBC) and stations in southern azimuths (DDR, MAT, YSS, BOD) with respect to South Sandwich Island: the latitude bias will be enhanced by a factor of two. The annual mean of these differences plotted for PKP as a function of time (Fig. 12) reveals a clear trend with a slope of 0.06 s/year, which corresponds to a mean shift of 22 km per 10 years, consistent with the observed position of the points on the mislocation band. This bias is large, but it reflects the evolution of the seismic networks, as will be discussed in the next section. It turns out that, for nearby events, only the doublet technique may efficently discriminate between location differences and inner core rotation. Even so, the method must be used with care if phase reversals are present. This is a real difficulty for the path SSI to COL, because most of the focal mechanisms at SSI are such that core phases are close to a nodal plane. Moreover, perfects teleseismic doublets are rare: This

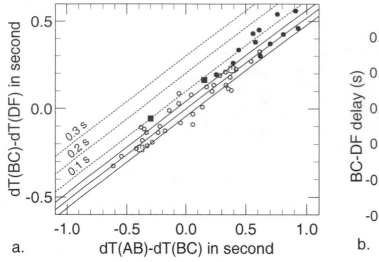
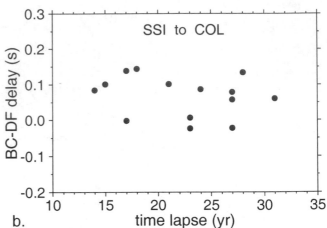

Figure 11. Results of the doublet technique applied to South Sandwich Island events recorded at COL. Circles correspond to the data of *Song and Richards* [1996] and *Poupinet et al.* [2000], squares to new high quality doublets extracted from the IRIS databank. a) Open symbols: differences in delays {dt(AB)- dt(BC)}, {dt(BC)- dt(DF)} for 17 doublets with events close in time (old events or recent events). They give the experimental "mislocation band".

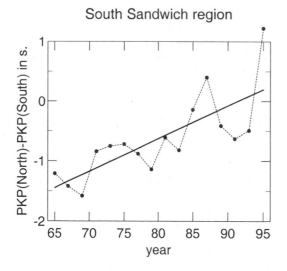

Figure 12. Difference of the mean PKP(DF) residuals for stations to northern azimuths with respect to South Sandwich Island (COL, IMA, INK, MBC), and stations to southern azimuths (DDR, MAT, YSS, BOD), as a function of time. The trend observed, with a mean slope of 0.06 s/year, reveals a mislocation bias toward the North which has decreased with time [*Poupinet and Souriau*, 2001].

would require the same rupture at the same place within the 30 years considered, a time lapse shorter than the time recurrence of the large events analysed in this study [*Song*, 2001; *Poupinet and Souriau*, 2001]. These various difficulties explain why different results have been obtained with the same data. Our results (Fig. 11a) show it remains

difficult to discriminate between depth mislocation and rotation, and they do not allow to completely discard a small rotation with rate below 0.2°/yr, slightly lower than the minimum rate of 0.3°/yr proposed by *Song* [2000a].

DISCUSSION AND CONCLUSION

The previous analyses show that the determination of an eventual inner core rotation is not an easy task. The most recent results (Table 2), which are based on different methods, seem to converge to either the absence of rotation, or to a slow eastward rotation, bounded to 0.2 °/yr, a value which is close to the present limit of detectability of the seismological methods. The main difficulties are: 1- the difficulty to identify an heterogeneity or an anisotropy variation which can be used to detect the rotation, 2- the biases due to mantle heterogeneities; 3- the biases due to event mislocations; 4- the limitations due to the availability and the quality of old records.

The investigations we performed on inner core structure reveal that the anisotropy is coherent over large regions, with no evidence of tilt of the symmetry axis with respect to the Earth rotation axis. The best identified heterogeneity is an hemispherical pattern in the inner core structure, with possibly a sharp transition between 175°W and 180°W longitude. A simple model with an external isotropic layer with thickness 100 km beneath the western hemisphere and 400 km beneath the quasi-eastern hemisphere (longitude 40°-180°), and an homogeneous anisotropic body with 3.5% anisotropy in the center (Fig. 4a) explains the general pattern of the PKP residuals and the geographic variations of the (BC-DF) ray parameter. The level of heterogeneity in

Table 2: Summary of the main results about inner core rotation. Rotation rate positive to the East (inner core faster than mantle). BC-DF: differential travel time PKP(BC)-PKP(DF), SSI: South Sandwich Island, NZ = Novaya Zemlya.

Reference	Method	Observation	Rotation rate
Souriau, 1989	PKiKP-PcP, Tuamotu to Australia	0.2 s in 10 yr	Probable rotation
Song and Richards, 1996	BC-DF, SSI to COL, Alaska (tilt of anisotropy axis) Tonga to GRF, Germany Kermadec to Norway	0.3 s in 30 yr ~ 0 - 0.2 s in 10 yr	1.1 °/yr (0.4 to 1.8 °/yr) No interpretation
Roudil, 1996	BC-DF, Tonga to France	0.3 s in 32 yr	No interpretation
Su et al., 1996	DF, worldwide	drift of the pole of the anisotropy axis 70° longitude in 25 yr	~3°/yr
Creager, 1997	BC-DF, SSI to COL (heterogeneities)	0.25 s in 30 yr	0.2-0.3 °/yr preferred (0.05 to 0.31 °/yr)
Souriau, 1998a, 1998c	BC-DF, NZ to DRV, Antarctica	~ 0.0 s in 23 yr	0°/yr
Ovtchinnikov et al., 1998	BC-DF, NZ to NVL, Antarctica	0.3 s in 28 yr	0.4 to 1.8 °/yr
Li et al., 1998	BC-DF, NZ to SBA, Antarctica to SPA, Antarctica	-0.09 ± 0.07 s in 19 yr 0.34 ± 0.66 s in 37 yr	Eastward rotation (rate not specified)
Richards et al., 1998	BC-DF, NZ to SBA, Antarctica	- 0.0053±0.0034 s/yr	Same
Sharrock et al., 1998	Splitting of eigenmodes	- 2.5 to - 0.5 °/yr	Westward rotation
Dziewonski and Su, 1998	DF, worldwide	large error bars	Rotation not detectable
Laske and Masters, 1999	Splitting of eigenmodes	-0.7 to 0.8 °/yr	0 ± 0.2 °/yr
Song and Li, 2000	BC-DF, Alaska to SPA	0.6 s in 37 yr	0.6 °/yr
Isse and Nakanishi, 2000	BC-DF, records at SYO, Antarctica	28 yr	0.0 °/yr
Souriau and Poupinet, 2000	BC-DF and polar DF, worldwide	32 yr	0 ± 1 °/yr
Vidale et al., 2000	Inner core scatterers, NZ to Montana	0.1 s in 3 yr	0.15 °/yr
Poupinet et al., 2000	Doublets of core phases, SSI to COL	< 0.1 s in 30 yr	< 0.2 °/yr
Song, 2000a	BC-DF pairs, SSI to Alaska	~ 0.45 s in 40 yr	0.3 to 1.1 °/yr
Collier and Hellfrich, 2001	BC-DF, South Pacific to UK		0.42±0.22 to 0.66±0.24 °/yr No monotonic variation
This study	Various studies, + synthesis		-0.2 to 0.2 °/yr

isotropic velocity is low at long and intermediate wavelengths. This does not rule out the possibility of anisotropy heterogeneities (variations in orientation or in anisotropy level), but the identification of such an heterogeneity is not an easy task, as it requires its sampling by rays with various orientations.

The perturbing effect of mantle heterogeneites is encountered in all the studies dealing with inner core rotation, in particular in those using differential travel times. This is well illustrated by the path from Tonga to France, or by the results of *Creager* [1997], who obtained rotation rates ranging from 0.05°/yr to 0.31°/yr for the path SSI to COL, depending on the level of heterogeneity at station side (see also *Creager* [2000]). Inner core sensitive eigenmodes also have a great sensitivity to mantle structures, and possibly to outer core structures [*Laske and Masters*, 1999; *Romanowicz and Bréger,* 2000]. Unfortunately they are not sensitive to patterns with an odd symmetry, thus they miss the most important feature of the inner core, i.e. the hemispherical pattern. The simultaneous use of several core sensitive modes partly allows to circumvent this problem [*Laske and Masters*, 1999]. For body waves, the simultaneous inversion of mantle structure, inner core structure and inner core rotation leads generally to an ill-posed inverse problem, because of the geometry of the ray paths, which are nearly parallel to each other. The best way to escape this problem is to use intercomparison of paths which are geographically close to each other, but distant in time: this approach is used in the doublet technique, but also in the analysis of inner core scattered waves [*Vidale et al.*, 2000].

The evolution with time of the mislocation biases are mostly related to the evolution of the seismological networks. They are difficult to correct, in particular for events in regions poorly sampled by seismic stations. Taking again the example of SSI events, the nearest station is for some events at a distance of ~ 20°, which is very large for hypocenter determinations, and which in addition may give travel times affected by P-triplication. Depth locations may be affected by large biases: a comparison of the locations given by USGS and the EHB-file [*Engdahl et al.*, 1998] shows that depth differences larger than 50 km are not rare (up to 110 km for SSI events). Relocating the events with the use of a joint epicenter determination method over the whole period of interest is impossible, as the different events are recorded by very few stations in common [*Souriau*, 1998c; *Poupinet et al.*, 2000]. Relocation for successive time intervals, as made by *Song* [2000a], resolves the location problem for the events inside each time lapse, but it does not remove the long term bias, i.e., the bias from one time lapse to the next one. The systematic use of doublets is clearly the best way to detect and minimize mislocation biases. A limitation of this method is however the need to find events which are not too far away from each other (less than about 100 km), as the doublet technique is very sensitive to local heterogeneities.

The present limitations in inner core detection are also inherent to the seismological data themselves. Old data of good quality are difficult to find, and their quality is often difficult to control. Moreover, the relevant paths are not necessarily well suited to core rotation studies. Readings reported in bulletins are not always very accurate, phase misidentifications may occur, and the temporal series in individual stations may exhibit fluctuations which have no relation with Earth structure [*Roehm et al.*, 1999; *Engdahl and Ritzwoller*, 2001]. The increasing number of stations with numerical data at high sampling rate will probably allow to reach a higher accuracy in a very near future. Some methods, as the use of scattered waves [*Vidale et al.*, 2000], will take benefit of these improvements.

An important question remains however: is a differential rotation possible from a physical point of view? Some of the dynamo models predict a differential rotation [e.g. *Glatzmaier and Roberts*, 1996, *Aurnou et al.*, 1996], whereas such a rotation is not required by other models

[*Jault*, 1996). On the other hand, the density heterogeneities in the mantle induce structure perturbations in the inner core and at inner core boundary [*Defraigne et al.*, 1996]. The gravitational coupling between mantle and inner core structures imposes a synchroneous rotation together with an inner core wobble [e.g. *Dehant et al.*, 1993; *Buffett*, 1996). A permanent deformation of the inner core by viscous deformation has been proposed to reconcile inner core rotation and gravitational constraints [*Buffett*, 1997]. However, it would be difficult to maintain, in such a permanently deforming inner core, some inner core structures which are independent of the mantle structure. Moreover, seismological observations must be compatible with the geodetic data [*Buffett and Creager*, 1999]. Dynamo models which takes into account gravitational and viscous coupling lead to inner core oscillations superimposed to, or alternating with, a slow super-rotation [*Kuang*, 1999; *Aurnou and Olson*, 2000; *Buffett and Glatzmaier*, 2000], but many important parameters of the problem remain poorly constrained. These oscillations, if real, will be difficult to detect with seismological data [*Collier and Helffrich*, 2001].

Finally, the inner core rotation has also implications for the distribution of the anisotropy inside the inner core. Whether the hemispherical inner core pattern is induced by the mantle gravitational coupling, or by magnetic forcing, remains an interesting question, as both the stress field and magnetic field [e.g. *Karato*, 1993, 1999; *Buffett and Wenk*, 2001] may play an important role in iron crystal orientation.

Acknowledgments. We thank John Vidale, Ken Creager and an anonymous reviewer for helpful criticisms of the manuscript. Financial support has been provided by INSU, program "Intérieur de la Terre".

REFERENCES

Aurnou, J., and P. Olson, Control of inner core rotation by electromagnetic, gravitational and mechanical torques, *Phys. Earth Planet. Inter.*, *117*, 111-121, 2000.

Aurnou, J., D. Brito, and P. Olson, Mechanisms of inner core super-rotation, *Geophys. Res. Lett.*, *23*, 3401-3404, 1996.

Bréger, L., B. Romanowicz, and H. Tkalcic, PKP(BC-DF) travel time residuals and short scale heterogeneity in the deep Earth, *Geophys. Res. Lett.*, *26*, 3169-3172, 1999.

Bréger, L., H. Tkalcic, and B. Romanowicz, The effect of D" on PKP(AB-DF) travel time residuals and possible implications for inner core structure, *Earth Planet. Sci. Lett.*, *175*, 133-143, 2000.

Buffett, B.A., A mechanism for decade fluctuations in the length of day, *Geophys. Res. Lett.*, *23*, 3803-3806, 1996.

Buffett, B.A., Geodynamic estimates of the viscosity of the Earth's inner core, *Nature, 338*, 571-573, 1997.

Buffett, B.A., and K.C. Creager, A comparison of geodetic and seismic estimates of inner core rotation. *Geophys. Res. Lett.*, *26*, 1509-1512, 1999.

Buffett, B.A., and G.A. Glatzmaier, Gravitational breaking of inner core rotation in geodynamo simulations, *Geophys. Res. Lett.*, *27*, 3125-3128, 2000.

Buffett, B.A., and H.-R. Wenk, Texturing the Earth's inner core by Maxwell stresses, *Nature*, *413*, 60-63, 2001.

Collier, J.D., and G. Helffrich, Estimate of inner core rotation rate from United Kingdom regional seismic network data and consequences for inner core dynamical behaviour, *Phys. Earth Planet. Inter.*, *193*, 523-537, 2001.

Creager, K.C., Anisotropy in the inner core from differential travel times of the phases PKP and PKIKP, *Nature, 356*, 309-314, 1992.

Creager, K.C., Inner core rotation rate from small scale heterogeneity and time-varying travel times, *Science*, *278*, 1284-1288, 1997.

Creager, K. C., Large-scale variations in inner core anisotropy, *J. Geophys. Res.*, *104*, 23127-23139, 1999.

Creager, K.C., Inner core anisotropy and rotation, in: *Mineral physics and seismic tomography*, S.I. Karato, L. Stixrude, R. Liebermann, G. Masters and A. Forte, Ed., AGU monograph, 2000.

Defraigne, P., V. Dehant, and J. Wahr, Internal loading of an inhomogeneous compressible Earth with phase boundaries, *Geophys. J. Int.*, *125*, 173-192, 1996.

Dehant, V., J. Hinderer, H. Legros, and M. Lefftz, Analytical approach to the computation of the Earth, the outer core and the inner core rotational motions, *Phys. Earth Planet. Inter.*, *76*, 259-282, 1993.

Durek, J.J., and B. Romanowicz, Inner core anisotropy inferred by direct inversion of normal mode spectra, *Geophys. J. Int.*, *139*, 599-622, 1999.

Dziewonski, A.M., and W.-J. Su, A local anomaly in the inner core, *Eos. Trans. AGU*, *79*, Spring Meeting Abstracts, S218, 1998.

Engdahl, R.E., R. van der Hilst, and R.P. Buland, Global teleseismic earthquake relocation with improved travel times and procedures for depth determination, *Bull. Seismol. Soc. Am.*, *88*, 722-743, 1998.

Engdahl, E.R., and M.H. Ritzwoller, Crust and upper mantle P- and S-wave delay times at Eurasian seismic stations, *Phys. Earth Planet. Inter.*, *123*, 205-219, 2001.

Garcia, R., Constraints on uppermost inner core structure from waveform inversion of core phases, *Geophys. J. Int.*, to be published, 2002.

Garcia, R., and A. Souriau, Inner core anisotropy and heterogeneitiy level, *Geophys. Res. Lett.*, *27*, 3121-3124, 2000.

Garcia, R., and A. Souriau, Correction to "Inner core anisotropy and heterogeneitiy level", *Geophys. Res. Lett.*, *28*, 85, 2001.

Glatzmaier, G.A., and P.H. Roberts, Rotation and magnetism of Earth's inner core, *Science*, *274*, 1887-1891, 1996.

Gubbins, D., Rotation of the inner core, *J. Geophys. Res.*, *86*, 11695-11699, 1981.

Isse, T., and I. Nakanishi, Inner core anisotropy beneath Autralia and differential rotation, *SEDI meeting*, Exeter (U.K.), 2000.

Jault, D., Magnetic field generation impeded by inner cores of planets, *C.R. Acad. Sc. Paris*, *323*, 451-458, 1996.

Karato, S.-I., Inner core anisotropy due to the magnetic field – induced preferred orientation of iron, *Science*, *262*, 1708-1711, 1993.

Karato, S.-I., Seismic anisotropy of the Earth's inner core resulting from flow induced by Maxwell stresses, *Nature*, *402*, 871-873, 1999.

Kennett, B.L.N., E.R. Engdahl, and R. Buland, Constraints on seismic velocities in the Earth from traveltimes, *Geophys. J. Int.*, *122*, 108-124, 1995.

Kuang, W., Force balance and convective state in the Earth's core, *Phys. Earth Planet. Inter.*, *116*, 65-79, 1999.

Laske, G., and G. Masters, Limits on differential rotation of the inner core from an analysis of the Earth's free oscillations, *Nature*, *402*, 66-68, 1999.

Li, A., P.G. Richards, and X. Song, Further measurements of changes in core travel times, to infer the rotation rate of the Earth's inner core. *EOS Trans. AGU*, *79*, Spring Meeting Abstracts, S218, 1998.

McSweeny, T.J., K.C. Creager, and R.T. Merrill, Depth extent of inner-core seismic anisotropy and implications for geomagnetism, *Phys. Earth Planet. Inter.*, *101*, 131-156, 1997.

Morelli, M., A.M. Dziewonski, and J.H. Woodhouse, Anisotropy of the inner core inferred from PKIKP travel times, *Geophys. Res. Lett.*, *13*, 1545-1548, 1986.

Niu, F., and L. Wen, Hemispherical variations in seismic velocity at the top of the Earth's inner core, *Nature*, *410*, 1081-1084, 2001.

Ouzounis, A., and K.C. Creager, Isotropy overlying anisotropy at the top of the inner core, *Geophys. Res. Lett.*, *28*, 4331-4334, 2001.

Ovtchinnikov, V.M., V. V. Adushkin, and V. A. An, About the velocity of differential rotation of the Earth's inner core (in Russian). *Dokl. Russ. Acad. Sci. Geophys.*, *362*, 683-686, 1998.

Poupinet, G., R. Pillet, and A. Souriau, Possible heterogeneity in the Earth's core deduced from PKIKP travel times, *Nature*, *305*, 204-206, 1983.

Poupinet, G., W.L. Ellsworth, and J. Fréchet, Monitoring velocity variation in the crust using earthquake doublet: an application to the Calaveras fault, California, *J. Geophys. Res.*, *89*, 5719-5731, 1984.

Poupinet, G., A. Souriau, and O. Coutant, The existence of an inner core super-rotation questioned by teleseismic doublets, *Phys. Earth Planet. Inter.*, *118*, 77-88, 2000.

Poupinet, G., and A. Souriau, Reply to Xiadong Song's comment on "The existence of an inner core super-rotation questioned by teleseismic doublets", *Phys. Earth Planet. Inter.*, *124*, 275-280, 2001.

Richards, P.G., X. Song, and A. Li, Detecting possible rotation of Earth's inner core *Science*, *282*, 1221, www.sciencemag.org, 1998.

Roehm, A.H.E., J. Trampert, H. Paulssen, and R.K. Snieder, Bias in reported seismic arrival times deduced from the ISC Bulletin, *Geophys. J. Int*, *137*, 163-174, 1999.

Romanowicz, B., X.-D. Li, and J. Durek, Anisotropy in the Inner Core: Could it be due to low-order convection?, *Science*, *274*, 963-966, 1996.

Romanowicz, B., and L. Bréger, Anomalous splitting of free oscillations: A reevaluation of possible interpretations, *J. Geophys. Res.*, *105*, 21,559-21,578, 2000.

Roudil, P., Etude sismologique de la structure du noyau terrestre au voisinage de la frontière noyau externe – noyau interne, *Thesis Univ. Paul Sabatier*, Toulouse, France, 126 pp., 1996.

Sharrock, D.S., and J.H. Woodhouse, Investigation of time

dependent inner core structure by the analysis of free-oscillation spectra, *Earth Planet Space, 50,* 1013-1018, 1998.

Shearer, P., Constraints on inner core anisotropy from PKP(DF) travel times. *J. Geophys. Res., 99,* 19647-19659, 1994.

Song, X., Anisotropy in central part of inner core, *J. Geophys. Res., 101,* 16, 089-16,097, 1996.

Song, X., Anisotropy of the Earth's inner core. *Rev. Geophys., 35,* 297-313, 1997.

Song, X., Joint inversion for inner core rotation, inner core anisotropy, and mantle heterogeneity, *J. Geophys. Res., 105,* 7931-7943, 2000a.

Song, X., Time dependence of PKP(BC)-PKP(DF) times: could this be an artifact of systematic earthquake mislocations?, *Phys. Earth Planet. Inter., 122,* 221-228, 2000b.

Song X., and P.G. Richards, Seismological evidence for differential rotation of the Earth's inner core, *Nature, 382,* 221-224, 1996.

Song, X., and A. Li, Support for differential inner core superrotation from earthquakes in Alaska recorded at South Pole station, *J. Geophys. Res., 105,* 623-630, 2000.

Song, X., Comment on "the existence of an inner core rotation questioned by teleseismic doublets" by G. Poupinet, A. Souriau and O. Coutant, *Phys. Earth Planet. Inter., 124,* 269-274, 2001.

Souriau, A., A search for time dependent phenomena inside the core from seismic data, *EGS meeting abstracts,* Barcelona, 1989.

Souriau, A., New seismological constraints on differential rotation of the inner core from Novaya Zemlya events recorded at DRV, Antarctica. *Geophys. J. Int., 134,* F1-F5, 1998a.

Souriau, A., Earth's inner core: Is the rotation real?, *Science, 281,* 55-56, 1998b.

Souriau, A., Response to P.G. Richards et al.'s comment : " Detecting possible rotation of Earth's inner core ", *Science, 282,* 1221, www.sciencemag.org, 1998c.

Souriau, A., and G. Poupinet, Inner core rotation : a test at the worldwide scale. *Phys. Earth Planet. Inter., 118,* 13-27, 2000.

Souriau, A., and B. Romanowicz, Anisotropy in the inner core: relation between P-velocity and attenuation, *Phys. Earth Planet. Inter., 101,* 33-47, 1997.

Souriau, A., P. Roudil, and B. Moynot, Inner core rotation : facts and artefacts, *Geophys. Res. Lett., 24,* 2103-2106, 1997.

Su, W.-J., and A.M. Dziewonski, Inner core anisotropy in three dimensions, *J. Geophys. Res., 100,* 9831-9852, 1995.

Su, W.-J., A.M. Dziewonski, and R. Jeanloz, Planet within a planet : Rotation of the inner core of the Earth, *Science, 274,* 1883-1887, 1996.

Tanaka, S., and H. Hamaguchi, Degree one heterogeneity and hemispherical variation of anisotropy in the inner core from PKP(BC)-PKP(DF) times, *J. Geophys. Res., 102,* 2925-2938, 1997.

Tkalcic, H., Romanowicz, B., and N. Houy, Constraints on D" structure using PKP(AB-DF), PKP(BC-DF) and PcP-P travel time data from broadband records, *Geophys. J. Int.,* submitted, 2001.

Vidale, J.E., and P.S. Earle, Fine-scale heterogeneity in the Earth's inner core, *Nature,* 404, 273-275, 2000.

Vidale, J.E., D.A. Dodge, and P.S. Earle, Slow differential rotation of the Earth's inner core indicated by temporal changes in scattering, *Nature, 405,* 445-448, 2000.

Vinnik, L., B. Romanowicz, and L. Bréger, Anisotropy in the center of the inner core, *Geophys. Res. Lett., 21,* 1671-1674, 1994.

Woodhouse, J.H., D. Giardini, and X.-D. Li, Evidence fo inner core anisotropy from free oscillations, *Geophys. Res. Lett., 13,* 1549-1552, 1986.

Georges Poupinet, LGIT/IRIGM, BP 53X, 38041 Grenoble Cedex, France ; Georges.Poupinet@obs.ujf-grenoble.fr
Annie Souriau, Observatoire Midi-Pyrénées, 14, Avenue Edouard Belin, 31400 Toulouse, France ; Annie.Souriau@cnes.fr

The three-dimensional phase diagram of iron

Orson L. Anderson

Institute of Geophysics and Planetary Physics, Department of Earth and Space Sciences, University of California, Los Angeles, California

The phase diagram of iron, constructed from experimental data, is displayed in the T-V and T-P planes. The P,T,V coordinates of the three triple-point boundaries, as well as the three slopes of the phase boundaries that join together at each triple point, are determined. The melting volume, ΔV_m, values are presented. The solidus of the hcp (ϵ) phase of iron is calculated from a formula for the melting slope, $\mathrm{d}T_\mathrm{m}/\mathrm{d}V$, called Gilvarry's rule. This formula requires the Grüneisen parameter arising from the variation in the Debye temperature with volume, called γ_vib (vib for atomic vibrations). Values of γ_vib and Θ measured up to 360 GPa are used in the calculation of $T_\mathrm{m}(V)$, which yields $T_\mathrm{m} = 5900 \pm 300$ K at the ICB pressure (330 GPa). To obtain the Grüneisen parameter appropriate for an adiabatic process, the contribution to specific heat from excited free electrons in iron must be determined. The free-electron correction to γ_vib yields an effective Grüneisen parameter, γ_eff, that is independent of pressure above 150 GPa and has a value of 1.5. In this range, γ_vib decreases substantially with P.

1. INTRODUCTION

A study of the physical properties of iron is a prerequisite to the study of the physical properties of the Earth's core, since the consensus heavily favors a model of the core as consisting of nearly pure iron for the inner solid core, with a mixture of iron and light impurities for the outer liquid core.

The cosmochemical argument in favor of a predominantly iron core relies on the assumption of approximate homogeneity in the solar system in so far as element distribution is concerned. If the element distribution is to be approximately maintained, the number of iron atoms should be close to the number of silicon atoms in the Earth [*Ganapathy and Anders*, 1976; *Ross and Aller*, 1976; *Wasson*, 1985]. Since iron is depleted relative to silicon in the Earth's mantle, it needs to be greatly enriched in the Earth's core. The volume of the mantle is much greater than the core. Consequently, to maintain the cosmic abundance ratio, the core has to be composed essentially of iron.

Experimental mineral physics has provided a great deal of data supporting the model of an iron core. The most important experimental evidence is from shockwave physics.

Birch [1963] showed that for solids, the bulk sound velocity, $v_b = \sqrt{K_S/\rho}$ (where K_S is the adiabatic bulk modulus and ρ is the density), is linear with ρ at constant mean atomic number. This linear curve is shifted along the ρ axis for higher mean atomic number. This means that a solid having a mean atomic number considerably higher than that of minerals in the Earth's mantle will have a much larger density for the same value of v_b. *Birch* [1963] found that shock-wave results showed that v_b of the core corresponds to a mean atomic number of iron (51), while v_b of the mantle corresponds to the average mean atomic number of minerals in the

Earth's Core: Dynamics, Structure, Rotation
Geodynamics Series 31
10.1029/31GD07

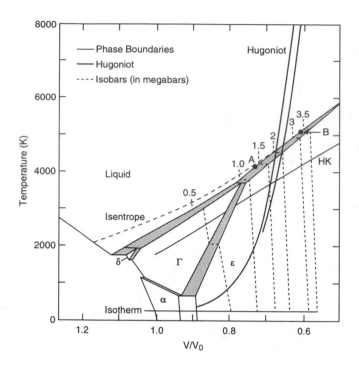

Figure 1. *Birch*'s [1972] conjectural phase diagram of epsilon iron in the *T-V* plane. There were few data at compressions smaller than $V/V_0 = 0.9$, so he assumed that all solidus curves increased linearly with volume. He calculated the slope of the *T-V* solidus from the measured dT_m/dP slope. The main feature is that the ΔV separating the liquidus from the solid is displayed. The crossing of the solidus and liquidus by the Hugoniot establishes experimentally the Hugoniot at 240 GPa. A represents the pressure for the CMB; B represents the pressure for the ICB.

ing the latest experimental information is the purpose of this paper.

In the construction of a phase diagram, the first problem is to decide the number of phases. Four crystalline phases are well established and were known to *Birch* [1972]. They are: (1) α, body-centered cubic (bcc) in structure, a ferromagnetic solid; (2) δ, also bcc, but non-magnetic; (3) Γ, face-centered cubic (fcc), the high-temperature phase; and (4) ϵ, hexagonal close-packed, the high-pressure phase. Two additional phases have been identified, but they are not well established, since there are scholarly papers based on experimental evidence that disputes their existence. These two controversial phases will be discussed in Section 10, but the phase diagram of iron in this paper will consist only of α, δ, Γ, and ϵ plus the liquid phase. There being five phases, three triple points exist.

2. BIRCH'S PHASE DIAGRAM OF IRON IN THE *T-V* PLANE

For thirty years the phase diagram (pd) of iron has been under construction, bit by bit. During this time, parts of the phase diagram were dependent on experimental information that either was guessed at, was in dispute, or was a matter of gross extrapolation from meager experiments. The required experimental information gradually accumulated because it depended on mastery of techniques for measuring iron properties at high pressure and temperature. For this reason, the historical phase diagram has changed through the years. Even though most of the experimental data were lacking, *Birch* [1972] laid out the guidelines using "speculative assumptions" that are still reasonably correct today for building the phase diagram. Birch's plan for the phase diagram includes the following points: 1) A phase diagram in the *T-V* plane is required as well as the usual phase diagram in the *T-P* plane; 2) values of ΔV and ΔS (the change in entropy) need to be evaluated at the triple points (tp) and the phase boundaries; ΔV versus *V* is conveniently displayed in the *T-V* plane; 3) slopes, dT/dP, of phase boundaries need to be evaluated; 4) the intersection of the Hugoniot with the melting line provides constraints on the melting temperature, T_m at high *P*; and 5) the ΔV between phases decreases as the pressure increases. These five guidelines are demonstrated in the hypothetical phase diagram that Birch presented in 1972, Figure 1.

Because of Point 5), the slope of the liquid melting line is less than the slope of the solid melting line. In physical chemistry, the liquid melting line is called the liquidus, and the solid melting line is called the solidus.

mantle (19). This finding disposed of the hypothesis that the Earth's core represents a high-pressure phase of upper mantle minerals because, by Birch's findings, a silicate core with the requisite density cannot have the v_b value found for the core by seismology.

Of all physical properties of iron, perhaps the most useful is the melting temperature, T_m, at high compression because this is the basis for understanding core temperatures. A corresponding problem is the relationship of the isentropes of compressed liquid iron to the melting curve. These problems are approached by finding T_m and the isentropes in iron at high pressure. Until recently, many properties were found by extrapolation of the existing measurements to higher pressure. However, since pure iron exists in four crystalline forms (or phases), it is necessary to define these phases with regard to each other so as to make extrapolations meaningful. This requires careful attention to the phase diagram. A presentation of the complete description of iron in all its phases by means of the phase diagram us-

I propose to follow this melting line notation. *Birch's* [1972] diagram (Figure 1) still gives a remarkably good picture of the general properties of the phase diagram of iron. We see, for example, from Figure 1 that the liquidus has a smaller slope than the solidus, and, although they tend to converge, there is still a significant ΔV at the inner core–outer core boundary pressure. At the two triple points along the melting line, the ΔV broadens, becoming equal to the sum of the two lower ΔV's. Where it crosses the melting lines, the Hugoniot is displaced by a change in volume. The Γ phase has a lower zero-pressure density than the α phase. Thus, the ΔV of the α-Γ phase is negative. This is consistent with the fact that dT/dP of the α-Γ transition boundary is negative.

In his phase diagram, *Birch* [1972] assumed that the slope of the solidus is invariant with V, because of the observation [*Kraut and Kennedy*, 1966] that an empirical linear relationship between melting temperature and ΔV along an isotherm holds for a number of metals (although the range of their measured pressure was small). As will be shown, the assumed linearity between T and ΔV does not hold for the phase diagram of iron, when the entire range of pressure in the phase diagram is considered.

3. THE THREE ESTABLISHED TRIPLE POINTS

Figure 2a,b shows the three triple points in the T-P plane, giving the slopes of the phase boundaries going into and leaving the triple points. The data from the lower triple points come from work in the late 1960's and 1970's [*Strong*, 1961; *Bundy and Strong*, 1962; *Bundy*, 1965; *Mao et al.*, 1967; *Strong et al.*, 1973; *Liu*, 1975]. The value of dT/dP in the lower branch of the α-Γ-ϵ triple point ($43°\mathrm{GPa}^{-1}$) comes from this paper. The fcc melting curve on the T-P plane (Figure 2b) comes from the experimental data: *Shen et al.* [1998] found the liquid-ϵ-Γ triple point to be at 60 GPa and 2800 K. *Saxena and Dubrovinsky* [2000] found from their measurements that this triple point is at 50 GPa, and the reported temperature was somewhat lower (2750 K). *Andrault et al.* [2000] reported the coordinates of the triple point to be in the vicinity of those reported by *Saxena and Dubrovinsky* [2000]. From the *Andrault et al.* [2000] paper, I read $P = 55$ GPa and $T = 2700$ K (see also our Figure 10). The small shifts in the coordinates of the three reports are well within the uncertainty in locating the phase boundary in melting determination experiments. These 3 triple points are a small cluster. I adopt the coordinates of the triple point used in

this paper as the center of the cluster, 55 ± 5 GPa and 2790 K, as shown in Figure 2b. The coordinates of this triple point comprise the lower bound of the hcp-solidus curve.

The Γ-ϵ-liquid triple point was reported by *Boehler* [1993] to be at 100 GPa and 2800 K, which is some distance from that reported by the three sets of investigators above. I did not choose the coordinates of this triple point as an anchor for the lower bound of the hcp-solidus curve. A choice must be made. I favored the triple point at 55 ± 5 GPa because this represents the results of three strong experimental laboratories (Washington, DC; Paris; and Uppsala). I judged the convergence of these three laboratories' results to one value to be compelling.

The historical search for the phase diagram at high compression has centered around the location of the hcp–fcc–liquid triple point. The location of this triple point is crucial in the determination of the properties of the phase at inner core conditions. The melting temperature of iron at ICB conditions depends greatly on where this triple point is located. A review of various locations of this important triple point is shown in Figure 3. The positions shown by *Birch* [1972], *Liu* [1975], and *Anderson* [1986] were located by estimates guided by the apparent slope of dT_m/dP emanating from the two lower triple points. There were no experimental data to confirm any of these suggestions. But *Boehler* [1993] was able to measure the melting curve of the fcc phase using the diamond cell measurement of both T and P. *Shen et al.* [1998], as well as *Saxena and Dubrovinsky* [2000] and *Andrault et al.* [2000], each traced the path of the fcc solidus found by *Boehler* [1993]. However, these three experimental papers found the triple point not at 100 GPa, but near 55 GPa. This is shown by the star in Figure 3, which represents the triple point used for the phase diagram in this paper. Adopting the 55 GPa triple point, the slope out of the ϵ-Γ-α triple point becomes $45.7°\mathrm{GPa}^{-1}$, Figure 2a,b.

From Figures 2a and 2b, the value of dT/dP can be determined anywhere in the phase diagram. The value of dT/dV, determined from dT/dP along the melting curve, can be calculated from a formula by *Birch* [1972]:

$$\left(\frac{dT}{dV}\right)_\mathrm{m} = -\frac{K_{T_\mathrm{m}}\,(dT/dP)_\mathrm{m}}{V\,(1 - \alpha K_T\,(dT/dP)_\mathrm{m})}, \qquad (1)$$

where m stands for melting, and K_{T_m} is the isothermal bulk modulus at melting.

It is difficult to find K_T at an arbitrary point in the phase diagram. To calculate it requires assumptions about the equation of state at high T. I shall bypass this

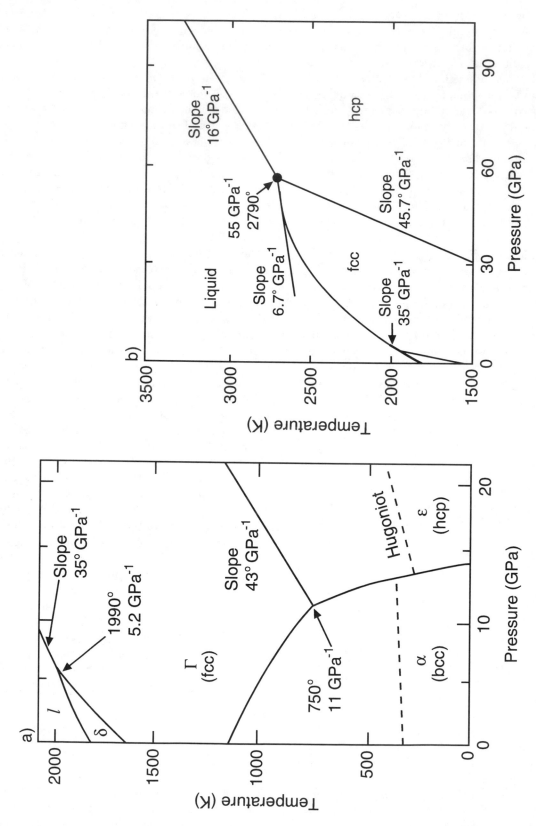

Figure 2. The three triple points in the *T-P* plane, showing the coordinates of the triple points, the value of the slope, dT/dP, and the phase boundaries.

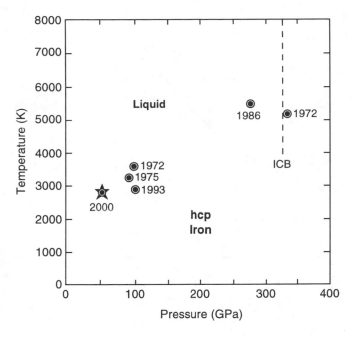

Figure 3. Historical record of the placement of the ϵ-Γ-liquid triple point. Until 1993, this triple point was found as the intercept between the slope of the extrapolated ϵ-Γ phase and the extrapolated Γ-liquid phase.

route and determine ΔV directly from the measured ΔV's at the phase boundaries.

4. THE VALUES OF ΔV AT PHASE BOUNDARIES AND TRIPLE POINTS

The ΔV values of the two lowest triple points were found experimentally in the 1960's and early 1970's. At that time, the piston cylinder high-pressure apparatus was popular because it allowed volume at the triple point to be accurately determined. The ΔV's at 11 GPa were reported by *Bundy* [1965]. The details of the 5.1 GPa triple point were reported by *Strong et al.* [1973]. Values of ΔV derived from these early measurements are shown in Figure 4 in the two lower triple points. Slight corrections to the early values of ΔV were made by *Anderson* [1986] (made while surveying these early experiments), to insure that $\sum \Delta V = 0$ around a triple point as required thermodynamically. *Anderson's* [1986] values are transferred to Figure 4. The original work did not give a definite value of dT_m/dP for the hcp-fcc phase boundary because at that time, the position of the hcp-fcc-liquid triple point had not been determined experimentally. But in Figure 4, where all three triple points are shown, we see that dT/dP for the hcp-fcc phase boundary is $45.7°\text{GPa}^{-1}$. Since the

Clapeyron law is $\Delta S/\Delta V = \Delta P/\Delta T$, then ΔS can be determined. All values of ΔS are listed in Table 1.

It is somewhat more difficult to obtain ΔV for the liquid-solid (hcp) phase boundary for compressions at 55 GPa and higher. We start with the inquiry about the ΔV's for the triple point at 55 GPa.

We know the value of dT/dP coming into 55 GPa: $6.7°\text{GPa}^{-1}$, so that the determination of ΔV depends on the value of ΔS since $\Delta V = (dT/dP)\,\Delta S$.

Along a melting line, the ΔS for phase transformation is

$$\Delta S_{\text{tran}} = \Delta S_{\text{config}} + \Delta S_{\text{volum}}, \qquad (2)$$

where the subscripts refer to transformation, configuration, and volumetric. The configuration term is significant for melting because the liquid has much more disorder of its atoms than is found in a lattice. The volumetric term arises because entropy diminishes with compression and rises with temperature, as shown in thermodynamic relationships.

The configurational entropy for monatomic substances with large coordination numbers has been measured by *Stishov* [1988]; he shows that as pressure increases, the value of $\Delta S \to 0.7R$, where R is the gas constant. This is very close to the theoretical value of $\Delta S = \ln 2R$. I shall use $\Delta S_{\text{config}} = 0.693R$ for $P = 55$ GPa.

The volumetric entropy is found from the Maxwell relationship [*Anderson, 1995*]

$$\left(\frac{\partial S}{\partial V}\right)_T = \left(\frac{\partial P}{\partial T}\right)_V = \alpha K_T, \qquad (3)$$

where α is the thermal expansivity, and K_T is the isothermal bulk modulus. Using the Grüneisen ratio relationship,

$$\gamma = \frac{\alpha K_T V}{C_V}, \qquad (4)$$

Equation (3) becomes

$$\left(\frac{\partial S}{\partial V}\right)_T = \frac{\gamma C_V}{V} \qquad (5)$$

or

$$\Delta S = \gamma C_V \left(\frac{\Delta V}{V}\right). \qquad (6)$$

We take the values $\gamma = 1.5$ and $C_V = 3.29R$ for ϵ iron at 60 GPa found in Table 2 of *Anderson and Isaak* [2000], which accounts for electronic specific heat. To find $\Delta V/V$ in Equation (6) requires an iterative approach, because ΔV is the final quantity we wish to calculate. However, since ΔS_{volum} is small compared to

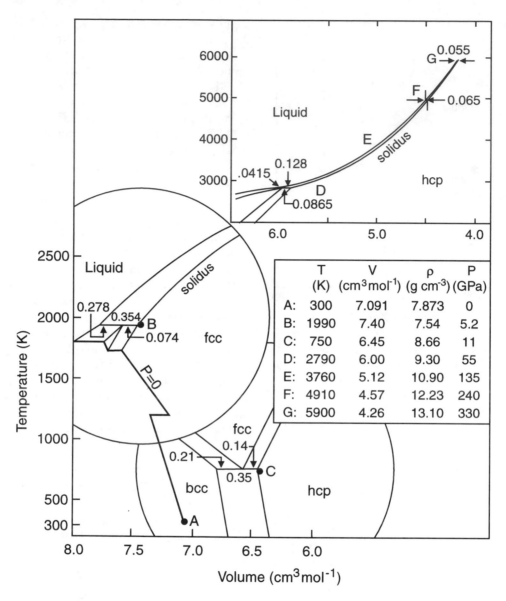

Figure 4. A plot of temperature versus volume that focuses on the three triple points. The values of ΔV at the three triple points and at phase boundaries are shown. The table inset gives the values of the coordinates of the triple points and two positions along the fcc solidus.

ΔS_{config}, the iteration process converges rapidly. Take $\Delta V / V = 0.01$. Then Equation (2) becomes

$$\Delta S_{\text{trans}} = (0.693 + 0.999)\, R$$
$$= 6.1923 \text{ J K}^{-1}\text{mol.}^{-1}. \qquad (7)$$

From Figure 2b we find the slope of the liquid-hcp boundary on the low-pressure side of the 55 GPa triple point to be $6.7°\text{GPa}^{-1} = 6.7 \times 10^{-3}\text{J cm}^{-3}$. Using the Clapeyron equation, $\Delta V = \Delta S_{\text{trans}}(\Delta T / \Delta P) = +0.0415 \text{ cm}^3\text{mol.}^{-1}$. This value of ΔV represents the

phase boundary coming from the 5.2 GPa triple point. There is another phase boundary to evaluate coming from the 11 GPa triple point. Interior phase boundaries are straight lines, and there is negligible entropy change along the boundary, so we reduce the ΔV from 14 cm^{3}mol.$^{-1}$ by the ratio of the V's at the two triple points. Thus, $\Delta V = -.0865 \text{ cm}^3\text{mol.}^{-1}$ for the solid-solid branch at the 55 GPa triple point. Since $\sum \Delta V = 0$ around a triple point, we have $\Delta V = 0.128 \text{ cm}^3\text{mol.}^{-1}$ on the higher-pressure side of the 55 GPa triple point (see Figure 4).

The value of ΔV at the ICB pressure (330 GPa) is needed. *Masters and Shearer* [1990] report that for the Earth, $\Delta \rho = 550$ kg m^{-3} coming from seismological measurements. Using the Earth's density at 330 GPa as 12,850 kg m^{-3}, we have $\Delta \rho / \rho = -0.0428$. The density change is composed of that due to the compositional change and that due to the difference between pure solid iron and liquid iron. Masters, as reported in *Gubbins et al.* [1979], suggests that $\Delta \rho_{\mathrm{m}} / \rho$ due to the solid-liquid phase change is at most -100 kg m^{-3}, leaving the majority of the observed change to compositional change. *Buffett et al.* [1996] propose -400 kg m^{-3} for the compositional change, leaving $\Delta \rho_{\mathrm{m}} = -150$ kg m^{-3} for the solid-liquid transformation. *Poirier* [1986] found $\Delta V_{\mathrm{m}} = 0.065$ cm^{3}mol^{-1} for hcp iron using dislocation-mediated melting theory, giving $\Delta \rho_{\mathrm{m}} = -220$ kg m^{-3}, apparently the largest value in the literature. *Stacey and Stacey* [1999] ascribe a small amount for the transformation, reporting $\Delta \rho_{\mathrm{m}} / \rho = -0.001$.

I shall take the median of the Poirier value and the Buffett value, $\Delta \rho_{\mathrm{m}} = -185$ kg m^{-3}, for the high-pressure transformation. Therefore, $\rho(330)$ of pure iron is 13,720 K m^{-3} and $(\Delta \rho_{\mathrm{m}} / \rho)_{\mathrm{trans}}$ is -0.01348. Thus, $\Delta V_{\mathrm{m}} / V$ for pure iron is $+0.01348$, or $\Delta V_{\mathrm{m}} = 0.0548$ cm^{3}mol^{-1} by using the volume for pure iron (330 GPa), 4.07 cm^{3}mol^{-1} [*Anderson*, 2002]. The values of ΔV_{m} found above are plotted in the upper right-hand corner of Figure 4. Thus, along the path of the hcp solidus, ΔV_{m} descends from 0.128 cm^{3}mol^{-1} to 0.0548 cm^{3}mol^{-1} going from $P = 55$ GPa to 330 GPa. An equation showing how ΔV_{m} changes with V for ϵ iron is needed.

Jeanloz [1985] derived an expression for $\Delta V_{\mathrm{m}} / V_1 = (V_2 - V_1) / V_1$, where subscript 1 refers to the solid and subscript 2 refers to the liquid

$$\frac{\Delta V_{\mathrm{m}}}{V_1} = \left. \frac{\Delta V_{\mathrm{m}}}{V_1} \right|_{P=0} + \frac{(V_2/V_{20}) - (V_1/V_{10})}{V_1/V_{10}}$$
$$+ \quad \text{high order terms.} \qquad (8)$$

This equation depends on the bulk modulus, K_T, ratio of the liquid K_2 and the solid K_1. When K_2/K_1 is much less than unity, the value of $\Delta V_{\mathrm{m}} / V_1$ falls rapidly with P/K_1. *Anderson and Ahrens* [1994] found $K_2 = 109$ GPa for liquid iron at $P = 0$ and $T = 1900$ K. The value for K_1 of ϵ iron at $P = 0$ and 300 K is 164 GPa [*Anderson et al.*, 2001]. Correcting for the temperature gives us $K_1 = 130$ GPa, so that $K_2/K_1 = 0.83$. By Equation (8) and Figure 5, $\Delta V_{\mathrm{m}} / V_1$ decreases rapidly with pressure for higher pressures.

Equation (8) could in principle be applied to find $\Delta V_{\mathrm{m}} / V_1$ at any P, but the EoS of liquid iron is not sufficiently well defined to extract the required information. From Figure 5, however, we see that $\Delta V_{\mathrm{m}} / V_1$ changes with P/K_1 as a straight line for most of the range P/K_1 when $K_2/K_1 = 0.83$. Equation (8) is usefully replaced by an empirical equation showing that ΔV_{m} changes with V with a linear law

$$\Delta V_{\mathrm{m}} = a + bV, \qquad (9)$$

where a and b are constants independent of V. The ending values are $\Delta V_{\mathrm{m}} = 0.128$ cm^{3}mol^{-1} at $V = 6.00$ cm^{3}mol^{-1} and $\Delta V_{\mathrm{m}} = 0.055$ cm^{3}mol^{-1} at $V = 4.34$ cm^{3}mol^{-1}. These volumes correspond to 55 GPa and 330 GPa, respectively, on the solidus (the equivalent volumes on the 300 K isotherm or lower). Using Equation (8), $a = -0.13586$ and $b = 0.04398$. Using Equation (9), ΔV_{m} is evaluated at 240 GPa where $V = 4.565$ cm^{3}mol^{-1}. The value found is $\Delta V_{\mathrm{m}} = 0.065$ cm^{3}mol^{-1}, giving $\Delta V_{\mathrm{m}} / V = 0.01423$. This gives $\Delta \rho_{\mathrm{m}}$ for iron $= -174$ kg m^{-3} at the crossing of the solidus by the Hugoniot. This value is comparable to the shock-wave measurements, as seen in Figure 6, which shows the data from *Al'tshuler et al.* [1962] for liquid iron. The solid Hugoniot data shown in the figure were gathered by *Marsh* [1980]. One can discern a shift in density at 243 GPa ($\rho = 12,234$ kg m^{-3}), where the Hugoniot crosses from the solid to the liquid. It is estimated to be -200 ± 50kg m^{-3}. This gives confidence in the use of Equation (9) to calculate ΔV_{m} throughout the whole range of pressure (30 GPa to 330 GPa). Values of ΔV_{m} at several values of P are listed in Table 2a.

5. THE ϵ IRON SOLIDUS MELTING CURVE

The solidus of pure iron at high pressure has received more attention than any part of the phase diagram because the solidus is the foundation for calculating the melting temperature, T_{m}, profile of the Earth's core. *Vocadlo et al.* [2000] found the most likely candidate for the stable structure of iron at core conditions to be the hcp phase [see also *Stixrude and Cohen,* 1995]. In the period between 1986 and 1993, various experimental data on T_{m} at high P were reported, but there was much disagreement. The scatter in data is shown in Figure 7 presented in 1997. The experimental situation was reviewed by *Anderson and Duba* [1997], who suggested that $T_{\mathrm{m}}(330)$ is limited to the range 5500 K to 6500 K, and they recommended $T_{\mathrm{m}}(300) = 6000$ K. They reported that the situation was clarified by the abstract of *Gallagher and Ahrens* 1995] and *Holland-Gallagher*

Table 1. Properties of the Three Triple Points

Property Clockwise	Dimensions	α-Γ-ϵ 11 GPa, 750 K	δ-Γ-liq. 5.2 GPa, 1990 K	ϵ-Γ-liq. 55 GPa, 2790 K
dT/dP	°GPa^{-1}	+43	+35	+16
dT/dP	°GPa^{-1}	−9.0	+38	+45.7
dT/dP	°GPa^{-1}	−29.5	+27.5	+6.7
ΔV	cm^{3}mol.$^{-1}$	+0.14	−0.354	−0.128
ΔV	cm^{3}mol.$^{-1}$	−0.35	+0.074	+0.087
ΔV	cm^{3}mol.$^{-1}$	+0.21	+0.278	+0.041
ΔS	J mol.$^{-1}$K^{-1}	$+3.25 \times 10^{-3}$	-10.11×10^{-3}	-8×10^{-3}
ΔS	J mol.$^{-1}$K^{-1}	$+3.88 \times 10^{-3}$	$+0.19 \times 10^{-3}$	$+1.9 \times 10^{-3}$
ΔS	J mol.$^{-1}$K^{-1}	-7.13×10^{-3}	$+9.92 \times 10^{-3}$	$+6.1 \times 10^{-3}$

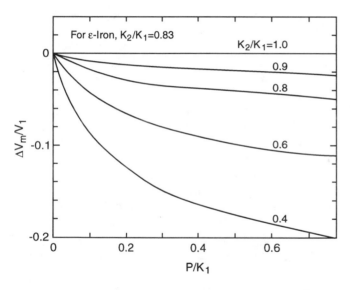

Figure 5. The variation of $\Delta V_m/V_1$, where V_1 is the volume of the solid, with pressure, calculated for various values of the ratio of the bulk modulus of the liquid to the bulk modulus of the solid, K_2/K_1 [after *Jeanloz*, 1985].

and Ahrens [1996], who reported that pioneering shock-wave data were probably in error where the temperatures are measured optically using the Wein displacement of frequency with temperature. This was due to an incorrect value of thermal diffusivity of the dielectric block Al_2O_3 supporting the sample. These reports placed doubt on the shock-wave temperature measurements of T_m of iron at outer core conditions, such as those of *Bass et al.* [1987] and *Yoo et al.* [1993] shown in Figure 7. Revised measurements of Hugoniot temperatures [*Ahrens et al.*, 1998] using a new, enhanced technique brought the values of T_m closer to the shock-

wave results of *Brown and McQueen* [1986] and to the temperature based on thermodynamic integration [*Anderson and Duba*, 1997].

The chief contenders for the experimental determination of T_m at core conditions are *Boehler* [1993], who measured $T_m(200) = 3900$ K using the diamond anvil cell, and *Brown and McQueen* [1986], whose shock-wave result gave $T_m(243) = 5000$ K. Both cannot be correct.

The thermodynamic method for finding T_m at core conditions is represented by the equation

$$\frac{d \ln T_m}{d \ln V} = f(\gamma), \tag{10}$$

where γ is the Grüneisen parameter. The best known representation of Equation (10) is the formula

$$\frac{d \ln T_m}{d \ln V} = -2 \left(\gamma - \frac{1}{3} \right). \tag{11}$$

The equation relating T_m to the Debye Θ from which Equation (11) is derived by differentiation has been incorrectly attributed to *Lindemann* [1910] by *Stacey* [1977], *Wolf and Jeanloz* [1984], *Anderson* [1986], *Anderson and Isaak* [2000], and others. The attribution belongs to *Gilvarry* [1956a,1956b] [see comments on this point by *Poirier*, 2000, Section 5.4.2]; in this paper, Equation (11) will be referred to as Gilvarry's rule. Gilvarry's derivation is based in part on the assumption of the Debye energy, contradicting Lindemann's assumption of the Einstein energy. *Gilvarry* [1957] also took certain assumptions from the theory of Debye and Waller for the thermal dependence of the intensity of x-rays diffracted by a crystal. Using this theory, *Gilvarry* [1956a,b] assumed that melting occurs when the root-

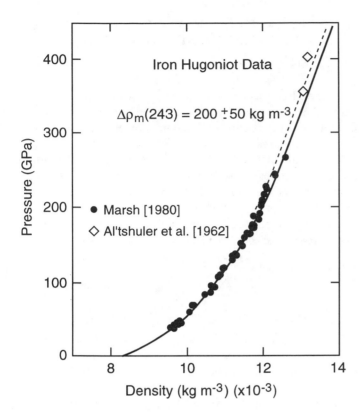

Figure 6. A plot of pressure versus density for the ϵ iron Hugoniot data. There is a jump in density at 200–240 GPa arising from the Hugoniot as it crosses the solidus. *Al'tshuler et al.* [1962] approached the crossover from the liquid state, while the compilation of *Marsh* [1980] shows the approach from the solid state. The jump in density from solid to liquid is estimated to be -200 ± 50 kg m^{-3}.

mean-squared amplitude of atomic vibrations exceeds a given fraction of the interatomic distance. Nowhere does the root-mean-squared amplitude enter into Lindemann's derivation [*Poirier,* 2000]. Gilvarry's assumption results in a volume-dependent Debye Θ, and thus, a volume-dependent γ, contradicting Lindemann's assumption that γ is independent of volume. In fact, Lindemann's theory of melting is only valid for $V = V_0$ [see *Poirier,* 1988 for a successful application of Lindemann's law at $P = P_0$].

Until recently, γ was not accurately pinned down at high compression. The recent measurement of γ for ϵ iron over a large range of pressure (30 GPa to 360 GPa) [*Anderson et al.,* 2001] removes a major roadblock. V, P, and Θ (the Debye temperature) have now been simultaneously measured at high pressure to better than three significant figures. The definition of γ from this experiment is

$$\gamma_{\text{vib}} = -\frac{\partial \ln \Theta}{\partial \ln V}. \tag{12}$$

Since Θ changes with V and Θ arises from the lattice vibrations, γ_{vib} is a function of V. It has now been determined to three significant figures over the entire pressure range and can be considered an experimental measurement. Using the experimental $\gamma_{\text{vib}}(V)$, which is needed in Equation (11), the variation of T_{m} with V can be calculated over the entire pressure range at small intervals of V.

Another form of Equation (10), based on the requirement that the Gibbs energy of a liquid equals that of

Table 2a. Properties of hcp Iron Along the Liquidus and Solidus, as a Function of Pressure

		Along the solidus				Along the liquidus		
P[a] (GPa)	T_{m}[b] (K)	ΔP_{TH}[c] (GPa)	P_0 (GPa)	V[a] (cm^{3}mol^{-1})	ρ (kg m^{-3})	ΔV_{m}[d] (cm^{3}mol^{-1})	V (cm^{3}mol^{-1})	ρ (kg m^{-3})
55	2790	33.00	22.00	5.85	9,550	0.128	5.98	9,340
100	3546	44.40	55.60	5.51	10,140	0.103	5.61	9,960
135	4018	52.76	82.27	5.22	10,700	0.090	5.31	10,520
150	4210	54.76	95.24	5.10	10,950	0.085	5.19	10,770
200	4790	64.16	135.8	4.83	11,560	0.073	4.90	11,390
240	5155	70.30	169.7	4.63	12,060	0.065	4.70	11,900
280	5465	75.50	204.5	4.48	12,470	0.058	4.54	12,310
330	5907	83.17	246.8	4.33	12,900	0.055	4.39	12,740

[a]Measured values.
[b]Solidus temperature calculated using Gilvarry's rule.
[c]Thermal pressure, see Section 7.
[d]Volume of melting.

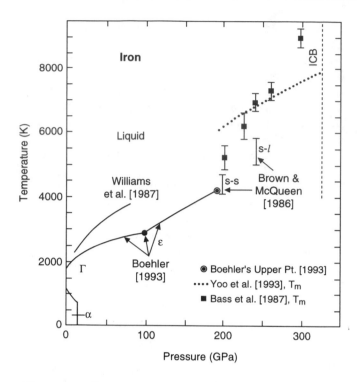

Figure 7. Experimental work contributing to the high-pressure regions of the phase diagram of iron (as of 1996). Below 200 GPa, all work on $T_m(P)$ is from diamond cell experiments (references in the figure). Work on the α-Γ and ϵ-Γ boundaries using the diamond cell was done by *Boehler* [1993]. Work on the Γ-liquid boundary using the diamond cell was done by *Williams et al.* [1987] and *Boehler* [1993]. Shock-wave results led to very high values of T_m by *Bass et al.* [1987] and *Yoo et al.* [1993]. *Brown and McQueen* [1986] reported two transitions on the Hugoniot (s-s and s-ℓ).

a solid (or the validity of the Clapeyron law) is [*Morse*, 1969]

$$\frac{\mathrm{d}\ln T_\mathrm{m}}{\mathrm{d}\ln V} = \frac{K_{T_m}\Delta V_\mathrm{m}}{L}, \qquad (13)$$

where K_{T_m} is the isothermal bulk modulus at melting, ΔV_m is the volume change of melting, and L is the latent heat of crystallization.

For use in Equation (13), *Stacey and Irvine* [1977] found f(γ) to be,

$$\mathrm{f}(\gamma) = -2(\gamma - 2\gamma^2\alpha T_\mathrm{m}), \qquad (14)$$

where α is the coefficient of thermal expansion at the solidus.

Since Equation (14) is based on the equality of the Gibbs free energy between solid and liquid, the gamma here is determined by the total energy: not only that from the vibration of the solid (as in Gilvarry's law), but also that arising from excited free electrons. This gamma is called the effective gamma, γ_eff; γ_eff is larger than γ_vib.

6. RELATIONSHIP BETWEEN γ_eff AND ELECTRONIC SPECIFIC HEAT

To obtain γ_eff for use in Equation (10), e.g., when Equation (13) is the basic assumption, one needs to consider free electrons in a metal, which provide energy that augments the value of γ_vib [*Bukowinski*, 1977]. Equation (13) requires consideration of the electronic density of states as well as the phonon density of states. The solidus of hcp iron lies in a temperature range in which the effect of electronic specific heat is large. Figure 8 shows the value of the electronic specific heat, C_{V_el}, in comparison with the value of vibrational (lattice) specific heat C_{V_vib}, as a function of T. In the presence of excited free electrons, γ is identified as γ_eff. The relationship between γ_eff and γ_vib is [*Bukowinski*, 1977]

$$\gamma_\mathrm{eff} = \frac{C_{V_\mathrm{el}}}{C_{V_\mathrm{total}}}\gamma_\mathrm{el} + \frac{C_{V_\mathrm{vib}}}{C_{V_\mathrm{total}}}\gamma_\mathrm{vib}, \qquad (15)$$

where $C_{V_\mathrm{total}} = C_{V_\mathrm{el}} + C_{V_\mathrm{vib}}$. Values of C_{V_total} and C_{V_el} for hcp iron at core conditions were found by *Anderson* [2000]. A plot of $\gamma_\mathrm{vib}(P)$ and $\gamma_\mathrm{eff}(P)$ is shown in Figure 9. For a more complete description of the derivation of γ_eff and C_{V_el}, leading to values in Figure 9, see *Anderson* [2000]. It is seen that γ_eff is virtually con-

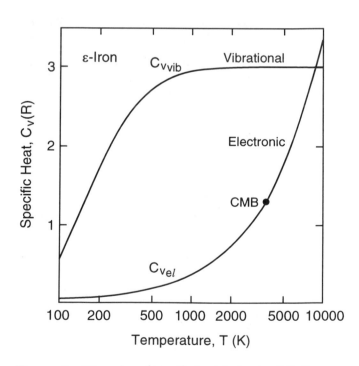

Figure 8. The value of specific heat in terms of R, the gas constant, versus T for the lattice, C_{V_vib}, and for the excited free electron specific heat, C_{V_el}. At CMB conditions, the electronic C_{V_vib} is 40% of C_{V_el}. At ICB conditions, C_{V_el} is only slightly smaller than C_{V_vib}.

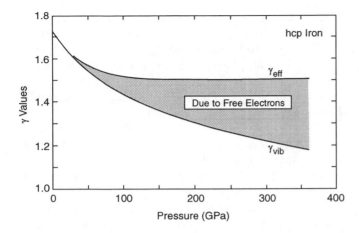

Figure 9. The variation of γ_{vib} and γ_{eff} with pressure along the solidus for ϵ iron. Throughout core conditions, γ_{eff} has an unchanging value of 1.5 and is virtually independent of pressure, whereas γ_{vib} decreases with increasing pressure.

stant throughout the outer core-pressure range and has a value of 1.5.

7. EVALUATION OF THE INTEGRATION CONSTANT AND THERMAL PRESSURE FOR $T_{\mathrm{m}}(P)$

Using the value of $\gamma_{\mathrm{vib}}(V)$ shown in Table 2b for Equation (11), $T_{\mathrm{m}}(V)$ and the corresponding $T_{\mathrm{m}}(P)$ are calculated once the integrating constant has been determined. Since the phase diagram has been well measured at low pressures, it is logical to use the well-understood part of the phase diagram to anchor the $T_{\mathrm{m}}(P)$ curve. The ϵ-γ-liquid triple point is used as the anchor point, even though this triple point has some uncertainty as to its exact position. The coordinates of this triple point are the constants of integration for Equation (10).

Three laboratories show the ϵ-γ-liquid triple point in the vicinity of 55 GPa. *Saxena and Dubrovinsky* [2000] give it at 50 GPa and 2760 K (see their Figure 2b). *Shen et al.* [1998] have a solution where the triple point is at 60 GPa and 2820 K. The phase diagram presented by *Andrault et al.* [2000] shows the triple point at 55 GPa and 2790 K. The solidus of the hcp phase is agreed upon in the above three experimental papers, so it is identical in its low-pressure region, but it terminates at three different triple points, as chosen by the experimentalists (see Figure 10). These three triple points effectively make a cluster. The integrating constant is applied here by terminating the integrated $T_{\mathrm{m}}(P)$ at the cluster of triple points (see Figure 10), given by $P = 55 \pm 5$ GPa and 2790 K.

The temperature associated with $V(330 \text{ GPa})$ will be T_{m}, and thus $V(330, T_{\mathrm{m}})$ will be larger than $V(330, 0)$. To find $V(330, T_{\mathrm{m}})$, the value of thermal pressure is required. At 300 K and 330 GPa, the measured volume is 4.07 cm^{3}mol^{-1} (see Table 2b). But at high T, the value of $P(330 \text{ GPa})$ evaluated at 300 K is augmented by the thermal pressure, P_{TH}, which is proportional to T. The value of P_{TH} is 66 GPa, which amounts to an increase of about 20% in P when $T_{\mathrm{m}} = 6000$ K. Thus, the volume corresponding to $P = 330$ GPa must be increased to about 4.26 cm^{3}mol^{-1}. This can be seen in Figure 11, in which the 330 GPa isobar begins at $V = 4.07$ cm^{3}mol^{-1} (300 K) and terminates on the solidus at $V = 4.26$ cm^{3}mol^{-1}. In the P,V,T relationship, the effect of ΔP_{TH} resulting from an increase in T is to decrease ρ at the same P. At 300 K, $\rho(330 \text{ GPa}) = 13.72 \times 10^3$kg m^{-3} (see Table 2b), but at the solidus, $\rho(330 \text{ GPa}) = 12.9 \times 10^3$ kg m^{-3} (see Table 2a).

Using the integrating constant (55 GPa and 2790 K), integration of Gilvarry's rule, Equation (11), yields val-

Table 2b. Properties of hcp Iron Along the 300 K Isotherm, as a Function of Pressure

P^{a} (GPa)	$\gamma_{\mathrm{vib}}{}^{\mathrm{a}}$	$\gamma_{\mathrm{eff}}{}^{\mathrm{b}}$	Θ^{a} (K)	$K_T{}^{\mathrm{c}}$ (GPa)	V^{a} (cm^{3}mol^{-1})	ρ^{a} (kg m^{-3})
55	1.51	1.73	582	438	5.52	10,120
100	1.42	1.65	660	644	5.07	11,020
135	1.38	1.62	707	801	4.83	11,560
150	1.36	1.60	725	868	4.74	11,780
200	1.30	1.57	777	1088	4.50	12,410
240	1.26	1.55	812	1264	4.34	12,870
280	1.23	1.53	844	1439	4.21	13,270
330	1.19	1.51	879	1658	4.07	13,720

[a]From measurements of *Anderson et al.*, 2001
[b]From calculations of *Anderson*, 2002.
[c]K_T calculated from an equation of state using values of V and P shown above.

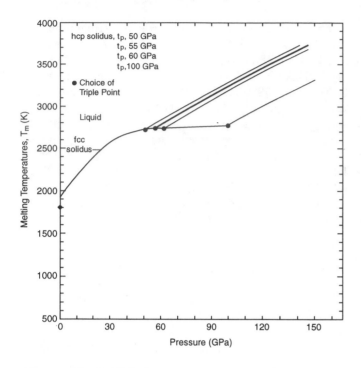

Figure 10. Published experimental values of the hcp-fcc-liquid triple point and the fcc-liquid boundary. The three lower-pressure triple points (at 50, 55, and 60 GPa) were reported by *Saxena and Dubrovinsky* [2000], *Andrault et al.* [2000], and *Shen et al.* [1998], respectively. The highest triple point (100 GPa) was reported by *Boehler* [1993]. The integration of $T_m(V)$ in Equation (11) (Gilvarry's rule) requires an integrating constant that is supplied by one of the triple points in this figure. The chosen triple point is indicated by a heavy line.

ues of the solidus, $T_m(P)$. In this paper, I use the experimental values of γ_{vib} listed in Tables 2a and 3 for the integration. Obtaining T_m values for many pressures (listed in Table 2a), the most pertinent result is $T_m(330) = 5907$ K.

Poirier [1986], using dislocation melting theory and the Clapeyron criterion (Equation 13) in his derivation of T_m versus V, found $T_m(330) = 5995$ K when the value of $\gamma = 1.5$, found by *Anderson* [2002], is used.

Finding $T_m(330)$ by means of the Stacey-Irvine formula, Equation (14), requires the use of γ_{eff} in place of γ. Exact information on $\alpha(V,T)$ is still lacking but enough is known for an educated guess of $T_m(330)$. At core pressure (300 GPa), $\alpha = 1 \times 10^{-5} \text{K}^{-1}$ and is independent of T [*Wasserman et al., 1996*]. Applying *Lindemann's* [1910] principle that αT_m is constant and assuming that this holds for variation in V, the value of $f(\gamma_{\text{eff}})$, Equation (10) can be approximated for every V. Upon integration, $T_m(330) \approx 5800$ K. Further work on the evaluation of $\alpha(V,T)$ needs to be done. Never-

theless, we see that three methods of thermal physics, using Equation (10), yield similar results for $T_m(P)$.

My recommendation is that $T_m(330) = 5900 \pm 300$ K be used for the phase diagram of ϵ iron. The calculated $T_m(V)$ for the entire range of V for the ϵ phase is shown in Figure 11. The calculated $T_m(P)$ is shown in Figure 12. Data on $T_m(P)$ are listed in Table 2a.

The value of $T_m(330) = 5900 \pm 300$ K for the melting of pure iron is near several other results obtained using different approaches, including those of *Steinle-Neumann et al.* [2001], 5700 K; *Spiliopoulos and Stacey* [1984], 6140 K; *Wasserman et al.* [1996] and *Stixrude et al.* [1997], 6150 K; *Brown and McQueen* [1982], 6200 K; and *Alfé et al.* [2002], 6250 ± 300 K. Values somewhat lower than 5700 K are those of: *Boehler* [1993], 4800 K and *Laio et al.* [2000], 5400 K. Higher values were reported by *Alfé et al.* [1999], 6700 ± 600 K, *Belonoshko et al.* [2000], 7000 K; and *Belonoshko and Ahuja* [1997], 7500 K.

8.   THE ASSEMBLED PHASE DIAGRAM

It is noted that in Figure 12, $T_m(P)$ is virtually linear in P above 135 GPa, e.g., through the core pressure range. Probably the behavior arises because $\gamma_{\text{eff}} = 1.5$, a constant throughout this range of compression. αK_T is independent of V, as shown in Table 1 of *Anderson* [2002]. Taking dT/dP as a constant, along a melting

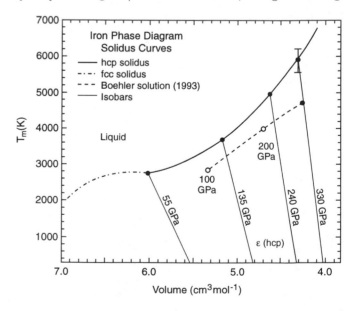

Figure 11. The variation of the solidus temperature with volume for epsilon iron in the T-V plane found using Equation (11). This figure uses the boundary conditions resulting from the 55 GPa triple point. The data from *Boehler* [1993] are plotted as open circles.

Table 3. Physical Parameters of ϵ-Iron at 300 K as a Function of Volume

V^a (cm^{3}mol^{-1})	P^a (GPa)	Θ^a (K)	$\gamma_{vib}{}^a$	$K_T{}^b$ (GPa)
6.73	0	422	1.71	164.0
5.9	29.96	524	1.59	318.8
5.8	35.56	539	1.57	346.1
5.7	41.75	554	1.55	375.8
5.6	48.62	569	1.53	408.3
5.5	56.22	585	1.51	444.0
5.4	64.64	602	1.49	483.2
5.3	73.99	618	1.47	526.2
5.2	84.38	636	1.45	573.7
5.1	95.92	654	1.43	626.0
5.0	108.76	673	1.41	638.8
4.9	123.05	692	1.39	747.8
4.8	139.00	712	1.37	818.9
4.7	156.80	733	1.35	897.8
4.6	176.71	754	1.32	985.7
4.5	198.99	776	1.30	1083.8
4.4	223.98	799	1.28	1193.5
4.3	252.06	822	1.25	1316.6
4.2	283.65	847	1.23	1454.9
4.1	319.26	872	1.20	1610.7
4.0	359.50	898	1.18	1786.8

[a]V, P, Θ, and γ_{vib} are measured values [*Anderson et al.*, 2001].

[b]K_T is calculated from an equation of state using values of V and P shown above.

curve, denoted as subscript m, Equation (1) can be replaced by

$$\left(\frac{dT}{dV}\right)_m = -\text{const.}\frac{K_T}{V} = -\text{const.}\frac{\alpha K_T}{\alpha V}. \qquad (16)$$

Since αK_T is independent of V, $(dT/dV)_m$ is inversely proportional to the negative of αV. Since the value of α descends rapidly as the value of V descends, then the value $(dT/dV)_m$ grows as V decreases (or as P increases). In Figure 11, we see that for the ϵ phase, the slope of T_m versus V increases as the volume decreases, just as Equation (16) predicts. I therefore conclude that straight line extrapolations of T_m with volume (or correspondingly, straight line extrapolations of T_m with density) are not valid for hcp iron, except when the extrapolation is limited to a narrow pressure range.

Assembling portions of Figures 4 and 11, the complete phase diagram in the T-V plane is given in Figure 13. The corresponding phase diagram in the T-P plane is shown in Figure 14.

The choice of location of the liquid-hcp-fcc triple point is crucial to the temperature profile phase diagram. For this reason, the four experimental plots from the four laboratories presenting the triple points are assembled and displayed in Figure 15. The reader may make his/her own decision on the value of P for this triple point.

9. IMPLICATIONS FOR THE EARTH'S CORE

9.1. The Outer Core Density Deficit

Figure 16 shows the comparison of the density of pure iron to that of the Earth's core as a function of pressure. The hcp curve labelled "solidus" is found from the pressure-volume measurements in Table 2a. The curve labelled "liquidus" (values listed in Table 2a) is corrected from the solidus by the value of ΔV_m computed from Equation (9).

We note that the density difference between hcp iron at 300 K and the outer core is 11 to 12%, according to Figure 16. This density difference model may have been the method of computing the outer core density deficit, because in the past, it was commonly held that the outer core density deficit was about 10%. However, I hold that the outer core density deficit is the difference between the density distribution along the liquidus and the seismically-determined density distribution of the outer core. From Figure 16, this difference is 5.4% using the 6000 K value for $T_m(330)$.

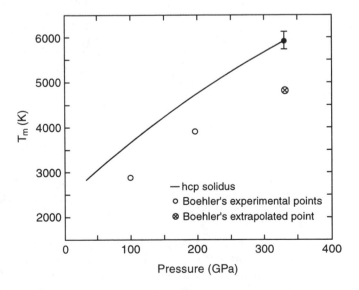

Figure 12. The variation of the solidus temperature with pressure in the T-P plane. The *Boehler* [1993] measured data for T_m of epsilon iron are plotted as circles. The circle containing a cross represents Boehler's extrapolation from the measured values.

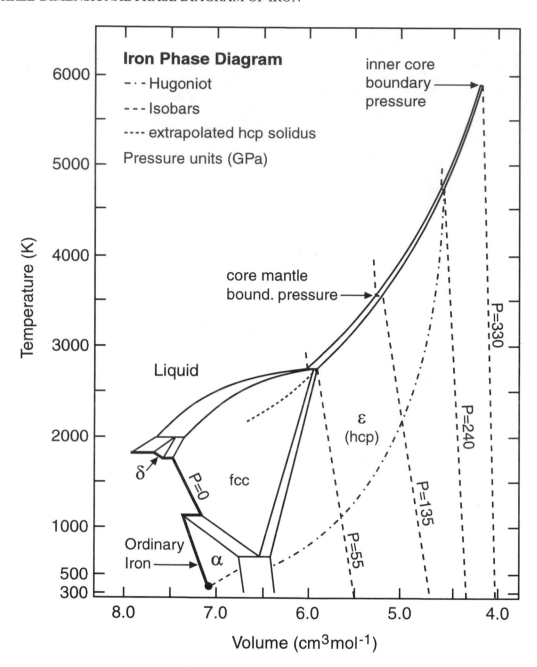

Figure 13. The phase diagram of iron in the T-V plane with isobars plotted as dashed lines. The Hugoniot intersects the solidus at $P = 240$ GPa. The hcp solidus intersects the fcc solidus at 55 GPa with a change in slope as shown by Figure 11.

Note that the density change is nearly constant along the majority of the outer core. The constancy of this outer core density deficit throughout most of the core indicates mixing due to vigorous convection, which overcomes the buoyancy of the light impurities.

At the interface with the CMB, the percentage drop is more than 5.4% (see Figure 16). This indicates that the effect of core convection is less able to cope with the buoyancy of the light elements at the interfaces. I

note that the core density deficit is 1.5% in the inner solid core as compared to 5.4% in the outer fluid core. This indicates that the amount of impurities in the inner core is about 30% of that in the outer core; this may be compared to the amount found by *Buffett et al.* [1996], who found 33%.

The value of P_{TH} for the solidus is rather high, as shown in Table 2a, being, for example, 62% of the total pressure at the ϵ-Γ-liquid triple point. The formula used

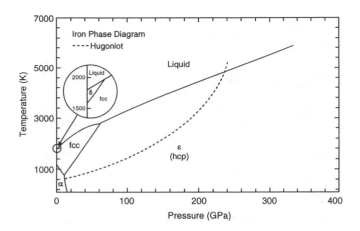

Figure 14. The phase diagram of epsilon iron in the T-P plane, showing the intersection of the Hugoniot with the solidus at 240 GPa. Note that the hcp solidus begins at the triple point located at 55 GPa and 2790 K.

for P_{TH} is $\alpha K_T(T_m - 300)$, where αK_T is independent of V [*Anderson*, 1999]. Any process that raises the value of either αK_T or T_m raises the percentage of P_{TH} in P.

Wasserman et al. [1996], using first principles methods, concluded that αK_T for hcp iron is linear in $(T_m - 300)$ in the core, and thus P_{TH} is quadratic in $(T_m - 300)$ in the core. *Wasserman et al.* [1996] pointed out that αK_T increases by about 40% going from 300 K to 6000 K for hcp iron. This number is used to prorate the temperature dependence of αK_T to solidus temperatures; the ΔP_{TH} for hcp iron thus found are listed in Table 2a.

A substantial increase in P_{TH} could arise from a large value of T_m at 330 GPa, which would reduce the outer core density deficit to unacceptably low levels. A value of $T_m = 7000$ K, for example, would lower the outer core density deficit to about 2.5% [see also Figure 4a in *Stixrude et al.*, 1997]. For further details on the core density deficit, see *Anderson and Isaak*, [2002].

9.2. Concentration of Light Elements

In earlier calculations of the melting point depression, it was often assumed that sulfur was the only impurity that needed to be considered, and therefore, the core's composition could be interpreted in terms of the Fe-FeS phase diagram [*Verhoogen*, 1980]. The problem posed was to find the variation of the eutectic T with P at core conditions. *Boehler* [1992,1996] measured the eutectic T of the Fe-FeS system up to 60 GPa, from which he suggested that Fe and FeS exhibit binary solid solution rather than eutectic behavior at core pressures and temperatures. This is probably true for other impurities as well. It is also generally understood that impurities are not just of one type. *Poirier* [1994] said,

"...It is not reasonable to look for one light element in the core." Therefore, we probably have a core of an iron alloy with several types of impurities (C, Si, S, H, O). It is noted that *Sherman* [1995,1997] found that oxygen is extremely unstable in Fe-FeO alloys. He found that the core is depleted in O compared to Si. I conclude that oxygen cannot be the dominant impurity.

In the cosmochemical literature, it is often argued that the concentration of sulfur in the core must be small. *Allégre et al.* [1995] and *O'Neill et al.* [1998] restrict the concentration of sulfur to about 6% by weight using cosmochemical abundance arguments. This restriction posed a severe problem when the core density deficit was thought to be about 10%. This problem led *O'Neill et al.* [1998] to conclude that "none of the cosmochemically abundant liquid elements appears to account for [the core density deficit]." However, the problem disappears when the difference between the density of the hcp liquidus and that of the outer core is 5.4%. A core which sulfur is the only light element would essentially satisfy the O'Neill et al. criterion.

Poirier [1994] provided equations and figures that relate the weight percent of cosmochemically-abundant light impurities to the core density deficit. His main result is redrawn and shown in Figure 17, which shows the relationship of the concentration of oxygen, sulfur, and silicon to the core density deficit. *Anderson and Isaak* [2002] found that for $T_m(330) = 6000$ K, the core density deficit is 5.4%. A horizontal line at 0.054 density deficit intersects the curves from which one can read the weight percent of the light element. From the intersection of the 0.054 density deficit line with the sulfur curve, one reads 5.2 for the weight fraction. In other words, 5.2 weight% sulfur satisfies the entire core density deficit. Similarly, a weight fraction of 9.2% silicon satisfies the entire core density deficit. *Poirier* [1994] calculated the weight percents of Si, S, and O in Fe (see his Figure 7), but he did not show the weight percents of C or H in Fe. This is probably because the equation of state of these two light elements in iron had not been agreed upon, and one needs the bulk modulus as well as the atomic mass for the calculations (the equations are given in the Appendix of *Poirier*, 1994). However, *Wood* [1993] showed that at 330 GPa, 0.3 weight% carbon with iron could form a stable phase of iron carbide, Fe_3C, and he proposed that this structure might be a cause of the core density deficit. *Scott et al.* [2001] reported an experimental determination of the bulk modulus of Fe_3C, 175 GPa, not very different from that of ϵ-Fe.

Okuchi [1997] suggested that H would be dissolved in the segregating core of the primitive Earth, rather

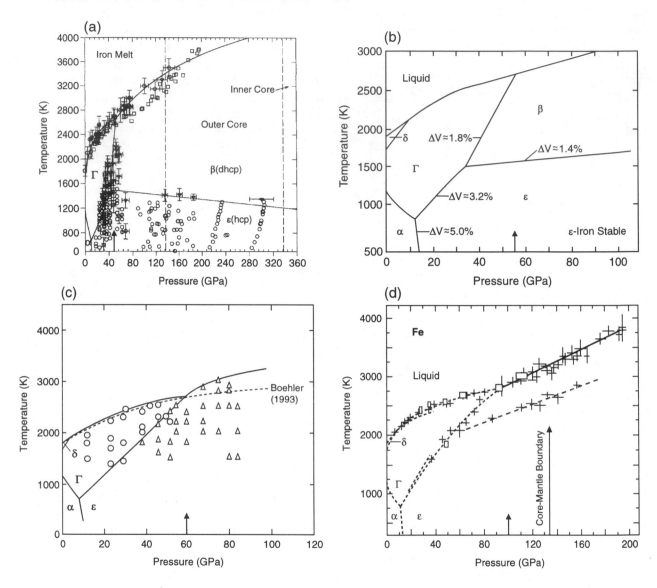

Figure 15. The original data from which the triple points described in Figure 10 were taken. The upper left curve is from *Saxena and Dubrovinsky* [2000]; the upper right curve is from *Andrault et al.* [2000]; the lower right curve is from *Boehler* [1993]; and the lower left curve is from *Shen et al.*[1998].

than escaping in the atmosphere. Thus, H would be stored in the deep interior, perhaps in the structure $FeH_{0.33}$. *Williams and Hemley* [2001] pointed out that iron hydride of the form FeH_x, where $x = 0.3$-0.4, is stable at core pressures and has a melting point about 600°C below that of pure iron.

In the calculation of the weight percent of the light element, as shown by *Poirier* [1994], there are two competing effects: atomic mass and bulk modulus. The masses of H and C are small compared to those of O and S. Because of the smallness of the masses of H and C, the weight percent appears far to the left in Figure 17

(shaded region). There is, however, some uncertainty in the value of the bulk modulus of the hydrogen-iron crystal (FeH_x).

Even a small concentration of either hydrogen or carbon (corresponding to a fraction of the core density deficit) can lead to a substantial effect on the temperature freezing point depression.

9.3. An Estimate of the Limiting Melting Point Depression

Following *Poirier* [1994], I assume that there is not one impurity, but several, and I take various combina-

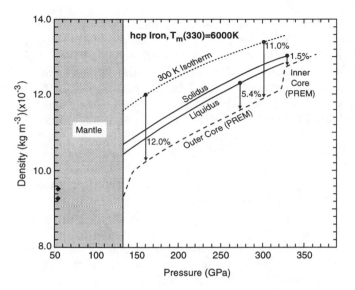

Figure 16. The densities of the solidus and the liquidus of hcp iron versus pressure are shown. Also shown is the density of the Earth's core according to PREM [Dziewonski and Anderson, 1981]. Note that for most of the region of the outer core, the density of the core is 5.4% less than that of the liquidus, and at the core-mantle boundary, the density difference is greater than 6%. Note that the density difference between the solidus of hcp iron and the inner core is 1.5%.

tions of Si, S, and O. As an example, assume that the 5.4% core deficit is made up of 3.4% Si, 1.8% S, and 0.2 % O, giving 5.0 wt.% Si, 1.8 wt.%S, and 0.3 wt.% O.

These concentrations are small enough to assume that the impurities form an ideal solid solution with iron. One equation used for calculating ΔT for the outer core [*Anderson*, 1998], when the approximation of an ideal solution is valid, is

$$\Delta T = T^* \sum_i \ln\left(1 - \chi\right), \tag{17}$$

where $T^* = 6000$ K (T_m of fcc iron at the ICB pressure), and $\chi(i)$ is the molar fraction of the various impurity elements. Evaluation of the above three impurities converting the wt.% to mole% gives $\chi(S) = 0.031$, $\chi(Si) = 0.091$, $\chi(O) = 0.006$. Substituting these values into Equation (17) gives $\Delta T = 831$ K. Other possible combinations of the cosmochemically abundant elements satisfying the core density deficit can be made. Trying out a number of combinations, the ΔT_m appears to vary between 800 K and 1100 K. If silicon were the only impurity, $\Delta T = -1100$ K.

I therefore take $\Delta T_m = 950 \pm 150$ K and consequently assign the value 5050 ± 300 K as T_m of the Earth at the ICB. *Braginsky and Roberts* [1995] estimated 5300 K;

Spiliopoulos and Stacey [1984], 5000 ± 900 K; *Brown and McQueen* [1986], 4800 K. The value of 5000 K for $T_m(330)$ used by many goes back to *Birch* [1972], who said "5000 K may have some standing as an upper limit."

9.4. The Outer Core Adiabat

Since the adiabat of the core and the melting temperature of the core cross at 330 GPa, we have $T_{ad}(330) = 5050 \pm 300$ K for the core. Since $\gamma_{eff} = 1.5$ for hcp iron (see Figure 9), we calculate γ_{eff} for the core by making a correction due to the smaller temperature and smaller density. *Anderson* [2002] found $\gamma_{eff}(core) = 1.45$ and constant throughout the core. The change in T_{ad} with ρ is found from the standard thermodynamic equation,

$$\left(\frac{\partial \ln T_{ad}}{\partial \ln \rho}\right)_S = \gamma_{eff}. \tag{18}$$

Since the value of γ_{eff} (not a function of ρ), is 1.45, the above equation reduces to

$$\frac{T_{ad}(1)}{T_{ad}(2)} = \left(\frac{\rho_1}{\rho_2}\right)^{1.45}. \tag{19}$$

The value of ρ_1/ρ_2 arising from the CMB and ICB is 0.824, from which T_{ad} at the CMB should be near 4100 ± 300 K.

Considering that the value of T at the base of the mantle is about 3000 K (where the non-adiabatic ΔT in the mantle is 400 K [*Shankland and Brown*, 1985], there is a drop of about 1200 K across the D″ layer.

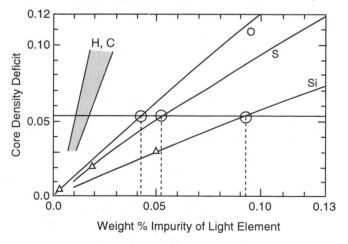

Figure 17. A plot showing the effect of sulfur and silicon impurities on the core density deficit. It is seen that a core density deficit of 5.4%, illustrated in Figure 16, would be accounted for by either 5.2 mass % impurity of sulfur or 9.2 mass % impurity of silicon.

This gives a gradient across D″ equal to about 6°km^{-1} (using 200 km for the thickness of D″) in comparison to the thermal gradient of the core at the CMB of about $-0.27°$km^{-1} [*Anderson*, 1998]. This is a reasonable thermal structure for D″ considering that the temperature and viscosity of D″ have to be higher, and lower, respectively, than the adjacent mantle in order that conditions of instability can be met for plume generation out of D″.

9.5. The Outer Core's Gamma is Invariant with Depth

The fact that γ_{eff} is essentially constant throughout the core deserves a further comment. Traditionally, the approach to thermodynamic problems of the core has been to assume that the value of γ, though independent of T, changes with volume according to a power law, such as

$$\frac{\gamma_1}{\gamma_2} = \left(\frac{V_1}{V_2}\right)^q, \qquad (20)$$

where q is some constant.

Equation (20) is approximately true for the γ_{vib} arising from lattice vibrations, but a metal, such as iron, has a counteracting energy effect on γ due to the conductive flow of free electrons. It turns out that for iron at core conditions, the free electron effect almost exactly compensates for the drop in γ with pressure found in lattice vibrations. For pure iron at $P = 0$, γ_0 is 1.72, but for impure iron, which constitutes the Earth's core, the $P = 0$ value is 1.6, yet throughout pressures of the core, γ is 1.45. Thus, scientists who used as an assumption that $\gamma = $ const. for core physics were fortuitously using a close approximation.

The large value of γ_{eff} (1.45) in the core makes the value of properties depending on γ, such as the thermal gradient, larger than previously calculated, where γ was thought to be smaller (~ 1.1) at core conditions. *Stacey* [1995], using theoretical methods, reported that γ at the CMB is close to 1.35, so there appears to be somewhat of a convergence to a high value both theoretically and experimentally.

10. CONSIDERATIONS CONCERNING TWO ADDITIONAL SOLID PHASES ADDED TO THE FIVE WELL-ESTABLISHED PHASES

10.1. A New Triple Point at 200 GPa?

It was suggested by *Brown and McQueen* [1986] that a solid-solid transition existed on the Hugoniot near 200 GPa as a result of sound velocity measurements along the Hugoniot. They suggested a γ-ϵ transition,

but now we know that the γ phase is terminated at the 55 GPa triple point. This idea was picked up by *Boehler* [1986], who named this suspected high-pressure solid phase as the Θ phase. The phase was thought to have a bcc structure from the theoretical work of *Ross et al.* [1990], but other theoretical work denied the stability of the bcc structure at core conditions [*Stixrude et al.*,1997]. *Matsui and Anderson* [1997], however, calculated the volume change for hcp to bcc at 200 GPa. *Nguyen and Holmes* [2000] repeated the early shockwave experiment of *Brown and McQueen* [1986], using sound velocity measurements, and failed to find evidence of a solid-solid transition at 200 GPa. They said that there is "no clear evidence from our work for a solid-solid phase transformation [at 200 GPa]..." and "there is no triple point in the vicinity of 200 GPa." *Brown* [2001], using new primary measured data on the Hugoniot of iron [*Brown et al.*, 2000] stated that [the 200 GPa density difference "is consistent with the density difference for transformation [between ϵ and the hypothetical bcc phase of iron.]" Thus, in recent reappraisals, two primary shock-wave laboratories have come to completely opposite conclusions. With all this evidence for and against a new phase on the solidus at about 200 GPa, the situation is too unsettled to incorporate such a phase in the present phase diagram.

Should the situation be resolved in favor of the extra phase at 200 GPa, then the phase diagram would show a fork in the liquidus above 200 GPa. It will appear much as in Figure 9 of *Anderson* [1993], who presented it at a time when this speculative 200 GPa triple point was much more in favor. Even if one adds an additional fork in the phase diagram at 200 GPa, there will be no change in the phase diagram below 200 GPa.

10.2. The β Phase

After the diamond-anvil cell was attached to the terminus of a synchrotron radiation beam line, the resulting intense x-rays produced high-quality, in-site diffraction patterns even at quite high temperatures. In the mid-nineties, the diamond-anvil cell on the synchrotron beam line became a tool to explore the structural details of the iron phase diagram. There followed a controversy over a new iron phase proposed by *Saxena et al.* [1993, 1995, 1996], which was named the β phase. Saxena and his colleagues from Uppsala reported that this phase was in the dhcp structure, a variant of the hcp structure. According to their mapping of the β phase, the γ-ϵ-liquid triple point is replaced by a γ-β-liquid triple point and a new triple point is added (ϵ-β-γ at about 35 GPa and 1500 K). The significance is that above about 1500 K and $P > 50$ GPa, the phase diagram of iron would be occupied by the β phase.

The controversy pitted the Uppsala group against the Lawrence Livermore National Laboratory and Geophysical Laboratory groups [*Yoo et al., 1995*], who instead found the standard hcp structure, where the Uppsala group had reported dhcp. There followed an announcement by the Paris group [*Andrault et al., 1997*] that they had verified the existence of a new phase where the Uppsala group had reported it to be, but they insisted that its crystallographic structure was orthorhombic rather than dhcp. This developed into a 3-sided controversy that has not yet been resolved, but was reinvigorated after *Shen et al.* [1998] from the Geophysical Laboratory reported that there was no evidence for the β phase.

Anderson and Isaak [2000] tried to resolve the controversy by searching for the values of the physical constants of the γ, ϵ, and β phases. After comparing the properties of ϵ with those of β, they concluded that "one can only assume that they [the β properties] are not substantially different from those of the hcp phase because ΔV for the ϵ-β boundary is very small." That is consistent with the report that dT/dP for the alleged boundary between ϵ and β is virtually horizontal (see Figures 15a and 15b). *Anderson and Isaak* [2000] concluded that the physical parameters of the suggested β phase and the ϵ phase are "so close in value that with the present state of knowledge, we need not consider these two phases separately."

With the present state of knowledge, it seems pointless to require that the β phase be placed in the phase diagram. The reader who wishes to know how the T-V phase diagram that includes the β phase would appear should see Figure 4 of *Anderson and Isaak* [2000] (see also Figures 15a and 15b).

Note added in press. The author notes with pleasure that in *Alfè et al.* [2002] it is reported that gamma "varies very little with pressure or temperature for $100 < p < 300$ GPa and $4000 < T < 6000$ K, and has a value of *ca* 1.5" (see also Figure 9 of this paper).

Acknowledgments. The author acknowledges a discussion of γ_{eff} with Ron Cohen that was of considerable assistance. Victoria Nelson assisted in the preparation of graphs and in calculations. Don Isaak was very helpful with discussions on the physics of melting and improved the manuscript considerably by locating and correcting inconsistent data. The works of J.-P. Poirier were very influential for this paper. Surendra Saxena, Denis Andrault, Lars Stixrude, and Leonid Dubrovinsky participated in discussions on iron melting that extend back seven years. Correspondence with Frank Stacey on the physical principles behind the Grüneisen parameter extends back twenty years. The aforementioned discussions and correspondence contributed to this paper. This research was supported by NSF EAR-0073989 and by the Office of Naval Research. IGPP # 5654.

REFERENCES

Ahrens, T. J., K. G. Holland, and C. Q. Chen, Shock temperatures and the melting point of iron, *AIP Conference Proceedings*, no. 429, 133–136, 1998.

Alfè, D., M. J. Gillan, and G. D. Price, The melting curve of iron at the pressures of the Earth's core from ab initio calculations, *Nature, 401,* 462–464, 1999.

Alfè, D., G. D. Price, and M. J. Gillan, Iron under Earth's core conditions: Liquid-state thermodynamics and high-pressure melting curve from ab-initio calculations, *Phys. Rev. B*, in press, 2002.

Allègre, C. J., J.-P. Poirier, E. Humler, and A. W. Hofmann, The chemical composition of the Earth, *Earth Planet. Sci. Lett., 134,* 515–526, 1995.

Al'tshuler, L. V., G. V. Sinakov, and R. F. Irunin, On the composition of the Earth's core, *Isv. Earth Phys., 1,* 1–3, 1962.

Anderson, O. L., Properties of iron at Earth's core conditions, *Geophys. J. R. Astron. Soc., 84,* 561–579, 1986.

Anderson, O. L., The phase diagram of iron and the temperature of the inner core, *J. Geomag. Geoelec., 45,* 135–1248, 1993.

Anderson, O. L., *Equations of State of Solids for Geophysics and Ceramic Science*, 405 pp., Oxford Univ. Press, New York, 1995.

Anderson, O. L., The Grüneisen parameter for iron at outer core conditions and the resulting conductive heat and power in the core, *Phys. Earth Planet. Inter., 109,* 177–199, 1998.

Anderson, O. L., The volume dependence of thermal pressure in perovskite and other minerals, *Phys. Earth Planet. Inter., 112,* 267–283, 1999.

Anderson, O. L., The Grüneisen ratio for the last 30 years, *Geophys. J. Int., 143,* 279–294, 2000.

Anderson, O. L., The power balance of the core-mantle boundary, *Phys. Earth Planet. Inter.*, in press, 2002.

Anderson, O. L., and A. Duba, The experimental melting curve of iron revisited, *J. Geophys. Res., 102,* 22,659–22,661, 1997.

Anderson, O. L., and D. G. Isaak, Calculated melting curves of iron, *Amer. Mineral., 83,* 376–383, 2000.

Anderson, O. L., and D. G. Isaak, Another look at the core density deficit of Earth's outer core, *Phys. Earth Planet. Inter.*, in press, 2002.

Anderson, O. L., L. Dubrovinsky, S. K. Saxena, and T. LeBihan, Experimental vibrational Grüneisen ratio values for ϵ-iron up to 330 GPa at 300 K, *Geophys. Res. Lett., 28,* 399–402, 2001.

Anderson, W. W., and T. J. Ahrens, An equation of state for liquid iron and implications for the Earth's core, *J. Geophys. Res., 99,* 4273–4284, 1994.

Andrault, D., G. Fiquet, M. Kunz, F. Visocekas, and D. Hausermann, The orthorhombic structure of iron: An in situ study at high-temperature and high-pressure, *Science, 278,* 831–834, 1997.

Andrault, D., G. Fiquet, T. Charpin, and T. LeBihan, Structural analysis and stability field of β iron at high P and T, *Amer. Mineral.*, *85*, 364–371, 2000.

Bass, J. D., B. Svendson, and T. Ahrens, The temperature of shock compressed iron, in *High Pressure Research in Mineral Physics, Geophys. Monogr. Ser.*, vol. 39, edited by M. H. Manghnani and Y. Syono, pp. 393–492, AGU, Washington, D. C., 1987.

Belonoshko, A. B., and R.. Ahuja, Embedded-atom molecular dynamic study of iron melting, *Phys. Earth Planet. Inter.*, *102*, 171–184, 1997.

Belonoshko, A. B., R. Ahuja, and B. Johansson, Quasi-ab initio molecular dynamic study of Fe melting, *Phys. Rev. Lett.*, *84*, 3638–3641, 2000.

Birch, F., Some geophysical applications of high pressure research, in *Solids Under Pressure*, edited by W. Paul and D. M. Warschauer, pp. 137–162, McGraw-Hill, New York, 1963.

Birch, F., The melting relations of iron, and temperatures in the Earth's core, *Geophys. J. R. Astron. Soc.*, *29*, 373–387, 1972.

Boehler, R., The phase diagram of iron to 430 kbar, *Geophys. Res. Lett.*, *13*, 1153–1156, 1986.

Boehler, R., Melting of the Fe-FeO and Fe-FeS systems at high pressure: Constraints on core temperatures, *Earth Planet. Sci. Lett.*, *111*, 217–227, 1992.

Boehler, R., Temperatures in the Earth's core from melting-point measurements of iron at high static pressures, *Nature*, *363*, 534–536, 1993.

Boehler, R., Fe-FeS eutectic temperatures to 620 kbar, *Phys. Earth Planet. Inter.*, *96*, 181–186, 1996.

Braginsky, S. I., and P. H. Roberts, Equations governing the convection in Earth's core and the geodynamo, *Geophys. Astrophys. Fluid Dyn.*, *79*, 1–97, 1995.

Brown, J. M., The equation of state of iron to 450 GPa: Another high pressure solid phase?, *Geophys. Res. Lett.*, *28*, 4339–4342, 2001

Brown, J. M., J. N. Fritz, and R. S. Hixon, Hugoniot data for iron, *J. Appl. Phys.*, *88*, 5496–5498, 2000.

Brown, J.M., and R.G. McQueen, The equation of state for iron and the Earth's core, in *Advances in Earth and Planetary Sciences*, edited by S. Akimoto and M. Manghnani, pp. 611–623, Center for Academic Publishing, Tokyo, 1982.

Brown, J. M., and R. G. McQueen, Phase transitions, Grüneisen parameter, and elasticity for shocked iron between 77 GPa and 400 GPa, *J. Geophys. Res.*, *91*, 7485–7494, 1986.

Buffett, B. A., H. E. Huppert, J. R. Lister, and A. W. Woods, On the thermal evolution of the Earth's core, *J. Geophys. Res.*, *101*, 7989–8006, 1996.

Bukowinski, M. S. T., A theoretical equation of state for the inner core, *Phys. Earth Planet. Inter.*, *14*, 333–344, 1977.

Bundy, F. P., Pressure-temperature phase diagram of iron to 200 kbar, 900°C, *J. Appl. Phys.*, *36*, 616–620, 1965.

Bundy, F. P., and H. M. Strong, Behavior of metals at high temperature and pressure, in *Solid State Physics,* edited by F. Seitz and D. Turnbull, pp. 81–143, Academic Press, New York, NY, 1962.

Dziewonski, A. M., and D. L. Anderson, Preliminary reference Earth model, *Phys. Earth Planet. Inter.*, *25*, 297–356, 1981.

Gallagher, K. G., and T. J. Ahrens, Analyses of shock temperature experiments of metals and the melting point of iron, *Eos Trans. AGU*, *76*(46), Fall Meet. Suppl., F553, 1995.

Ganapathy, R., and E. Anders, Bulk compositions of the Moon, Earth estimated from meteorites, *Geochem. Cosmochim. Acta*, Suppl., *5*, 1181–1206, 1976.

Gilvarry, J. J., The Lindemann and Grüneisen laws, *Phys. Rev.*, *102*, 307–316, 1956a.

Gilvarry, J., Equation of the fusion curve, *Phys. Rev.*, *102*, 325–331, 1956b.

Gilvarry, J. J., Temperature in the Earth's interior, *J. Atm. Terrestr. Phys.*, *10*, 84–95, 1957.

Gubbins, D., T. G. Masters, and J. A. Jacobs, Thermal evolution of the Earth's core, *Geophys. J. Roy. Astron. Soc.*, *59*, 57–99, 1979.

Holland–Gallagher, K. G., and T. J. Ahrens, Ultra high pressure thermal diffusivity and shock temperature experiments on metals and the phase diagram of iron, *Eos Trans. AGU*, *77*(22), West. Pac. Geophys. Meet. Suppl., W132, 1996.

Jeanloz, R., Thermodynamics of phase transitions, in *Microscopic to Macroscopic: Atomic Environments to Mineral Thermodynamics*, Reviews in Mineralogy, vol. 14, edited by S. W. Kieffer and A. Navrotsky, pp. 389–428, Mineralogical Society of America, Chelsea, MI, 1985.

Kraut, E. A., and G. Kennedy, New melting law at high pressures, *Phys. Rev. Lett.*, *16*, 608–609, 1966.

Laio, A., S. Bernard, G. L. Chiarotti, S. Scandolo, and E. Tosatti, Physics of iron at Earth's core conditions, *Science*, *287*, 1027–1030, 2000.

Lindemann, F. A., Uber die Berechnung molecular eigenfrequnzen, *Phys. Z.*, *11*, 609–612, 1910.

Liu, L., On the (γ,ϵ,ℓ) triple point of iron and the Earth's core, *Geophys. J. R. Astron. Soc.*, *43*, 697–705, 1975.

Mao, H. K., W. A. Bassett, and T. Takahashi, Effect of crystal structure and lattice parameters of iron up to 300 kbar, *J. Appl. Phys.*, *38*, 272–276, 1967.

Marsh, S. P. (Ed.), *LASL Shock Hugoniot Data*, 668 pp., Univ. of Calif. Press, Berkeley, 1980, p. 89.

Masters, T. G., and P. M. Shearer, Seismological constraints on the structure of the Earth's core, *J. Geophys. Res.*, *95*, 21,691–21,695, 1990.

Matsui, M., and O. L. Anderson, The case for a body centered cubic phase (α') for iron at inner core conditions, *Phys. Earth Planet. Inter.*, *103*, 55–62, 1997.

Morse, P. M., *Thermal Physics,* 2nd ed., 431 pp., Benjamin, New York, 1969.

Nguyen, J. H., and N. C. Holmes, Iron sound velocities in shock wave experiments, in *Shock Compression of Condensed Matter*, AIP Conference Proceedings, Snowbird, Utah, pp. 84–84, 2000.

Okuchi, T., Hydrogen partitioning into molten iron at high pressure: Implications for Earth's core, *Science*, *278*, 1781–1784, 1997.

O'Neill, H. St. C., D. Canil, and D. C. Rubie, Oxide-metal equilibria to 2500°C and 25 GPa: Implications for core formation and the light component in the Earth's core, *J. Geophys. Res.*, *103*, 12,239–12,260, 1998.

Poirier, J.-P., Dislocation-mediated melting of iron and temperature of the Earth's core, *Geophys. J. R. Astron. Soc.*, *83*, 313–328, 1986.

Poirier, J.-P., Lindemann law and the melting temperatures of perovskites, *Phys. Rev. Lett., 54,* 364–369, 1988.

Poirier, J.-P., Light elements in the Earth's outer core: a critical review, *Phys. Earth Planet. Inter., 85,* 319–337, 1994.

Poirier, J.-P., *Introduction to the Physics of the Earth's Interior,* 2nd ed., 312 pp., Cambridge University Press, 2000.

Ross, T. E., and L. H. Aller, The chemical composition of the Sun, *Science, 191,* 1223–1229, 1976.

Ross, M., D. A. Young, and R. Grover, Theory of iron phase diagram at earth-core conditions, *J. Geophys. Res., , 95,* 21,713–21,716, 1990.

Saxena, S., and L. Dubrovinsky, Iron phases and temperatures, phase transitions and melting, *Amer. Mineral., 85,* 372–375, 2000.

Saxena, S. K., L. S. Dubrovinsky, P. Haggkvist, Y. Ccrcnius, G. Shen, and H. K. Mao, Synchrotron x-ray study of iron at high pressure and temperature, *Science, 269,* 1703–1704, 1995.

Saxena, S. K., L. S. Dubrovinsky, and P. Häggkvist, X-ray evidence for the new phase β-iron at high temperature and high pressure, *Geophys. Res. Lett., 23,* 2441–2444, 1996.

Saxena, S. K., G. Shen, and P. Lazor, Experimental evidence for a new iron phase and implications for the Earth's core, *Science, 260,* 1312–1314, 1993.

Scott, H. P., Q. Williams, and E. Knittle, Stability and equation of state of Fe_3C to 73 GPa: Implications for carbon in the Earth's core, *Geophys. Res. Lett., 28,* 1875–1878, 2001.

Shankland, T. J. and J. M. Brown, Homogeneity and temperatures in the lower mantle, *Phys. Earth Planet. Inter., 38,* 51–58, 1985.

Shen, G., H. K. Mao, R. J. Hemley, T. S. Duffy, and M. L. Rivers, Melting and crystal structure of iron at high pressures and temperatures, *Geophys. Res. Lett., 25,* 373–376, 1998.

Sherman, D. M., Stability of possible Fe–FeS and Fe-FeO alloy phases at high pressure, *Earth Planet. Sci. Lett., 132,* 87–98, 1995.

Sherman, D. M., The composition of the Earth's core: Constraints on S and Si to temperatures, *Earth Planet. Sci. Lett., 153,* 149–155, 1997.

Spiliopoulos, S., and F. D. Stacey, The Earth's thermal profile: Is there a mid-mantle thermal boundary layer?, *J. Geodyn., 1,* 61–77, 1984.

Stacey, F. D., A thermal model of the Earth, *Phys. Earth Planet. Inter., 15,* 341–348, 1977.

Stacey, F. D. Theory of thermal and elastic properties of the lower mantle and core, *Phys. Earth Planet. Inter., 89,* 219–245, 1995.

Stacey, F.D., and R. D. Irvine, Theory of melting: Thermodynamic basis of Lindemann's law, *Aust. J. Phys., 30,* 631–640, 1977.

Stacey, F. D. and C. H. B. Stacey, Gravitational energy of core evolution: Implications for thermal history and geodynamo power, *Phys. Earth Planet. Inter., 110,* 83–93, 1999.

Steinle-Neumann, G., L. Stixrude, R. E. Cohen, and O. Gürlseren, Elasticity of iron at the temperatures of the Earth's inner core, *Nature, 413,* 57–60, 2001.

Stishov, S. M., Entropy, disorder, melting, *Sov. Phys. USP, 31,* 52–54, 1988.

Stixrude, L., and R. E. Cohen, Constraints on the crystalline structure of the inner core-mechanical instability of bcc iron at high pressure, *Geophys. Res. Lett., 22,* 125–128, 1995.

Stixrude, L., E. Wasserman, and R. E. Cohen, Composition and temperature of the Earth's inner core, *J. Geophys. Res., 102,* 24,729–24,739, 1997.

Strong, H. M., Melting and other phase transformations at high pressure, in *Progress in Very High Pressure Research,* edited by F. P. Bundy, W. R. Hibbard, and H. M. Strong, p. 182, Wiley, New York, 1961.

Strong, H. M., R. E. Tuft, and R. E. Hannemann, The iron fusion curve and the γ-δ-ℓ triple point, *Metall. Trans., 4,* 2657–2661, 1973.

Verhoogen, J., *The Energetics of the Earth,* 139 pp., National Academy of Sciences, Washington, D.C., 1980.

Vocadlo, L., J. Brodholt, D. Alfè, M. J. Gillian, and G. D. Price, Ab initio free energy calculations on the polymorphs of iron at core conditions, *Phys. Earth Planet. Inter., 117,* 123–137, 2000.

Wasserman, E., L. Stixrude, and R. E. Cohen, Thermal properties of iron at high pressures and temperatures, *Phys. Rev. B Condens. Mattcr, 53,* 8296–8309, 1996.

Wasson, J., *Meteorites: Their Record of Early Solar-System History,* 267 pp., W. H. Freeman and Co., New York, 1985.

Williams, Q., and R.J. Hemley, Hydrogen in the deep Earth, *Ann. Rev. Earth Planet. Sci., 29,* 365–418, 2001.

Williams, Q., R. Jeanloz, J. Bass, B. Svendson, and T. J. Ahrens, T. J., Melting curve of iron to 250 GPa: A constraint on the temperature of the earth's center, *Science, 236,* 181–183, 1987.

Wolf, G. H. and R. Jeanloz, Lindemann melting law: Anharmonic correction and test of its validity for minerals, *J. Geophys. Res., 89,* 7821–7835, 1984.

Wood, B.J., Carbon in the core, *Earth Planet. Sci. Lett., 117,* 593–607, 1993.

Yoo, C. S., J. Akella, A. J. Campbell, H. K. Mao, and R. J. Hemley, Phase diagram of iron by in situ x-ray diffraction: Implications for the Earth's core, *Science, 270,* 1473–1475, 1995.

Yoo, C. S., N. C. Holmes, M. Ross, D. Webb, and C. Pike, Shock temperatures and melting of iron at Earth core conditions, *Phys. Rev. Lett., 70,* 3931–3934, 1993.

O. L. Anderson, Institute of Geophysics and Planetary Physics, University of California, Los Angeles, CA 90095-1567. (email: olanderson@adam.igpp.ucla.edu)

Solidification of the Earth's Core

Michael I. Bergman

Physics Department, Simon's Rock College, Great Barrington, MA

As the Earth cools, the inner core solidifies from the fluid outer core. The outer core is about 6 - 10 % less dense than pure iron at core pressures, the inner core perhaps 2 - 3 % less dense. The exact makeup of the less dense alloying components remains uncertain, though some combination of metallic iron oxide and iron sulphide are likely candidates. It is not known whether the iron alloy forms a eutectic or a solid solution. If the former, the core has a composition on the iron side of the eutectic, *i.e.,* the core is iron-rich; if the latter, pure iron has a higher melting temperature than the dominant alloying component, at core pressures.

In either case, it has been predicted that the inner-outer core boundary is several hundred times supercritical for a morphological instability. The instability is due to constitutional supercooling resulting from melting point suppression by the alloying component. The inner core thus solidifies with a dendritic structure, and at least the top of the inner core is a mushy zone where interdendritic melt pockets enriched in the alloy component remain. With an estimated mushy zone Rayleigh number at least one thousand times supercritical, there is a large exchange of fluid between the dendritic mushy zone and the outer core. This results in a solid fraction within the inner core that increases rapidly with depth, so that the inner core presents a sharp interface on the seismic body wave lengthscale of a few kilometers. Nevertheless, fluid pockets may be an explanation for the possible high attenuation of seismic body waves near the top of the inner core.

Extrapolations from laboratory solidification experiments and evidence from meteorites that have retained their solidification structure predict an inner core grain size of about a kilometer. This large grain size agrees with that found by assuming that the inner core seismic attenuation anisotropy is due to scattering off grain boundaries of a partially textured inner core, such as can occur as a result of solidification. If the inner core is undergoing recrystallization at a stress level of 1 Pa, as occurs due to longitudinal Lorentz forces, the grain size could also be as large as a kilometer. Larger inner core stresses have also been suggested, which would result in a grain size smaller than that predicted by solidification.

Earth's Core: Dynamics, Structure, Rotation
Geodynamics Series 31
Copyright 2003 by the American Geophysical Union
10.1029/31GD08

Because dendrites tend to grow most rapidly in the direction of heat flow, and because they grow in a particular crystallographic direction, solidification results in a lattice preferred orientation. Because of rotational constraints, convective transport of heat and solute in the outer core is primarily in the cylindrically radial direction. There is therefore a tendency for dendrites to grow in this direction. Thus, the inner core solidifies with a texture with a tendency towards cylindrical symmetry, as is observed seismically. For geometrical reasons the anisotropy along a ray path can actually increase with depth. However, it is not clear whether a simple solidification texture model can explain the particular depth dependence or longitudinal variations of the elastic anisotropy. Moreover, the elastic constants of iron under inner core conditions remain uncertain, and the effects of deformation on the solidification texture are also not known.

The presence of less dense alloying components also provides for a source of compositional convection in the outer core. Convection resulting from solidification commonly occurs as channel convection, where fluid rises out of the dendritic mushy zone in narrow, isolated, and fast jets. The mushy zone Rayleigh number in the core is highly supercritical, so that one might expect channel convection. However, the Lorentz force in the mushy zone at the top of the inner core is large enough to supplant the Darcy viscous force as the primary retardant to flow. Channels do not then form, so that the compositional buoyancy flux thus varies only smoothly, if at all, at the inner-outer core boundary. The solidification texture is not, however, affected by channel convection.

1. INTRODUCTION

The Earth's solid iron inner core has been described, because of its possible super-rotation with respect to the Earth's mantle and crust, as a 'planet within a planet' (*Su, et al., 1996*). Although such a fanciful description of the inner core may involve some scientific license, the inner core is in many ways as remote as the most distant planet in our solar system. Indeed, the presence of the inner core at the center of our planet was not discovered until six years after Tombaugh's discovery of Pluto in 1930. Then, Lehmann realized that compressional waves arriving in the core shadow zone had amplitudes too large to be attributed to diffraction, but rather, represented reflections off a body with a higher seismic velocity than the fluid outer core (*Lehmann, 1936*). Bullen (*1946*) is credited with discovering the rigidity of the inner core, though detection of shear waves in the inner core, providing solid evidence for a rigid inner core, has been difficult and uncertain (*Julian, et al., 1972; Deuss, et al., 1998; Okal & Cansi, 1998*). Birch (*1940; 1952*) recognized that the inner core is solidifying from the outer core, and that the inner-outer core boundary represents a phase transition. The solidification occurs at smaller radii r, despite higher temperatures, because the change in the melting temperature of iron $(dT_m/dp)(dp/dr)$ is steeper than the temperature gradient dT/dr present in the core, where p is the pressure.

Verhoogen (*1961*) first recognized that solidification of the core results in a release of latent heat that could be important for the heat budget of the Earth. He suggested that this heat could help drive the convective motions in the outer core that are thought to sustain the geodynamo. Braginsky (*1963*) further suggested that partitioning of an iron alloy during solidification of the core provides a source of compositional buoyancy that could also help drive convective motions. Considerable theoretical work has followed in an effort to understand the role of the inner core for the thermodynamics of the Earth. Much of this work has been directed towards understanding the heat budget of the core and the relative importance of thermal versus compositional buoyancy (*Loper, 1978a; Gubbins, et al., 1979; Buffett, et al., 1992*). Estimates vary because of uncertainty on the amount of heat that can be conducted down the core adiabat, the heat flux across the core-mantle boundary, and the partitioning of light elements into the inner core, but compositional convection is likely to contribute at least 50 - 80 % of the dynamo power (*Lister & Buffett, 1995; Stacey & Stacey, 1999*). Other theoretical and laboratory studies have been directed towards investigating the structure of solidifying metallic alloys such as the inner core, and the nature of solidification-driven compositional convection in the outer core (topics covered in some depth in this review).

Observational advances concerning the inner core lagged behind those on the rest of the Earth, understandable in light of the noisy filter that is the mantle. Many seismological studies of the core involved efforts to determine the radial dependence of the speed of compressional waves. In particular, one group of studies (*Haddon & Cleary, 1974; Bolt, 1982*) discounted the existence of the proposed 'F-layer' above the inner-outer core boundary, where the compressional wave speed had been thought to decrease with depth (*Jeffreys & Bullen, 1935*). Many other studies focused on attenuation in the inner core (*Doornbos, 1974; Qamar & Eisenberg, 1974*).

With improvements in data and data analysis the inner core has received more interest from seismologists in the past fifteen years. Some surprising inferences have been made, such as an elastic anisotropy in which the inner core is 3-4 % faster parallel to the rotation axis than perpendicular to the rotation axis (*Morelli, et al., 1986; Woodhouse, et al., 1986*), an attenuation anisotropy in which the fast direction exhibits more damping and more complex waveforms (*Souriau & Romanowicz, 1996*), a frequency dependent quality factor (*Cormier, 1981; Widmer, et al., 1991*), and, of course, the prograde rotation of the inner core relative to the Earth's surface (*Song & Richards, 1996*). While there is some, though not universal, agreement that these inferences are generally correct (*Creager, 1992; Tromp, 1993; Creager, 1997; Souriau, et al., 1997; Laske & Masters, 1999; Vidale, et al., 2000*), the details are hotly disputed, and the physical mechanisms causing them, possibly including solidification, remain contentious as well.

In addition to its thermodynamic role and seismic properties, many aspects of the inner core are now the subject of research. For instance, mineral physicists are investigating the phase and elastic properties of iron under inner core conditions (*Anderson, this volume; Steinle-Neumann, et al., this volume*). Although there is both experimental and theoretical evidence that the hexagonal closest packed (hcp) ε phase is the most stable (*Yoo, et al., 1995; Vocadlo, et al., 1999*), there are suggestions that a double hcp phase may exist (*Saxena, et al., 1993*). There are also suggestions that the presence of silicon in the inner core may stabilize a body centered cubic (bcc) phase to co-exist with the hcp phase (*Lin, et al., 2002*), though the amount of silicon in the core is unlikely to be enough for the bcc phase to be dominant. In any case, the elastic properties remain uncertain (*Stixrude & Cohen, 1995; Bergman, 1998; Mao, et al., 1998; Steinle-Neumann, et al., 2001; this volume*). Meanwhile, fluid dynamicists are studying the inner core's role in generating the geomagnetic field (*Hollerbach & Jones, 1993; Olson & Aurnou, 1999*). This intense study of the inner core has not always the case. Analytical studies on the difficult geodynamo problem have understandably ignored the presence of the inner core (*Roberts (1988)* described the inclusion of an inner core as a peripheral issue!), but, for instance, the possibility that the inner core helps stabilize the geomagnetic field against reversals has been demonstrate (*Hollerbach & Jones, 1993*).

This review does not cover in depth the thermodynamics, seismology, mineral physics, or fluid dynamics of the core.

Rather, it concentrates on those aspects of all of these subjects that bear on the solidification of the Earth's core. Section 2 reviews the essentials of alloy phase diagrams, and Section 3 examines the composition and phase diagram of the core. Section 4 covers directional solidification and dendritic growth of alloys, and Section 5 examines the evidence for such growth in the inner core. Section 6 studies further the solidification structure of the inner core, in particular, its possible relevance to the seismic anisotropy of the inner core. Section 7 discusses work on compositional convection driven by solidification, relating the results to the outer core. Although there is little doubt that the inner core is solidifying from the outer core, the connection between solidification of the core and surface observations is less clear, and it is the intent of this review to examine this connection.

2. REVIEW OF ALLOY PHASE DIAGRAMS

As discussed in the next section, there is evidence that the core is not pure iron. Before turning to the composition and phase diagram of the core, it is thus worthwhile to review briefly phase diagrams for solidifying alloys. For a binary system (A and B) where each pure solid exists only in a single solid phase (crystal structure), and where A and B are miscible in the liquid phase, three types of phase diagrams are possible. (Metals are usually miscible in the liquid phase, and although there have been some doubts about the miscibility of iron oxide in iron, recent calculations by Alfe, *et al.* (*1999*) suggest full miscibility under outer core conditions.) Which type of phase diagram exists for a given system is determined by the minimum in Gibbs free energy G = H - TS, where H is the enthalpy, T is the temperature, and S is the entropy.

If the crystal structures are the same, and the atomic sizes are similar, then A and B may be completely miscible in the solid state as well as the liquid state ($\Delta H = 0$ upon mixing). Liquid and solid are then both ideal solutions, and only two phases, liquid and solid, are possible in the temperature-composition field (Figure 1). An example is the copper-nickel system.

If the change in enthalpy upon mixing in the solid state is non-zero, then the solid is a real solution. If $\Delta H > 0$, atoms A and B are incommensurate, either because of a differing crystal structure, atomic size, or both. In this case it is energetically favorable for the mixture to remain a liquid to a temperature beneath that of the pure solid melting temperatures (Figure 2). Accompanying the melting point depression is a miscibility gap in the solid at low temperatures such that two solid phases co-exist. If ΔH is large enough, the miscibility gap can extend up to the liquid phase (*Porter & Easterling, 1992*). In this case, the lower entropy associated with two separate phases, one A-rich and the other B-rich, rather than randomly mixed atoms in a single phase, is made up for by the lower enthalpy that results from A and B existing

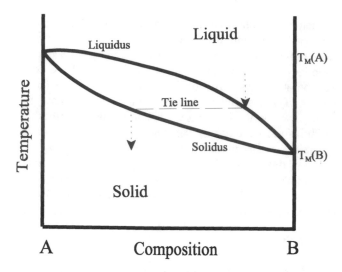

Figure 1. Phase diagram for a binary system (A,B) exhibiting complete solubility in both the liquid and solid phases (the enthalpy of mixing, ΔH, equals zero), so that both the liquid and solid phases are ideal solutions. The liquidus represents the temperature at which liquid of a given composition solidifies. A horizontal tie line at that temperature connects to the solidus, which gives the composition of the solid. The space between the curves represents the width of the phase loop, which is a measure of the compositional difference between the liquid and solid.

primarily in separate phases. Such a system exhibits a eutectic phase diagram (Figure 3); the alloy forms a eutectic. Due to entropy, even in a eutectic system B will have non-zero solubility in the A-rich solid phase α, and vice versa, but such a solid solution is not ideal. Examples are ubiquitous, including model laboratory systems sodium chloride-water, ammonium chloride-water, lead-tin, and zinc-tin.

In a binary system with $\Delta H > 0$ it is also possible for stable solid phases other than those of pure A and pure B to exist. When such intermediate phases are present (as is common with iron alloys), a more complex phase diagram results (Figure 4). Other solidification reactions such as peritectic solidification can result. However, the essential solidification structure of the alloy is similar to that of a eutectic system (*Porter & Easterling, 1992*).

If $\Delta H < 0$, it is energetically favorable for a solid phase to form, so that a maximum in the melting temperature occurs for some intermediate composition (Figure 5). Such a solid is known as an ordered alloy because the atoms arrange themselves in a particular structure known as a superlattice. Ordered alloys often occur only within a certain compositional range, for instance, a particular type of brass, β-brass. There is no experimental evidence that any of the possible core alloy components form an ordered alloy with iron under inner core conditions, so I will not consider further such phases.

3. COMPOSITION AND PHASE DIAGRAM OF THE CORE

The composition of the core remains uncertain. Based on the average density of the Earth, the solar abundance of the elements, and the presence of iron meteorites thought to be remnants of planetoid cores, it is clear that the core contains elemental iron (*Birch, 1952*). Based on iron meteorites (*Brown & Patterson, 1948; Buchwald, 1975*), it is likely that the core also contains about 8 % nickel by mass. Little theoretical or experimental work has been done on the partitioning of nickel upon solidification under core conditions. Because nickel's density is similar to that of iron (*Stacey, 1992*), it is difficult to detect seismically, and it would seem to play little role in core thermodynamics.

Under much more scrutiny has been the composition of the less dense alloying components in the core, and the partitioning upon solidification. Early compression experiments with extrapolation to core conditions (*Bridgman, 1949; Birch, 1952*) and shock compression experiments (*McQueen & Marsh, 1966*) have shown that iron-nickel alloys under core pressures are about 6-10 % more dense than the outer core (*Bullen, 1949*), so that less dense alloying components must be present. Recent calculations (*Laio, et al., 2000*) using first principles and molecular dynamics confirm

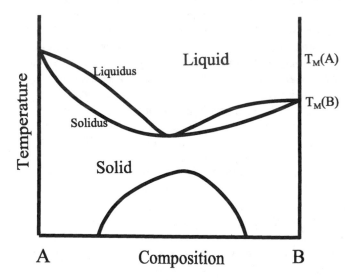

Figure 2. Phase diagram for a binary system exhibiting complete solubility in the liquid phase, but where $\Delta H > 0$ in the solid phase, so that the solid phase is a real solution. Because $\Delta H > 0$ in the solid phase, the system remains liquid to lower temperatures than the melting temperature of pure A or pure B ($T_M(A)$ or $T_M(B)$). Also, at low temperatures where the entropy contribution to the free energy is less, A and B tend to segregate into two solid phases, one A-rich and one B-rich.

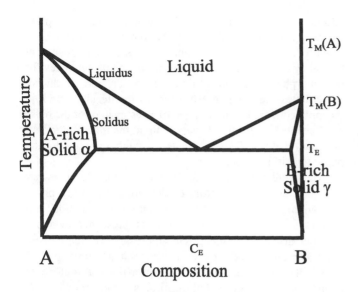

Figure 3. As for Figure 2, but where ΔH is sufficiently positive that the miscibility gap in the solid phase extends up to the liquid phase. Such a phase diagram represents a eutectic system. Two solid phases, A-rich α and B-rich γ, exist at all temperatures beneath the eutectic temperature T_E, the lowest temperature at which liquid can be in equilibrium with solid. At T_E, the composition of the liquid is C_E.

that liquid iron at 330 GPa and 5400 K is about 6 % more dense than given by PREM (*Dziewonski & Anderson, 1981*). Iron meteorites contain iron sulphide (troilite) inclusions, suggesting that iron sulphide may alloy with iron in the core (*Mason, 1966*). Also, experiments show that sulphur has an affinity for liquid iron-nickel alloys, at least in the range 2 - 25 GPa and 2073 - 2623 K (*Li & Agee, 2001*). Moreover, the mantle is depleted in sulphur relative to its chondritic abundance by two orders of magnitude, so that the deficit could reside in the core (*Murthy & Hall, 1970*). However, since sulphur is volatile, that line of argument could be suspect (*Ringwood, 1977*).

Iron oxide is not very soluble in iron at atmospheric pressure: much less than .1 % by mass in solid iron, and only a few tenths percent just above the liquidus (*Ringwood, 1977*). However, its solubility increases with temperature, and with pressure, as it becomes metallic. Thus, although iron oxide is not present within iron meteorites, given oxygen's abundance in the solar system, and in the mantle, iron oxide could be present in the core (*Ringwood & Hibberson, 1990*). Other possible alloying components include iron silicate, iron carbide, and iron hydride (*MacDonald & Knopoff, 1958; Wood, 1993; Fukai, 1984; Okuchi, 1997*), all likely to be metallic under core pressures. Abundance and volatility during condensation from the nebula, and solubility in iron during core formation, determined the less dense elements in the core. However, the process of core formation, and hence the

temperature and pressure history of the iron alloy, is not known. In particular, it is not known whether the core formed from a planetary magma ocean or via percolation from a solid silicate matrix (*Stevenson, 1990; Minarik, et al., 1996; Shannon & Agee, 1996; Karato & Murthy, 1997; Bruhn, et al., 2000*).

Earth models indicate that the inner core is more dense than the outer core, by about 600 kg/m³ in PREM (*Dziewonski & Anderson, 1981*), and 550 +/- 50 kg/m³ according to Masters & Shearer (*1990*). Part of this difference, perhaps about 200 kg/m³, can be attributed to iron being more dense in the solid phase (*Laio, et al., 2000*). The rest is attributed to the less dense alloying components partitioning into the outer core during solidification of the core. This has profound consequences for the thermodynamics, fluid dynamics, and metallurgy of the core, as Braginsky (*1963*) realized. The partitioning results in a radial compositional gradient, which can lead to a convective instability that can stir up the fluid outer core and drive the geodynamo. It can also result in a morphological instability that can influence the structure of the solid inner core. Braginsky proposed that the concentration of iron in the core is less than the eutectic composition, and that the inner core has the eutectic composition. In this scenario, crystals of the less dense, iron-depleted phase rise out of a slurry layer directly above the inner core (Jeffreys and Bullen's F-layer) to drive convection. However, a slurry layer is not in accord with the metallurgical evidence for the way in which metallic alloys solidify (*Rutter,*

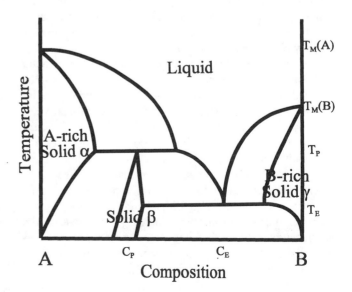

Figure 4. Phase diagram for a binary system exhibiting complete solubility in the liquid phase, where ΔH > 0 in the solid phase, and where an intermediate phase β is present. At the peritectic temperature T_P, liquid and solid phase α form solid phase β of composition C_P.

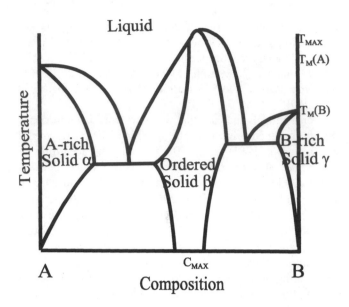

Figure 5. Phase diagram for a binary system exhibiting complete solubility in the liquid phase, but where $\Delta H < 0$ in the solid phase. Because $\Delta H < 0$ in the solid phase, the system forms an ordered phase β, which has a maximum melting temperature T_{MAX} at a composition C_{MAX}.

1958; Chalmers, 1964). Also, Loper (*1978b*) and Fearn and Loper (*1983*) showed thermodynamically that an iron-poor core is not capable of driving the geodynamo. Moreover, the seismic evidence for the F-layer, which was part of the original motivation for Braginsky's model, has been reinterpreted in terms of scattering off the core-mantle boundary (*Haddon & Cleary, 1974*).

The seismically inferred density increase at the inner-outer core boundary, above that due to the phase transition, can be explained if the concentration of iron in the core is greater than the eutectic composition. An iron-rich core also provides for a more viable source of gravitational energy to drive the geodynamo. However, an iron-rich core does not preclude some fraction of less dense alloying components in the inner core, which depends on the slope of the solidus (Figure 1 or 3). In fact, Jephcoat and Olson (*1987*) extrapolated the density of pure iron to inner core conditions and found that it is too dense to explain the seismic data. Similarly, Laio, *et al.* (*2000*) has calculated that solid iron under the conditions at the inner-outer core boundary is 2 - 3 % more dense than given by PREM (*Dziewonski & Anderson, 1981*).

Jephcoat and Olson extrapolated that the inner core contains some 3-7 % sulphur by mass. Stixrude, *et al.* (*1997*) calculated similar results, but found that the mass fraction of the less dense element depends on which less dense element and alloy is present. (For instance, 1 % mass fraction oxygen as FeO at 7000 K, 2 % sulphur as FeS_2 at 5500K, and >8 % sulphur as Fe_9S at unrealistically low inner core temperatures

<3500 K, all as ideal solid solutions with iron, can all explain the inner core density.) To the extent that the inner core does contain less dense alloying components the compositional buoyancy available to drive convection, and the tendency for the morphological instability to occur, are reduced. Nevertheless, reasonable estimates for the density deficit in the inner core are less than the 6-10 % estimated for the outer core, accounting for the density jump at the inner-outer core boundary that is 400 kg/m³ greater than that expected for the phase transition alone.

Along with the composition of the core, considerable work has been done in an effort to determine the alloy phase diagram of the core. At low pressures iron-iron sulphide form a eutectic, but Boehler (*1996a*) has presented evidence showing that the melting point suppression decreases with increasing pressure. His experiments reach pressures up to 62 GPa, still much lower than the 330 Gpa at the inner-outer core boundary. This suggests that iron-iron sulphide may form an ideal solid solution at inner core pressures. This does not preclude an iron-enriched inner core, provided the liquidus and solidus are sufficiently separated in composition (the phase loop in Figure 1 is wide), and the melting temperature of pure iron is higher than that of iron sulphide, at core pressures. The latter may well be the case (*Boehler, 1992; Anderson & Ahrens, 1996*). However, Alfe, *et al.* (*2000*) found from *ab initio* calculations that the concentration of sulphur in the solid state is very nearly that of the liquid state. Such a similarity in composition cannot explain the density jump at the inner-outer core boundary. It would also not provide the driving force for the compositional convective and morphological instabilities. (The similarity indicates that the liquidus and solidus are close in composition. While this suggests an ideal solid solution, a eutectic system could still be a possibility if a significant fraction of the inner core has not yet cooled to the eutectic temperature (*Fearn, et al., 1981*)).

Just as iron sulphide alone cannot explain the density jump at the inner-outer core boundary, so too there have been arguments that iron oxide alone may not be able to explain the presence of a less dense component in the inner core. Sherman (*1995*), also using first principles calculations, showed that the concentration of oxygen in the solid state is very low. On the other hand, Alfe, *et al.* (*1999*) suggested that the concentration could be higher for other assumed crystal structures, and Stixrude, *et al.* (*1997*) were able to explain the inner core density with FeO present in solid solution with pure iron rather than as a mixture of separate phases. It seems entirely plausible that the core is not a binary system, and at present the dominant less dense alloying components and their enthalpy of mixing remain uncertain, though FeO and perhaps FeS_2 are likely candidates. Similarly, the presence of nickel and its possible effect on the phase diagram of the core is not known. Without knowing the phase diagram appropriate to the core, the temperature and thermal history of the core will also remain uncertain (see Jeanloz (*1990*), Poirier (*1994*), and Boehler (*1996b*) for reviews). The latter also reviews the pressure-temperature phase diagram of pure iron (*i.e.*, the existence of various solid phases and the melting temperature). Although knowledge of the stable phase and elastic properties of iron under inner core conditions is essential for understanding the role of solidification in causing the seismic anisotropy of the inner core (Section 6), progress can nevertheless be made towards understanding the directional solidification of the core.

4. REVIEW OF DIRECTIONAL SOLIDIFICATION AND CRYSTAL GROWTH OF METALLIC ALLOYS

Solidification studies of the core have typically assumed a binary eutectic phase diagram and have not been concerned with the exact identity of the less dense component(s). The justification for this is that provided the core is iron-rich (if the core forms a eutectic) or the melting point of pure iron is greater than the dominant less dense component (if the inner core forms an ideal solid solution), then an iron-enriched phase in the inner core will solidify. The rejection of less dense solute into the liquid upon solidification can then drive the compositional convective and morphological instabilities that have been of much interest. Although the core is almost certainly more complicated, an iron-rich, binary eutectic system thus captures the essential features necessary to understand the solidification of the Earth's core. An additional complication in the Earth, not reproducible in the laboratory, is that the pressure range in the core is sufficiently large to affect the phase diagram (which of course is why the core is solidifying outwards). The phase diagram becomes three-dimensional (*Fearn & Loper, 1983*), with the three coordinates being temperature, composition, and pressure (or radius).

After it was realized in the late 1970's that an iron-poor core was thermodynamically incapable of driving the geodynamo (*Loper, 1978b; Fearn & Loper, 1983*), focus turned to an iron-rich core. It is then easy to see qualitatively the origin of compositional convection driven by solidification. As an iron-enriched phase in the inner core solidifies, the fluid rejected upon solidification is enriched in the less dense solute (Figure 1 or 3). The solute boundary layer can become gravitationally unstable underneath the fluid with the bulk composition of the outer core. This compositional convective instability may help power the geodynamo. I will not discuss further the relative contributions of thermal and compositional buoyancy to the power budget of the core, but Section 7 discusses details of the style of compositional convection driven by solidification.

In addition to the compositional convective instability there is the morphological instability that can result during directional (*e.g.*, radially outwards) solidification. Consider the case where solidification is slow enough that solid state diffusion is important, and where there is no convection in the fluid. Rejection of solute during solidification leads to a solute boundary layer in the fluid with a scale thickness D/V, where D is the solute diffusivity and V is the growth velocity. As can be seen in Figure 6, the equilibrium freezing temperature T_L (the liquidus) is a function of the composition of the fluid. The equilibrium freezing range ΔT is the temperature difference between the liquidus and solidus at the initial composition of the fluid (like the composition difference it is also a measure

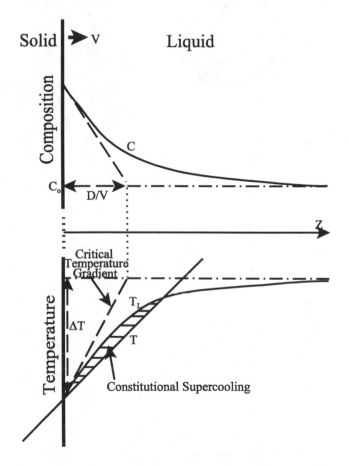

Figure 6. Dendritic growth in directionally solidifying alloys can occur when the liquid is constitutionally supercooled. Adjacent to the solid is a solute boundary layer of scale thickness D/V, where D is the solute diffusivity and V is the growth rate of the solid. The solute concentration C in the boundary layer is enriched relative to its value C_o far from the solid. This enrichment depresses the equilibrium freezing temperature T_L by an amount ΔT from its value far from the solid. If the temperature T in the liquid is less than T_L, then solid is predicted ahead of the solid/liquid interface, a condition known as constitutional supercooling. The criterion for constitutional supercooling is that the temperature gradient in the liquid, dT/dz < ΔT/(D/V). After Porter & Easterling (*1992*).

Figure 7. Constitutional supercooling results in the growth of solid perturbations into the liquid. The initial instability leads to cellular growth, and side branching resulting from further instabilities leads to dendritic growth. Dendrites form the mushy zone, where the temperature and composition are linked by the liquidus curve. Shown here is a mushy zone in an ammonium chloride-rich aqueous solution. The solid fraction in mushy zones in metallic alloys is typically higher than in aqueous salt solutions because of their higher thermal conductivity (*Hellawell, et al., 1993*). Courtesy of M.G. Worster.

of the width of the phase loop). If the gradient of the local freezing temperature, $dT_L/dz = \Delta T/(D/V)$, exceeds the actual temperature gradient in the fluid, dT/dz, then freezing is predicted ahead of the flat solid-fluid interface. This condition is known as constitutional supercooling (*Rutter, 1958*). Essentially, a perturbation of solid into the fluid may grow because, although the temperature increases with distance away from the solid, the product of the liquidus slope times the change in solute concentration decreases more so.

The breakdown of a planar solid-fluid interface is known as the morphological instability (*Mullins & Sekerka, 1963; 1964*). In the first stage solid perturbations grow into the fluid, with lateral solute rejection resulting in local equilibrium freezing temperature depression between protrusions. These protrusions develop into cellular structures, with the cells that grow out of the same crystal parallel. Those cells elongated close to the direction of heat flow grow most rapidly. In the second stage the cells themselves may become morphologically unstable, with transverse secondary arms forming for essentially the same reason (solute rejection) that the initial solid perturbations grew from the planar solid-fluid

interface. With yet further instability, tertiary solid arms form, leading to a fully dendritic structure (*Porter & Easterling, 1992*). Directionally solidified metallic alloys typically exhibit columnar crystals, elongated in the direction of dendritic growth.

The region of constitutional supercooling is known as a mushy zone, and is a mixed solid-fluid region. Mushy zones are ubiquitous features in directionally solidified metallic alloys, organic systems, and aqueous salt solutions on both sides of the eutectic (Figure 7). Dendritic growth and the structure of mushy zones are studied widely in materials science because of their practical importance (*Chalmers, 1964; Flemings, 1974; Langer, 1980; Kurz & Fisher, 1992*). With their relatively high ratio of latent heat of fusion to melting temperature, solid-fluid interfaces in silicates tend to be atomically flat, as opposed to in metals where they are more diffuse (*Jackson, 1958*). It has been observed (*Miller & Chadwick, 1969*) that materials with atomically flat solid-fluid interfaces usually solidify with low index close-packed facets. Perhaps because of the low entropy associated with faceted dendrites, dendritic mushy zones are less common in silicates.

The criterion for instability of a planar interface during solidification with solid state diffusion, $\Delta T/(D/V) > dT/dz$, shows that a large melting point depression ΔT, a large growth velocity V, and a small temperature gradient dT/dz all favor a mushy zone forming. The lack of solid state diffusion during solidification increases the tendency for instability by increasing the effective freezing range ΔT by further depleting the solid and enriching the fluid in solute. Solid state diffusion is probably not important during solidification at the inner-outer core boundary because the diffusional timescale of the inner core, L^2/D, where L is the radius of the inner core, far exceeds even the age of the Earth, using a typical value for substitutional diffusivity D in metals, 10^{-10} m^2/s (*Porter & Easterling, 1992*).

Loper and Roberts (*1981*) extended the condition for morphological instability to include the effect of pressure variations on the liquidus. By comparing the pressure variation of the liquidus with that of the actual core temperature (the latter from the rate at which latent heat of fusion can be conducted away from the inner-outer core boundary), they estimated that the inner core growth rate is nearly 500 times supercritical. The morphological instability at the inner-outer core boundary is plausible because the variation of the liquidus with pressure (the melting point suppression) is proportional to the ratio of the liquidus to the latent heat of fusion (the Clapeyron slope), which is usually large in metals (*Jackson, 1958*). In other words, the condition for morphological instability, $\Delta T/(D/V) > dT/dz$, can be satisfied because, although V is very small, dT/dz is also small, while ΔT is large. Fearn, *et al.* (*1981*) argued further that the mushy zone, with its non-zero liquid fraction, could extend to the center of the Earth because the temperature at the center of the Earth is likely to exceed the eutectic temperature. These two papers introduced the idea that the dendritic growth common to the metallurgical laboratory may also be found in the core. Loper and Roberts (*1978; 1980*) had earlier worked out the thermodynamics of a slurry, which is a general mixed phase region, but they were apparently not yet aware of dendritic growth in alloys. Their theory is general, but the emphasis concerned an inner core that formed by precipitation of heavy, iron-rich particles downwards. This is contrary to the way in which directionally solidified metallic alloys grow by dendritic growth upwards.

Morse (*1986; 2001*) has questioned the prediction of a mushy inner core because of the assumption that heat and solute near the inner-outer core boundary are removed only by diffusion. Loper and Roberts (*1981*) ignored the effect of convection, arguing that it must become small near the boundary. On the other hand, Morse argued that convection at the base of the outer core reduces the solute buildup, making the inner core growth rate nearly five orders of magnitude less than that needed for morphological instability. However,

although it has been observed that convection can thin a mushy zone, in part due to solute transport and in part due to broken dendrite arms serving as new nucleation sites (*Flemings, 1974*), convection has not been observed to eliminate a mushy zone, presumably due to the existence of a solute boundary layer.

5. CRYSTAL GROWTH IN THE CORE

The observational evidence for a mushy inner core, though hardly conclusive, comes from seismology and from meteoritics. The evidence for the F-layer at the base of the outer core (*Jeffreys & Bullen, 1935*), which was part of the original motivation for Braginsky proposing compositional convection driven by a slurry (*1963*), was later reinterpreted in terms of diffraction off the core-mantle boundary (*Haddon & Cleary, 1974*). However, there is evidence for high attenuation, as measured by a low quality factor Q in both compression and shear, throughout the inner core. Several studies suggest that compressional wave Q_α varies from about 200 in the shallow inner core to about 1000 deep in the inner core (*Doornbos, 1974; 1983; Cormier, 1981; Choy & Cormier, 1983; Masters & Shearer, 1990*). Moreover, Mao, *et al.*, (*1998*) found that the aggregate shear wave speed of hcp iron extrapolated from 220 GPa and room temperature to inner core conditions is about 15 % greater than that in the inner core (*Dziewonski & Anderson, 1981*), suggesting near-melting softening.

High compressional attenuation can be interpreted in terms of unsolidified, interdendritic liquid pockets. Moreover, one expects a larger liquid fraction closer to the top of an inner core mushy zone (*Doornbos, 1974; Stiller, et al., 1980; Loper & Fearn, 1983*). However, some seismic work suggests that although the overall attenuation in the inner core is high, the depth dependence of the attenuation is not resolved (*Bhattacharyya, et al., 1993*). Shear wave attenuation, as measured by Q_β from normal modes, may also be high, though there are some discrepencies with the body wave data that might be explained by a frequency dependence due to an absorbtion band (*Widmer, et al., 1991*). There is little depth resolution in the normal mode data. On the other hand, some studies (*Jackson, et al., 2000; Laio, et al., 2000; Steinle-Neumann, et al., 2001*) find a Poisson's ratio for hcp iron under inner core conditions comparable to that from PREM, so that partial melt may be unnecessary to explain the seismic results.

Perhaps more difficult to understand is how a mushy inner core with a liquid fraction that decreases smoothly with depth can yield sharp reflections (the very ones that led to Lehmann's discovery of the inner core) off an interface resolved to less than a few kilometers by 1 Hz body waves (*Masters & Shearer, 1990*). The resolution to this may involve

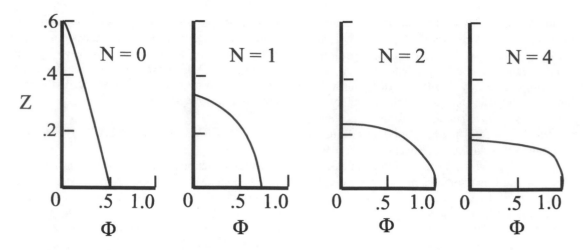

Figure 8. The solid fraction Φ versus non-dimensional height z for various values of the number density of channels N. As N increases, corresponding to increasing mass flux between the liquid and the mushy zone, the solid fraction increases more rapidly with depth into the mushy zone. After Worster (*1991*), where details of the numerical calculations can be found.

convection. Compositional convection driven by solidification of alloys typically occurs not just in the fluid, but rather, since buoyant fluid is released throughout the mushy zone, in the mushy zone as well (*Chen & Chen, 1991*). The buoyant, solute-enriched fluid rises out of the mushy zone in fast, narrow plumes known as chimneys in geophysics (*Roberts & Loper, 1983*), freckles in metallurgy (*Copley, et al., 1970*), and brine channels in oceanography (*Bennington, 1963; Wettlaufer, et al., 1997; Cole & Shapiro, 1998*). The return flow, which is depleted in the solute, and is therefore more readily solidified, is slow and broad. Such flows, and their relevance to convection in the outer core, are considered in more detail in Section 7.

Using a Taylor expansion in the solid fraction, Loper (*1983*) showed analytically that, except in the narrow and isolated chimneys, the solid fraction increases rapidly with depth when the broad return flow speed greatly exceeds the growth velocity of the solid. This is easily understood: convection enhances the solidification rate, but if the interface between the mushy zone and the fluid is advancing slowly, the solid fraction must become large even very close to the interface. Loper estimated that for parameter values reasonable for the core the solid fraction rises to order one within a depth of a few hundred meters, smaller than seismic body wavelengths. Hence, sharp reflections are possible off of a mushy inner core. He also pointed out that the expansion is not strictly valid when the expansion parameter, the solid fraction, becomes order one, so that one cannot necessarily treat the inner core as fully solid beneath a few hundred meters. More recent numerical calculations by Worster (*1991*) also exhibit this dependence of the solid fraction with depth

into the mushy zone as a function of the vigor of convection (Figure 8, see also Section 7). The mushy zone Rayleigh number is a measure of the vigor of convection between the mushy zone and the melt. Bergman and Fearn (*1994*) estimated it for the inner core, and found it to be highly supercritical, supporting Loper's results.

The meteoritic evidence, though also ambiguous because of the relative lack of data and the large extrapolation from a meteorite to a planetary core, lends support for the inner core solidifying dendritically. Iron meteorites such as those from the Cape York shower exhibit compositional gradients that are too large to result from the general fractionation of a planetoid core, but rather are more likely due to microsegregation between secondary and tertiary dendrite arms (*Esbensen & Buchwald, 1982; Haack & Scott, 1992*). Moreover, the iron sulphide (troilite) nodules are elongated and oriented, suggesting interdendritic pockets of melt during directional solidification (*Esbensen & Buchwald, 1982*).

Extrapolation (by some six orders of magnitude!) of laboratory data to planetoid core cooling rates gives a planetoid primary dendrite arm spacing of a few tens of meters (*Esbensen & Buchwald, 1982*). This is consistent with a typical spacing of ten centimeters between meteoritic troilite nodules, which are believed to represent pockets between tertiary dendrite arms. Columnar dendritic crystals that are typical of directionally solidified metallic alloys have the structure shown in Figure 9. Each crystal is composed of about ten parallel primary dendrites, suggesting a characteristic grain width d in a planetoid core of a few hundred meters. Primary dendrites grow in a particular crystallographic direction, so that dendrites within a given

Figure 9. A photograph of a directionally solidified zinc-rich tin alloy. Columnar crystals are composed of zinc-rich dendrites, which are visible on the ingot surface due to solidification shrinkage. Dendrites grow in a particular crystallographic direction, so that they are parallel within a single grain. A selection mechanism allows dendrites parallel to the direction of heat flow to grow most easily, so that dendrites are nearly parallel between grains. This is the origin of solidification texturing. The tin-rich or eutectic phase forms as crystallites between dendrites.

grain must be parallel. Moreover, due to their thermosolutal origin as discussed in the previous section, dendrites growing close to the direction of heat flow have a growth advantage. Hence, the direction of heat flow becomes a preferred crystallographic direction. This is the origin of solidification texturing, discussed more fully in the next section.

High attenuation in the inner core, especially in the body wave frequency range, could be due to interdendritic fluid. In addition, there have also been observations of an inner core attenuation anisotropy (*Creager, 1992; Souriau & Romanowicz, 1996; 1997*), with 10 - 30 km wavelength (λ) body waves that propagate in the direction parallel to the rotation axis exhibiting more complex waveforms and smaller amplitudes. One way to explain this attenuation anisotropy is by scattering off partially textured grains, an effect that has been observed in directionally solidified hcp zinc alloys (*Bergman, et al., 2000*). For instance, as discussed in the next section, if the a-axes of the inner core hcp iron crystals are aligned in directions perpendicular to the rotation axis, but the c-axes are randomly oriented in the planes transverse to these directions, then there will be more scattering due to impedance contrasts between grains for waves propagating parallel to the rotation axis.

Assuming that the seismic attenuation anisotropy (between a one-third and one-fifth difference in amplitude, *Creager, 1992; Souriau & Romanowicz, 1996*) is due to scattering, and assuming a maximum single crystal anisotropy of 7 % (see the next paragraph), Bergman (*1998*) estimated an inner core columnar grain width of a few hundred meters. This is in agreement with the crude laboratory extrapolation. Similarly, Cormier, *et al.* (*1998*), not assuming the inconsistent high frequency limit where $\lambda \ll d$, obtained a characteristic grain width of a couple of kilometers. This result is based on a maximum 10-12 % impedance contrast between crystals (there is a tradeoff between the maximum impedance contrast and the number of scatterers, which is inversely proportional to the grain width d). These calculations involved only the relative amplitudes of east-west versus north-south propagating waves, and do not bear directly on Q in the inner core. However, scatterers with a characteristic lengthscale of a few kilometers might help explain the discrepancy between Q as inferred from body waves versus that from normal modes. Vidale & Earle (*2000*) have also found evidence for scatterers in the inner core with a lengthscale of a couple of kilometers, with an impedance contrast of a couple percent.

A characteristic grain width of a kilometer, and much longer in the growth direction, is not surprising given the very slow cooling rate of the core. Indeed, iron meteorites with cross-sections of several square meters typically exhibit a Widmanstatten pattern with the same orientation across the entire meteorite, indicating solid state precipitation from a single crystal (*Buchwald, 1975*). Further, based on the similarity between the magnitude of the inner core elastic anisotropy, about 3-4 % (*Creager, 1992*), and the 0 K first principles calculations of the single crystal elastic anisotropy of hcp iron, about 4 %, Stixrude & Cohen (*1995*) suggested that the inner core might be a single crystal. However, in analog materials such as atmospheric pressure titanium (*Bergman, 1998*), and in first principles calculations of hcp iron at inner core temperatures (*Steinle-Neumann, et al., 2001; this volume*), the effect of temperature is to increase the single crystal anisotropy from about 4 % to over 10 %, so that the rationale for the inner core behaving as a single crystal might not exist. X-ray diffraction studies of polycrystalline hcp iron at 220 GPa and room temperature suggest a single crystal elastic anisotropy as large as 25% (*Mao, et al., 1998*). Nevertheless, if the inner core retains its solidification structure the grains are very likely much larger than metallurgists, or even geologists, are accustomed to.

On the other hand, if the inner core is undergoing significant deformation as it solidifies (like a giant slush puppy), the grain size could be smaller. Buffett (*1997*) argued that for the inner core to super-rotate at 1 deg/yr (*Song & Richards, 1996*) in spite of a gravitational torque with the mantle, the inner core viscosity must be less than 10^{16} Pa s. If the mechanism for plastic flow in the inner core is high temperature diffusion creep (*Poirier, 1985*), then a grain size of less than 5 mm may be necessary to achieve such a low viscosity (*Frost & Ashby, 1982*). Stresses σ on the order of 10^4 Pa have been suggested to arise from a variety of sources: inner core thermal convection (*Jeanloz & Wenk, 1988*), solid state flow due to misalignment between the gravitational equipotential and the thermodynamical equilibrium figure of the inner core (*Yoshida, et al., 1996*), longitudinal inner core flow driven by mantle inhomogeneities (*Buffett, 1997*), and radial inner core flow driven by Lorentz stresses (*Karato, 1999*). During dynamic recrystallization the grain size is given empirically by $d = bK (\sigma/G)^{-r}$, where b is the magnitude of the Burgers vector (approximately 10^{-10} m), G is the rigidity (2 x 10^{11} Pa, *Dziewonski & Anderson, 1981*), and K and r are

empirical constants approximately equal to 100 and 1.2 (*Poirier, 1985*). This yields a grain size of about one meter. More recently, Buffett and Bloxham (*2000*) suggested that inner core radial flow is inhibited by stable stratification, and Buffett and Wenk (*2001*) argued that longitudinal inner core flow driven by Lorentz stresses as small as 1 Pa could be responsible for inner core anisotropy. In this case, the recrystallized grain size is of the order of a few kilometers, comparable to the size suggested by solidification studies. If the grain size is larger than 5 mm, and if the inner core super-rotation rate is less than 1 deg/yr, then the inner core viscosity could be greater than 10^{16} Pa s.

6. SOLIDIFICATION TEXTURING AND THE SEISMIC ANISOTROPY OF THE INNER CORE

In the mid-1980s seismologists inferred from both body waves and normal modes that the inner core is elastically anisotropic (*Morelli, et al., 1986; Woodhouse, et al., 1986*), with the direction parallel to the rotation axis 3-4 % fast. This inference was confirmed (*Creager, 1992; Tromp, 1993*), and considerable effort has been spent to understand the detailed pattern of anisotropy (see *Song, 1997* for a review). There is evidence that the anisotropy is weak in the upper 100 km (*Shearer, 1994; Song & Helmberger, 1995*), and that it increases with depth (though the resolution of normal modes in the inner core is not good, and the AB - DF differential travel times used to study anisotropy deep in the inner core could be biased by the lower mantle). There is also evidence that the anisotropy is weaker in the eastern hemisphere (*Tanaka & Hamaguchi, 1997*). The level of lateral heterogeneity in the inner core is still uncertain, ranging from very little (*Creager, 1999*) to .8 % (*Niu & Wen, 2001*).

The fast axis of symmetry appears to be tilted with respect to the rotation axis (*Su & Dziewsonski, 1995*). This has allowed seismologists to infer a 1 deg/yr prograde super-rotation of the inner core over the past thirty years (*Song & Richards, 1996*). Some studies (*Su, et al., 1996*) reported a prograde super-rotation as fast as 3 deg/yr. However, these inferences have been doubted because of questions about our ability to reliably detect the tilt of the symmetry axis (*Souriau, et al., 1997*), and because source mislocations can result in an overestimate of the rotation rate (*Poupinet, et al., 2000*). A normal mode study finds a preferred super-rotation rate of 0.0 +/- 0.2 deg/yr (*Laske & Masters, 1999*). This is consistent with some body wave studies that find a prograde super-rotation rate in the range 0.1 - 0.3 deg/yr (*Creager, 1997; Vidale, et al., 2000*), so there is not yet consensus on whether a rotation of the inner core has been observed. Some (*Romanowicz & Breger, 1999*) have even questioned the inference of elastic anisotropy of the inner core. However, an alternate explanation for the normal mode splitting requires putting lateral structure in the fluid outer core, which is unlikely (*Piersanti, et al., 2001*). As discussed, there have also been suggestions for an inner core attenuation anisotropy, with the fast direction parallel to the rotation axis exhibiting more complex waveforms and smaller amplitudes (*Creager, 1992; Souriau & Romanowicz, 1996; 1997*).

Many hypotheses have been put forth to explain the elastic anisotropy, all involving texturing of the hcp iron crystals that comprise the inner core. However, all proposed texturing mechanisms suffer some shortcoming in being able to fully explain the seismic inferences (*Buffett, 2000*). The hypotheses fall broadly into two categories: those involving solidification, and those involving post-solidification deformation and/or recrystallization. Those that involve solidification include texturing due to anisotropic paramagnetic susceptibility (*Karato, 1993*), the inner core being a single crystal (*Stixrude & Cohen, 1995*), and texturing due to directional solidification (*Bergman, 1997*). Those that involve post-solidification modification include inner core thermal convection (*Jeanloz & Wenk, 1988*), solid state flow due to misalignment between the gravitational equipotential and the thermodynamical equilibrium figure of the inner core (*Yoshida, et al., 1996*), radial flow due to Lorentz stresses (*Karato, 1999*), and longitudinal flow due to Lorentz stresses (*Buffett & Wenk, 2001*).

As no single hypothesis has yet been generally accepted, and given the uncertainties in the seismic inferences, it seems sensible to explore further each of the hypotheses, perhaps in conjunction with others. Here I focus on the possible role of solidification. Yoshida, *et al.* (*1996*) suggested that because convection in a rapidly rotating fluid such as the outer core is more efficient at transporting heat in the direction perpendicular to the rotation axis (*Roberts, 1968; Busse, 1970; Cardin & Olson, 1992*), the inner core will tend to solidify more oblately than the gravitational equipotential. This results in a solid state flow (Figure 10), and the stress can lead to a recrystallization texture. However, as the driving force for recrystallization is quite small, the timescale for the texture development may be as long or longer than the age of the Earth. On the other hand, a depth dependence can result as deeper parts of the inner core have had more time to recrystallize. Bergman (*1999*) showed experimentally that fluid convection can result in ellipsoidal solidification of salt water at the center of a rapidly rotating hemispherical shell, though the timescale of the experiments did not allow for examination of subsequent solid state flow.

However, the analysis by Yoshida, *et al.* (*1996*) assumed an inner core that is isotropic before a recrystallization texture develops. Bergman (*1997*) suggested that prior to deformation, a solidification texture is frozen-in. Because convection in a rapidly rotating fluid favors a cylindrically radial transport of heat, it was suggested that dendritic growth

might tend to have cylindrical rather than spherical symmetry. Since dendrites grow in a particular crystallographic direction (<112'0> for hcp alloys), a solidification texture with cylindrical symmetry arises (Figure 11). However, the texturing is incomplete: in the absence of fluid flow during solidification the c-axes (<0001>) are oriented randomly in the plane transverse to the local cylindrically radial growth direction. Surprisingly, this geometry can lead to a depth dependence of the elastic anisotropy, when integrated along a seismic ray path. For instance, for rays that turn on the equatorial plane the anisotropy increases with the $\sin(\pi r/2R)$, where r is the ray turning depth and R is the inner core radius.

The experiments on salt water solidifying at the center of a rapidly rotating hemispherical shell (*Bergman, 1999*) support the hypothesis that the inner core is composed of columnar crystals with a tendency towards cylindrical symmetry. If solidification texturing is at least in part responsible for the inner core elastic anisotropy, there must either be little deformation in the inner core, or that deformation must not destroy the pre-existing fabric. Clearly, with cylindrical growth symmetry there must be at least some flow, in order to redistribute solid from the equator to the poles, perhaps as suggested by Yoshida, *et al.* (*1996*). Since the driving force for recrystallization in this model is small, the solidification texture may not be destroyed, but further

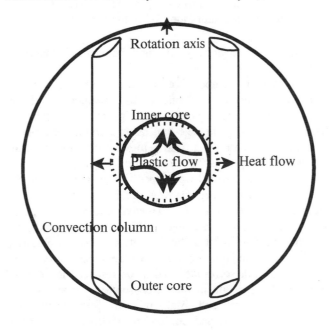

Figure 10. Convection in a rapidly rotating, spherical fluid shell such as the outer core is more efficient at transporting heat perpendicular to the rotation axis. This leads to an inner core that solidifies more oblately than the gravitational equipotential. A solid-state flow results, and the stress may lead to a recrystallization texture. After Yoshida, *et al.* (*1996*).

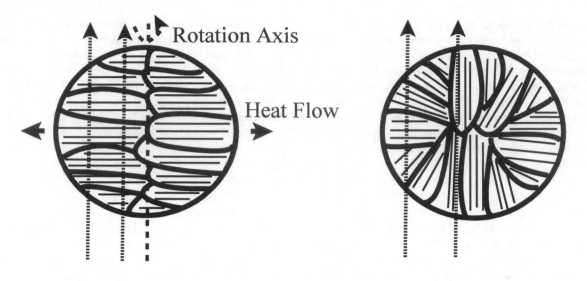

Figure 11. Heat flow perpendicular to the rotation axis leads to dendritic growth in the cylindrically radial direction. The heavier lines represent columnar crystals, the lighter lines primary dendrites. The left panel represents a longitudinal cross section, the right panel an equatorial cross section. North-south seismic rays (left panel), represented by the dotted arrows, are always perpendicular to the growth direction of dendrites. The component of rays perpendicular to the rotation axis (right panel) that is parallel to the growth direction of dendrites increases with turning depth in the inner core. This is the origin of the depth dependence associated with solidification texturing of the inner core. (Such geometric depth dependence becomes less strong for rays not turning on the equatorial plane.) After Bergman (*1997*).

experiments are needed to understand the slow deformation of actively solidifying systems.

Both the low temperature elastic constants of hcp iron (*Stixrude & Cohen, 1995*) and the high temperature elastic constants of metals analogous to (high pressure) hcp iron (*Bergman, 1998*) yield the c-axis faster than the basal plane. This results from the c/a ratio being less than the ideal for closest packed spheres. The experiments of Mao, *et al.* (*1998*) show the compressional wave speed to be a maximum for propagation near 45 degrees from the c-axis. Most recently, Steinle-Neumann, *et al.* (*2001*) have computed the high temperature elastic constants of hcp iron, finding that the c/a ratio becomes greater than the ideal at elevated temperatures, so that the c-axis is slower than the basal plane. The differences in these results need to be resolved in order to understand the origin of the inner core elastic anisotropy. More of the difficult high pressure experiments on polycrystalline iron, and perhaps iron alloys, and further testing of the finite temperature numerical methods on hcp metals that have been well-studied experimentally, should give progress. It is interesting to note that if Morse (*2001*) is correct about the inner core growing via adcumulus growth rather than dendritically, then a texture with c-axes in the cylindrically radial growth direction would result (*Bergman,*

et al., 2000). Such a texture is qualitatively similar to the texture resulting from flow due to longitudinal Maxwell stresses (*Buffett & Wenk, 2001*).

As with all proposed texturing mechanisms, it is difficult to understand how solidification can result in hemispherical variations in the anisotropy. One possibility is long term mantle control over the fluid dynamics of the outer core (*Bloxham & Gubbins, 1987; Sumita & Olson, 1999*), which can influence the solidification of the inner core. However, this explanation requires that the inner core is locked to the mantle (*Buffett, 1996*), implying that seismologists are inferring an inner core oscillation rather than a rotation. Future observations should resolve this issue. However, since the longitudinal variations appear to extend at least 500 km deep into the inner core (*Tanaka & Hamaguchi, 1997*), presumably representing at least 500 hundred million years during which the mantle has been convecting, it is still difficult to understand the origin of hemispherical variations in the anisotropy.

It was recently suggested (*Song & Helmberger, 1998*) that the nearly isotropic zone near the top of the inner core, some 100 km in the western hemisphere and thicker in parts of the eastern hemisphere, might represent a distinct layer with a variable thickness. However, the physical origin of a distinct, variable thickness layer in the inner core is unclear. The evidence for this layer comes from broadening of long period waveforms and reflections at short periods of north-south PKIKP waves. A different interpretation for these observations is that they represent scattering off particularly misoriented grains. As discussed in the previous section, such scattering has been suggested as the cause for the inner core attenuation anisotropy (*Bergman, 1998; Cormier, et al., 1998*). If the inner core texture is such that grains are well aligned in the cylindrically radial direction, but the c-axes are randomly oriented in the transverse planes (as might be the case for solidification texturing (*Bergman, 1997*), Figure 11), then an anisotropic apparent attenuation results.

Two other recent studies may bear on the role of solidification and inner core anisotropy. Bergman, *et al.* (*2001*) found that solidification-driven convective flow in salt water causes a texture in which the c-axes do not lie randomly in the plane transverse to growth, but rather, lie at 90 degrees to the flow direction. This is in contrast to previous studies on solidifying salt water, which show that an externally forced flow such as an ocean current (as opposed to a solidification-driven convective flow) causes the c-axes to lie in the direction of the flow (*Weeks & Gow, 1978*). These flow effects are thought to arise from the platelet nature of hcp dendrites, common to sea ice and possibly iron under inner core conditions. They indicate that the texture that results from solidification may be more complicated than that due only to directional solidification (*Bergman, 1997*). Whether the subtleties of solidification texturing are reflected in the texture of the inner core is not yet clear, but if they are, then the inner core has been recording the flow in the outer core in the vicinity of the inner core. Finally, it has been suggested that aligned fluid pockets, such as those between dendrites, can, depending on their volume fraction and shape, strengthen or weaken a crystalline preferred orientation already present (*Singh, et al., 2000*). However, due to solid state flow, *i.e.*, dendritic compaction (*Sumita, et al., 1996*), it is not clear whether the 3-10 % liquid fraction they claim is needed to simultaneously explain the compressional wave anisotropy, the low shear wave velocity, and the high body wave attenuation is reasonable, except in the very upper part of the inner core.

7. COMPOSITIONAL CONVECTION IN THE OUTER CORE DRIVEN BY SOLIDIFICATION

Solidification affects convection in the outer core as well as the structure in the inner core. Although the possible thermodynamical importance of compositional convection in the outer core was recognized by Braginsky (*1963*), it has often been assumed that the style of compositional convection will be similar to that of thermal convection. For instance, numerical calculations (*Glatzmaier & Roberts, 1995; Kuang & Bloxham, 1997*) typically assume a uniform flux of thermal buoyancy at the inner-outer core boundary, which is intended to also simulate a flux of compositional buoyancy. Cardin & Olson (*1992*) have shown experimentally that the primary difference between compositional and thermal convection when a uniform flux along the inner sphere is prescribed is that the features of compositional convection are smaller scale, owing to the material diffusivity being smaller than the thermal diffusivity. When both sources of buoyancy are present, the convection resembles more nearly thermal convection.

However, there is considerable experimental evidence that compositional buoyancy resulting from solidification in a mushy zone is not introduced as a uniform flux along the freezing interface. Rather, as discussed briefly in Section 5, the buoyancy flux out of the mushy zone occurs via fast, narrow channel flow, with a slower and broader return flow. Copley, *et al.* (*1970*) first recognized that defects in nickel-base castings, known as freckles, are a result of convection through the mushy zone during solidification. They demonstrated this using a transparent analog, an ammonium chloride-rich aqueous solution. A gas turbine blade, operating at high temperatures, is stronger when it is a single crystal, because at high temperatures the primary creep mechanism relies on diffusion from grain boundaries (*Flemings, 1974*). During the slow, directional solidification required to produce a single crystal alloy blade, freckles can form, which render the blade weak. Channels have since been observed in a wide range of metallic and non-metallic alloy systems, and progress has been made in understanding why they have not been observed in certain systems (*Worster & Kerr, 1994*). The literature on channel flow is immense, in part because of the practical metallurgical interest, in part because of the oceanographic and geophysical interest, and in part because channel flow is an interesting fluid dynamical phenomenon. I will concentrate on why channels form, and their possible relevance to the core.

In essence, channel flow is a result of a non-linear process with positive feedback (Figure 12, *Tait, et al., 1992*). To fix ideas, consider upwards solidification, with a less dense solute, as is common in laboratory experiments. Fluid rejected upon solidification in the mushy zone is cold, but enriched in the solute. The combined thermal and solutal expansivity may be such that the rejected fluid rises through the mushy zone. (In the core, the fluid rejected upon solidification is both thermally and compositionally buoyant, but this does not affect the essential mechanism by which channels form.) Because heat diffuses more rapidly than does solute, the rising fluid warms up to its surroundings but remains enriched in the solute. To remain in local equilibrium it thus melts dendrites and dendrite sidebranches as it rises (Figure 13). Because the

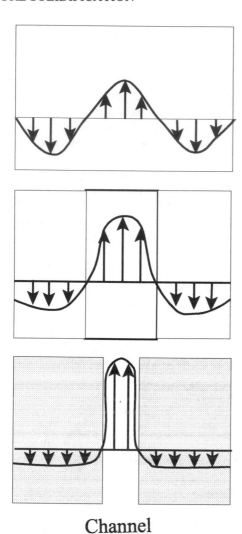

Channel

Figure 12. Consider upwards solidification, with a less dense solute. At linear stability for convective motion in a mushy zone, the solid fraction is uniform and there is symmetry between flow into and out of the mushy zone (top panel). Because heat diffuses more rapidly than does solute, cold, solute enriched fluid rising out of the mushy zone heats up faster than solute can diffuse away. The rising fluid thus melts dendrites, leading to a lower solid fraction in its path (middle panel). This in turn decreases the Darcy frictional force, leading to a larger flow speed away from the solid, and a correspondingly slower flow speed towards the solid. This is the non-linear process thought to focus the flow in solid-free channels (bottom panel). The downwards flow leads to further coating of dendrites with solid. After Tait, *et al.* (*1992*).

viscous drag in the mushy zone, which can be treated as a porous medium, can be parameterized as Darcy friction, which depends on the permeability, the rising fluid encounters less of a retarding force as it melts dendrites. It thus rises quickly, re-enforcing the original flow. In the laboratory, a typical channel encompasses a few dendrites across. By continuity, the downwards return flow of solute-depleted fluid occurs over a broad area, and more slowly. This solute-depleted fluid coats the dendrites, leading to a higher solid fraction with more vigorous convection, as Loper (*1983*) realized.

Channel convection has been studied extensively both experimentally and theoretically. See Hellawell, *et al.* (*1993*) and Worster (*1997*) for detailed reviews. Analytical work has concentrated on the limits of well-developed channel convection (*Roberts & Loper, 1983; Worster, 1991*), and linear stability (*Fowler, 1985; Worster, 1992*). The linear stability studies have uncovered a critical value for mushy zone Rayleigh number, $Ra_m = \beta|\Delta C|g\Pi H/\nu\kappa$, for the onset of channel convection. Here β is the combined thermosolutal expansivity, $|\Delta C|$ is the difference in concentration between the initial melt and the eutectic, g is gravity, Π is the characteristic permeability of the mushy zone, H is the mushy zone thickness, ν is the kinematic viscosity of the fluid, and κ is the thermal diffusivity. Experiments (*Chen & Chen, 1991; Bergman, et al., 1997; Beckermann, et al., 2000*) have confirmed the existence of a critical mushy zone Rayleigh number close to the theoretical value, and the usefulness of Ra_m as a predictor of the vigor of channel convection.

The mushy zone Rayleigh number is very much like the usual fluid Rayleigh number, with the permeability replacing the square of the lengthscale associated with the usual viscous force. This follows from the viscous force being parameterized by the Darcy frictional force, $\nu v/\Pi$, where v is the flowspeed. Worster (*1992*) showed two possible modes of instability: one with a larger lengthscale comparable to the depth of the mushy zone and associated with convection in the mushy zone, and a second with a smaller lengthscale comparable to the thickness of the solutal boundary layer adjacent to the mushy zone and associated with convection originating in the solutal boundary layer. Using a weakly non-linear analysis, Amberg & Homsy (*1993*) showed that the first mode exhibits a sub-critical bifurcation that leads to variations in the liquid fraction in the mushy zone, and hence is clearly associated with channel convection. The second mode, sometimes known as a salt-finger instability, has also been observed (*Chen & Chen, 1991*), and is often the most unstable until the mushy zone has reached sufficient depth.

Bergman & Fearn (*1994*) estimated Ra_m for channel convection in the mushy zone near the top of the inner core, finding it to be at least one thousand times supercritical. In essence, the high supercriticality at the inner-outer core boundary results from the large interdendritic spacing (tens of centimeters for tertiary arms), and hence the large permeability, as compared with laboratory experiments. The large Ra_m supports Loper's (*1983*) prediction that when the rate of mass interchange between the melt and mushy zone far exceeds the growth rate of the solid, the solid fraction increases rapidly with depth into the mushy zone.

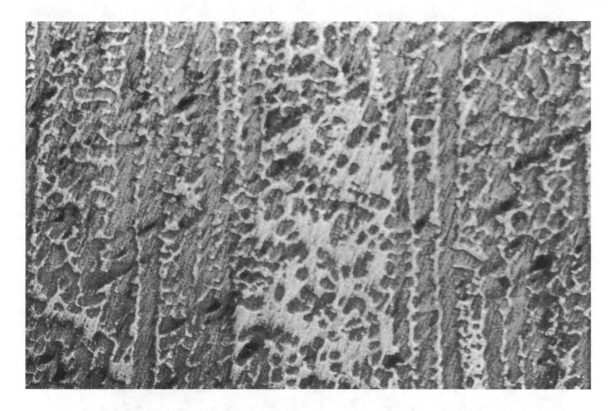

Figure 13. A micrograph of a longitudinal cut of a channel in a lead-rich tin alloy. The dendrites are lead-rich; the channel, having solidified last, is a mixture of lead-rich and tin-rich phases. Sidebranches that have melted off primary dendrites are also visible in the channel (due to melting point suppression in the solute boundary layer the sidebranches tend to narrow where they connect to the primary dendrite, and so can melt off without completely melting). Width of micrograph is 5 mm.

The likelihood of channel convection in the core has led to investigations of fluid convection driven by narrow plumes or blobs emanating from the boundary (*Loper & Moffatt, 1993; Moffatt & Loper, 1994; Shimizu & Loper, 1997; Classen, et al., 1999*), in an effort to see how such convection might differ from that driven by a uniform buoyancy flux. Because of the small lengthscale (as compared with the radius of the core) associated with interdendritic spacing, the Coriolis force is unimportant within the mushy zone, even in the very slowly solidifying inner core (*Bergman & Fearn, 1994*). However, in the fluid outer core the Coriolis force is important, and Classen, *et al.* (*1999*) found experimentally that plumes of buoyant fluid quickly become unstable and break into blobs (Figure 14). Numerical investigations including both the Coriolis and Lorentz forces (*St Pierre, 1996*) have shown that these blobs are also unstable. If plumes do not maintain their integrity, then the style of convection in the outer core driven by chimneys from the inner core is in fact not likely to be very different from that driven by a uniform buoyancy flux.

Moreover, Bergman & Fearn (*1994*) have argued that the Lorentz force in the mushy zone in the upper part of the inner core is larger than the Darcy frictional force. Since the Lorentz force is independent of the permeability, the positive feedback

that leads to channels forming may no longer be important. Laboratory experiments on lead-tin alloys (*Bergman, et al., 1999*) confirmed that a magnetic field can prevent channels from forming when the primary resistance to flow is the Lorentz force rather than Darcy frictional force. The experiments also showed that the longitudinal macrosegregation, which is a measure of the vigor of convection during solidification, is unchanged by the absence of channels. This suggests that fluid can be exchanged between the mushy zone and the melt even in the absence of channels. The observed rapid increase in the solid fraction with depth into the inner core is thus still possible.

These results make channel convection near the Earth's inner-outer core boundary less likely, though the phenomenon remains a fascinating fluid dynamical problem with unquestioned metallurgical importance. If channels do not even form, then irrespective of plume behavior, the compositional buoyancy flux from the inner-outer core boundary is likely to be uniform over any regional patch of the boundary, though there could be latitudinal variations on a global scale (*Yoshida, et al., 1996; Bergman, 1999; Karato, 1999*). Finally, it is important to note that the presence (or absence!) of isolated, localized channels does not affect the

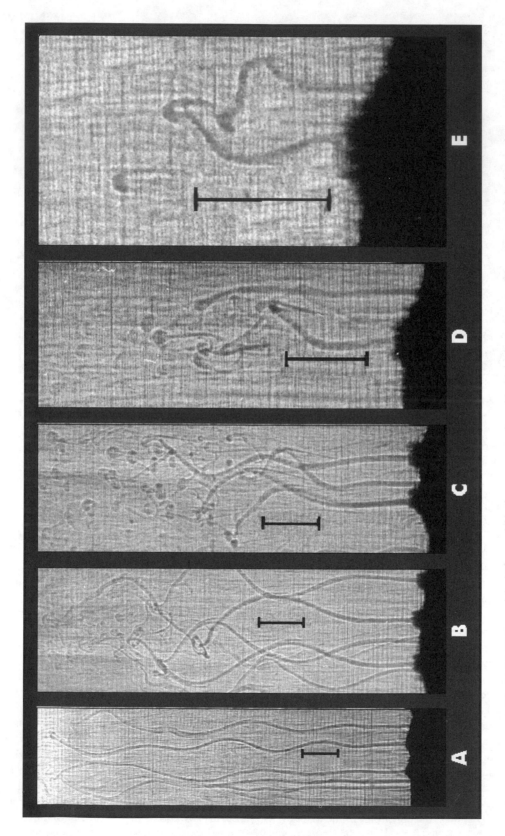

Figure 14. Water-enriched plumes rising from an ammonium chloride-rich mushy zone. The sequence of shadowgraphs shows the effects of decreasing Ekman number (faster rotation): the plumes become unstable and break into blobs. The scale in all figures is 10 mm. *Courtesy of Classen, et al. (1999).*

overall solidification texture, which is determined by the dendritic growth.

8. CONCLUSIONS

It is reasonably certain that the Earth is cooling, and that as it does so, the inner core is solidifying from the outer core. It is also reasonably certain that the inner core is iron-enriched. There is considerably less consensus on the exact makeup and partitioning of the less dense alloying components, and the phase diagram of the core. In spite of our not knowing the less dense components, nor even whether the inner core is a solid solution or forms a eutectic, mere knowledge that the core is a directionally solidifying alloy has led to the prediction that the inner core is growing dendritically. Such growth is common in metallic alloys, as opposed to silicates, because solid-fluid interfaces in metals tend to be atomically diffuse.

Because of the large mushy zone Rayleigh number associated with solidification of the inner core, the sharp interface between the fluid outer core and solid inner core on the seismic body wave lengthscale is not inconsistent with a mushy inner core. Moreover, although there are discrepancies between different seismic datasets and between different mineral physics experiments and numerical calculations on the anelasticity of high pressure iron, possible high attenuation of short period body waves near the top of the inner core might be explained by an inner core mushy zone.

Comparison between laboratory solidification experiments and iron meteorites suggest iron meteorites have retained their solidification structure. Extrapolated laboratory data and meteorites have then provided an estimate for the inner core primary dendrite spacing of some tens of meters, and for a typical grain size of about a kilometer. This grain size is consistent with estimates made by assuming that the inner core seismic attenuation anisotropy (not to be confused with the directionally averaged attenuation discussed in the preceeding paragraph) is due to scattering off grain boundaries of a partially textured inner core. Such a large grain size may also be possible if the inner core is undergoing recrystallization at a stress level of 1 Pa, which has been suggested to result from longitudinal Lorentz stresses, though larger stresses from other causes, leading to a smaller grain size, have also been suggested.

Due to the dominant effects of rotation on convection in the fluid outer core, convective heat and solute transport in the outer core tends to be in the cylindrically radial direction. This may lead to cylindrical, columnar dendritic growth in the inner core. Such growth leads to a solidification texture with cylindrical symmetry. (Even in the absence of dendritic growth, a solidification texture occurs. Although the texture is different, it also has a tendency towards cylindrical symmetry to the extent that transport is in the cylindrically radial direction.) For geometrical reasons, a solidification texture can actually give an elastic anisotropy that increases with depth. However, it is not clear whether the particular observed depth dependence or the longitudinal variations of the seismic anisotropy can be satisfactorily explained by a simple solidification texturing model. Moreover, imperfect knowledge about the stable phase and elastic properties of iron under inner core conditions, and incomplete seismic data, make precise predictions and comparisons difficult and untrustworthy. In addition, since it is likely that the inner core is deforming as it solidifies, studies of deformation during solidification are necessary to assess properly the role of solidification in causing the seismic observations.

Channel convection in the dendritic mushy zone is an interesting fluid dynamical phenomenon with metallurgical importance. However, because of the effect of the Lorentz force, it may not occur at the inner-outer core boundary. Moreover, even if it did, the resulting plume would quickly become unstable and not maintain a sharp density contrast with the surrounding fluid. Thus, compositional convection in the outer core is likely to be driven by a buoyancy flux that is uniform on a regional scale. The lack of channel convection does not, however, modify the general dendritic structure of the mushy zone, nor the resulting elastic anisotropy.

With advances in quantum mechanical calculations and high pressure mineral physics we will better understand the material properties of iron alloys under core conditions. Advances in seismic data processing will give us a better map of the elastic and attenuation properties of the inner core. Laboratory experiments may also continue to uncover new phenomena relevant to the solidification of the core, such as the role played by melt flow during crystallization. Because of the differences in lengthscale and timescale between the Earth and laboratory experiments on solidification and deformation, it will continue to be necessary to make large extrapolations from the laboratory to inner core conditions. In spite of such difficulties a better understanding of the solidification of the Earth's core may be necessary to help us to better understand the unusual properties of this most remote part of our planet.

Acknowledgements. Jeremy Bloxham, Dave Cole, Dan DuVall, David Fearn, Louis Giersch, Michael Hinczewski, Valerie Izzo, Jackson Jones, Shun-Ichiro Karato, Eric Kramer, Dan Nielsen, Peter Olson, Nancy Perron, Peggy Shannon, Frans Spaepen, and Stephen Zatman have contributed in various ways to this work. The continued support of the Research Corporation and the NSF is gratefully acknowledged.

REFERENCES

Alfe, D., Price, G.D., & Gillan, M.J., Oxygen in the Earth's core: a first-principles study, *Phys. Earth Planet. Int., 110*, 191-210, 1999.

Alfe, D., Gillan, M.J., & Price, G.D., Constraints on the composition of the Earth's core from *ab initio* calculations, *Nature, 405*, 172-175, 2000.

Amberg, G. & Homsy, G.M., Nonlinear analysis of buoyant convection in binary solidification with application to channel formation, *J. Fluid Mech., 252*, 79-98, 1993.

Anderson, W.W. & Ahrens, T., Shock temperatures and melting in iron sulfides at core pressures, *J. Geophys. Res., 101*, 5627-5642, 1996.

Beckermann, C., Gu, J.P., & Boettinger, W.J., Development of a freckle predictor via Rayleigh number method for single crystal nickel-base superalloy castings, *Metall. Trans.,31A*, 2545-2557, 2000.

Bennington, K.O., Some crystal growth features of sea ice, *J. Glaciol., 4*, 669-688, 1963.

Bergman, M.I., Measurements of elastic anisotropy due to solidification texturing and the implications for the Earth's inner core, *Nature, 389*, 60-63, 1997.

Bergman, M.I., Estimates of the Earth's inner core grain size, *Geophys. Res. Lett., 25*, 1593-1596, 1998.

Bergman, M.I., Experimental studies on the solidification of the Earth's inner core, *EOS, Trans. AGU, F18*, 1999.

Bergman, M.I. & Fearn, D.R., Chimneys on the Earth's inner-outer core boundary?, *Geophys. Res. Lett., 21*, 477-480, 1994.

Bergman, M.I., Fearn, D.R., Bloxham, J., & Shannon, M., Convection and channel formation in solidifying Pb-Sn alloys, *Metall. Trans., 28A*, 859-866, 1997.

Bergman, M.I., Fearn, D.R., & Bloxham, J., Suppression of channel convection in solidifying Pb-Sn alloys via an applied magnetic field, *Metall. Trans., 30A*, 1809-1815, 1999.

Bergman, M.I., Giersch, L., Hinczewski, M., & Izzo, V., Elastic and attenuation anisotropy in directionally solidified (hcp) zinc, and the seismic anisotropy in the Earth's inner core, *Phys. Earth Planet. Int., 117*, 139-151, 2000.

Bergman, M.I., Cole, D.M., & Jackson, J.R., Preferred crystal orientations due to melt convection during solidification, in press, *J. Geophys. Res.*, 2002.

Bhattacharyya, J., Shearer, P., & Masters, G., Inner core attenuation from short-period PKP(BC) versus PKP(DF) waveforms, *Geophys. J. Int., 114*, 1-11, 1993.

Birch, F., The alpha-gamma transformation of iron at high pressures, and the problem of the Earth's magnetism, *Amer. J. Sci., 238*, 192-211, 1940.

Birch, F., Elasticity and constitution of the Earth's interior, *J.Geophys. Res., 57*, 227-286, 1952.

Bloxham, J. & Gubbins, D., Thermal core-mantle interactions, *Nature, 325*, 511-513, 1987.

Boehler, R., Melting of the Fe-FeO and Fe-FeS systems at high pressures: constraints on core temperatures, *Earth Planet. Sci. Let., 111*, 217-227, 1992.

Boehler, R., Fe-FeS eutectic temperatures to 620 kbar, *Phys. Earth Planet. Int., 96*, 181-186, 1996a.

Boehler, R., Melting temperature of the Earth's mantle and core: Earth's thermal structure, *Annu. Rev. Earth Planet. Sci., 24*, 15-40, 1996b.

Bolt, B.A., The constitution of the core: seismological evidence, *Phil. Trans. R. Soc. Lond. A306*, 11-20, 1982.

Braginsky, S.I., Structure of the F-layer and reasons for convection in the Earth's core, *Doklady Akad. Nauk SSSR, 149*, 8-10, 1963.

Bridgman, P.W., Linear compression to 30,000 kg/cm^2, including relatively incompressible substances, *Proc. Amer. Acad. Arts Sci., 77*, 187-234, 1949.

Brown, H. & Patterson, C., The composition of meteoritic matter. III-Phase equilibria, genetic relationships, and planet structures, *J. Geol., 56*, 85-111, 1948.

Bruhn, D., Groebner, N., & Kohlstedt, D.L., An interconnected network of core-forming melts produced by shear deformation, *Nature, 403*, 883-886, 2000.

Buchwald, V.F., *Handbook of Iron Meteorites, Volume 1*, University of California Press, Berkeley, 1975.

Buffett, B.A., Gravitational oscillations in the length of day, *Geophys. Res. Lett., 23*, 2279-2282, 1996.

Buffett, B.A., Geodynamic estimates of the viscosity of the Earth's inner core, *Nature, 388*, 571-573, 1997.

Buffett, B.A., Dynamics of the Earth's core, in *Earth's Deep Interior: Mineral Physics and Tomography from the Atomic to the Global Scale*, AGU, Washington, DC, 2000.

Buffett, B.A. & Bloxham, J., Formation of Earth's inner core by electromagnetic forces, *Geophys. Res. Lett., 27*, 4001-4004, 2000.

Buffett, B.A., Huppert, H.E., Lister, J.R., & Woods, A.W., Analytical model for solidification of the Earth's core, *Nature, 356*, 329-331, 1992.

Buffett, B.A. & Wenk, H.-R., Texturing of the Earth's inner core by Maxwell stresses, *Nature, 413*, 60-63, 2001.

Bullen, K.E., A hypothesis on compressibility at pressures of the order of a million atmospheres, *Nature, 157*, 405, 1946.

Bullen, K.E., Compressibility-pressure hypothesis and the Earth's interior, *Mon. Not. R. Astr. Soc., Geophys. Sup., 5*, 355-368, 1949.

Busse, F.H., Thermal instabilities in rapidly rotating systems, *J. Fluid Mech., 44*, 441-460, 1970.

Cardin, P. & Olson, P., An experimental approach to thermochemical convection in the Earth's core, *Geophys. Res. Lett., 19*, 1995-1998, 1992.

Chalmers, B., *Principles of Solidification*, Wiley, New York, 1964.

Chen, C.F., Experimental study of convection in a mushy layer during directional solidification, *J. Fluid Mech., 293*, 81-98, 1995.

Chen, C.F. & Chen, F., Experimental study of directional solidification of aqueous ammonium chloride solution, *J. Fluid Mech., 227*, 567-586, 1991.

Choy, G.L. & Cormier, V.F., The structure of the inner core inferred from short period and broadband GDSN data, *Geophys. J. R. Astr. Soc., 72*, 1-21, 1983.

Classen, S., Heimpel, M., & Christensen, U., Blob instability in rotating compositional convection, *Geophys. Res. Lett., 26*, 135-138, 1999.

Cole, D.M. & Shapiro, L.H., Observations of brine drainage networks and microstructure of first-year sea ice, *J. Geophys. Res., 103*, 21,739-21,750, 1998.

Copley, S.M., Giamei, A.F., Johnson, S.M., & Hornbecker, M.F., The origin of freckles in unidirectionally solidified castings, *Metall. Trans., 1*, 2193-2204, 1970.

Cormier, V.F., Short period PKP phases and the anelasticity mechanism of the inner core, *Phys. Earth Planet. Int., 24*, 291-301, 1981.

Cormier, V.F., Xu, L., & Choy, G.L., Seismic attenuation of the inner core: viscoelastic or stratigraphic?, *Geophys.Res.Lett., 25*, 4019-4022, 1998.

Creager, K.C., Anisotropy of the inner core from differential travel times of the phases PKP and PKIKP, *Nature, 356*, 309-314, 1992.

Creager, K.C., Inner core rotation rate from small-scale heterogeneity and time-varying travel times, *Science, 278*, 1284-1288, 1997.

Creager, K.C., Large-scale variations in inner core anisotropy, *J. Geophys. Res., 104*, 23,127-23,139, 1999.

Deuss, A., Woodhouse, J., Paulssen, H., & Trampert, J., Observations of inner core shear waves, *EOS, Trans. AGU, F79*, 1998.

Doornbos, D.J., The anelasticity of the inner core, *Geophys. J. R. Astr. Soc., 38*, 397-415, 1974.

Doornbos, D.J., Observable effects of the seismic absorption band in the Earth, *Geophys. J. R. Astr. Soc., 75*, 693-711, 1983.

Dziewonski, A.M. & Anderson, D.L., Preliminary reference Earth model, *Phys. Earth Planet. Int., 25*, 297-356, 1981.

Esbensen, K.H. & Buchwald, V.F., Planet(oid) core crystallization and fractionation-evidence from the Agpalilik mass of the Cape York iron meteorite shower, *Phys. Earth Planet. Int., 29*, 218-232, 1982.

Fearn, D.R. & Loper, D.E., The evolution of an iron-poor core I. Constraints on the growth of the inner core, in *Stellar and Planetary Magnetism*, ed. Soward, A.M., 351-370, Gordon & Breach, London, 1983.

Fearn, D.R., Loper, D.E., & Roberts, P.H., Structure of the Earth's inner core, *Nature, 292*, 232-233, 1981.

Flemings, M.C., *Solidification Processing*, McGraw-Hill, New York, 1974.

Fowler, A.C., The formation of freckles in binary alloys, *IMA J. Appl. Math., 35*, 159-174, 1985.

Frost, H.J. & Ashby, A.F., *Deformation Mechanism Maps*, Pergamon Press, Oxford, 1982.

Fukai, Y., The iron-water reaction and the evolution of the Earth, *Nature, 308*, 174-175, 1984.

Glatzmaier, G.A. & Roberts, P.H., A three-dimensional self-consistent computer simulation of a geomagnetic field reversal, *Nature, 377*, 203-209, 1995.

Gubbins, D., Masters, T.G., & Jacobs, J.A., Thermal evolution of the Earth's core, *Geophys. J. R. Astr. Soc., 59*, 57-99, 1979.

Haack, H. & Scott, E.R.D., Asteroid core crystallization by inward dendritic growth, *J. Geophys. Res., 97*, 14,727-14,734, 1992.

Haddon, R.A.W. & Cleary, J.R., Evidence for scattering of PKP waves near the core-mantle boundary, *Phys. Earth Planet. Int., 8*, 211-234, 1974.

Hellawell, A., Sarazin, J.R., & Steube, R.S., Channel convection in partly solidified systems, *Phil. Trans. R. Soc. Lond. A345*, 507-544, 1993.

Hollerbach, R. & Jones, C.A., Influence of the Earth's inner core on geomagnetic fluctuations and reversals, *Nature, 365*, 541-543, 1993.

Jackson, I., Fitzgerald, J.D., & Kokkonen, H., High temperature viscoelastic relaxation in iron and its implication for the shear modulus and attenuation of the Earth's inner core, *J. Geophys. Res., 105*, 23605-23634, 2000.

Jackson, K.A., Mechanism of growth, in *Liquid Metals and Solidification*, ed. Maddin, R., *et al.*, 174-186, American Society for Metals, Cleveland, 1958.

Jeanloz, R., The nature of the Earth's core, *Annu. Rev. Earth Planet. Sci., 18*, 357-386, 1990.

Jeanloz, R. & Wenk, H.R., Convection and anisotropy of the inner core, *Geophys. Res.Lett., 15*, 72-75, 1988.

Jeffreys, H. & Bullen, K.E., Times of transmission of earthquake waves, *Bur. Centr. Seism. Internat. A, Fasc. 11*, 1935.

Jephcoat, A. & Olson, P., Is the inner core of the Earth pure iron?, *Nature, 325*, 332-335, 1987.

Julian, B.R., Davies, D., & Shepard, R.M., PKJKP, *Nature, 235*, 317-318, 1972.

Karato, S.I., Inner core anisotropy due to the magnetic field-induced preferred orientation of iron, *Science, 262*, 1708-1711, 1993.

Karato, S.I., Seismic anisotropy of the Earth's inner core resulting from flow induced by Maxwell stresses, *Nature, 402*, 871-873, 1999.

Karato, S.I. & Murthy, V.R., Core formation and chemical equilibrium in the Earth – 1. Physical considerations, *Phys. Earth Planet. Int., 100*, 61-79, 1997.

Kuang, W.J. & Bloxham, J., An earth-like numerical dynamo model, *Nature, 389*, 371-374, 1997.

Kurz, W. & Fisher, D.J., *Fundamentals of Solidification*, Trans-Tech, Switzerland, 1992.

Laio, A., Bernard, S., Chiarotti, G.L., Scandolo, S., & Tosatti, E., Physics of iron at Earth's core conditions, *Science, 287*, 1027-1030, 2000.

Langer, J., Instabilities and pattern formation in crystal growth, *Rev. Mod. Phys., 52*, 1-28, 1980.

Laske, G. & Masters, G., Limits on differential rotation of the inner core from an analysis of the Earth's free oscillations, *Nature, 402*, 66-69, 1999.

Lehmann, I., P', *Bur. Centr. Seism. Internat. A14*, 3-31, 1936.

Li, J. & Agee, C.B., Element partitioning constraints on the light element composition of the Earth's core, *Geophys. Res. Lett., 28*, 81-84, 2001.

Lin, J.-F., Heinz, D.L., Campbell, A.J., Devine, J.M., & Shen, G., Iron-silicon alloy in Earth's core?, *Science, 295*, 313-315, 2002.

Lister, J.R. & Buffett, B.A., The strength and efficiency of thermal and compositional convection in the geodynamo, *Phys. Earth Planet. Int., 91*, 17-30, 1995.

Loper, D.E., The gravitationally powered dynamo, *Geophys. J. R. Astr. Soc., 54*, 389-404, 1978a.

Loper, D.E., Some thermal consequences of a gravitationally powered dynamo, *J. Geophys. Res., 83*, 5961-5970, 1978b.

Loper, D.E., Structure of the inner core boundary, *Geophys. Astrophys. Fluid Dynamics, 25*, 139-155, 1983.

Loper, D.E. & Fearn, D.R., A seismic model of a partially molten inner core, *J.Geophys.Res., 88*, 1235-1242, 1983.

Loper, D.E. & Moffatt, H.K., Small-scale hydromagnetic flow in the Earth's core: rise of a vertical buoyant plume, *Geophys. Astrophys. Fluid Dynamics, 68*, 177-202, 1993.

Loper, D.E. & Roberts, P.H., On the motion of an iron-alloy core containing a slurry I. General theory, *Geophys. Astrophys. Fluid Dynamics, 9*, 289-321, 1978.

Loper, D.E. & Roberts, P.H., On the motion of an iron-alloy core containing a slurry II. A simple model, *Geophys. Astrophys. Fluid Dynamics, 16*, 83-127, 1980.

Loper, D.E. & Roberts, P.H., A study of conditions at the inner core boundary of the Earth, *Phys. Earth Planet. Int., 24*, 302-307, 1981.

MacDonald, G.J.F. & Knopoff, L., The chemical composition of the outer core, *J.Geophys, 1*, 1751-1756, 1958.

Mao, H.K., Shu, J., Shen, G., Hemley, R.J., Li, B., & Singh, A.K., Elasticity and rheology of iron above 220 GPa and the nature of the Earth's inner core, *Nature, 396*, 741-743, 1998.

Mason, B., Composition of the Earth, *Nature, 211*, 616-618, 1966.

Masters, T.G. & Shearer, P.M., Summary of seismological constraints on the structure of the Earth's core, *J. Geophys. Res., 95*, 21,691-21,695, 1990.

Miller, W.A., & Chadwick, G.A., The equilibrium shapes of small liquid droplets in solid-liquid phase mixtures: metallic h.c.p. and metalloid systems, *Proc. Roy. Soc. A312*, 257-276, 1969.

Minarik, W.G., Ryerson, F.J., & Watson, E.B., Textural entrapment of core-forming melts, *Science, 272*, 530-533, 1996.

Moffatt, H.K. & Loper, D.E., The magnetostrophic rise of a buoyant parcel in the Earth's core, *Geophys. J. Int., 117*, 394-402, 1994.

Morelli, A., Dziewonski, A.M., and Woodhouse, J.H., Anisotropy of the inner core inferred from PKIKP travel times, *Geophys. Res. Lett., 13*, 1545-1548, 1986.

Morse, S.A., Adcumulus growth of the inner core, *Geophys. Res. Lett., 13*, 1557-1560, 1986.

Morse, S.A., No mushy zones in the Earth's core, submitted, *Geochimica et Cosmochimica Acta*, 2001.

McQueen, R.G. & Marsh, S.P., Shock-wave compression of iron-nickel alloys and the earth's core, *J. Geophys. Res., 71*, 1751-1756, 1966.

Mullins, W.W. & Sekerka, R.F., Stability of a planar interface during solidification of a dilute binary alloy, *J. Appl. Phys., 35*, 444-451, 1964.

Mullins, W.W. & Sekerka, R.F., Morphological stability of a particle growing by diffusion or heat flow, *J. Appl. Phys., 34*, 323-329, 1963.

Murthy, V.R. & Hall, H.T., The chemical composition of the Earth's core: possibility of sulfur in the core, *Phys. Earth Planet. Int., 2*, 276-282, 1970.

Niu, F. & Wen, L., Hemispherical variations in seismic velocity at the top of the Earth's inner core, *Nature, 410*, 1081-1084, 2001.

Okal, E.A., & Cansi, Y., Detection of PKJKP at intermediate periods by progressive multi-channel correlation, *Earth Planet. Sci. Lett., 164*, 23-30, 1998.

Okuchi, T., Hydrogen partitioning into molten iron at high pressures: implications for Earth's core, *Science, 278*, 1781-1784, 1997.

Olson, P. & Aurnou, J., A polar vortex in the Earth's core, *Nature, 402*, 170-173, 1999.

Piersanti, A., Boschi, L., & Dziewonski, A.M., Estimating lateral structure in the Earth's outer core, *Geophys. Res. Lett., 28*, 1659-1662, 2001.

Poirier, J.-P., *Creep of Crystals*, Cambridge University Press, Cambridge, 1985.

Poirier, J.-P., Light elements in the Earth's outer core: a critical review, *Phys. Earth Planet. Int., 85*, 319-337, 1994.

Porter, D.A. & Easterling, K.E., *Phase Transformations in Metals and Alloys*, Chapman & Hall, London, 1992.

Poupinet, G., Souriau, A., & Coutant, O., The existence of an inner core super-rotation questioned by teleseismic doublets, *Phys. Earth Planet. Int., 118*, 77-88, 2000.

Ringwood, A.E., Composition of the core and implications for the origin of the Earth, *Geochem. J., 11*, 111-135, 1977.

Ringwood, A.E. & Hibberson, W., The system Fe-FeO revisited, *Phys. Chem. Minerals, 17*, 313-319, 1990.

Roberts, P.H., On the thermal instability of a rotating fluid sphere containing heat sources, *Phil. Trans. R. Soc. Lond. A263*, 93-117, 1968.

Roberts, P.H., Future of geodynamo theory, *Geophys. Astrophys. Fluid Dynamics, 44*, 3-31, 1988.

Roberts, P.H. & Loper, D.E., Towards a theory of the structure and evolution of a dendrite layer, in *Stellar and Planetary Magnetism*, ed. Soward, A.M., 329-349, Gordon & Breach, London, 1983.

Romanowicz, B.A. & Breger, L., Anomalous splitting of core sensitive normal modes: is inner core anisotropy the cause?, *EOS, Trans. AGU, F17*, 1999.

Rutter, J.W., Imperfections resulting from solidification, in *Liquid Metals and Solidification*, ed. Maddin, R., *et al.*, 243-262, American Society for Metals, Cleveland, 1958.

Qamar, A. & Eisenberg, A., The damping of core waves, *J. Geophys. Res., 79*, 758-765, 1974.

St Pierre, M.G., On the local nature of turbulence in Earth's outer core, *Geophys. Astrophys. Fluid Dynamics, 83*, 293-306, 1996.

Saxena, S.K., Shen, G., & Lazor, P., Experimental evidence for a new iron phase and implications for Earth's core, *Science, 260*, 1312-1314, 1993.

Shannon, M.C. & Agee, C.B., High pressure constraints on percolative core formation, *Geophys. Res. Lett., 23*, 2717-2720, 1996.

Shearer, P.M., Constraints on inner core anisotropy from PKP(DF) travel times, *J. Geophys. Res., 99*, 19,647-19,659, 1994.

Sherman, D.M., Stability of possible Fe-FeS and Fe-FeO alloy phases at high pressure and the composition of the Earth's core, *Earth Planet. Sci. Lett., 132*, 87-98, 1995.

Shimizu, H. & Loper, D.E., Time and length scale of buoyancy-driven flow structures in a rotating hydromagnetic fluid, *Phys. Earth Planet. Int., 104*, 307-329, 1997.

Singh, S.C., Taylor, M.A.J., & Montagner, J.P., On the presence of liquid in Earth's inner core, *Science, 287*, 2471-2474, 2000.

Song, X., Anisotropy of the Earth's inner core, *Rev. Geophys., 35*, 297-313, 1997.

Song, X. & Helmberger, D.V., Depth dependence of anisotropy of Earth's inner core, *J. Geophys. Res., 100*, 9805-9816, 1995.

Song, X. & Helmberger, D.V., Seismic evidence for an inner core transition zone, *Science, 282*, 924-927, 1998.

Song, X. & Richards, P.G., Seismological evidence for differential rotation of the Earth's inner core, *Nature, 382*, 221-224, 1996.

Souriau, A. & Romanowicz, B., Anisotropy in inner core attenuation: a new type of data to constrain the nature of the solid core, *Geophys.Res.Lett., 24*, 2103-2106, 1996.

Souriau, A. & Romanowicz, B., Anisotropy in the inner core: relation between P-velocity and attenuation, *Phys. Earth Planet. Int., 101*, 33-47, 1997.

Souriau, A., Roudil, P., & Moynot, B., Inner core differential rotation: facts and artifacts, *Geophys. Res. Lett., 24*, 2103-2106, 1997.

Stacey, F.D., *Physics of the Earth, 3rd ed.*, Brookfield Press, Brisbane, 1992.

Stacey, F.D. & Stacey, C.H.B., Gravitational energy of core evolution: implications for thermal history and geodynamo power, *Phys. Earth Planet. Int., 110*, 83-93, 1999.

Steinle-Neumann, G., Stixrude, L., Cohen, R.E., & Gulseren, O., Elasticity of iron at the temperature of the Earth's inner core, *Nature, 413*, 57-60, 2001.

Stevenson, D.J., Fluid dynamics of core formation, in *Origin of the Earth*, ed. Newsom, H.E. & Jones, J.H., 231-250, Oxford Press, New York, 1990.

Stiller, H., Franck, S., & Schmit, U., On the attenuation of seismic waves in the Earth's core, *Phys. Earth Planet. Int., 22*, 221-225, 1980.

Stixrude, L. & Cohen, R.E., High-pressure elasticity of iron and anisotropy of Earth's inner core, *Science, 267*, 1972-1975, 1995.

Stixrude, L., Wasserman, E., & Cohen, R.E., Composition and temperature of Earth's inner core, *J. Geophys. Res., 102*, 24,729-24,739, 1997.

Su, W. & Dziewonski, A.M., Inner core anisotropy in three dimensions, *J. Geophys. Res., 100*, 9831-9852, 1995.

Su, W., Dziewonski, A.M., & Jeanloz, R., Planet within a planet: rotation of the inner core of Earth, *Science, 274*, 1883-1887, 1996.

Sumita, I., Yoshida, S., Kumazawa, M., & Hamano, Y., A model for sedimentary compaction of a viscous medium and its application to inner-core growth, *Geophys. J. Int., 124*, 502-524, 1996.

Sumita, I. & Olson, P., A laboratory model for convection in Earth's core driven by a thermally heterogeneous mantle, *Science, 286*, 1547-1549, 1999.

Tait, S., Jahrling, K., & Jaupart, C., The planform of compositional convection and chimney formation in a mushy layer, *Nature, 359*, 406-408, 1992.

Tananka, S. & Hamaguchi, H., Degree one heterogeneity and hemispherical variation of anisotropy in the inner core from PKP(BC)-PKP(DF) travel times, *J. Geophys. Res., 102*, 2925-2938, 1997.

Tromp, J., Support for anisotropy of the Earth's inner core from free oscillations, *Nature, 366*, 678-681, 1993.

Verhoogen, J., Heat balance of the earth's core, *Geophys. J. R. Astr. Soc., 4*, 276-281, 1961.

Vidale, J.E. & Earle, P.S., Fine-scale heterogeneity in the Earth's inner core, *Nature, 404*, 273-275, 2000.

Vidale, J.E., Dodge, D.A., & Earle, P.S., Slow differential rotation of the Earth's inner core indicated by temporal changes in scattering, *Nature, 405*, 445-448, 2000.

Vocadlo, L., Brodholt, J., Alfe, D., Price, G.D., & Gillan, M.J., The structure of iron under the conditions of the Earth's inner core, *Geophys. Res. Lett., 26*, 1231-1234, 1999.

Weeks, W.F. & Gow, A.J., Preferred crystal orientations in the fast ice along the margins of the Arctic Ocean, *J.Geophys.Res., 83*, 5105-5121, 1978.

Wettlaufer, J.S., Worster, M.G., & Huppert, H.E., Natural convection during solidification of an alloy from above with application to the evolution of sea ice, *J. Fluid Mech., 344*, 291-316, 1997.

Widmer, R., Masters, G., & Gilbert, F., Spherically symmetric attenuation within the Earth from normal mode data, *Geophys. J. Int., 104*, 541-553, 1991.

Wood, B.J., Carbon in the core, *Earth Planet. Sci. Lett., 117*, 593-607, 1993.

Woodhouse, J.H., Giardini, D., & Li, X.-D., Evidence for inner core anisotropy from free oscillations, *Geophys.Res.Lett., 13*, 1549-1552, 1986.

Worster, M.G., Natural convection in a mushy layer, *J. Fluid Mech., 224*, 335-359, 1991.

Worster, M.G., Instabilities of the liquid and mushy regions during solidification of alloys, *J. Fluid Mech., 237*, 649-669, 1992.

Worster, M.G., Convection in mushy layers, *Annu. Rev. Fluid Mech., 29*, 91-122, 1997.

Worster, M.G. & Kerr, R.C., The transient behavior of alloys solidified from below prior to the formation of chimneys, *J. Fluid Mech., 269*, 23-44, 1994.

Yoo, C.S., Akella, J., Campbell, A.J., Mao, H.K., & Memley, R.J., Phase diagram of iron by in situ X-ray diffraction: implications for Earth's core, *Science, 270*, 1473-1475, 1995.

Yoshida, S., Sumita, I., & Kumazawa, M., Growth model of the inner core coupled with the outer core dynamics and the resulting elastic anisotropy, *J.Geophys.Res., 101*, 28085-28103, 1996.

Michael I. Bergman, Physics Department, Simon's Rock College, 84 Alford Road, Great Barrington, MA 01230 USA. bergman@simons-rock.edu

Thermodynamics of Epsilon Iron at Core Physical Conditions

Surendra K. Saxena

Center for the Study of Matter at Extreme Conditions, Florida International University, Miami, Florida

Experimental data on epsilon or the HCP-iron to a pressure of 305 GPa and temperatures to 1300 K (Dubrovinsky et al., 2001) along with other available data on pressure-volume-temperature have been used to determine equation of state parameters. The parameters have been constrained in such a way that when used in conjunction with thermochemical data on enthalpy, entropy and heat capacity, they form a complete set of parameters with which molar Gibbs free energy can be calculated for iron at core conditions. The thermochemical data on HCP iron is calculated from: ΔH°_{f} (J/mol.) 7700, S° (Gibbs/mol.) 34.40 and C_p J.mol^{-1} = 52.275 -.355155E-03 T + 790710.86 T^{-2} + -619.07 T$^{-\frac{1}{2}}$. The EoS parameters for the high-temperature Birch-Murnaghan equation (Saxena et al., 1993b) are: K_0 174 (9.152), K' 5.400 (.155), $\delta K/\delta T$ -.016 (.0011), K_1 2.00 x 10^{-5} (1.011x 10^{-4}), α_0 2.10e-5 (1.776 x 10^{-5}), α_1 5.0 x 10^{-9} (1.42 x 10^{-8}) and V_o 6.69 (.01). The thermal expansion has large errors and new data at high pressures are needed. The high pressure-high temperature heat capacities, entropy, enthalpy and Gruneisen parameter are calculated by differentiating the molar Gibbs free energy equation. The calculated densities within the range of experimental data show a trend that clearly attests to the density of iron being too high relative to the core density of the PREM model.

INTRODUCTION

Iron is one of the most abundant elements in Earth with a suitable density and therefore it is recognized as the principal component of the core. There is intensified interest in study of iron at physical conditions of the core because of new developments in high pressure-temperature techniques. Iron occurs in four distinct polymorphs. Three (BCC body centered cubic, δ-BCC, and FCC) are stable at one atmosphere and the fourth (HCP) phase is stable at high pressures. Saxena et al. (1993a) suggested the occurrence of a fifth phase, which they called β. During the last

six years we have conducted a series of experiments on establishing the identity and stability field of this phase (e.g. Saxena et al., 1995, Dubrovinsky et al., 1998). While the exploration of the β-phase continues, it is believed that it may not differ significantly in its physical properties at core physical conditions from that of the highly studied HCP-iron. This paper deals with the analysis of the pressure-volume-temperature experimental data on iron; the emphasis is to use the data in a form that not only permits us obtain the physical parameters as exemplified in the recent studies of Anderson et al. (2001) and Dubrovinsky et al. (2000a) but also to obtain thermo-chemical data as a function of temperature and pressure. The goal is to avoid initially the use of any theoretical assumptions (e.g. the applicability of the Debye model) and obtain thermochemical and evaluation of state data over the range of an experimentally achieved pressure and temperature range. Fernandez-Guillermot and Gustafsson (1985) assessed the

Earth's Core: Dynamics, Structure, Rotation
Geodynamics Series 31
Copyright 2003 by the American Geophysical Union
10.1029/31GD09

thermochemical data on BCC, δ-BCC, FCC and melt at 1 bar and calculated a phase diagram to a pressure of 20 GPa. While there are several excellent papers on systematics of the physical properties of iron as mentioned above, the integration of these properties with thermochemical data has not been attempted explicitly.

Analysis of the Experimental Data

Experimental data. In our recent experiments, we have attained some of the highest pressures (to ~300 GPa) and temperatures (~1300 K) without using laser-heating. There is a large number of data of various kinds available on iron on phase equilibrium (Boehler, 1986, Boehler at al., 1990, Boehler, 1993, Liu and Bassett, 1975, Mao et al., 1987, Bundy, 1965, Nasch and Manghnani, 1994, Shen et al., 1993, Saxena et al., 1993a) and on the physical property (Huang et al., 1987, Manghnani et al., 1987, Boehler et al., 1990, Funamori et al., 1996) in addition to the data already used by Fernández Guillermot and Gustafson (1985). There is no attempt to use all these data but care is taken to ascertain that the present results are consistent with the existing data. In the present study, let us consider only the data on HCP-iron, which is obtained either at ambient temperatures or by the use of external heating of a diamond-anvil cell (Dubrovinsky et al., 1998, 2000b). The reasons for this choice are the difficulties associated with laser heating of a sample in-situ. There is significant axial as well as front to back temperature gradient in the spot being studied by x-ray which may not be as important in phase transition studies as it could be for the pressure-volume measurements. An alternate method is to use the multianvil cell press but here one is limited to pressures below 25 GPa. In addition to the studies of Huang et al (1987) and Funamori et al. (1996), recently Uchida et al. (2001) measured the pressure-volume-temperature (PVT) data on iron to a pressure of 18 GPa and 873 K and considered the data of sufficient use for extrapolation to core conditions. The present study is based on all the avail-able data with the range of pressure extending to over 300 GPa and to temperatures of 1300 K. There is significant extrapolation involved in temperature (~6000 K) if one wants to consider the inner core conditions; this is discussed later in the paper.

Equation of state. Among the various equations of state available, the most commonly used is the third or fourth order Birch-Murnaghan equation. Cohen et al (2000) have recommended the use of the Vinet equation. The isobaric thermal expansion αp and isothermal compressibility β_T, respectively, are given by:

$$\alpha_p = \frac{1}{V_T}\left(\frac{\partial V}{\partial T}\right)_p \qquad (1)$$

$$\beta_T = \frac{1}{V_T}\left(\frac{\partial V}{\partial P}\right)_T \qquad (2)$$

Many different polynomial equations have been used for fitting the measured data (Saxena and Zhang, 1989). In this work, the following equations are used:s

$$\alpha_p = a_O + a_1 T + a_2 T^{-1} + a_3 T^{-2} + \ldots\ldots \qquad (3)$$

$$K_T = 1/(\beta_O + \beta_1 T + \beta_2 T^2 + \beta_3 T^3 + \ldots) \qquad (4)$$

where α are the coefficients in the expansion for isobaric thermal expansion, and β_i the coefficients of compressibility.

The Birch-Murnaghan equation of state is given by:

$$P_{B-M} = \frac{3}{2}K_{T,0}\left[\left(\frac{V^0}{V}\right)^{7/3} - \left(\frac{V^0}{V}\right)^{5/3}\right]\left\{1 - \frac{3}{4}(4 - K'_{T,0})\left[\left(\frac{V^0}{V}\right)^{2/3} - 1\right] + \ldots\right\} \qquad (5)$$

where K_T and K'_p $(= [\delta K_T/\delta p]_T)$ are the isothermal bulk modulus and its pressure derivative at 298 K (or 0 K depending on modeling parameters), respectively.

With all the information on the temperature dependence of α and K available, one may proceed to apply the isothermal form of the BM equation (see Saxena and Zhang, 1989) at different temperatures as follows.

The VdP is calculated by adopting the third order Birch-Murnaghan equation of state [equation (5)] where the temperature dependence of the isothermal bulk modulus is included and V^0/V is replaced by $V(1,T)/V(P,T)$. The temperature dependence of all variables, except *of the pressure derivative* K' is known from the data systematization at conditions of 1 bar and T. By using the experimental data on in-situ P-V-T determin-ations or with the help of the phase equilibrium experi-mental data (less preferably the shock-wave data, see Saxena and Zhang, 1990 for the method), we may determine the temperature dependence of the pressure derivative K'_p by expressing $(\delta K_T/\delta P)_T$ with an appropriate polynomial, e.g.

$$(\delta K_T/dP)_T = K' + K_i(T)ln(T) \qquad (6)$$

where K' is the pressure derivative in the Birch - Murnaghan equation and K_1 the temperature coefficient (not to be confused with K" the second derivative of the bulk modulus). The temperature term "$T\,ln(T)$" has been established empirically to facilitate differentiation of the Gibbs energy function in the thermodynamic analysis. Saxena et al. (1993b) called this model as *the high-temperature Birch-Murnaghan (HTBM)* model. It is possible to consider the Vinet model similarly modified for use with high temperature PVT data.

Table 1. EoS parameters at room temperature in various studies

EoS	KT,GPa	K'	Vo,cm^3 mol	Reference
BM (Third order)	155.8	5.81	6.73	Anderson et al. (2001)
BM (Fourth order)	164 (1.7)	5.35 (0.12) (constrained)	6.73 cm^3	Anderson et al. (2001)
Vinet (Third order)	156.2 (1.8)	6.08 (0.12)	6.72 (0.2)	Anderson et al. (2001)
BM (Third order)	166 (5.8)	5.51 (.84)	6.69 (.265)	Fitted to Mao et al. (1990)
BM (Third order)	166.8 (2.4)	5.4 (.004)	6.721 (.009)	Fitted to Dubrovinsky et al. (2000)

*K" = 1.34 (0.12) GPa^{-1}

Table 2. Result of regression analysis with constraints

Parameter	Value	Standard. Error.
K'	5.400	1.554e-1
K_0	166.79	9.152
dK/dT	-6.5e-3	1.071e-2
K_1	-2.500e-5	1.011e-4
α_0	1.90e-5	1.776e-5
α_1	5.0e-9	1.421e-8
V_o	6.73	8.12e-2

Thermodynamics, results of data analysis and the equation of state. Both the Vinet and the BM equations were tried with the pressure-volume data. Although Cohen et al. (2001) advocate the use of the Vinet equation and Anderson et al. (2001) determined EoS parameters well (Table 1), in this study, it gave physically unrealistic parameters with the data. In view of the previously used EoS parameters for the high temperature BM equation, the use of the third order BM equation was continued in this study.

An unconstrained fit usually led to parameters that were unusable in thermodynamic analysis. For example the calculated entropy or Gruneisen parameter became negative at high pressures, which was clearly a result of overfitting but could also be due to the inappropriate model adopted in this work. Therefore, several constraints had to be used; these are 1) the calculated entropy should not become negative and should decrease with increasing pressure, 2) heat capacities and thermal expansion were assumed to be either constant or to increase with temperature and decrease with pressure and 3) bulk modulus should decrease with temperature.

The regression analysis was used to determine the parameters of equation of state as required in equations (3) to (6) and resulted in the data presented in Table 2. It is clear that the parameters with temperature dependence are quite unreliable with high uncertainties. However, as seen in Fig.1, their use does improve the fit of the data to the model significantly. In the procedure for regression analy-sis, there is no actual data available on the thermochemical data as mentioned above. Starting from the best fit parameters, a test was made to see if the constraints of thermochemistry were satisfied. The procedure was to calculate the molar Gibbs free energy as a function of pressure and temperature by combining G (1 bar,T) and the VdP integral and then obtain other thermochemical data. The molar Gibbs free energy(with respect to standard reference state) given by

$$\Delta G(P,T) = \Delta H_T^0 - T\Delta S_T^0 + \int_1^P \Delta V(P,T)dP \qquad (7)$$

where ΔH^0_T and ΔS^0_T are the standard enthalpy and entropy of the phase, respectively, at temperature T and 1 bar given by

$$\Delta H_T^0 = \Delta H_{298}^0 + \int_{298}^T \Delta C_P dT \qquad (8)$$

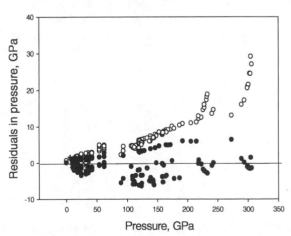

Figure 1. The result of the fit constrained as described in the text. In spite of the high uncertainties in the temperature parameters of the equation, their use increases the fit demonstrably. The circles are the data from Dubrovinsky et al. (2000); the crosses are from Uchida et al. (2001) and not visible on the scale of the pressure range.

and

$$\Delta S_T^0 = \Delta S_{298}^0 + \int_{298}^T \frac{\Delta C_P}{T} dT \qquad (9)$$

where EH^o_{298} and ES^o_{298} are the standard enthalpy and entropy of the phase at 298.15 K, ΔC_P is the heat capacity difference between products and reactants, and $\Delta V(P,T)$ is the volume change for the phase. For computational convenience PdV may be calculated from equation (5), instead of from VdP. The relation between PdV and VdP is given by

$$\int_1^P V dP = \int_{V(P,T)}^{V(1,T)} PdV + V(P-1) \qquad (10)$$

where

$$\int_{V(P,T)}^{V(1,T)} PdV =$$

$$\frac{3}{2} K_T V(1,T) \left[\frac{3}{4}(1+2x)(Y^{4/3}-1) - \frac{3}{2}(1+x)(Y^{2/3}-1) - \frac{1}{2}x(Y^2-1) \right] \quad (11)$$

where

$$x = \frac{3}{4}\left[4 - \left(\frac{\partial K_T}{\partial P} \right)_T \right] \quad \text{and} \quad Y = \frac{V(1,T)}{V(P,T)}$$

We can then calculate all other thermochemcal data at any pressure, volume or temperature by suitably differentiating the molar Gibbs free energy as presented below.

The change in Gibbs free energy (ΔG) as a function of temperature (T) and pressure (P) can be obtained from the relations:

$$\left(\frac{\partial^2 \Delta G}{\partial T^2} \right)_P = -\frac{C_P(T)}{T} \qquad (12)$$

and

$$\left(\frac{\partial \Delta G}{\partial P} \right)_T = V(P,T) \qquad (13)$$

where $C_P(T)$ is heat capacity at constant P and at a temperature T and $V(P,T)$ is the molar volume at a temperature T and a pressure P.

In general, C_P for a solid can be expressed as

$$C_P = C_V + \alpha^2 V K_T T + C_P' \qquad (14)$$

where C_V is the heat capacity of a crystal at constant volume. a is the coefficient of thermal expansion, V is the molar volume, and K_T is the isothermal bulk modulus, all at temperature T. C_P' is the contribution from cation disordering and anharmonicity (other than those incorporated in the $\alpha^2 V K_T T$ term). We may calculate the thermodynamic Gruneisen parameter (γ) from $\alpha V K_T / Cv$.

The thermochemical data on four iron phases BCC (δ-BCC), FCC, HCP and melt at 1 bar were assessed by Fernández Guillermot and Gustafson (1985) where the assessment method and errors in the data are described. The thermochemical data at 1 bar from their 1985 study is used here are as follows: Equations for thermal expansion and compressibility are (10) and (11) respectively where β parameters are obtained from the inverse of K_T.

ΔH^o_f (J/mol.) 7700.00, S^o (Gibbs/mol.) 34.40

$C_p = 52.275\ -.355155E\text{-}03\ T\ + 790710.86\ T^{-2} + -619.07\ T^{-\frac{1}{2}}$

The thermochemical data for HCP-iron is listed in Tables 3 and 4.

DISCUSSION AND APPLICATION

This is the first attempt to integrate thermochemical and physical data on iron at high pressure that does not involve any theoretical assumptions except the constraints listed earlier. However to maintain objectivity, this also requires that we refrain from any significant extrapolations. Fig.2 shows all the experimental data that is available and with which the consistency of the present parameters can be demonstrated. It should be emphasized that calculations of all thermodynamically and physical parameters at high pressure are done by first calculating the molar Gibbs energy of the phase through equation (7)

$$\Delta G(P,T) = \Delta H_T^0 - T\Delta S_T^0 + \int_1^P \Delta V(P,T)dP$$

and then by suitable differentiations. For example equation (12) would yield C_p and then C_v is determined by

$$C_v = C_p - \alpha^2 V K_T T.$$

The determination of thermodynamic Gruneisen parameter (γ) then follows from

$$\gamma = \alpha K_T V / C_v.$$

Entropy. The need for such a calculation procedure is obvious. In all questions of phase transformation and phase

Table 3. Thermodynamic data on iron at various pressures at 300 K

P, GPa	Mol. Vol. cm³	C_p, J.mol⁻¹	C_v J.mol⁻¹	S, gibbs mol⁻¹	ΔG, KJ mol⁻¹	Density, gm.cm³	α x 10^5	K_T, GPa	γ
1.0e-4	6.69	25.21	25.04	34.56	-2.62	8.35	2.26	169.16	1.02
20.00	6.10	25.03	24.91	32.19	124.74	9.15	1.57	277.16	1.06
40.00	5.73	24.89	24.79	30.55	242.80	9.75	1.24	385.16	1.10
60.00	5.45	24.75	24.66	29.29	354.50	10.24	1.03	493.16	1.12
80.00	5.24	24.60	24.53	28.27	461.36	10.66	0.89	601.16	1.14
100.00	5.06	24.45	24.39	27.41	564.34	11.03	0.78	709.16	1.15
120.00	4.92	24.30	24.24	26.68	664.10	11.36	0.69	817.16	1.15
140.00	4.79	24.14	24.09	26.05	761.08	11.67	0.62	925.16	1.14
160.00	4.67	23.98	23.94	25.49	855.66	11.95	0.56	1033.16	1.14
180.00	4.57	23.82	23.78	24.99	948.11	12.21	0.51	1141.16	1.13
200.00	4.48	23.65	23.61	24.55	1038.65	12.46	0.47	1249.16	1.11
220.00	4.40	23.48	23.45	24.15	1127.47	12.69	0.43	1357.16	1.10
240.00	4.33	23.31	23.28	23.79	1214.71	12.91	0.40	1465.16	1.08
260.00	4.26	23.13	23.10	23.46	1300.51	13.12	0.37	1573.16	1.06
280.00	4.19	22.95	22.93	23.16	1384.97	13.32	0.34	1681.16	1.04
300.00	4.13	22.77	22.75	22.89	1468.20	13.52	0.31	1789.16	1.02
320.00	4.08	22.59	22.57	22.64	1550.28	13.70	0.29	1897.16	0.99
340.00	4.02	22.41	22.39	22.42	1631.28	13.88	0.27	2005.16	0.97
360.00	3.97	22.23	22.21	22.21	1711.26	14.05	0.25	2113.16	0.94

equilibrium, the minimization of the Gibbs free energy of an assemblage of phases coexisting together in equilibrium is required. Similarly high-pressure data on entropy and enthalpy are needed for all thermo-physical calculations. Furthermore, by not making this important connection, we may calculate the correct densities but the data may not be compatible with other physical parameters. An example is shown in Fig.3 where the calculated molar entropy is plotted as a function of pressure. It can be seen that the entropy is calculated to be negative at a pressure of 80 GPa if we use the data in Anderson et al. (2001). The calculated data from Uchida et al. (2001) remains positive in the experimental range but any significant extrapolation will lead to negative entropy.

Heat capacities. Fig. 5 shows the data on heat capacities as produced in this study. The heat capacities decrease with

Table 4. Physical data on HCP-iron at a pressure of 360 GPa as a function of temperature

T, K	Mol. Vol. cm³	C_p, J.mol⁻¹	C_v J.mol⁻¹	S, gibbs mol⁻¹	ΔG, KJ Mol⁻¹	Density, gm.cm³	Density, gm.cm³	K_T, GPa	γ
300	25.21	25.04	34.56	6.69	2.62	8.35	2.26	169.16	1.02
400	6.71	26.12	25.88	41.90	6.46	8.33	2.30	167.59	1.00
500	6.72	27.57	27.26	47.89	10.96	8.31	2.36	166.01	0.97
600	6.74	28.98	28.60	53.04	16.01	8.29	2.40	164.42	0.93
700	6.75	30.24	29.78	57.61	21.55	8.27	2.45	162.82	0.91
800	6.77	31.34	30.79	61.72	27.51	8.25	2.51	161.22	0.89
900	6.79	32.30	31.66	65.46	33.88	8.23	2.55	159.62	0.87
1000	6.80	33.13	32.41	68.91	40.60	8.21	2.60	158.01	0.86
1100	6.82	33.87	33.05	72.11	47.65	8.19	2.65	156.40	0.86
1200	6.84	34.53	33.61	75.08	55.01	8.16	2.69	154.79	0.85
1300	6.86	35.11	34.08	77.87	62.66	8.14	2.75	153.18	0.85
1400	6.88	35.64	34.49	80.49	70.58	8.12	2.80	151.58	0.85
1500	6.90	36.11	34.85	82.96	78.75	8.10	2.85	149.97	0.84
1600	6.92	36.54	35.16	85.31	87.17	8.07	2.90	148.37	0.85
1700	6.94	36.93	35.42	87.54	95.81	8.05	2.95	146.78	0.85
1800	6.96	37.29	35.65	89.66	104.67	8.03	3.00	145.18	0.85
1900	6.98	37.62	35.84	91.68	113.74	8.00	3.05	143.60	0.85
2000	7.00	37.92	36.01	93.62	123.01	7.98	3.10	142.02	0.86

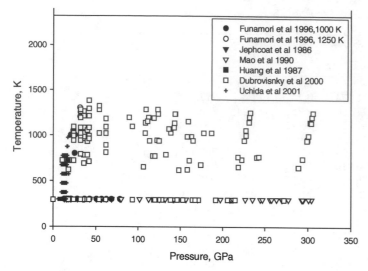

Figure 2. Experimental pressure-volume-temperature data on HCP-iron considered in this study.

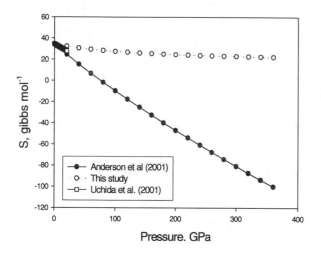

Figure 3. Calculated entropy from different sets of EoS parameters plotted as a function of pressure. None of the sets are satisfactory but the parameters of the present study do keep the molar entropy positive over a large range of pressure.

pressure and increase with temperature. The calculated data are consistent with these assumptions in the experimental range. However, the calculated data at a pressure of 360 GPa shows the need of further improvement in the present equation of state parameters. Above 1200 K, the isobaric heat capacities decrease with increasing temperature.

Gruneisen parameter. While the entropy data did not seem to contradict any rule if the Uchida et al. (2001) data (α = 3.98E-5+5.07E-8T, K_T = 135, K'=6 and $\delta K/\delta T$ =-4.48E-2), are used, the Gruneisen parameter in Fig. 4 is clearly at odds with all other studies. The data calculated

here yields somewhat lower values matching with those of Jeanloz (1979) at the highest compression (see Dubrovinsky et al., 2001) and not far off our previous data (Anderson et al., 2001) or with those of Stacey's (1995).

Thermal expansion. The data in Tables 3 and 4 show that the thermal expansion at a pressure of 360 GPa approaches a value of 2.5×10^{-6} at 300 K and 3.1×10^{-5} at 2000 K. Thus we may expect thermal expansion of iron to be close to this value at high temperatures between 5000 to 6000 K. The thermal expansion is significantly low compared to other data (Uchida et al. 2001, Anderson et al., 2001). But we

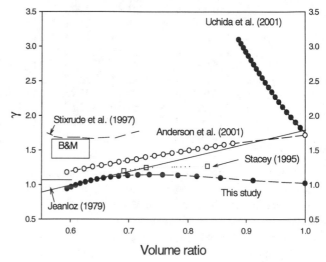

Figure 4. Gruneisen parameter plotted as a function of compression. The thermodynamic parameter as calculated in this study is found to be consistently smaller than all previous studies in the low compression range.

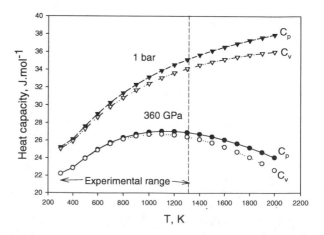

Figure 5. Calculated heat capacities at pressures of 1 bar and 360 GPa. In the experimental range of temperature, the heat capacities increase with temperature. Above a temperature of 1200 K the slope changes and would lead to unacceptable low heat capacities at core temperatures.

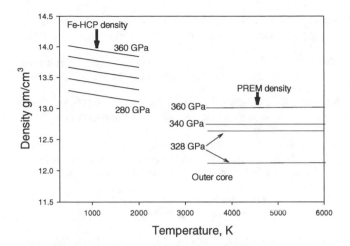

Figure 6. A comparison of the PREM inner core density with that of iron. No significant extrapolation is used in drawing the density profiles at pressures ranging from 280 to 360 GPa between 500 to 2000 K. The density trends clearly show that iron is too dense even at very high temperatures relative to the PREM model.

have already noted that the high thermal expansion cannot be used while fitting the data to the highest pressure experimental data (Dubrovinsky et al., 2000b) and that if used with thermochemical data gives contrasting data on the Gruneisen parameter or the entropy.

Core density. Fig.6 shows the calculated density at various pressures between 280 to 360 GPa and at temperatures ranging from 500 to 2000 K. As is clear there is no significant extrapolation involved here and yet the figure clearly shows that the density of iron at core temperatures and the PREM model density would be significantly different. The high density of iron relative to the PREM core density was modeled before by Jephcoat and Olsen (1987), Stixrude et al. (1997) and recently by Uchida et al. (2001) who used the Mie-Gruneisen-Debye EOS. As Fig. 2 shows, the data used in this study extends to core pressures and should firmly establish the significantly higher density of iron relative to the modeled core density. This conclusion is based on a lower value of thermal expansion than that of Uchida et al. (2001). Within the errors produced by the selection of parameters in this study, the latter data fits well (Fig.2).

If we assume the inner core temperature to be around 6000 K, then the difference between the iron and core (PREM) density would be about 7 to 8%. If nickel is to be considered as usually present with iron, this density difference will be further enhanced. There are many diluents proposed for alloying with iron to make it light and appropriate for the outer core. They could also be considered for the inner core. It is unnecessary to consider incompatible elements such as Si which require a highly reduced environment and incomprehensible method of incorporation in to the core. A suitable substance that is already present in abundance in the mantle is wuestite (Fe_xO). Wuestite is known to undergo a phase transition between 60 to 70 GPa. From the available data, one may estimate the density of Fe_xO to be around 9 gm^3/mol at the core conditions. The core density could be matched with a solid solution of 91% iron and 9% wuestite. These are very rough estimates.

Acknowledgments. I thank O. Anderson for several inspiring discussions during his visit to Uppsala and Miami. The research was supported by NSF and generous support from the Division of Sponsored Research at FIU. Ms. Debby Arnold helped in preparing the final version of the manuscript.

REFERENCES

Anderson, O. L. (1993) The phase diagram of iron and the temperature of the inner core; *J. Geomag. Geoelectr.*, 2, 145-156.

Anderson, O .L., Dubrovinsky, L. S., Saxena, S. K., LeBihan, T. (2001) Experimental vibrational Gruneisen ratio values for ε-iron up to 330 GPa at 300 K, *Geophys. Res. Lett.*, 28, 399-402.

Boehler, R. (1986) The phase diagram of iron to 430 kbar, *Geophys. Res. Lett.*, 13, 1153-1156.

Boehler, R. (1993) Temperatures in the Earth's core from melting-point measurements of iron at high static pressures, *Nature*, 363, 534-536.

Boehler, R., von Bargen, N. and Chopelas, A. (1990) Melting, thermal expansion, and phase transition of iron at high pressures, *J. Geophys. Res.*, 95, 21731-21736.

Boness, D. A. and Brown, J. M. (1990) The electronic band structures of iron, sulfur, and oxygen at high pressures and the Earth's core, *J. Geophys. Res.*, 95, 21721-21730.

Brown, J.M. and McQueen, R. G. (1986) Phase transitions, Grüneisen parameters and elasticity for shocked iron between 77 GPa, *J. Geophys. Res.*, 91, 7485-7494.

Bundy, F.P. (1965) Pressure-temperature phase diagram of iron to 200 kbar, 900° C, *J. Appl. Phys.*, 36:2, 616-620.

Cohen, R.E., Gulseren, O. and Hemley, R. J. (2000) Accuracy of equation-of-state formulations, *Amer. Mineralogist*, 85, 338-344.

Dubrovinsky, L. S., Saxena, S. K., and Lazor, P. (1997) X-ray study of iron with in-situ heating at ultra high pressures, *Geophys. Res. Lett.*,24, 1835-1838.

Dubrovinsky, L. S., Saxena, S. K., and Lazor, P. (1998) High-pressure and high-temperature in situ X-ray diffraction study of iron and corundum to 58 GPa using an internally heated diamond anvil cell, *Phys. Chem. Miner.*, 25, 434-441.

Dubrovinsky, L. S., Saxena, S K.,Dubrovinskaia, N.A., Rekhi, S.,and LeBihan, T. (2000a) Gruneisen parameter of e-iron up to 300 GPa from in-situ X-ray study, *Amer. Mine.*, 85, 1-3.

Dubrovinsky, L.S., Saxena, S.K., Tutti, F., and Rekhi, S. (2000b) In situ x-ray study of thermal expansion and phase transition in iron at multimegabar pressure, *Phys. Rev. Lett*, 84, 1720-1723.

Fernández Guillermot, A. and Gustafson, P. (1985) An assessment of the thermodynamic properties and the (p,T) phase diagram of iron, *High Temp.-High Press.*, 16, 591.

Funamori, N., Funamori, M., Jeanloz, R., and Hamaya, N. (1997)

Broading of X-ray powder diffraction lines under Nonhydrostatic stress, *J. Applied Physics*, 82, 142-146.

Huang, E., Bassett, W. A. and Tao, P. (1987) Pressure-temperature-volume relationship for hexagonal close packed iron determined by synchrotron radiation, *J. Geophys. Res.*, 92, 8129-8135.

Jeanloz, R. (1979) Properties of iron at hiagh pressure and the state of the core, *J. of Geophys. Res*, 84, 6059-6069.

Jephcoat, A.P. and Olsen, P. (1987) Is the inner core of the Earth pure iron?, *Nature*, 325, 331-335.

Lazor, P., Shen, G. and Saxena, S.K. (1993) Laser-heated diamond anvil cell experiments at high pressure: Melting curve of nickel up to 700 kbar, *Phys. Chem. Minerals*, 20, 86-90.

Liu, L. and Bassett, W.A. (1975) The melting of iron to 200 kbar, *J. Geophys. Res.*, 80, 3777-3782.

Manghnani, M.H., Ming, L.C. and Nakagiri, N. (1987) Investigation of the α-Fe⇔ε-Fe phase transition by synchrotron radiation, in *High-Pressure Research in Mineral Physics*, Eds.

M. H. Manghnani and Y. Syono, pp. 155-163, TERRAPUB, Tokyo/AGU, Washington DC.

Mao, H.K., Bell, P.M. and Hadidiacos, C. (1987) Experimental phase relations of iron to 360 kbar, 1400°C, determined in an internally heated diamond anvil apparatus, in *High-Pressure Research in Mineral Physics*, Eds. M. H. Manghnani and Y. Syono, pp. 135-140, TERRAPUB, Tokyo/AGU, Washington DC.

Nasch, Ph. M. and Manghanani, M. H. (1994) A modified ultrasonic interferometer for sound velocity measurements in molten metals and alloys, *Re. Sci. Instrum*, 65, 682-688

Saxena, S.K. and Zhang, J. (1989) Assessed high-temperature thermochemical data on some solids, *J. Phys. Chem. Solid*, 50, 723-727.

Saxena, S.K., Shen, G., Lazor, P. (1993a) Experimental evidence for anew iron phase and implications for Earth's core, *Science*, 260, 1312-1314.

Saxena, S.K., Chatterjee, N., Fei, Y. and Shen, G. (1993b) *Thermodynamic data on oxides and silicates*. Springer-Verlag, Heidelberg, 428 pp.

Saxena, S.K., Dubrovinsky, L.S., Häggqvist, P., Cerenius, Y., Shen, G. and Mao, H.K. (1995) Synchrotron x-ray study of iron at high pressure and temperature, *Science*, 269, 1703-1704.

Shen, G., Lazor, P. and Saxena, S.K. (1993) Melting of wüstite and iron up to pressures of 600 kbar, *Phys. Chem. Minerals*, 20, 91-96.

Stixrude, L. and Wasserman, E. and Cohen, R. (1997) Composition and temperature of Earth's inner core, *J. Geophys Res.*, 102, 24,729-24,739.

Uchida, T., Wang, Y., Rivers, M.L. and Sutton, S.R. (2001) Stability field and thermal equation of state of ε-iron determined by synchrotron X-ray diffraction in a multianvil apparatus, *J Geophys. Res.*, 106, 21,799-21,810.

Surendra K. Saxena, Center for the Study of Matter at Extreme Conditions (CeSMEC), VH-150, Florida International University, Miami, FL 33199

Physical Properties of Iron in the Inner Core

Gerd Steinle-Neumann[1] and Lars Stixrude

Department of Geological Sciences, University of Michigan, Ann Arbor

R. E. Cohen

Geophysical Laboratory, Carnegie Institution of Washington, Washington, DC

The Earth's inner core plays a vital role in the dynamics of our planet and is itself strongly exposed to dynamic processes as evidenced by a complex pattern of elastic structure. To gain deeper insight into the nature of these processes we rely on a characterization of the physical properties of the inner core which are governed by the material physics of its main constituent, iron. Here we review recent research on structure and dynamics of the inner core, focusing on advances in mineral physics. We will discuss results on core composition, crystalline structure, temperature, and various aspects of elasticity. Based on recent computational results, we will show that aggregate seismic properties of the inner core can be explained by temperature and compression effects on the elasticity of pure iron, and use single crystal anisotropy to develop a speculative textural model of the inner core that can explain major aspects of inner core anisotropy.

1. INTRODUCTION

The presence and slow growth of Earth's inner core is one of the most significant manifestations of the dynamics in the interior of our planet. As it is inaccessible to direct observation, an understanding of the physical state of the inner core requires an integrative approach combining results from many fields in the geosciences. Seismology, geo- and paleomagnetism, geo- and cosmochemistry, geodynamics, and mineral physics have ad-

vanced our knowledge of the structure and processes in the inner core, revealing many surprises.

Foremost among these have been the discoveries of anisotropy and heterogeneity in the inner core. Long assumed to be a featureless spherically symmetric body, a higher number and higher quality of seismic data revealed that the inner core is strongly anisotropic to compressional wave propagation [*Morelli et al.*, 1986; *Woodhouse et al.*, 1986]. Generally, seismic waves travel faster along paths parallel to the Earth's polar axis by 3-4% compared to equatorial ray paths [*Creager*, 1992; *Song and Helmberger*, 1993]. The presence of anisotropy is significant because it promises to reveal dynamical processes within the inner core.

Anisotropy is usually attributed to lattice preferred orientation, which may develop during inner core growth [*Karato*, 1993; *Bergman*, 1997], or by solid state deformation [*Buffett*, 2000]. The source of stress that may be responsible for deformation of the inner core is

---

[1] Now at the Geophysical Laboratory, Carnegie Institution of Washington, Washington, DC.

Earth's Core: Dynamics, Structure, Rotation
Geodynamics Series 31

unknown, although several mechanisms have been proposed [*Jeanloz and Wenk, 1988; Yoshida et al., 1996; Buffett 1996; 1997; Karato, 1999; Buffett and Bloxham 2000*].

An understanding of the origin of inner core anisotropy will require further advances in our knowledge of the physical properties of iron at inner core conditions, and may rely critically on further observations of the detailed structure of the inner core. For example, recent observations indicate that the magnitude of the anisotropy may vary with position: heterogeneity has been observed on length scales from 1-1000 km [*Creager, 1997; Tanaka and Hamaguchi, 1997; Vidale and Earle, 2000*]. Inner core structure may change with time as well. *Song and Richards* [1996] interpreted apparent changes in travel times of inner core sensitive phases in terms of super-rotation of the inner core with respect to the mantle. Some recent studies have argued for a much slower rotation rate than that advocated originally, or questioned the interpretation of time dependent structure [*Souriau, 1998; Laske and Masters, 1999; Vidale and Earle, 2000*].

The inner core also plays an essential role in the dynamics of the overlying outer core. The anisotropy and long magnetic diffusion time of the inner core may alter the frequency and nature of reversals, and influence the form of the time-averaged field [*Hollerbach and Jones, 1993; Clement and Stixrude, 1995*]. Moreover, important energy sources driving the geodynamo process are associated with solidification of the inner core: the density contrast across the inner core boundary is due to the phase transition from the liquid to the solid, and chemical differentiation during the incongruent freezing of the inner core. Both of these processes provide energy for the dynamo through the release of latent heat [*Verhoogen, 1961*] and the generation of chemical buoyancy [*Braginsky, 1963*]. Other energy sources for magnetic field generation are secular cooling of the Earth, gravitational energy from thermal contraction of the core, radioactive heat generation, and precession [*Verhoogen, 1980; Buffett et al, 1996*].

Both thermal and compositional contributions to the buoyancy depend on the thermal state of the core. The more viscous mantle controls the cooling time scale of the Earth and facilitates the formation of a thermal boundary layer at the core mantle boundary. The heat flux out of the core controls the rate of inner core growth and light element partitioning during this process [*Buffett et al, 1996*]. Conversely, a reliable estimate on temperature in the Earth's core would advance our understanding of the current thermal state and evolution of the Earth [*Jeanloz and Morris, 1986; Yukutake, 2000*]

with important implications for the dynamics of the Earth.

Because the inner core is inaccessible, the study of model systems by theory and experiments is essential. Here we consider the ways in which mineral physics may lend deeper insight into inner core processes and to the origin of its structure, extending previous reviews by *Jeanloz* [1990] and *Stixrude and Brown* [1998]. We begin with geophysical background on the inner core including recent seismological advances, constraints on the composition, thermal state, and dynamics of the inner core. As our subsequent discussions draw on various experimental and theoretical approaches in mineral physics we then give an overview of recent developments in methods in the following section, focusing on computational mineral physics (a recent review focusing on advances in experiments has been given by *Hemley and Mao* [2001]). To the extent that the inner core is composed of nearly pure iron, physical properties of this element at high pressure and temperature govern the behavior of the inner core; we consequently review advances in our knowledge of the high pressure physical properties of iron, focusing on crystalline structure, equation of state, and elasticity at both static condition and high temperature. In the final section we examine the implications of these results for inner core temperature, and integrate aspects of elasticity with considerations of the dynamics in the inner core to develop a simple speculative model of polycrystalline structure that explains major aspects of its anisotropy.

2. GEOPHYSICAL BACKGROUND

2.1. Aggregate Seismic Properties

Lehmann [1936] discovered the inner core by recognizing weak arrivals of PKiKP within the P-wave shadow zone of the core. The amplitudes of these arrivals were sufficient to invoke a discontinuous seismic boundary in the Earth's core. The P-wave contrast across this boundary was soon established; *Birch* [1940] and *Bullen* [1946] argued that the inner core must be solid based on this estimate. The best evidence for inner core solidity comes from studies of inner core sensitive normal modes [*Dziewonski and Gilbert, 1971*]: Earth models with finite shear modulus of the inner core provide a significantly better fit to eigenfrequency observations than those with a liquid inner core. Recent observations of body wave phases involving a shear wave in the inner core (PKJKP, SKJKP, and pPKJKP) [*Okal and Cansi, 1998; Deuss et al., 2000*] support solidity, but are still controversial. The inferred shear wave velocity v_S of the inner core is remarkable: it is low compared to the compressional wave velocity v_P, a property which can also be expressed in terms of the Poisson's ratio σ. The value of $\sigma=0.44$ for the inner core is nearly that of a liquid (0.5), leading to speculation that this region may be partially molten [Singh et al., 2000].

In principle density ρ, v_P, and v_S also depend on depth. However, constraints on the depth dependence of ρ and v_S are weak. Seismic observations are consistent with an inner core in a state of adiabatic self-compression.

2.2. Anisotropy

First evidence for deviations from a spherically symmetric structure came from the observation that eigenfrequencies of core sensitive normal modes are split much more strongly than predicted by ellipticity and rotation of the Earth alone [Masters and Gilbert, 1981]. Anomalies in PKIKP travel times were initially interpreted as topography on the inner core [Poupinet et al., 1983]. Morelli et al. [1986] and Woodhouse et al. [1986] interpreted similar observations of eigenfrequencies and travel times as inner core anisotropy. Observation of differential travel times PKIKP-PKP$_{BC}$ [Creager, 1992; Song and Helmberger, 1993] and a reanalysis of normal mode data [Tromp, 1993] confirmed that the inner core displays a hexagonal (cylindrical) pattern of anisotropy with a magnitude of 3-4% and symmetry axis nearly parallel to Earth's rotation axis. For example, PKIKP arrives 5-6 s earlier along polar paths than predicted from radially symmetric Earth models such as PREM [Dziewonski and Anderson, 1981]. It is worthwhile pointing out that some of the travel time differences could be due to mantle structure not accounted for in the reference model [Bréger et al., 1999; Ishii et al., 2002a; 2002b]. In particular, small scale heterogeneity in the lowermost mantle could be sampled preferentially for select body wave core paths [Bréger et al., 1999; Tromp, 2001; Ishii et al., 2002b].

In further investigation deviations from first order anisotropy have been put forward, for example lateral variations in v_P of the inner core on length scales ranging from hemispherical differences [Tanaka and Hamaguchi, 1997; Creager, 1999; Niu and Wen, 2001], to hundreds of kilometers [Creager, 1997], down to a few kilometers [Vidale and Earle, 2000]. Radial variations may also exist: weak anisotropy may be present in the uppermost inner core (to a depth of 50-100 km) [Shearer, 1994; Song and Helmberger, 1995; Su and Dziewonski, 1995] and strong uniform anisotropy in its inner half [Song and Helmberger, 1995; Creager, 1999]. Seismological studies of the inner core are discussed in more detail elsewhere in this volume.

2.3. Composition

Seismically determined properties of the core may be compared to laboratory measurements under high compression. Measurements of the equation of state show that only elements with an atomic number close to that of iron satisfy the seismic constraints [Birch, 1964]. Additional arguments are necessary to uniquely implicate iron [Jeanloz, 1990]: iron is one of the most abundant elements in stars and meteorites, much more so than in the portions of the Earth that are directly observable [Brown and Mussett, 1993]; and a conducting liquid is necessary in the outer core to explain the existence of a long lived dynamo process that creates Earth's magnetic field [Merrill et al., 1996].

To the degree that we are certain about the main constituent of the core we are also sure that the core contains other lighter elements: pure iron can not satisfy the seismological constraints for both portions of the core. Liquid iron is about 10% too dense to satisfy both the density and bulk modulus in the outer core [Birch, 1964; Jeanloz, 1979; Brown and McQueen, 1986] and while solid iron can explain the bulk modulus of the inner core for reasonable temperatures it overestimates the density even for very high temperature (8000 K) [Jephcoat and Olson, 1987; Stixrude et al., 1997]. The identity and amount of the light element is still uncertain, but based on cosmochemical arguments hydrogen, carbon, oxygen, magnesium, silicon, and sulfur have been proposed [Poirier, 1994], with oxygen and sulfur being the most popular. To infer information on the composition of the core from geochemistry, two questions are of central importance: did the core form in chemical equilibrium [Karato and Murthy, 1997] and what are the physical conditions of the core forming event, as pressure and temperature critically determine the partition coefficient of various elements between silicate and metallic melt [Ito et al., 1995; Li and Agee, 1996; Okuchi, 1997].

Alternatively, one may use the available seismological information on the current physical state of the outer and inner core (ρ, v_P, v_S) and compare to the physical properties of candidate iron alloys at the appropriate pressure and temperature condition. The compositional space of Fe-X with X any light element has been sparsely sampled in shock wave experiments at the conditions relevant for the core. Only binary compounds in the Fe-S system (pyrrhotite $Fe_{0.9}S$ [Brown et al., 1984], troilite FeS [Anderson and Ahrens, 1996], and pyrite Fe_2S [Ahrens and Jeanloz, 1987; Anderson and Ahrens, 1996]) as well as wüstite FeO [Jeanloz and Ahrens, 1980; Yagi et al., 1988] have been exposed to shock. The data has been extrapolated to inner core

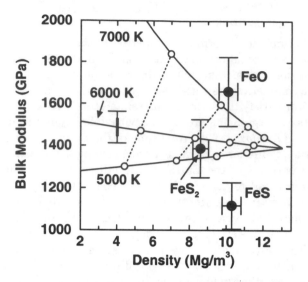

Figure 1. Properties of the alloy fraction that are required to match the seismically observed properties of the inner core [*Stixrude et al., 1997*]. For a given temperature (solid lines) the required effective bulk modulus is plotted as a function of required effective density. The dashed lines connect points of common alloy fraction for any light element X (2, 5, 10, and 20% from left to right). Estimated uncertainties in the alloy fractions required are indicated with the error bars on the curve corresponding to 6000 K (1% in density and 5% in bulk modulus). Superimposed are extrapolations of shock wave experimental estimates for FeO [*Jeanloz and Ahrens, 1980; Yagi et al., 1988*], FeS [*Anderson and Ahrens, 1996*], and Fe₂S [*Ahrens and Jeanloz, 1987; Anderson and Ahrens, 1996*] at 345 GPa and 6000 K (estimated uncertainties are 5% in density and 10% in bulk modulus).

conditions [*Stixrude et al., 1997*] and compared to the required elastic parameters (Fig. 1). Fig. 1 This analysis indicates that small amounts of either S or O (few atomic percent) would be sufficient to match the properties of the inner core.

Alfè et al. [2000a; 2000b; 2002] combined the geophysical approach with a chemical argument. They evaluated the liquid-solid partition coefficients of candidate light elements assuming thermodynamic equilibrium at the inner core boundary. The results show that neither S, Si, nor O alone can satisfy the observed density contrast at the inner core boundary and that a ternary or higher mixture of small amounts of S or Si with O is required.

2.4. Thermal State

Like the composition, the temperature of the inner core cannot be determined by direct observation. Assuming that the inner core is growing in equilibrium from incongruent freezing of the outer core liquid a

knowledge of the melting behavior of iron-rich systems at the pressure of the inner core boundary (330 GPa) would yield an important fixed-point temperature for the construction of whole Earth geotherms. Because the core is not a pure system, the temperature at the inner core boundary should differ from the melting point of pure iron. Freezing point depression in an eutectic system with no solid solution is given by the van Laar equation [*Brown and McQueen, 1982*] which yields a value of 800 K for a melting point of iron of 6000 K and 10% mole fraction of the light element. The value for the freezing point depression must be viewed as highly uncertain, however, since solid solution almost certainly exists at the high temperatures of the core. An independent estimate of the temperature of the core may be obtained by comparing the elastic properties of iron with those seismologically determined. We describe this approach as applied to the inner core below.

Seismic observations do provide constraints on some aspects of the thermal state of the core. In the outer core the compressional wave velocity v_P equals the bulk sound velocity $v_B = \sqrt{K_S/\rho}$. In a homogeneous, convecting system, v_B and ρ are related by adiabatic self-compression. Deviations from this state are characterized by the *Bullen* [1963] inhomogeneity parameter η being different from one. η is defined as

$$\eta = -\frac{v_B^2}{\rho g}\frac{\partial \rho}{\partial r}, \qquad (1)$$

where g the gravitational acceleration, and r the radius. η for the outer core is constrained by seismology to be 1 ± 0.05 [*Masters, 1979*]. This is consistent with (but does not uniquely require) a vigorously convecting outer core, and a resulting geotherm close to an adiabat, characterized by the gradient:

$$\frac{\partial T}{\partial r} = -\frac{\gamma g}{v_B^2}T, \qquad (2)$$

where γ is the Grüneisen parameter. Adopting the value measured for liquid iron at core conditions (γ=1.5) [*Brown and McQueen, 1986*] and a temperature at the inner core boundary of 6000 K, one finds a temperature contrast of 1500 K across the outer core.

The temperature contrast in the inner core is likely to be small. We can place an upper bound on it by assuming that the inner core is a perfect thermal insulator. If the inner core grows through freezing of the outer core its temperature profile will follow the solidus temperature. Based on this assumption *Stixrude et al.* [1997] estimated the total temperature difference across the inner core to be less than 400 K. Conductive or convective heat loss will further reduce this temperature gradient relative to the insulating case. Conduction is likely to be a very effective way to extract heat from the inner core, such that the temperature profile may fall below an adiabat [*Yukutake*, 1998; *Buffett*, 2000].

2.5. Dynamics

Song and Richards [1996] found that the differential travel time of PKIKP-PKP$_{BC}$ for earthquakes in the South Sandwich islands recorded in Alaska increased by 0.3 s over a period of three decades, and concluded that the inner core rotates relative to the mantle by 1°/year, a finding that was confirmed qualitatively using global data sets [*Su et al.*, 1996]. *Creager* [1997] showed that part of the signal could be explained by lateral heterogeneity in the inner core, and reassessed the rotation rate to a lower value. Recent years have seen body wave [*Souriau*, 1998] and free oscillations studies [*Laske and Masters*, 1999] that can not resolve inner core rotation, and put close bounds on rotation rate.

If it is present, differential inner core rotation would provide one of the few opportunities for direct observations of the dynamics in Earth's deep interior. Moreover, differential rotation could have a significant effect on the angular momentum budget of the Earth yielding an explanation of decadal fluctuations in the length of day [*Buffett*, 1996; *Buffett and Creager*, 1999]. Its origin is not fully understood, but geodynamo simulations produce super-rotation by electro-magnetic coupling with the overlying outer core [*Glatzmaier and Roberts*, 1996; *Kuang and Bloxham*, 1997; *Aurnou et al.*, 1998]. Gravitational stresses, arising from mass anomalies in the mantle, are also expected to act on the inner core [*Buffett*, 1996; 1997]. These tend to work against super-rotation by gravitationally locking the inner core into synchronous rotation with the mantle. In detail, the interplay between forces driving and resisting super-rotation depend on the rheology of the inner core, which is currently unknown. If the viscosity is sufficiently low, super-rotation may take place, with the consequence that the inner core undergoes continuous viscous deformation in response to the gravitational perturbations.

The interaction of super-rotation with gravitational stresses is just one of many proposed sources of internal deformation in the inner core. The subject has received substantial attention because solid state flow in the inner core can result in lattice-preferred orientation, thought to be essential for producing seismically observed anisotropy.

Other proposed sources of stress in the inner core include:

(a) Coupling with the magnetic field generated in the overlying outer core [*Karato*, 1999; *Buffett and Bloxham*, 2000; *Buffett and Wenk*, 2001]. Karato [1999] considered the radial component of the Lorentz force (F_r) at the inner core boundary which is caused by the zonal magnetic field (B_ϕ). This is typically the strongest contribution to the Lorentz force in geodynamo models [*Glatzmaier and Roberts*, 1995; *Kuang and Bloxham*, 1997]. *Buffett and Bloxham* [2000] argue that the inner core adjusts to F_r in a way to minimize steady solid state flow: only weak flow in the inner core is induced that is largely confined to the outermost portion. Considering additional terms to the Lorentz force $\vec{F}$ by including radial components of the magnetic field B_r they conclude that the azimuthal term F_ϕ which is proportional to $B_r B_\phi$ induces a steady shear flow throughout the inner core.

(b) Thermal convection [*Jeanloz and Wenk*, 1988; *Wenk et al.*, 2000a]. As in any proposed model of inner core flow, the viscosity of the inner core remains an important uncertainty, as does the origin and magnitude of heat sources required to drive the convection.

(c) Aspherical growth of the inner core [*Yoshida et al.*, 1996]. Fundamental considerations, based on the expected cylindrical symmetry of flow in the outer core, and detailed geodynamo simulations [*Glatzmaier and Roberts*, 1995] indicate that heat is transported more efficiently in the equatorial plane than along the poles, leading to an inhomogeneous growth rate of the inner core, and internal viscous relaxation. A key question is whether the magnitude of the effect is sufficient to produce lattice preferred orientation. In particular, the resulting strain rates are very small, and may not be sufficient to generate significant polycrystalline texture via recrystallization.

It has also been proposed that polycrystalline texture in the inner core may be acquired during solidification [*Karato*, 1993; *Bergman*, 1997]. However, if the inner core does experience solid state deformation, by one or more of the mechanisms described above, it is unclear to what extent the texture acquired during solidification would be preserved. It is possible that texture in the outermost portions of the inner core is dominated by the solidification process, whereas lattice preferred orientation in the bulk of the inner core is produced by deformation.

Further progress in our understanding of the composition, temperature, dynamics, and origin of anisotropy in the inner core is currently limited by our lack of knowledge of the properties of iron and iron alloys at high pressures and temperatures. A better understanding of elastic and other properties of iron at inner core conditions can provide a way to test hypotheses concerning the state and dynamics of the inner core.

3. MINERAL PHYSICS METHODS

As elasticity plays a central role in deep Earth geophysics we will emphasize aspects of mineral physics that are directly related to the determination of elastic properties. To gain deeper insight into complex elastic behavior, such as anisotropy, we need to know the full elastic constant tensor at the conditions in the Earth's center. We will focus on methods based on first-principles quantum mechanical theory, but also briefly review experimental progress, as it relates to comparison and validation of theory. A full review of experimental work has been given recently by *Hemley and Mao* [2001].

3.1. Experimental Progress

Determination of the elastic constants of metals in the diamond cell remains a particular challenge as now-standard techniques, such as Brillouin spectroscopy, cannot readily be used for opaque materials. A variety of alternative methods have been developed and applied to study iron at high pressure and ambient temperature. In the lattice strain technique [*Singh et al.,* 1998a; *Singh et al.,* 1998b] X-ray diffraction is used to study the strain induced in a polycrystal by uniaxial stress. A full determination of the elastic constant requires the measurement of d-spacing for many (h, k, l) lattice planes, or additional assumptions such as homogeneity of the stress field in the sample [*Mao et al.,* 1998]. Other efforts have exploited the relationship between phonon dispersion in the long-wavelength limit and elastic wave propagation: measurements of phonon dispersion have been used to estimate the average elastic wave velocity [*Lübbers et al.,* 2000; *Mao et al.,* 2001], and the longitudinal wave velocity [*Fiquet et al.,* 2001]. An approximate calibration has been explored in which the zone-center Raman active optical mode is related to the c_{44} shear elastic constant by a Brillouin-zone folding argument [*Olijnyk and Jephcoat,* 2000]. Finally, *Anderson et al.* [2001] by analyzing pressure induced changes in the intensity of X-ray diffraction patterns from hcp iron, extracted a Debye temperature Θ_D, and thus average elastic wave velocity, which they equated with v_S.

Whereas diamond anvil cell experiments most readily measure properties at ambient temperature, shock wave experiments achieve pressure and temperature conditions similar to those of the core through dynamic compression. By varying the speed of the driver impacting the sample, a set of different thermodynamic conditions are accessed, along a curve in pressure-density space called the Hugoniot. Temperature is not determined directly by the Rankine-Hugoniot equations, it must be measured using special techniques, such as optical pyrometry [*Yoo et al.,* 1993] or calculated on the basis of a thermodynamic model [*Brown and McQueen,* 1986]. Using temperature and Grüneisen paramter an adiabatic bulk modulus (K_S) on the Hugoniot can be determined. The impact of the driver plate on the sample not only sets up a shock in the sample but also in the plate itself. When the shock wave reflects off the back of the impactor, pressure is released and a longitudinal (compressional) sound wave is set up traveling forward through the system of impactor and sample. This has been exploited to determine v_P; combining v_P with K_S the corresponding v_S can be calculated [*Brown and McQueen,* 1986].

3.2. Computational Mineral Physics

With the sparse probing of thermodynamic conditions relevant for Earth's inner core by the experimental methods discussed in the previous section, and the difficulty to obtain information on single crystal elasticity, first-principles material physics methods provide an ideal supplement to experimental study, with all of thermodynamic space accessible, and various approaches to determine elasticity at hand. In the following sections we will introduce the basic principles of calculating such properties.

3.3. Total Energy Methods

Density functional theory [*Hohenberg and Kohn,* 1964; *Kohn and Sham,* 1965] provides a powerful and in principle exact way to obtain the energetics of a material with N nuclei and n interacting electrons in the groundstate (for a review see *Lundqvist and March* [1987]), with the electronic charge density $\rho_e(\vec{r})$ being the fundamental variable. It can be shown [*Hohenberg and Kohn,* 1964] that ground state properties are a unique functional of $\rho_e(\vec{r})$ with the total (internal) energy

$$E[\rho_e(\vec{r})] = T[\rho_e(\vec{r})] + U[\rho_e(\vec{r})] + E_{xc}[\rho_e(\vec{r})]. \quad (3)$$

Here T is the kinetic energy of a system of non-inter-/acting electrons with the same charge density as the interacting system, and U is the electrostatic (Coulomb) energy containing terms for the electrostatic interaction between the nuclei, the electrons, and nuclei-electron interactions. The final term E_{xc} is the exchange-corre/-lation energy accounting for many body interactions between the electrons. Density functional theory allows one to calculate the exact charge density $\rho_e(\vec{r})$ and hence the many-body total energy from a set of n single-particle coupled differential equations [*Kohn and Sham*, 1965]

$$\{-\nabla^2 + V_{KS}[\rho_e(\vec{r})]\}\psi_i = \varepsilon_i \psi_i, \qquad (4)$$

where ψ_i is the wave function of a single electronic state, ε_i the corresponding eigenvalue, and V_{KS} the effective (Kohn-Sham) potential that includes the Coulomb and exchange-correlation terms from (3). The Kohn-Sham equations are solved self-consistently by iteration. Density functional theory has been generalized to spin polarized (magnetic) systems [*Singh*, 1994].

While density functional theory is exact in principle the exact solution of the Kohn-Sham equations requires the knowledge of the universal form of the exchange-correlation potential which is yet unknown. Approximations for V_{xc} however have been very successful. The local density approximation (LDA) [*Lundqvist and March*, 1983] replaces V_{xc} at every point in the crystal with the value of a homogeneous electron gas with the same local charge density. This lowest order approximation yields excellent agreement with experiment for a wide variety of materials, but fails for some metals. Most prominently for iron LDA wrongly predicts hcp as the ground state structure for iron at ambient pressure [*Stixrude et al.*, 1994]. Generalized gradient approximations (GGA) include a dependence on local gradients of the charge density in addition to the charge density itself [*Perdew et al.*, 1996]. GGA yields the correct ground state of iron at ambient pressure and predicts the phase transition from bcc to hcp iron at the experimentally determined pressure [*Asada and Terakura*, 1992; *Stixrude et al.*, 1994].

In addition to total energy it is possible to calculate directly first derivatives of the total energy with first-principles methods. This allows one to determine forces acting on the nuclei and stresses acting on the lattice [*Nielsen and Martins*, 1985].

All-electron, or full potential methods make no additional essential approximations to density functional theory. Computational methods such as the Linearized Augmented Plane Wave (LAPW) method provide an important standard of comparison. All-electron methods are very costly (slow), and are currently impractical for many problems of interest. More approximate computational methods have been developed, which, when applied with care, can yield results that are nearly identical to the all-electron limit.

In the pseudopotential approximation the nucleus and core electrons are replaced inside a sphere of radius r_c (cut-off radius) with a simpler object that has the same scattering properties (for a review see *Pickett* [1989]). The pseudopotential is much smoother than the bare Coulomb potential of the nuclei, and the solution sought is only for the pseudo-wavefunctions of the valence electrons that show less rapid spatial fluctuations than the real wavefunction in the core region or those of the core electrons themselves. The construction of the pseudopotential is non-unique and good agreement with all-electron calculations must be demonstrated.

Iron provides a particular challenge. For example, all-electron results show that pressure-induced changes in the $3p$ band are important for the equation of state [*Stixrude et al.*, 1994], and so should be treated fully as valence electrons in a pseudopotential approach.

For our work on the high temperature elasticity of hcp iron [*Steinle-Neumann et al.*, 2001] we have constructed a Troullier-Martins [*Troullier and Martins*, 1991] type pseudopotential for iron in which $3s$, $3p$, $3d$, and higher electronic states are treated fully as valence electrons. Agreement with all-electron calculations of the equation of state and elastic constants is excellent: for hcp iron the pressure at inner core densities is within 1 % of all-electron (LAPW) results (Fig. 2), Fig. 2 and the elastic constant tensor at inner core density is within 2% rms.

The predictions of density functional theory can be compared directly with experiment, or with geophysical observations. For example, by computing the total energy as a function of volume, one obtains the equation of state. This equation of state is static, that is the effects of thermal vibrations are absent. This athermal state is one that is not attainable in the lab where zero-point motion cannot be eliminated. Static properties are often directly comparable to experimental measurements taken at ambient temperature since the effect of 300 K is small for properties such as the density, or the elastic constants. However, for comparison with the earth's core, thermal effects are essential. Calculation of the effects of temperature from first-principles is more involved because one must calculate the energies associated with atomic displacements, including those that break the symmetry of the lattice.

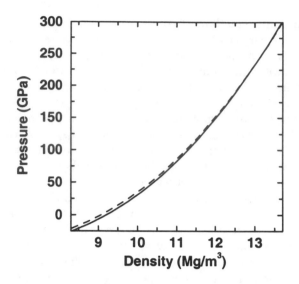

Figure 2. Equation of state for hcp iron obtained from all electron results (dashed line) [*Steinle-Neumann et al.*, 1999] and the pseudopotential used in the calculation of high temperature thermoelasticity (solid line) [*Steinle-Neumann et al.*, 2001].

3.4. High Temperature Methods

Statistical mechanics provides the tools to deal with material properties at high temperature. The thermodynamic behavior of any physical system is uniquely defined by the so-called fundamental relation, which, for a non-magnetic or Pauli-paramagnetic solid in the canonical ensemble (particle number N, volume V, and temperature T held constant) takes the form

$$F(V,T) = E(V,T) - TS_{el}(V,T) + F_{vib}(V,T) \quad (5)$$

where F is the Helmholtz free energy. The total energy $E(V,T)$ is now a function of T as well, because we explicitly have to account for thermal excitation of electrons according to Fermi-Dirac statistics. S_{el} is the entropy associated with this excitation of the electrons [*McMahon and Ross*, 1977], and F_{vib} is the vibrational part of the free energy. F_{vib} is derived from the partition function for a system with N atoms, which in the classical limit, appropriate at high temperature conditions (significantly above Θ_D) is

$$Z_{vib} = \frac{1}{N!\Lambda^{3N}} \cdot \qquad (6)$$
$$\int d\vec{R}_1 d\vec{R}_2 \ldots d\vec{R}_N \exp\left[-\frac{F_{el}(\vec{R}_1, \vec{R}_2, \ldots, \vec{R}_N; T)}{kT}\right]$$

and

$$F_{vib} = -kT \ln Z_{vib}. \qquad (7)$$

Z_{vib} is a 3N dimensional integral over the coordinates of the nuclei located at $\vec{R}_i$ with the electronic free energy $F_{el} = E - TS_{el}$ uniquely defined by the coordinates of

the atoms and T. $\Lambda = h/\sqrt{2\pi mkT}$ is the de Broglie wavelength with h the Planck and k the Boltzmann constant, and m the nuclear mass.

A naïve attempt to evaluate the integral (7) fails because of the large dimensionality, and because most configurations contribute little to the integral. What is required is a search of configuration space that is directed towards those configurations that have relatively low energy. In the particle-in-a-cell (PIC) method and the lattice dynamics method described next, atoms are restricted to vibrations about their ideal crystallographic sites, that is diffusion is neglected. This is not a severe approximation to equilibrium thermodynamic properties at temperatures below the premelting region. Molecular dynamics, described last, in principle permits diffusion, although in practice computationally feasible dynamical trajectories are sufficiently short that special techniques are often required to study non-equilibrium processes.

3.4.1. Particle-in-a-Cell. Here the basic approximation motivates the division of the lattice into non-overlapping sub-volumes centered on the nuclei (Wigner-Seitz cell Δ_{WS}) with the coordinates of each atom restricted to its cell. A second basic assumption in the PIC model is that the motions of the atoms are uncorrelated. We can expect this approximation to become increasingly valid with rising temperature above Θ_D and below melting.

If the energy change resulting from moving one particle be independent of the vibrations of the other atoms, the partition function factorizes, and the $3N$ dimensional integral is replaced by the product of N identical 3-dimensional integrals, which reduces the computational burden tremendously

$$Z_{vib} = \frac{1}{\Lambda^{3N}} \left(\int_{\Delta_{WS}} d\vec{R}_w \exp\left[\frac{-\Delta F_{el}(\vec{R}_w, T)}{kT}\right] \right)^N \quad (8)$$

with

$$\Delta F_{el}(\vec{R}_w, T) = F_{el}(\vec{R}_w, T) - F_{el}(\vec{R}_{w0}, T). \quad (9)$$

($N!$ no longer appears in the prefactor of (8) as the atoms are distinguished by their lattice site.) One evaluates the energetics of moving one atom w (the so-called wanderer) with the equilibrium lattice position $\vec{R}_{w0}$ in the potential of the otherwise ideal lattice; this is a mean field approach to the vibrational free energy.

Because large displacements of w are included in the integral (up to about 1/2 nearest neighbor distance) the PIC model treats anharmonicity of the vibrations explicitly. The method is computationally very efficient, and can be sped up even more by minimizing the total number of calculations involved in evaluating the integral (8). The angular integrations can be performed efficiently by developing a quadrature which requires evaluation of the integrand along a small number of special directions that are determined by the point symmetry of the lattice site [*Wasserman et al.*, 1996].

The cell model has been used for calculations on iron for the thermodynamics of both the hcp and fcc phase [*Wasserman et al.*, 1996], and for high temperature elasticity of hcp iron [*Steinle-Neumann et al.*, 2001]. It has also been successfully applied to thermoelasticity of tantalum [*Cohen and Gülseren*, 2001; *Gülseren and Cohen*, 2002].

3.4.2. Lattice Dynamics. The calculation of forces on the atoms allows one directly to compute the vibrational frequencies of the material. The dynamical matrix may be calculated row by row by displacing one atom by a small amount from its equilibrium site in a supercell and evaluating the resulting forces on the other atoms. F_{vib} is then calculated by performing the appropriate summation over wavevector and phonon branches.

A fundamental approximation in lattice dynamics is that the vibrations of the atoms about their equilibrium are harmonic, only terms that are quadratic in atomic displacements are retained in the expression for the total energy. However, anharmonicity might become important for conditions of the inner core where the temperature is just below the melting point of the material, and considerable effort must be put into anharmonic corrections.

Lattice dynamics has been used extensively over the past few years to address the thermodynamics of hcp iron [*Alfè et al.*, 2001], studies on stability of various phases [*Vočadlo et al.*, 2000], melting [*Alfè et al.*, 1999], and studies on core composition [*Alfè et al.*, 2000a; 2000b; 2002].

3.4.3. Molecular Dynamics. While lattice dynamics and the PIC model essentially evaluate ensemble averages for the thermodynamics of a system, molecular dynamics explores the time evolution of a single realization of the system. In this case thermodynamic properties are calculated as time averages by appealing to the ergodic hypothesis. To obtain the time evolution of the system the forces acting on the atoms are coupled with Newton's second law. The set of N coupled differential equations is then integrated. A large supercell (100 atoms) and long time series (thousands of time steps) are required for convergence of equilibrium properties. Molecular dynamics has been applied to the study of high temperature properties of solid iron including melting [*Laio et al.*, 2000] by using a clever

hybrid scheme for the electronic structure, as well as for liquid iron [*de Wijs et al.*, 1998]. *Laio et al.*, [2000] combined first-principles total energy and force calculations for a limited number of time steps with a semi-empirical potential fit to the first-principle results.

3.5. Elastic Constants

Seismic wave propagation is governed by the elastic constants, and the density. In most cases, seismic frequencies are sufficiently high that the adiabatic elastic constants, c_{ijkl}^S, are relevant. First-principles calculations of the type described here yield the isothermal elastic constants, c_{ijkl}^T, most directly, and a conversion must be applied [*Davies*, 1974]

$$c_{ijkl}^S = c_{ijkl}^T + \frac{T}{\rho C_V}\lambda_{ij}\lambda_{kl}, \qquad (10)$$

where C_V is the specific heat at constant volume and

$$\lambda_{ij} = \sum_{l,k\leq l}\alpha_{kl}c_{ijkl}^T. \qquad (11)$$

The thermal expansivity tensor α_{ij} for a hexagonal system has only entries in the diagonal with

$$\alpha_{11} = \alpha_{22} = 1/a\cdot(\partial a/\partial T)_P \quad, \quad \alpha_{33} = 1/c\cdot(\partial c/\partial T)_P, \qquad (12)$$

the linear thermal expansivity of the a- and c-axis, respectively.

Under conditions of isotropic pre-stress, elastic wave propagation is governed by the so-called stress-strain coefficients which are defined by

$$\sigma_{ij} = c_{ijkl}^T\varepsilon_{kl}, \qquad (13)$$

with σ_{ij} the stress and ε_{ij} the strain. Other definitions of the elastic constants have appeared in the literature [*Barron and Klein*, 1965]: one may define elastic constants as the second strain derivatives of the free energy, which are not equivalent to (13) in general. If the pre-stress is isotropic, and if the applied strain is isochoric to all orders, then the free energy may be directly related to the stress-strain coefficients

$$F(V,\varepsilon_{ij}',T) = F(V,0,T) + \frac{1}{2}c_{ijkl}^T(V,T)\varepsilon_{ij}'\varepsilon_{kl}'. \qquad (14)$$

where the primes indicate the deviatoric strain. This relationship has been used to calculate the elastic constants from first-principles calculations of the total energy alone [*Cohen et al.*, 1997; *Steinle-Neumann et al.*, 1999]. The elastic constants may also be calculated by appealing to the dissipation-fluctuation theorem which relates the c_{ijkl}^T to fluctuations in the shape of the crystal at constant stress. This provides one means of calcu-

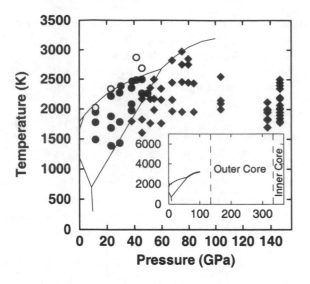

Figure 3. Phase diagram of iron from diamond anvil cell experiments. Stable crystalline phases are bcc at ambient conditions, fcc (solid circles) at high temperature, and hcp (diamonds). Liquid iron is shown in the open symbols. bcc also has a small phase stability field at low pressure immediately below melting. Data are from *Shen et al.* [1998] and *Shen and Heinz* [1998]. The inset shows the pressure-temperature range relevant for a study of the inner core.

lating the elastic constants in molecular dynamics simulations [*Parrinello and Rahman*, 1982; *Wentzcovitch*, 1991].

Elastic wave velocities are related to the elastic constants by the Christoffel equations

$$\left(c_{ijkl}^S n_j n_k - \rho v^2 \delta_{il}\right) u_i = 0, \qquad (15)$$

where $\vec{n}$ is the propagation direction and $\vec{u}$ the polarization of the wave, v the phase velocity, and δ_{il} is the Kronecker delta function. Solving (15) for a given propagation direction yields three velocities, one with quasi-longitudinal polarization (v_P), and two with quasi-transverse polarizations (v_S). From the full elastic constant tensor, we may also determine the bulk (K_S) and shear (G) moduli using Hashin-Shtrikman bounds [*Watt and Peselnick*, 1980] which give tighter bounds on G than the usually used Voigt-Reuss-Hill averages [*Hill*, 1963]. The isotropically averaged aggregate velocities v_P and v_S can then be calculated by

$$v_P = \sqrt{\frac{K_S + \frac{4}{3}G}{\rho}} \quad , \quad v_S = \sqrt{\frac{G}{\rho}} \qquad (16)$$

Usually the Voigt notation is used to represent elastic constants, replacing the fourth rank tensor c_{ijkl} with a 6×6 pseudomatrix in which pairs of indices are replaced

by a single index utilizing the symmetry of the stress and strain tensors:

$$11 \to 1; \qquad 22 \to 2; \qquad 33 \to 3; \qquad (17)$$
$$23, 32 \to 4; \qquad 13, 31 \to 5; \qquad 12, 21 \to 6.$$

In this notation the five single crystal elastic constants for a hexagonal system are: the longitudinal elastic constants c_{11} and c_{33}, the off-diagonal elastic constants, c_{12} and c_{13}, and a shear constant c_{44}. In the following discussions we also refer to another linearly dependent shear constant for comparison with c_{44}: $c_{66} = 1/2(c_{11} - c_{12})$. To calculate the five elastic constants with a first-principles total energy method, we must evaluate the effect of five different strains on the free energy (14). The bulk modulus and the change in the equilibrium c/a ratio with compression provide two pieces of information, yielding two independent combinations of elastic constants. The application of three isochoric strains, of hexagonal, orthorhombic, and monoclinic symmetry, yield the other three pieces of information necessary to obtain the full elastic constant tensor [*Steinle-Neumann et al.*, 1999].

4. PHYSICAL PROPERTIES OF DENSE IRON

4.1. Phase Diagram

To the extent that the inner core is composed of pure iron (or nearly pure iron), the phase diagram of iron determines the crystalline structure of the inner core. Despite considerable progress in experimental determination of the phase diagram and melting at pressures approaching those of the core, the stable phase of iron at inner core conditions can not yet uniquely be identified.

This issue is of great geophysical and geochemical importance. First, it is central in our understanding of inner core anisotropy. Different phases of iron show a distinctly different single crystal anisotropy both in magnitude and symmetry [*Stixrude and Cohen*, 1995b]. The enthalpy of various phases and hence the amount of latent heat released at the inner core boundary may depend strongly on the crystalline phase. Finally, the ability to incorporate impurities may be determined by the structure.

Three phases of iron have been unambiguously identified (Fig. 3): Fig. 3 the ambient condition ferromagnetic body center cubic phase (bcc, α) is stable to about 13 GPa and up to 1200 K; at higher temperature the cubic close packed (fcc, γ) phase exists, with bcc reappearing in a narrow stability field (δ) just below melting. The magnetic ground state of the fcc phase is a complex spin

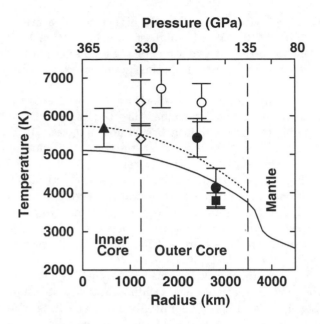

Figure 4. Melting temperatures of iron and estimates of the geotherm in Earth's core, with the geotherm from *Stacey* [1992] in the solid line. Experiments on melting of iron are shown with a square from static diamond anvil cell experiments [*Boehler*, 1993] and with circles for melting along the Hugoniot (solid from *Brown and McQueen* [1986] and open from *Yoo et al.* [1993]). The two points from *Brown and McQueen* [1986] show the uncertainty in the detection of melting: both points represent discontinuities in acoustic velocity along the Hugoniot, and the occurrence of melting is ambiguous. Two points from *Yoo et al.* [1993] bracket melting as observed with optical pyrometry. Diamonds show theoretical estimates of the melting point of iron at the pressure of the inner core boundary by *Alfè et al.* [2002b] (upper symbol) and *Laio et al.* [2000] (lower symbol). Inner core temperature estimated from a comparison of inner core elasticity with that of iron [*Steinle-Neumann et al.*, 2001] is shown with a solid triangle. The dashed line is an adiabat through the core, based on the latter result.

density wave [*Tsunoda et al.*, 1993; *Uhl et al.*, 1994]; in the stability field of fcc iron the local moments appear not to be ordered, however. The hexagonal close packed (hcp, ε) phase is the high pressure phase, stable to at least 300 GPa at room temperature [*Mao et al.*, 1990]. Theory predicts this phase to be non-magnetic (Pauli paramagnet) at core pressures [*Söderlind et al.*, 1996; *Steinle-Neumann et al.*, 1999]. Magnetism is not observed experimentally in the hcp phase [*Taylor et al.*, 1991], but magnetic moments on the atoms are predicted on the basis of first-principles theory [*Steinle-Neumann et al.*, 1999] up to 50 GPa. Moreover, magnetism is observed in epitaxially grown overexpanded lattices of hcp iron [*Maurer et al.*, 1991], consistent with theoretical predictions. The possible presence of

magnetic states in hcp iron is important for understanding the equation of state (see below) and the phase diagram in the sub-megabar range. The competition between magnetic and non-magnetic contributions to the internal energy, differences in vibrational and magnetic entropy, and differences in volume all contribute to phase stability in iron [*Moroni et al.*, 1996].

Two experimental lines of evidence suggest additional stable polymorphs of iron at high pressure and temperature. First, in the shock wave experiment by *Brown and McQueen* [1986] (see also *Brown et al.* [2000] and *Brown* [2001]) there are two discontinuities in v_P, one at 200 GPa, the other 243 GPa (Fig. 4). Fig. 4 While the one at higher pressure was associated with melting of the sample, the lower one was originally attributed to the hcp to fcc phase transition. With a better characterization of the phase diagram at lower temperature and pressure today this is an unlikely scenario: the fcc-hcp phase transition ends with a triple point at much lower pressure (Fig. 3) [*Shen et al.*, 1998; *Boehler*, 2000]. Re-appearance of the bcc phase of iron has been suggested as an explanation of the apparent solid-solid phase transformation at 200 GPa [*Matsui and Anderson*, 1997]. However, first-principles theory shows that the bcc phase is mechanically unstable at high pressure and is unlikely to exist as a stable phase [*Stixrude and Cohen*, 1995a; *Söderlind et al.*, 1996; *Vočadlo et al.*, 2000]. An alternative interpretation attributes the first discontinuity to the onset of melting, with melting completed only at 243 GPa [*Boehler and Ross*, 1997]. A recent repetition of the experiment of *Brown and McQueen* [1986] has been unable to resolve whether or not a phase transformation in addition to melting occurs on the Hugoniot [*Nguyen and Holmes*, 2001].

Second, additional phases of iron have been proposed on the basis of static high pressure experiments. X-ray diffraction patterns measured in laser heated diamond anvil cell experiments have been argued to be incompatible with any known iron polymorph [*Saxena et al.*, 1995; *Yoo et al.*, 1995; *Andrault et al.*, 1997]. The anomalous signal is subtle, and the proposed structures, a double hexagonal dhcp [*Saxena et al.*, 1995; *Yoo et al.*, 1995] and an orthorhombic structure [*Andrault et al.*, 1997], are closely related to hcp. Using in-situ X-ray diffraction *Shen et al.* [1998] found only fcc and hcp phases at pressure and temperature, while the dhcp phase was observed in temperature quenched samples. *Andrault et al.* [1997] used high strength pressure media which could induce non-hydrostatic conditions [*Boehler*, 2000]. *Vočadlo et al.* [2000] examined the relative stability of proposed high pressure phases using computational *ab-initio* methods, and found that the

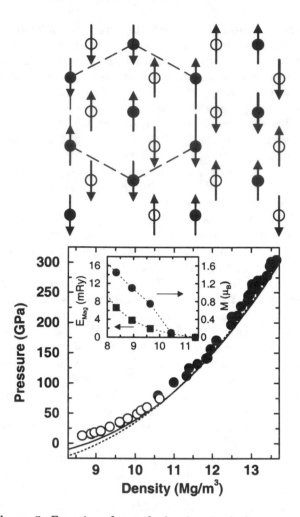

Figure 5. Equation of state for hcp iron in the lower panel. Static theoretical results are compared to room temperature diamond anvil cell experiments by *Jephcoat et al.* [1986] (open circles) and *Mao et al.* [1990] (solid circles). The dashed line shows non-magnetic results [*Stixrude et al.*, 1994], the solid line the density-pressure relation for an antiferromagnetic structure (afmII) [*Steinle-Neumann et al.*, 1999]. The inset shows the magnetic moment (circles) and associated magnetic stabilization energy (squares) for the afmII structure. The afmII structure itself with atoms at z=1/4 in solid, z=3/4 in open circles is displayed in the upper panel. Arrows indicate the spin polarization of the atoms.

orthorhombic structure is mechanically unstable, and that the dhcp phase is energetically less favored than hcp.

Most measurements of the melting temperature of iron from static experiments show reasonable agreement up to a pressure of 100 GPa, where the melting point is 2800-3300 K [*Shen and Heinz*, 1998; *Boehler*, 2000; *Hemley and Mao*, 2001]; the data of *Williams et*

al. [1987] yield a significantly higher temperature at 100 GPa (4100 K). The highest pressure datum from diamond anvil cell experiments is at 200 GPa where *Boehler* [1993] finds the melting point at 3800 K (Fig. 4).

As mentioned above, in shock wave experiments the temperature is not determined directly; based on their dynamic compression data *Brown and McQueen* [1986] calculated the temperature at the Hugoniot melting point (243 GPa) to be in the range of 5000 to 5700 K (Fig. 4), consistent with subsequent theoretical calculations of the Hugoniot temperature [*Wasserman et al.*, 1996]. Measurements of the temperature by optical pyrometry yield a melting point between 235 and 300 GPa with temperatures of 6350 and 6720 K, respectively [*Yoo et al.*, 1993].

Two *ab-initio* calculations of the melting curve of iron have been carried out, yielding inconsistent results (Fig. 4). *Alfè et al.* [1999] initially found a melting temperature of 6700 ± 600 K at the inner core boundary by comparing Gibbs free energies of solid and liquid, a subsequent reanalyzation of their results [*Alfè et al.*, 2002b] did yield a lower temperature of approximately 6350 K. *Laio et al.* [2000] determined a considerably lower temperature of 5400 ± 400 K. The origin of these discrepancies are not clear, but may be related to the quality of the anharmonic corrections in the study of *Alfè et al.* [1999], the semi-empirical potential used to augment first-principles calculations in the study of *Laio et al.* [2000], or different statistical sampling and runtime adopted in these two studies. In this context, it is worth pointing out that theoretical calculations of the melting temperature are extremely demanding as they involve the comparison of two large numbers (Gibbs free energies of solid and liquid) which must both be calculated to high precision.

4.2. Static Equation of State

Experimental measurements and theoretical predictions of the equation of state of non-magnetic hcp iron agree well at core pressures (Fig. 5). Fig. 5 At relatively low pressures, however, a discrepancy develops that is larger than that for the bcc phase and what is typical for other transition metals [*Körling and Häglund*, 1992; *Steinle-Neumann et al.*, 1999] suggesting that fundamental aspects of the physics of hcp iron may not be well understood to date.

Although experiment has so far not detected magnetism in hcp iron, recent first-principles theoretical calculations have found stable magnetic states [*Steinle-Neumann et al.*, 1999]. These are more stable than the non-magnetic state by more than 10 mRy at low pressure. The most stable magnetic arrangement found so far is one of antiferromagnetic ordering (afmII, Fig. 5) which retains a finite moment up to 50 GPa, well into the pressure region where hcp iron is stable [*Steinle-Neumann et al.*, 1999]. Because magnetism tends to expand the lattice, the presence of magnetism reduces the discrepancy between experimental and theoretical equations of state considerably. It is likely that still more stable magnetic states will be found, and that the ground state is a more complex magnetic structure involving spin-glass like disorder, incommensurate spin density waves, non-collinear magnetism, or a combination of these [*Cohen et al.*, 2002].

4.3. Static Elastic Constants

A comparison shows considerable disagreement of the single crystal elastic constants (Fig. 6) Fig. 6 between experiment [*Mao et al.*, 1998; *Singh et al.*, 1998b] and theory [*Stixrude and Cohen*, 1995b; *Söderlind et al.*, 1996; *Cohen et al.*, 1997; *Steinle-Neumann et al.*, 1999]. The difference in the longitudinal constants c_{11} and c_{33} decreases from ~50% at low pressure to a little more than 10% at high compression. As pronounced is the discrepancy for the shear elastic constants: the difference in c_{44} increases with compression to 30%, and c_{66} differs as much as 40% at high pressure, and by a factor of two at the low density data point.

Because the full elastic constant tensor of hcp iron has been measured only by the new lattice strain technique, the large discrepancy between experiment and theoretical prediction prompted a stringent comparison of both methods for a well characterized hcp metal. For the $5d$ transition metal rhenium, ultrasonic measurements of elastic constants were not only performed at ambient condition, but also up to 0.5 GPa, constraining the initial pressure slope (Fig. 6) [*Manghnani et al.*, 1974]. Calculated elastic constants at zero pressure and their compression dependence [*Steinle-Neumann et al.*, 1999] show excellent agreement with the ultrasonic data. In contrast, lattice strain experiments [*Duffy et al.*, 1999] do not compare favorably with either result. One of the longitudinal elastic constants (c_{33}) is considerably smaller (20%), and as in the case for iron the shear constant c_{44} shows the largest discrepancy, being larger than the ultrasonic value by 50%. This comparison indicates that additional factors other than elasticity influence the measurements in lattice strain experiments. Subsequently it has been found that strong lattice preferred orientation developed in the experiments for hcp iron [*Wenk et al.*, 2000b; *Matthies et al.*, 2001], which

may cause the assumption of stress homogeneity to be violated.

For aggregate v_P and v_S, the discrepancies between theory and experiment are smaller, but still significant (Fig. 7). Fig. 7 Part of the discrepancy between theory and experiment may be attributed to the equation of state. Since theory overestimates the density of hcp iron

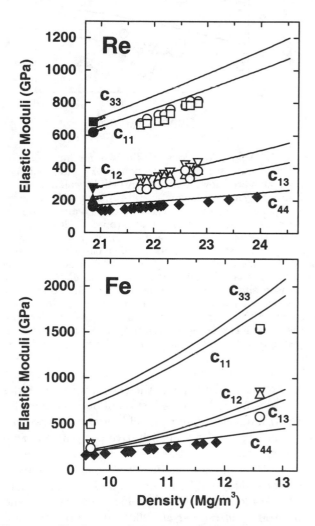

Figure 6. Single crystal elastic constants for hcp rhenium (upper panel) and iron as a function of compression. The solid lines are Eulerian finite strain fits to computational results [*Steinle-Neumann et al.*, 1999]. Open symbols show lattice strain experiments from *Duffy et al.* [1999] for rhenium and *Mao et al.* [1998] for iron: c_{33} squares, c_{11} circles, c_{12} triangles down, c_{13} triangles up, and c_{44} circles. The shear elastic constant c_{44} as inferred from Raman frequency measurements are shown with diamonds; data for rhenium are from *Olijnyk et al.* [2001], for iron from *Merkel et al.* [2000]. For rhenium ultrasonic measurements are available at low pressure [*Manghnani et al.*, 1974] (filled symbols as above, with initial pressure dependence indicated).

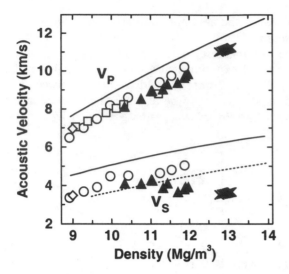

Figure 7. Aggregate acoustic velocities for hcp iron and the inner core (crosses). Compressional (v_P) and shear (v_S) wave velocity are shown from first-principles calculations in solid lines for the non-magnetic state [*Steinle-Neumann et al.*, 1999]. Experimental results at ambient temperature are based on the phonon density of states (open circles) [*Mao et al.*, 2001], the longitudinal acoustic phonon frequency (open squares) [*Fiquet et al.*, 2001], ultrasonic measurements (open diamonds) [*Mao et al.*, 1998], and the intensity of x-ray diffraction peaks (dashed line) [*Anderson et al.*, 2001]. For comparison shock wave experimental data in the stability field of hcp iron is included (filled triangles) [*Brown and McQueen*, 1986]: for the shock wave results temperature increases with compression resulting in a temperature of approximately 4500 K at the highest density point in the figure.

at low pressure, one expects the elastic wave velocities to be overestimated. The reduction of bulk modulus for the afmII magnetic state will yield lower compressional wave velocity, but for a quantitative comparison of v_P and v_S information on the full elastic constant tensor for this orthorhombic structure (with nine independent elastic constants) will be needed. The LA phonon data (v_P) by *Fiquet et al.* [2001] at high compression and the v_S from *Anderson et al.* [2001] at low pressure appear to be anomalous as they fall below the shock wave data [*Brown and McQueen*, 1986] which, due to thermal effects, would be expected to yield smaller sound velocities than room temperature experiments.

4.4. Thermal Equation of State

The Hugoniot provides the strongest constraint on the equation of state of iron at core conditions. First-principles theoretical calculations of the Hugoniot [*Wasserman et al.*, 1996, *Laio et al.*, 2000, *Alfè et al.*, 2001] have typically found excellent agreement with that experimentally measured [*Brown and McQueen*, 1986].

A direct comparison of solid state properties between shock wave experiments and computations at inner core pressures is unfortunately not possible, as the Hugoniot is in the stability field for the liquid above 250 GPa as discussed above. We have performed a detailed comparison of the properties of iron at inner core conditions obtained from several first-principles theoretical calculations [*Laio et al.*, 2000; *Alfè et al.*, 2001; *Steinle-Neumann et al.*, 2001] and from static [*Dubrovinsky et al.*, 2000] and dynamic compression experiments (Fig. 8). Fig. 8 The properties of iron determined from these sources agree with each other to within 2% in pressure and to within 10% in bulk modulus at 13 Mg/m^3 and 6000 K. This comparison supports the conclusion that the inner core is not pure iron, but that it must contain a small amount of lighter elements. Iron is consistently found to be denser than the inner core, even at a temperature of 7000 K. The bulk modulus of iron, while showing considerably more scatter, appears to be consistent with that of the inner core at a temperature near 6000 K.

4.5. Structure

The axial ratio c/a of the hexagonal unit cell is important for understanding the elastic anisotropy. Among transition metals at ambient conditions, the value of c/a is found to be correlated with the longitudinal wave anisotropy, which can be characterized by the ratio c_{33}/c_{11}. Large values of c/a are associated with a relatively small c_{33} and slow P-wave propagation along the c-axis [*Simmons and Wang*, 1971]. A change in axial ratio through compression or temperature could change the single crystal anisotropy considerably.

Similar to most ambient condition hcp transition metals iron crystallizes with an axial ratio c/a slightly below the ideal value (~ 1.6) [*Jephcoat et al.*, 1986; *Stixrude et al.*, 1994]. It changes little with compression, experiments show a slight decrease with pressure, theory a minor increase.

At ambient pressure, hcp transition metals typically show significant but small changes in c/a up to high homologous temperature [*Eckerlin and Kandler*, 1971]. The ratio c_{33}/c_{11} shows correspondingly small changes [*Simmons and Wang*, 1971]. The largest change in this ratio is exhibited by titanium which shows a change of 13%. Experiments on hcp iron at higher pressure observed a small increase in c/a with temperature in the pressure range of 15-30 GPa and temperatures up to 1200 K (Fig. 9) Fig. 9 [*Huang et al.*, 1987; *Funamori et al.*, 1996; *Uchida et al.*, 2001]. Similarly, *Dubrovinsky et al.* [1999] report a $c/a=1.623$ at 61 GPa and 1550 K. However, these results can be questioned on

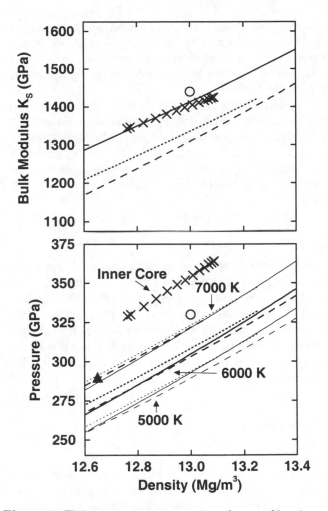

First-principles theoretical calculations have predicted a significant temperature induced increase in the c/a ratio of hcp iron [*Wasserman et al.*, 1996; *Steinle-Neumann et al.*, 2001], which is apparently consistent with the majority of the existing experimental data. For an inner core density of 13.04 Mg/m³ the axial ratio increases from the static value close to 1.6 to about 1.7 at 6000 K (Fig. 10). Fig. 10 This implies that at constant density the c-axis grows at the expense of the a-axis. The linear thermal expansivities α_{11} and α_{33} at constant pressure provide another way to represent the change in structural properties. Over a wide range of thermodynamic conditions, the a-axis is predicted to compress slightly with increasing temperature (Fig. 11). Fig. 11

The temperature-induced change in c/a appears to depend on the absolute temperature rather than homologous temperature: the reason that c/a of hcp iron reaches such large values at inner core conditions appears to be that higher sub-solidus temperatures may be reached. The origin of the temperature induced increase in c/a can be traced to the vibrational entropy,

Figure 8. Finite temperature equation of state of hcp iron. The lower panel shows a comparison for pressure-density relation in the inner core (crosses) and for hcp iron at temperatures of 5000 K, 6000 K, and 7000 K. Experimental data (long dashed lines) from *Dubrovinsky et al.* [2000] are extrapolated to inner core conditions. Two sets of calculations are from *Steinle-Neumann et al.* [2001] (solid lines) and from *Alfè et al.* [2001] (dashed lines). For comparison included are pressure and density on the Hugoniot at 7000 K (triangle) [*Brown and McQueen*, 1986], and results by *Laio et al.* [2000] at 5400 K (open circles). The upper panel compares the corresponding adiabatic bulk moduli along 6000 K isotherms with those of the inner core (same symbols). For the theoretical results by *Alfè et al.* [2001] and *Steinle-Neumann et al.* [2001] K_S is calculated self-consistently. K_T from static experiments [*Dubrovinsky et al.*, 2000] and the result from *Laio et al.* [2000] are converted using thermodynamic parameters from theory.

the grounds that non-hydrostatic stress may have influenced the temperature dependence of c/a, a contention supported by hysteresis on a heating and cooling cycle seen in the data of *Funamori et al.* [1996], and data by *Dubrovinsky et al.* [2000] at 185 GPa and 1115 K with a small c/a (1.585).

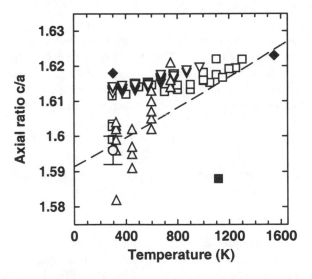

Figure 9. Axial ratio c/a of hcp iron as a function of temperature. The dashed line shows theoretical results for a density of 12.52 Mg/m³ corresponding to a static pressure of 195 GPa [*Steinle-Neumann et al.*, 2001]. The symbols show a polybaric set of experimental data at lower pressures, with measurements in the pressure range of 15-20 GPa from *Uchida et al.* [2001] (triangles down) and *Huang et al.* [1987] (triangles up). Open squares from *Funamori et al.* [1996] are in the pressure range of 23-35 GPa. Higher pressure data are from *Dubrovinsky et al.* [1999] (61 GPa, filled diamonds) and *Dubrovinsky et al.* [2000] (185 GPa, square). For comparison the room temperature axial ratio from *Mao et al.* [1990] (197 GPa, open circle) is included. The experimental uncertainty shown for *Mao et al.* [1990] can be taken as representative.

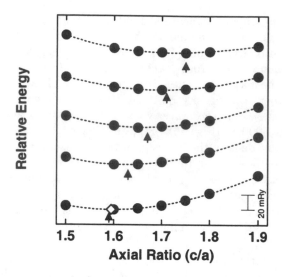

Figure 10. Relative energies (F) for hcp iron at an inner core density of 13.04 Mg/m^3 as a function of axial ratio c/a for temperature increments of 2000 K, starting with the static results at the bottom [*Steinle-Neumann et al.*, 2001]. Arrows indicate the minima on the lines, showing significant increase in equilibrium c/a with temperature. The open diamond is the experimentally determined c/a at room temperature and 270 GPa [*Mao et al.*, 1990].

other contributions to the vibrational and electronic free energy have little effect (Fig. 12). Fig. 12 Our calculations show that the entropy increases substantially with increasing c/a. The entropic contribution to the free energy depends on the absolute temperature

$$F_{vib} = E_{vib} - T S_{vib} \qquad (18)$$

so that large values of c/a become increasingly more favorable energetically at high temperature.

4.6. High Temperature Elasticity

As a consequence of the increase in c/a the longitudinal anisotropy changes radically with temperature, c_{11} becomes larger than c_{33} (Fig. 13), Fig. 13 with compressional wave propagation in the basal plane being faster than along the c-axis (Fig. 14). Fig. 14 As the c-axis expands it becomes more compressible, and the corresponding longitudinal modulus, c_{33}, softens. c_{11} in turn increases slightly. The off-diagonal elastic constants are also affected by the temperature-induced change in structure: c_{12} increases rapidly with temperature because the basal plane shrinks with increasing temperature at constant density.

The shear constants c_{44} and c_{66} show a strong temperature dependence and decrease almost by a factor of four at 6000 K and change order as well (Fig. 13).

The velocity of shear waves is considerably smaller at high temperature and the sense of shear anisotropy is reversed (Fig. 14), with the propagation of the (001) polarized shear wave becoming faster along the c-axis than in the basal plane.

Our calculations imply a shear instability in hcp iron at very high temperature, c_{66} approaches zero at 7000 K yielding an upper bound on its mechanical stability or melting point. This is in qualitative agreement with the results by *Laio et al.* [2000] who predict a shear instability at somewhat lower temperature.

5. PROPERTIES OF THE INNER CORE

5.1. Aggregate Elasticity

Our current understanding of the physical properties of iron shows that the high Poisson's ratio of the inner core can be explained by solid-phase elasticity [*Laio et al.*, 2000; *Steinle-Neumann et al.*, 2001]. The shear elastic constants are predicted to decrease rapidly with increasing temperature - by a factor of four at 6000 K. As a result, G becomes rapidly smaller with increasing temperature (Fig. 15), Fig. 15 leading to a Poisson's ratio of hcp iron that is in quantitative agreement with seismic models of the inner core. These results confirm inferences on the basis of shock wave measurements of v_P and estimates of v_S at core conditions [*Brown and McQueen*, 1986]. It does not seem necessary to invoke additional mechanisms to explain the high Poissons ratio of the inner core such as viscoelastic dispersion [*Jackson et al.*, 2000] or the presence of partial melt [*Singh et al.*, 2000].

5.2. Temperature

Knowledge of the elasticity of iron permits an estimate of inner core temperature that is independent of the iron melting curve and the associated uncertainties related to freezing point depression. The temperature of the inner core is estimated to be 5700 ($\pm$ 500) K. At this temperature Poisson's ratio, K_S, and G of iron simultaneously agree with the properties of seismological models of the inner core (Fig. 15). We infer that the additional light elements that must be present in the inner core have a greater effect on the density through a decrease in mean atomic weight, than they do on the elastic moduli. Further investigations of the elasticity of iron-light element alloys will be needed to test this hypothesis.

Based on an estimate of the Earth's central temperature, we may construct a core geotherm. We assume that the temperature distribution in the inner and outer core are adiabatic and adopt $\gamma = 1.6$ [*Wasserman*

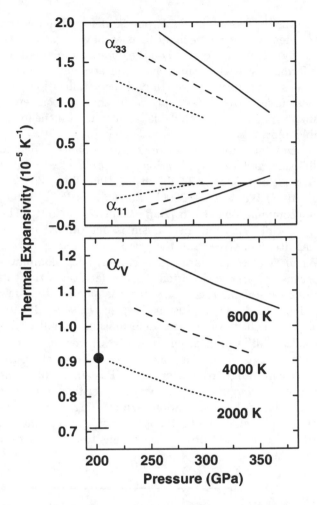

Figure 11. Thermal expansivity of iron at high pressure. The lower panel shows the volume thermal expansivity α_V of hcp iron at temperatures of 2000 K, 4000 K, and 6000 K. The experimental datum is from *Duffy and Ahrens* [1993] for an average thermal expansivity over the range of 300-5200 K. The upper panel shows the corresponding linear thermal expansivities for the *a*- (α_{11}) and *c*-axes (α_{33}).

et al., 1996; *Alfè et al.*, 2001] for the inner core, and $\gamma = 1.5$ for the outer core (Fig. 4). The latter value is consistent with shock wave data [*Brown and McQueen*, 1986]. With this choice the temperature is 5750 K in the center of the Earth, 5500 K at the inner core boundary, and 4000 K at the core mantle boundary (Fig. 4).

Other estimates of the temperature at the inner core boundary fall in the range of 4500-6000 K, depending on melting point estimate and degree of melting point depression. Our temperature is somewhat higher than the geotherm given by *Brown and McQueen* [1986] with 5000 K at the inner core boundary, and also higher than the extrapolation from static experiments (< 5000 K) [*Boehler*, 2000] or that based on the melting point from *Laio et al.* [2000] (5400 K). The melting point

from *Alfè et al.* [2002b] combined with a melting point depression of 700 K [*Alfè et al.*, 2002a] yields an inner core boundary temperature at 5600-5700 K, in remarkable agreement with our result. Melting points obtained with pyrometric measurements in shockwave experiments [*Williams et al.*, 1987; *Yoo et al.*, 1993] are considerably higher.

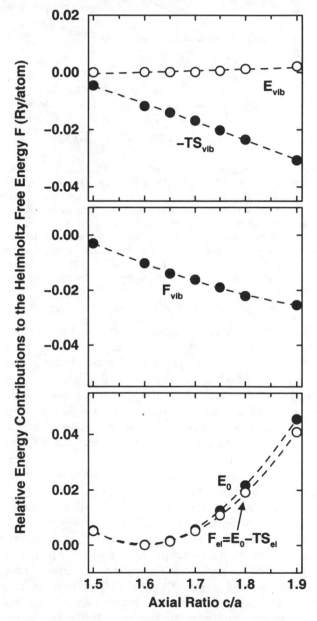

Figure 12. Energy contributions to the total Helmholtz free energy F as a function of axial ratio for $\rho = 13.04$ Mg/m^3 and $T = 4000$ K. The lower panel shows the static energy E (filled symbols) and the electronic free energy F_{el} (opaque symbols). The middle panel shows the vibrational free energy F_{vib} which is divided into internal vibrational energy E_{vib} (open symbols) and a vibrational entropy term $-TS_{vib}$ (filled symbols) in the upper panel.

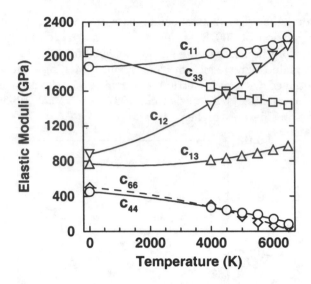

Figure 13. Single crystal elastic constants of hcp iron at a density of 13.04 Mg/m³ [*Steinle-Neumann et al.*, 2001]. Static values are connected to high temperature results with a quadratic fit (lines): the longitudinal elastic constants c_{11} and c_{33} are shown with circles and squares, the off-diagonal constants c_{12} and c_{13} with triangles down and up, for the shear elastic constant c_{44} and c_{66} circles and diamonds are used. $c_{66} = 1/2(c_{11} - c_{12})$ (dashed line) is not an independent elastic constant but included for comparison with c_{44}.

5.3. Simplified Model of Texture and Anisotropy

The sense of anisotropy that is found at high temperature changes our view of the polycrystalline texture of the inner core and the dynamic processes that may produce it. On the basis of first-principles calculations of the elastic constants, we propose a simple model of the polycrystalline texture of the inner core that explains the main features of its anisotropy [*Steinle-Neumann et al.*, 2001]. We find that if 1/3 of the basal planes are aligned with Earth's rotation axis in an otherwise randomly oriented medium, compressional wave travel time anomalies are well explained (Fig. 16). Fig. 16 This model is almost certainly over-simplified. The key element is the tendency for the fast crystallographic direction (*a*) to be aligned with the observed symmetry axis of inner-core anisotropy (approximately polar). It is probable that the actual direction and degree of crystallographic alignment will vary with geographic location. Such variations may account for seismological observations of heterogeneity [*Creager*, 1997; *Tanaka and Hamaguchi*, 1997; *Vidale and Earle*, 2000].

Important remaining questions include the origin of polycrystalline texture in the inner core, which may have been acquired during solidification, or may have developed subsequently as a result of plastic deforma-

tion, as discussed above. If plastic deformation is the prevalent texturing mechanism as we have argued, crystal alignment must depend on the dominant microscopic deformation mechanism in hcp iron at inner core conditions or patterns of growth and recrystallization, and the source of the stress field. While these are currently unknown, our simple model is consistent with the available information.

Candidates for the deformation mechanism include diffusion and recrystallization [*Yoshida et al.*, 1996; *Stixrude and Cohen*, 1995b], dislocation glide [*Wenk et al.*, 1988; *Wenk et al.*, 2000b; *Poirier and Price*, 1999], or a combination of both [*Buffett and Wenk*, 2001]. The type of deformation mechanism is determined by the magnitude of stress and the grain size of the material. When stresses are large, dislocation creep dominates because a large dislocation density facilitates the slip along crystal planes; if the stress is small with a low dislocation density diffusion creep prevails as diffusion of point defects dominates deformation. Small grains facilitate diffusion, larger grain sizes favor dislocation glide. Estimates on the grain size of the crystals in the inner core range from a few mm [*Buffett*, 1996] to the km scale [*Bergman*, 1998]. With the critical grain size determining the deformation regime being in the meter range [*Yoshida et al.*, 1996] the large uncertainty in grain size does not provide the means to favor one over the other.

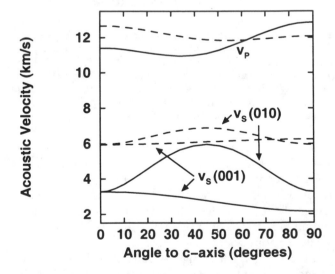

Figure 14. Single crystal anisotropy in hcp iron from static calculations (dashed lines) and at 6000 K (solid lines) [*Steinle-Neumann et al.*, 2001]. The wave propagation velocities for the P-wave (v_P) and the two polarizations of the S-wave (v_S, with polarizations given in the parentheses) are shown as a function of the angle with respect to the *c*-axis.

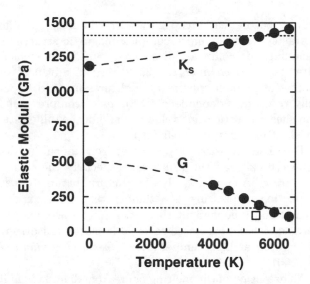

Figure 15. Aggregate acoustic properties of iron calculated as a function of temperature [*Steinle-Neumann et al.*, 2001] in comparison to the inner core. The adiabatic bulk (K_S) and shear moduli (G) at 13.04 Mg/m^3 are shown as a function of temperature (solid circles) with the corresponding values of the inner core at the same density (dotted lines). The open square shows a previous computational result by *Laio et al.* [2000].

Yoshida et al. [1996] argue that diffusion and recrystallization would result in a texture with the elastically stiffest (fastest) axis coinciding with the direction of the flow, minimizing the strain energy; for iron at high temperature this would tend to align basal planes with the dominant pattern of flow. Among the crystallographic slip planes that may participate in dislocation glide, the basal plane is the slip system that is most easily activated at high pressure according to recent theoretical and experimental work [*Poirier and Price*, 1999; *Wenk et al.*, 2000b]. The predicted high axial ratio at inner core conditions would probably further enhance basal slip as it is typical for ambient hcp metals with large c/a. As crystals deform in an external stress field basal planes would rotate in the direction of maximum shear, yielding a texture with basal planes aligned with the direction of flow as well. Active recrystallization during slip would tend to enhance the resulting fabric.

Above we have discussed possible sources for stress in the inner core, which all yield distinct flow patterns. Many of them share a flow that is dominant in the polar direction [*Yoshida et al.*, 1996; *Karato*, 1999] which would yield a texture of basal planes aligned with Earth's rotation axis. The shear flow invoked by *Buffett and Bloxham* [2000] also yields a texture with the general characteristics of basal planes aligned with the polar direction [*Buffett and Wenk*, 2001].

To obtain the seismic characteristics of such a textural model from single crystal results we convert the elastic anisotropy of iron to differential travel time anomalies. We start by considering an aggregate in which all a-axes of hcp iron are aligned with the polar axis of the inner core but otherwise randomly oriented. We obtain the elastic properties of this aggregate by averaging elasticity over the solid angle about the a-axis [*Stixrude*, 1998] which again yields an aggregate elasticity of hexagonal (cylindrical) symmetry. Using the Christoffel equations (15) we can calculate v_P as function of the angle ξ between the P-wave propagation direction and the pole. Defining the amplitude of the anisotropy as

$$\delta v_P(\xi)/v_{P,av} = (v_P(\xi) - v_{P,av})/v_{P,av} \qquad (19)$$

($v_{P,av}$ is the average v_P) we obtain an amplitude of 10% at $\xi = 0°$. This is about a factor of three to five larger than global seismic anisotropy models for the inner core [*Song*, 1997; *Ishii et al.*, 2002a]. It is unlikely that crystallographic alignment in the inner core is perfect. Consequently, in our simple model, the degree of alignment is reduced by a factor of three in order to match the gross amplitude of the seismically observed anisotropy.

To compare directly with seismic observations we compute differential PKIKP travel time anomalies due to inner core anisotropy by

$$\delta t(\Delta, \xi) = -t(\Delta)\delta v_P(\xi)/v_{P,av}, \qquad (20)$$

where Δ is the angular distance from source to receiver, and t the travel time of the PKIKP phase through the inner core. PKIKP and PKP$_{BC}$ are seen together over a narrow range of distances near $\Delta = 150°$ (t=124 s) which we use as a reference distance [*Stixrude and Cohen*, 1995b]. The resulting differential travel time differences are in good agreement both in amplitude and angular dependence with seismic data (Fig. 16).

While we refer to plastic deformation explicitly in the development of this textural model, it is also consistent with other classes of structure. The general character of solidification texture from dentritic growth for pure zinc (like iron at high temperature, a transition metal with high c/a) [*Bergman et al.*, 2000] is in agreement with the model we propose.

6. CONCLUSIONS AND OUTLOOK

The seismically observed complexity in inner core structure and importance of inner core crystallization for geodynamo processes have initiated considerable interest in the physical state and dynamics of this inner-

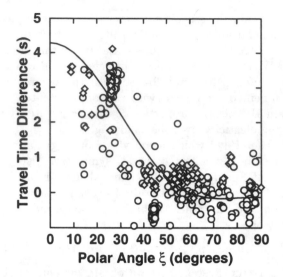

Figure 16. Differential travel time differences of PKIKP-PKP (BC-DF) due to inner core anisotropy as a function of propagation direction. The solid line shows the results based on a model of the inner core in which 1/3 of the crystals have their basal planes aligned with the rotation axis in conjunction with the high temperature elastic constants for hcp iron from *Steinle-Neumann et al.* [2001]. Seismological observations are from Song and Helmberger [1993] (diamonds) and Creager [1999; 2000] (circles).

most portion of our planet. Advances in experimental and theoretical methods in mineral physics have made it possible to address important questions regarding core composition, temperature, crystalline structure, and elasticity in the inner core.

Combining geophysical information on core structure and chemical constraints for the partitioning of elements between the solid and liquid promising steps have been made to characterize the light element composition in the core [*Alfè et al.*, 2000a; *Alfè et al.*, 2000b; *Alfè et al.*, 2002a] with first results indicating that at least a ternary mixture is needed to satisfy all data. The effect of the light element on inner core elastic properties is expected to be minor, except for density, as evidenced by the ability of pure iron to reproduce inner core aggregate elastic properties [*Laio et al.*, 2000; *Steinle-Neumann et al.*, 2001].

Discrepancies in static properties of the high pressure phase of iron, hcp, between experimental data and theoretical predictions suggest that some aspect of the physics of this phase is not well understood to date. Theory indicates the possible presence of magnetic moments on the atoms in the hcp phase [*Steinle-Neumann et al.*, 1999]. Experimental and theoretical efforts should be targeted towards a better characterization of magnetic properties at pressures below 100

GPa. Theoretical investigation should focus on the characterization of more complex magnetic structures including non-collinear, disordered, and incommensurate states [*Cohen et al.*, 2002]. Also, advances in studies of phonon spectra at high pressure will hopefully result in independent estimates of compressional and shear acoustic wave velocities and of the full elastic constant tensor in the near future.

The thermoelastic properties of iron appear to depend critically on the c/a ratio, especially the elastic anisotropy. A careful study of structural parameters as a function of pressure and temperature using experimental and independent theoretical methods could be used to test the predicted reversal in elastic anisotropy of hcp iron at high temperature [*Steinle-Neumann et al.*, 2001].

An extension of the melting curve from diamond anvil cell experiments to higher pressures could help to lower current uncertainties in the temperature of the Earth's core. However, present uncertainties in the melting point of iron are probably comparable to uncertainties in the freezing point depression. Further investigation of the phase stability and elastic properties of iron light element alloys will be important.

Acknowledgments.
We greatly appreciate helpful communication of results and preprints from D. Alfè, B. Buffett, L. Dubrovinsky, O. Gülseren, M. Ishii, J. Nguyen, and J. Tromp. This work was supported by the National Science Foundation under grants EAR-9980553 (LS) and EAR-998002 (REC), and by DOE ASCI/ASAP subcontract B341492 to Caltech DOE W-7405-ENG-48 (REC). Computations were performed on the Cray SV1 at the Geophysical Laboratory, supported by NSF grant EAR-9975753 and by the W. M. Keck Foundation.

REFERENCES

Ahrens, T. J., and R. Jeanloz, Pyrite shock compression, isentropic release, and composition of the Earth's core, *J. Geophys. Res., 92,* 10363-10375, 1987.

Alfè, D., M. J. Gillan, and G. D. Price, The melting curve of iron at the pressures of the Earth's core from *ab- initio* calculations, *Nature, 401,* 462-464, 1999.

Alfè, D., M. J. Gillan, and G. D. Price, Constraints on the composition of the Earth's core from *ab- initio* calculations, *Nature, 405,* 172-175, 2000a.

Alfè, D., M. J. Gillan, and G. D. Price, Thermodynamic stability of Fe/O solid solution at inner-core conditions, *Geophys. Res. Lett., 27,* 2417-2420, 2000b.

Alfè, D. , G. D. Price, and M. J. Gillan, Thermodynamics of hexagonal close packed iron under Earth's core conditions, *Phys. Rev. B Condens. Matter, 64,* 045123, 2001.

Alfè, D., M. J. Gillan, and G. D. Price, Composition and temperature of Earth's core constrained by combining *ab-initio* calculations and seismic data, *Earth Planet. Sci. Lett., 95,* 91-98, 2002.

Alfè, D., G. D. Price, and M. J. Gillan, Iron under Earth's core conditions: Liquid-state thermodynamics and high-pressure melting curve from ab initio calculations, *Phys. Rev. B Condens. Matter, 65,* 165118, 2002.

Anderson, O. L., L. Dubrovinsky, S. K. Saxena, and T. LeBihan, Experimental vibrational Grüneisen ratio values for ε-iron up to 330 GPa at 300 K, grl *28,* 399-402, 2001. Correction, *Geophys. Res. Lett., 28,* 2359, 2001.

Anderson, W. W., and T. J. Ahrens, Shock temperature and melting in iron sulfides at core pressure, *J. Geophys. Res., 101,* 5627-5642, 1996.

Andrault, D., G. Fiquet, M. Kunz, F. Visocekas, and D. Häusermann, The orthorhombic structure of iron: An in situ study at high-temperature and high-pressure, *Science, 278* 831-834, 1997.

Asada, T, and K. Terakura, Cohesive properties of iron obtained by use of generalized gradient approximation, *Phys. Rev. B Condens. Matter, 46,* 13599-13602, 1992. Correction, *Phys. Rev. B Condens. Matter, 48,* 17649, 1993.

Aurnou, J.M., D. Brito, and P. L. Olson, Anomalous rotation of the inner core and the toroidal magnetic field, *J. Geophys. Res., 103,* 9721-9738, 1998.

Barron, T. H.K., and M. L. Klein, Second-order elastic constants of a solid under stress, *Proc. Phys. Soc. London, 85,* 523-532, 1965.

Bergman, M. I., Measurements of elastic anisotropy due to solidification texturing and the implications for the Earth's inner core, *Nature, 389,* 60-63, 1997.

Bergman, M. I., Estimates of the Earth's inner core grain size, *Geophys. Res. Lett., 25,* 1593-1596, 1998.

Bergman, M. I., L. Giersch, M. Hinczewski, and V. Izzo, Elastic and attenuation anisotropy in directionally solidified (hcp) zinc, and the seismic anisotropy in the Earth's inner core, *Phys. Earth Planet. Inter., 117,* 139-151, 2000.

Birch, F., The α-γ transformation of iron at high pressures, and the problem of Earth's magnetism, *Amer. J. Sci., 238,* 192-211, 1940.

Birch, F., Density and composition of mantle and core, *J. Geophys. Res., 69,* 4377-4388, 1964.

Boehler, R., Temperatures in the Earth's core from melting point measurements of iron at high static pressures, *Nature, 363,* 534-536, 1993.

Boehler, R., High-pressure experiments and the phase diagram of lower mantle and core materials, *Rev. Geophys., 38,* 221-245, 2000.

Boehler, R., and M. Ross, Melting curve of aluminum in a diamond cell to 0.8 Mbar: implications for iron, *Earth Planet. Sci. Lett., 153,* 223-227, 1997.

Braginsky, S. I., Structure of the F layer and reasons for convection in the Earth's core, *Dokl. Akad. SSSR, 149,* 1311-1314, 1963.

Bréger, L., B. Romanovicz, and H. Tkalčić, PKP(BC-DF) travel time residuals and short scale heterogeneity in the deep Earth, *Geophys. Res. Lett., 26,* 3169-3172, 1999.

Brown, G. C., and A. E. Mussett *The inaccesible Earth,* 276 pp., Chapman and Hall, London, 1993.

Brown, J. M., The equation of state of iron to 450 GPa: Another high pressure solid phase, *Geophys. Res. Lett., 28,* 4339-4342, 2001.

Brown, J. M., and R. G. McQueen, The equation of state for iron and the Earth's core, in *High-Pressure Research in Geophysics,* edited by S. Akimoto and M. H. Manghnani, pp. 611-623, Reidel, Dordrecht, 1982.

Brown, J. M., and R. G. McQueen, Phase transitions, Grüneisen parameter, and elasticity for shocked iron between 77 GPa and 400 GPa, *J. Geophys. Res., 91,* 7485-7494, 1986.

Brown, J. M., T. J. Ahrens, and D. L. Shampine, Hugoniot data for pyrrhotite and the Earth's core, *J. Geophys. Res., 89,* 6041-6048, 1984.

Brown, J. M., J. N. Fritz, and R. S. Hixson, Hugoniot data for iron, *J. Appl. Phys., 88* 5496-5498, 2000.

Buffett, B. A., A mechanism for decade fluctuations in the length of day, *Geophys. Res. Lett., 23,* 3803-3806, 1996.

Buffett, B. A., Geodynamic estimates of the viscosity of the Earth's inner core, *Nature, 388,* 571-573, 1997.

Buffett, B. A., Dynamics of Earth's core, in *Earth's Deep Interior: Mineral Physics and Tomography from the Atomic to the Global Scale, Geophysical Monograph, vol. 117,* edited by S.-I. Karato et al., pp. 37-62, American Geophysical Union, Washington, DC, 2000.

Buffett, B. A., and J. Bloxham, Deformation of Earth's inner core by electromagnetic forces, *Geophys. Res. Lett., 24,* 4001-4004, 2000.

Buffett, B. A., and K. C. Creager, A comparison between geodetic and seismic estimates of inner core rotation, *Geophys. Res. Lett., 26,* 1509-1512, 1999.

Buffett, B. A., and H.-R. Wenk, Texturing of the inner core by Maxwell stresses, *Nature, 412,* 60-63, 2001.

Buffett, B. A., H. E. Huppert, J. R. Lister, and A. W. Woods, On the thermal evolution of the Earth, *J. Geophys. Res., 101,* 7989-8006, 1996.

Bullen, K. E., A hypothesis on compressibility at compressionals of the order a million atmospheres, *Nature, 157,* 405, 1946.

Bullen, K. E., An index of degree of chemical inhomogeneity in the earth, *Geophys. J. R. Astron. Soc., 7,* 584-592, 1963.

Clement, B. M., and L. Stixrude, Inner-core anisotropy, anomalies in the time-averaged paleomagnetic field, and polarity transition paths, *Earth Planet. Sci. Lett., 130,* 75-85, 1995.

Cohen, R. E., and O. Gülseren, Thermal equation of state of tantalum, *Phys. Rev. B Condens. Matter, 63,* 224101, 2001.

Cohen, R. E., L. Stixrude, and E. Wasserman, Tight-binding computations of elastic anisotropy of Fe, Xe, and Si under compression, *Phys. Rev. B Condens. Matter, 56,* 8575-8589, 1997.

Cohen, R. E., S. Gramsch, S. Mukherjee, G. Steinle-Neumann, and L. Stixrude, Importance of magnetism in phase stability, equations of state, and elasticity, in *Proc. Int. School of Physics "Enrico Fermi", vol. 147,* edited by R. Hemley et al., (in press), 2002.

Creager, K. C., Anisotropy of the inner core from differential travel times of the phases PKP and PKIKP, *Nature, 356,* 309-314, 1992.

Creager, K. C., Inner core rotation rate from small-scale heterogeneity and time varying travel times, *Science, 278,* 1284-1288, 1997.

Creager, K. C., Large-scale variations in inner core anisotropy, *J. Geophys. Res., 104,* 23127-23139, 1999.

Creager, K. C., Inner core anisotropy an rotation, in *Earth's*

Deep Interior: Mineral Physics and Tomography from the Atomic to the Global Scale, Geophysical Monograph, vol. 117, edited by S.-I. Karato et al., pp. 89-115, American Geophysical Union, Washington, DC, 2000.

Davies, G. F., Effective elastic-moduli under hydrostatic stress- I. Quasi-harmonic theory, *J. Phys. Chem. Solids, 35,* 1513-1520, 1974.

Deuss, A., J. H. Woodhouse, H. Paulsen, and J. Trampert, The observation of inner core shear waves, *Geophys. J. Int., 112,* 67-73, 2000.

Dubrovinsky, L. S., P. Lazor, S. K. Saxena, P. Häggkvist, H. P. Weber, T. LeBihan, and D. Häusermann, Study of laser heated iron using third generation synchrotron X-ray radiation facility with image plate at high pressure, *Phys. Chem. Minerals, 26,* 539-545, 1999.

Dubrovinsky, L. S., S. K. Saxena, F. Tutti, S. Rekhi, and T. LeBihan, In situ X-ray study of thermal expansion and phase transition of iron at multimegabar pressure, *Phys. Rev. Lett., 84,* 1720-1723, 2000.

Duffy, T. S., and T. J. Ahrens, Thermal-expansion of mantle and core materials at very high-pressures, *Geophys. Res. Lett., 20,* 1103-1106, 1993.

Duffy, T. S., G. Shen, D. L. Heinz, J. Shu, Y. Ma, H.-K. Mao, R. J. Hemley, and A. K. Singh, Lattice strains in gold and rhenium under nonhydrostatic compression to 37 GPa, *Phys. Rev. B Condens. Matter, 60,* 15063-15073, 1999.

Dziewonski, A. M., and D. L. Anderson, Preliminary reference Earth model, *Phys. Earth Planet. Inter., 25,* 297-356, 1981.

Dziewonski, A. M., and F. Gilbert, Solidity of the inner core of the Earth inferred from normal mode observations, *Nature, 234,* 465-466, 1971.

Eckerlin, P., and H. Kandler, Structure data of elements and intermetallic alloys, *Landolt-Börnstein, Group III, Vol. 6,* Springer, Berlin, 1971.

Fiquet, G., J. Badro, F. Guyot, H. Requardt, and M. Krisch, Sound velocities of iron to 110 Gigapascals, *Science, 291,* 468-471, 2001.

Funamori, N., T. Yagi, and T. Uchida, High-pressure and high-temperature *in situ* x-ray diffraction study of iron to above 30 GPa using MA8-type apparatus, *Geophys. Res. Lett., 23,* 953-956, 1996.

Glatzmaier, G. A., and P. H. Roberts, A three-dimensional convective dynamo solution with rotating and finitely conducting inner core and mantle, *Phys. Earth Planet. Inter., 91,* 63-75, 1995.

Glatzmaier, G. A., and P. H. Roberts, Rotation and magnetism of Earth's inner core, *Science, 274,* 1887-1891, 1996.

Gülseren, O., and R. E. Cohen, High pressure and temperature elasticity of bcc tantalum, *Phys. Rev. B Condens. Matter, 65,* 064103, 2002.

Hemley, R. J., and H.-K. Mao, *In-situ* studies of iron under pressure: new windows on the Earth's core, *Int. Geology Rev., 43,* 1-30, 2001.

Hill, R., The elastic behaviour of a crystalline aggregate, *Proceedings of the Physical Society, London, A, 65,* 349-355, 1952.

Hohenberg, P., and W. Kohn, Inhomogeneous electron gas, *Phys. Rev., 136,* B864-871, 1964.

Hollerbach, R., and C. A. Jones, Influence of the Earths inner-core on geomagnetic fluctuations and reversals, *Nature, 365,* 541-543, 1993.

Huang, E., W. A. Bassett, and P. Tao, Pressure-temperature-volume relationship for hexagonal close packed iron determined by synchroton radiation, *J. Geophys. Res., 92,* 8129-8135, 1987.

Ishii, M., J. Tromp, A. M. Dziewonski, and G. Ekström, Joint inversion of normal-mode and body-wave data for inner core anisotropy: 1. Simple inner-core models and mantle heterogeneity, submitted to *J. Geophys. Res.,* (2002a).

Ishii, M., A. M. Dziewonski, J. Tromp, and G. Ekström, Joint inversion of normal-mode and body-wave data for inner core anisotropy: 2. Possible complexity within the inner core and mantle, submitted to *J. Geophys. Res.,* (2002b).

Ito, E., K. Morooka, O. Ujike, and T. Katsura, Reactions between molten iron and silicate melts at high pressure: implications for the chemical evolution of the Earth's core, *J. Geophys. Res., 100,* 5901-5910, 1995.

Jackson, I., J. D. F. Gerald, and H. Kokkonen, High-temperature viscoelastic relaxation in iron and its implication for the shear modulus and attenuation for the Earth's inner core, *J. Geophys. Res., 105,* 23605-23634, 2000.

Jeanloz, R., Properties of iron at high pressures and the state of the core, *J. Geophys. Res., 84,* 6059-6069, 1979.

Jeanloz, R., The nature of the Earth's core, *Annu. Rev. Earth Planet. Sci., 18,* 357-386, 1990.

Jeanloz, R., and T. J. Ahrens, Equations of state of FeO and CaO, *Geophys. J. R. Astron. Soc., 62,* 505-528, 1980.

Jeanloz, R., and S. Morris, Temperature distribution in the crust and mantle, *Annu. Rev. Earth Planet. Sci., 14,* 377-415, 1986.

Jeanloz, R., and H.-R. Wenk, Convection and anisotropy of the inner core, *Geophys. Res. Lett., 15,* 72-75, 1988.

Jephcoat, A. and P. Olson, Is the inner core of the Earth pure iron, *Nature, 325,* 332-335, 1987.

Jephcoat, A. P., H.-K. Mao, and P. M. Bell, Static compression of iron to 78 GPa with rare gas solids as pressure-transmitting media, *J. Geophys. Res., 91,* 4677-4684, 1986.

Karato, S.-I. Inner core anisotropy due to the magnetic field induced preferred orientation of iron, *Science, 262,* 1708-1710, 1993.

Karato, S.-I. Seismic anisotropy of the Earth's inner core resulting from flow induced by Maxwell stresses, *Nature, 402,* 871-873, 1999.

Karato, S.-I., and V. R. Murthy, Core formation and chemical equilibrium in the Earth – I. Physical considerations, *Phys. Earth Planet. Inter., 100,* 61-79, 1997.

Kohn, W., and L. J. Sham, Self-consistent equations including exchange and correlation effects, *Phys. Rev., 140,* A1133-1138, 1965.

Körling, M., and J. Häglund, Cohesive and electronic properties of transition metals: the generalized gradient method, *Phys. Rev. B Condens. Matter, 45,* 13293-13297, 1992.

Kuang, W., and J. Bloxham, An Earth-like numerical dynamo model, *Nature, 389,* 371-374, 1997.

Laio, A., S. Bernard, G. L. Chiarotti, S. Scandolo, and E.

Tosatti, Physics of iron at Earth's core conditions, *Science*, *287*, 1027-1030, 2000.

Laske, G., and G. Masters, Rotation of the inner core from a new analysis of free oscillations, *Nature*, *402*, 66-69, 1999.

Lehmann, I., P', *Bur. Cent. Seismol. Int. A*, *14*, 3-31, 1936.

Li, J., and C. B. Agee, Geochemistry of mantle-core differentiation at high pressure, *Nature*, *381*, 686-689, 1996.

Lübbers, R., H. F. Grünsteudel, A. I. Chumakov, and G. Wortmann, Density of phonon states in iron at high pressure, *Science*, *287* 1250-1253, 2000.

Lundqvist, S., and N. H. March, *Theory of the inhomogeneous electron gas*, 395 pp., Plenum Press, New York, 1983.

Manghnani, M. H., K. Katahara, and E. S. Fisher, Ultrasonic equation of state of rhenium, *Phys. Rev. B Condens. Matter*, *9*, 1421-1431, 1974.

Mao, H.-K., Y. Wu, L. C. Chen, J. F. Shu, and A. P. Jephcoat, Static compression of iron to 300 GPa and $Fe_{0.8}Ni_{0.2}$ alloy to 260 GPa: Implications for the composition of the core, *J. Geophys. Res.*, *95*, 21737-21742, 1990.

Mao, H.-K., J. Shu, G. Shen, R. J. Hemley, B. Li, and A. K. Sing, Elasticity and rheology of iron above 220 GPa and the nature of the Earth's inner core, *Nature*, *396*, 741-743, 1998. Correction, *Nature*, *399*, 280, 1999.

Mao, H.-K., J. Xu, V. V. Struzhkin, J. Shu, R. J. Hemley, W. Sturhahn, M. Y. Hu, E. E. Alp, L. Vočadlo, D. Alfè, G. D. Price, M. J. Gillan, M. Schwoerer-Böhning, D. Häusermann, P. Eng, G. Shen, H. Giefers, R. Lübbers, and G. Wortmann, Phonon density of states of iron up to 153 Gigapascals, *Science*, *292*, 914-916, 2001.

Masters, G., Observational constraints on the chemical and thermal structure of the Earths deep interior, *Geophys. J. R. Astron. Soc.*, *57*, 507-534, 1979.

Masters, G., and F. Gilbert, Structure of the inner core inferred from observations of its spheroidal shear modes, *Geophys. Res. Lett.*, *8*, 569-571, 1981.

Matsui, M., and O. L. Anderson, The case for a body-centered cubic phase (α') for iron at inner core conditions, *Phys. Earth Planet. Inter.*, *103*, 55-62, 1997.

Matthies, S., S. Merkel, H.-R. Wenk, R. J. Hemley, and H.-K. Mao, Effects of texture on the determination of elasticity of polycrystalline ε-iron from diffraction measurements, *Earth Planet. Sci. Lett.*, *194*, 201-212, 2001.

Maurer, M., M. Piecuch, M. F. Ravet, J. C. Ousset, J. P. Sanchez, C. Aaron, J. Dekoster D. Raoux, A. Deandres, M. Desantis, A. Fontaine, F. Baudelet, J. L. Rouviere, and B. Dieny, Magnetism and structure in hexagonal Fe/Ru superlattices with short periodicity, *J. Magn. Magn. Mater.*, *93*, 15-24, 1991.

McMahon, A. K., and M. Ross, High-temperature electron-band calculations *Phys. Rev. B Condens. Matter*, *15*, 718-725, 1977.

Merkel, S., A. F. Goncharov, H.-K. Mao, P. Gillet, and R. J. Hemley, Raman spectroscopy of iron to 152 GPa: Implications for Earth's inner core, *Science*, *288*, 1626-1629, 2000.

Merrill, R. T., M. W. McElhinny, and P. L. McFadden, *The magnetic field of the Earth*, 531 pp., Academic Press, San Diego, CA, 1996.

Morelli, A., A. M. Dziewonski, and J. H. Woodhouse, Anisotropy of the inner core inferred from PKIKP travel times, *Geophys. Res. Lett.*, *13*, 1545-1548, 1986.

Moroni, E. G., G. Grimvall, and T. Jarlborg, Free energy contributions to the hcp-bcc transformation in transition metals, *Phys. Rev. Lett.*, *76*, 2758-2761, 1996.

Nguyen, J. H., and N. C. Holmes, Iron sound velocity and its implications for the iron phase diagram, *submitted to Science*, 2001.

Nielsen, O. H., and R. M. Martin, Quantum-mechanical theory of stress and force, *Phys. Rev. B Condens. Matter*, *32*, 3780-3791, 1985.

Niu, F., and L. Wen, Hemispheral variations in seismic velocity at the top of the Earth's inner core, *Nature*, *410*, 1081-1084, 2001.

Okal, E. A., and Y. Cansi, Detection of PKJKP at intermediate periods by processing multichannel correlation, *Earth Planet. Sci. Lett.*, *164*, 23-30, 1998.

Okuchi, T., Hydrogen partitioning into molten iron at high pressure: implications for Earth's core, *Science*, *278*, 1781-1784, 1997.

Olijnyk, H., and A. P. Jephcoat, Optical zone-center phonon modes and macroscopic elasticity in hcp metals, *Solid State Communications*, *115*, 335-339, 2000.

Olijnyk, H., A. P. Jephcoat, and K. Refson, On optical phonons and elasticity in the hcp transition metals Fe, Ru and Re at high pressure, *Europhysics Letters*, *53*, 504-510, 2001.

Parrinello, M., and A. Rahman, Strain fluctuations and elastic constants, *J. Chem. Phys.*, *76*, 2662-2666, 1982.

Perdew, J. P., K. Burke, and M. Ernzerhof, Generalized Gradient Approximation Made Simple, *Phys. Rev. Lett.*, *77*, 3865-3868, 1996. Correction, *Phys. Rev. Lett.*, *78*, 1396, 1997.

Pickett, W. E., Pseudopotential methods in condensed matter applications, *Comp. Phys. Rep.*, *9*, 115-197, 1989.

Poirier, J. P., Light elements in the Earth's outer core - a critical review, *Phys. Earth Planet. Inter.*, *85*, 319-337, 1994.

Poirier, J. P., and G. D. Price, Primary slip system of ε-iron and anisotropy of the Earth's inner core, *Phys. Earth Planet. Inter.*, *110*, 147-156, 1999.

Poupinet, G., R. Pillet, and A. Souriau, Possible heterogeneity of the Earth's core deduced from PKIKP travel times, *Nature*, *305*, 204-206, 1983.

Saxena, S. K., L. S. Dubrovinsky, P. Haggkvist, Y. Cerenius, G. Shen, and H.-K. Mao, Synchrotron X-ray study of iron at high-pressure and temperature, *Science*, *269*, 1703-1704, 1995.

Shearer, P. M., Constraints on inner core anisotropy from ISC PKP(DF) travel-times, *J. Geophys. Res.*, *99*, 19647-19659, 1994.

Shen, G., and D. L. Heinz, High pressure melting of deep mantle and core materials, *Reviews in Mineralogy*, *37*, 369-396, 1998.

Shen, G., H.-K. Mao, R. J. Hemley, T. S. Duffy, and M. L. Rivers, Melting and crystal structure of iron at high pressures and temperatures, *Geophys. Res. Lett.*, *25*, 373-376, 1998.

Simmons, G., and H. Wang, *Single crystal elastic constants and calculated aggregate properties: a handbook*, 370 pp., M. I. T. Press, Cambridge, Mass., 1971.

Singh, A. K., C. Balasingh, H.-K. Mao, J. Shu, and R. J. Hemley, Analysis of lattice strains measured under nonhydrostatic pressure, *J. Appl. Phys.*, *83*, 7567-7575, 1998a.

Singh, A. K., H.-K. Mao, J. Shu, and R. J. Hemley, Estimation of single-crystal elastic moduli from polycrystalline X-ray diffraction at high pressure: application to FeO and iron, *Phys. Rev. Lett., 80*, 2157-2160, 1998b.

Singh, D. J., *Planewaves, pseudopotentials and the LAPW method* 115 pp., Kluwer Academic Publishers, Dordrecht, 1994.

Singh, S. C. , M. A. J. Taylor, and J. P. Montagner, On the presence of liquid in Earth's inner core, *Science, 287*, 2471-2474, 2000.

Söderlind P., J. A. Moriarty, and J. M. Wills, First-principles theory of iron up to Earth-core pressures: Structural, vibrational, and elastic properties, *Phys. Rev. B Condens. Matter, 53*, 14063-14072, 1996.

Song, X. D., Anisotropy of Earth's inner core, *Rev. Geophys., 35*, 297-313, 1997.

Song, X. D., and D. V. Helmberger, Anisotropy of the Earth's inner core, *Geophys. Res. Lett., 20*, 2591-2594, 1993.

Song, X. D., and D. V. Helmberger, Depth dependence of anisotropy of Earth's inner core, *J. Geophys. Res., 100*, 9805-9816, 1995.

Song, X. D., and D. V. Helmberger, Seismic evidence for an inner core transition zone, *Science, 282*, 924-927, 1998.

Song, X. D., and P. G. Richards, Seismological evidence for differential rotation of the Earth's inner core, *Nature, 382*, 221-224, 1996.

Souriau, A., New seismological constraints on differential rotation rates of the inner core from Novaya Zemlya events recorded at DRV, Antarctica, *Geophys. J. Int., 134*, F1-5, 1998.

Stacey, F. D., *Physics of the Earth,* 513 pp., Brookfield Press, Brisbane, 1992.

Steinle-Neumann, G., L. Stixrude, and R. E. Cohen, First-principles elastic constants for the hcp transition metals Fe, Co, and Re at high pressure, *Phys. Rev. B Condens. Matter, 60*, 791-799, 1999.

Steinle-Neumann, G., L. Stixrude, R. E. Cohen, and O. Gülseren Elasticity of iron at the temperature of Earth's inner core, *Nature, 413*, 57-60, 2001.

Stixrude, L., Elastic constants and anisotropy of MgSiO$_3$ perovskite, periclase, and SiO$_2$ at high pressure, in *The Core-Mantle Boundary Region, Geodynamic Series 28*, edited by M. Gurnis et al., pp. 83-96, American Geophysical Union, Washington, DC, 1998.

Stixrude, L., and J. M. Brown, The Earth's core, *Rev. Mineralogy, 37*, 261-280, 1998.

Stixrude, L., and R. E. Cohen, Constraints on the crystalline structure of the inner core - mechanical instability of bcc iron at high-pressure, *Geophys. Res. Lett., 22*, 125-128, 1995a.

Stixrude, L., and R. E. Cohen, High-pressure elasticity of iron and anisotropy of Earth's inner-core, *Science, 267*, 1972-1975, 1995b.

Stixrude, L., R. E. Cohen, and D. J. Singh, Iron at high pressure: Linearized-augmented-plane-wave computations in the generalized-gradient approximation, *Phys. Rev. B Condens. Matter, 50*, 6442-6445, 1994.

Stixrude, L., E. Wasserman, and R. E. Cohen, Composition and temperature of Earth's inner core, *J. Geophys. Res., 102*, 24729-24739, 1997.

Su, W.-J., and A. M. Dziewonski, inner core anisotropy in three dimensions, *J. Geophys. Res., 100*, 9831-9852, 1995.

Su, W.-J., A. M. Dziewonski, and R. Jeanloz, Planet within a planet: Rotation of the inner core of the Earth, *Science, 274*, 1883-1887, 1996.

Tanaka, S., and H. Hamaguchi, Degree one heterogeneity and hemispheral variation of anisotropy in the inner core from PKP(BC)-PKP(DF) times, *J. Geophys. Res., 102*, 2925-2938, 1997.

Taylor, R. D., M. P. Pasternak, and R. Jeanloz, Hysteresis in the high-pressure transformation of bcc-iron to hcp-iron *J. Appl. Phys., 69*, 6126-6128, 1991.

Tromp, J., Support for anisotropy of the Earth's inner core from free oscillations, *Nature, 366*, 678-681, 1993.

Tromp, J., Inner-core anisotropy and rotation, *Annu. Rev. Earth Planet. Sci., 29*, 47-69, 2001.

Troullier, N., and J. L. Martins, Efficient pseudopotentials for plane-wave calculations, *Phys. Rev. B Condens. Matter, 43*, 1993-2006, 1991.

Tsunoda, Y., Y. Nishioka, and R. M. Nicklow, Spin fluctuations in small γ-Fe precipitates, *J. Magn. Magn. Mater., 128*, 133-137, 1993.

Uchida, T., Y. Wang, M. L. Rivers, and S. R. Sutton, Stability field and thermal equation of state of ε-iron determined by synchrotron X-ray diffraction in a multianvil apparatus, *J. Geophys. Res., 106*, 21799-21810, 2001.

Uhl, M., L. M. Sandratskii, and J. Kübler, Spin fluctuations in γ-Fe and Fe$_3$Pt Invar from local-density-functional calculations, *Phys. Rev. B Condens. Matter, 50*, 291-301, 1994.

Verhoogen, J., Heat balance of the Earth's core, *Geophys. J. R. Astron. Soc., 4*, 276-281, 1961.

Verhoogen, J., *Energetics of the Earth,* 139 pp., National Academy of Sciences, Washington, DC, 1980.

Vidale, J. E., and P. S. Earle, Fine scale heterogeneity in the Earth's inner core, *Nature, 406*, 273-275, 2000.

Vočadlo, L., J. Brodholt, D. Alfè, M. J. Gillan, and G. D. Price, Ab initio free energy calculations on the polymorphs of iron at core conditions. *Phys. Earth Planet. Inter., 117*, 123-137, 2000.

Wasserman, E., L. Stixrude, and R. E. Cohen, Thermal properties of iron at high pressures and temperatures, *Phys. Rev. B Condens. Matter, 53*, 8296-8309, 1996.

Watt, J. P., and L. Peselnick, Clarification of the Hashin-Shtrikman bounds on the effective elastic moduli of polycrystals with hexagonal, trigonal, and tetragonal symmetry, *J. Appl. Phys., 51*, 1525-1531, 1980.

Wenk, H.-R., T. Takeshita, R. Jeanloz, and G. C. Johnson, Development of texture and elastic-anisotropy during deformation of hcp metals, *Geophys. Res. Lett., 15*, 76-79, 1988.

Wenk, H.-R., J. R. Baumgardner, R. A. Lebensohn RA, and C. N. Tomé, A convection model to explain anisotropy of the inner core, *J. Geophys. Res., 105*, 5663-5677, 2000a.

Wenk, H.-R., S. Matthies, R. J. Hemley, H.-K. Mao, and J. Shu, The plastic deformation of iron at pressures of the Earth's inner core, *Nature, 405*, 1044-1046, 2000b.

Wentzcovitch, R. M., Invariant molecular-dynamics approach to structural phase transitions, *Phys. Rev. B Condens. Matter, 44*, 2358-2361, 1991.

de Wijs, G. A., Kresse, G., Vočadlo, L., Dobson, D., Alfè,

D., Gillan, M. J., and G. D. Price, The viscosity of liquid iron at the physical conditions of the earth's core, *Nature, 392*, 805-807, 1998.

Williams, Q., R. Jeanloz, J. Bass, B. Svendsen, T. J. Ahrens, The melting curve of iron to 250 gigapascals - a constraint on the temperature at Earth's center, *Science, 236*, 181-182, 1987.

Woodhouse, J. H., D. Giardini, and X. D. Li, Evidence for inner core anisotropy from free oscillations, *Geophys. Res. Lett., 13*, 1549-1552, 1986.

Yagi, T., K. Fukuoka, H. Takei, and Y. Syono, Shock compression of wüstite, *Geophys. Res. Lett., 15*, 816-819, 1988.

Yoo, C. S., N. C. Holmes, M. Ross, D. J. Webb, and C. Pike, Shock temperatures and melting of iron at Earth core conditions, *Phys. Rev. Lett., 70*, 3931-3934, 1993.

Yoo, C. S., J. Akella, A. J. Campbell, H.-K. Mao, and R. J. Hemley, Phase-diagram of iron by in-situ x-ray-diffraction - implications for Earth core, *Science, 270*, 1473-1475, 1995.

Yoshida, S., I. Sumita, and M. Kumazawa, Growth model of the inner core coupled with outer core dynamics and the resultant elastic anisotropy. *J. Geophys. Res., 101*, 28085-28103, 1996.

Yukutake, T., Implausibility of thermal convection in the Earth's solid inner core, *Phys. Earth Planet. Inter., 108*, 1-13, 1998.

Yukutake, T., The inner core and the surface heat flow as clues to estimating the initial temperature of Earth's core, *Phys. Earth Planet. Inter., 121*, 103-137, 2000.

G. Steinle-Neumann and R. E. Cohen, Geophysical Laboratory, Carnegie Institution of Washington, 5251 Broad Branch Road, NW, Washington, DC 20015-1305 (e-mail: g.steinle-neumann@gl.ciw.edu; cohen@gl.ciw.edu)

L. Stixrude, Department of Geological Sciences, University of Michigan, 245 E. University Avenue, Ann Arbor, MI 48109-1063, (e-mail: stixrude@umich.edu)

Thermal Core-Mantle Interactions: Theory and Observations

David Gubbins

School of Earth Sciences, University of Leeds, UK

The Earth's core and mantle convect on very different time scales but they interact strongly through their influence on each other's boundary conditions. The high heat flow from the core must produce a thermal boundary layer in the mantle that is often associated with the seismic layer D''. Boundary layers in the core are very thin and lateral variations in heat flow through D'' can control core convection and the geodynamo directly. This control is constant on the time scale of core dynamics and the geodynamo and large in spatial scale. It may even stabilise the geodynamo by forcing large scales when core dynamics otherwise demands smaller length scales and more rapid time variations.

Lateral variations in seismic velocity and density within D'' may arise from variations in temperature or composition. It now seems likely that a cold ring exists beneath subduction zones of the Pacific. This can force downwelling in the core and influence the geomagnetic field by concentrating field lines. Evidence for this comes from the longitudinal distribution of the modern field, the time-average of the field over the last 5 Myr, and the transition field during polarity reversals. While each of these observations are open to criticism because they are sensitive to errors in the paleomagnetic data, the existence of a unifying theory that explains all three is convincing.

INTRODUCTION

The core-mantle boundary (CMB) is the largest discontinuity in physical properties on Earth. For geodynamic purposes, we regard this interface to be between two very different fluids. The core is made of liquid iron, which flows like water and conducts heat and electricity very well. The mantle is not really fluid at all, it is made of rock that conducts heat and electricity poorly but it flows by creep processes on very long time scales. Their densities and seismic properties also differ enormously.

The lowermost mantle, the region called D'', is anomalous. First proposed by K. E. Bullen on the basis of an apparent change in the radial gradient of the P-wave velocity, it is now better known as a region of strong lateral variations with an uneven but in some places rather sharp upper seismic discontinuity [*Haddon and Cleary*, 1974; *Lay and Helmberger*, 1983]. *Haddon and Cleary* [1974] used the ability of seismic arrays to measure the incoming direction of seismic waves to establish that well-known precursors to the phase PKIKP, the reflection from the inner core surface, were in fact scattered energy from the lower mantle. *Lay and Helmberger* [1983] discovered shear wave reflections from a hundred or more kilometers above the CMB, and P-waves reflections have also been observed [*Weber and Davis*, 1990].

D'' is probably a boundary layer, just as the lithosphere is a boundary layer at the top of the mantle. Vigorous lower mantle convection will require a thermal boundary layer, and the sharp seismic discontinu-

Earth's Core: Dynamics, Structure, Rotation
Geodynamics Series 31
Copyright 2003 by the American Geophysical Union
10.1029/31GD11

ities required to reflect short period P-waves suggest a compositional component to the boundary layer. This is not surprising in view of the proximity of materials with very different composition. D″ is thicker than the lithosphere, averaging at about 250 km, but thinner than the upper mantle. Seismic reflections from the top of D″ suggest a change in properties at least as strong as those at the 650 km discontinuity at the base of the upper mantle, or those at 450 km discontinuity.

Huge differences in fluid velocity and viscosity between the core and mantle lead to fundamental differences in their fluid mechanics. The core is strongly influenced by Earth's rotation and its magnetic field, while the mantle is not. The heat flux per unit area across the CMB is large, comparable with that at the Earth's surface. This is likely to lead to plume formation and these may be linked to hot spots at the Earth's surface, although direct seismic evidence of lower mantle plumes is still lacking. Plumes may carry a significant amount of heat away from the CMB in small scale features that have so far remained undetected. Another discovery of seismology, ultra-low velocity zones [*Garnero et al.*, 1998], suggest partial melt of the mantle and incursion of iron from the core. They are very thin features (depth about 20 km) with exceptionally low shear wave velocity.

Seismic tomography shows increasingly convincing evidence of plates sinking through the 650 km discontinuity into the lower mantle [*Hilst et al.*, 1998]. They may reach the CMB, and may even form part of D″ [*Kendall*, 2000]. *Lay et al.* [1998] have suggested that cold plates reaching the CMB will produce a heat flux anomaly and stimulate plume formation. D″ may therefore be a very complex boundary layer.

Core convection must inevitably be influenced by lateral variations in the overlying mantle, specifically in the boundary layer or D″. The important question is whether the influence is strong, and whether it produces any observable effects. The clearest evidence we have of mantle control of the core is the change in frequency of geomagnetic reversals. The detailed record of reversals revealed on the ocean floor includes the Cretaceous superchron, a long period when there were no reversals, and an apparent gradual increase in reversal frequency throughout the subsequent 80 Myr [*Lowrie*, 1982]. The time scale is too long for any core process, but is typical of lower mantle convection.

Core-mantle interaction could take the form of changes in the overall heat through the CMB, changes in the lateral variations of heat flux from place to place around the CMB [*Bloxham and Gubbins*, 1987], of electrical conductivity [*Herrero-Bervera and Runcorn*, 1997], or

CMB topography [*Bell and Soward*, 1996; *Hide*, 1967]. Such lateral variations have the potential for a seismological signature. Seismic tomography and normal mode splitting give good maps of variations in P and S-wave velocity averaged through the lowermost 150 km of the mantle [*Masters et al.*, 1996]; their interpretation is complicated because they could arise from temperature or compositional variations, or a combination of the two. Topography on the CMB is found in principle from times of reflection, but the estimate is contaminated by delays introduced by unknown mantle anomalies. Rather low topography, below that resolvable by seismology, could have an influence on the core [*Hide*, 1967]. D″ structure is highly variable and our knowledge of it is patchy [*Kendall*, 2000; *Kendall and Shearer*, 1994], possibly because of the presence of plates that have sunk all the way to the CMB.

Geomagnetism and paleomagnetism give useful constraints on the heat flow across the CMB [*Gubbins et al.*, 2001; *Labrosse, Poirier and LeMouël*, 1997]. The morphology of the modern field illustrates clearly the influence of rotation (through its primary axial dipole nature) and the influence of the inner core, with most of the magnetic field appearing just outside the circle where the tangent cylinder, the cylinder enclosing the inner core with its axis parallel to the rotation axis, meets the core surface [*Gubbins and Bloxham*, 1987]. The tangent cylinder is known to be important for the dynamics of rotating convection, and some recent numerical dynamo simulations have most of the field generated near the tangent cylinder throughout the volume of the core, in agreement with observation.

Any long-term departure from axial symmetry suggests the influence of the mantle, because a perfectly spherically symmetric core would support a dynamo with no longitude preference. The modern field contains some relatively stationary features [*Bloxham and Gubbins*, 1985] that might point to a link with the mantle, and secular variation is weak in the Pacific hemisphere.

The historical record is too short to be sure that these effects are genuinely the result of core-mantle interactions, and we must turn to paleomagnetism, with much less accurate magnetic measurements and poor dating, to look for longer term effects. The time-averaged paleomagnetic field, paleosecular variation, and reversal transition paths, all hint at departures from axial symmetry.

THEORY

Boundary Conditions

Fluid motion in the core is very rapid compared with motion in the mantle. Recent changes in the Earth's

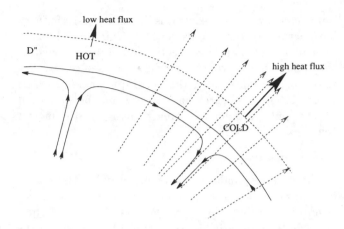

Figure 1. A simple illustration of thermal core-mantle interaction. The overlying boundary layer is cold where heat flux is high, core fluid downwells, and magnetic field lines are compressed giving a strong vertical field.

magnetic field suggest speeds of 10–50 km/yr [*Bloxham and Jackson*, 1991], five orders of magnitude larger than typical plate motions, and therefore several orders of magnitude larger than speeds of lower mantle convection. This rapid, and probably chaotic, motion of a very good thermal conductor will make the CMB very close to isothermal.

To obtain an estimate of the lateral temperature variation around the CMB consider the force balance at the top of the core. Flow driven by buoyancy forces resulting from a temperature variation ΔT will be comparable with other forces acting, by far the largest of which is the Coriolis force arising from rotation. A very rough balance of these two forces gives

$$2\Omega v \approx g\alpha\Delta T \qquad (1)$$

where $\Omega = 7.272 \times 10^{-5}$ s^{-1} is Earth's angular rotation frequency, v is a typical core flow speed, $g = 10.7$ ms^{-2} is the acceleration due to gravity, and $\alpha \approx 1.3 \times 10^{-5}$ K^{-1} is the thermal expansion coefficient. A typical core flow speed of 1 mm/s or 30 km/yr then gives

$$\delta T = 10^{-3} \text{ K}$$

This is an overerestimate because most of the Coriolis force is balanced by the pressure. On very small scales it may be an underestimate because of the effects of turbulence, but the calculation is sufficient to show that the CMB is at a very uniform temperature.

Very small temperature fluctuations are important for core dynamics but not for mantle dynamics; thermal convection in the lower mantle should therefore be modelled with an isothermal lower boundary. The existence theorem for the convection equations requires

either isothermal or constant heat flux boundary conditions: specifying both overdetermines the system. The heat flux across the CMB must inevitably vary from place to place, depending on the details of convection in the mantle.

Core convection may be modelled by a prescribed variable heat flux boundary condition, the precise nature of which depends on convection in the mantle. The two fluid systems are coupled and therefore depend upon each other, but the enormous disparity of time scales between core and mantle convection makes the dependency rather easy to evaluate: the core experiences an imposed boundary heat flux that varies so slowly in time that we may treat it as stationary, while the mantle experiences a uniform lower boundary temperature.

Mantle Convection

Heat passes across the CMB by conduction through thermal boundary layers. The boundary layer in the core is likely to be very thin but the boundary layer on the mantle side will be thick because the fluid flow is very much slower than in the core. (The anonymous reviewer points out that the boundary layer may not exist at all if the top of the core is stably stratified, or may have a negative temperature gradient if compositional convection takes heat downwards as *Loper* [1978] has suggested.) Conducted heat flow is proportional to the temperature gradient, and therefore the temperature gradient in the boundary layer will be steep in regions where heat flow is high and shallow where heat flow is low. The CMB temperature is uniform everywhere, so the mean temperature in the boundary layer will be relatively cold where the gradient is steep and heat flux high, and relatively hot where the gradient is shallow and the heat flux low. This situation is illustrated in Figure 1. It is somewhat counter-intuitive to have high heat flux where the mantle is cold, but it is only like holding an ice cube—the ice is cold and therefore draws heat out of your hand.

How thick is this thermal boundary layer? An estimate can be obtained if we assume the convecting parts of the core and mantle are close to adiabatic. This will be true in the core unless it is partially stratified; it is also a good approximation in the mantle away from mineral phase changes. Estimates of the adiabatic gradients in the core and mantle are available from properties of iron and silicates at high pressure and temperature. The temperature throughout the outer core may then be found by integrating up from the inner core boundary, which is assumed to be at the solidus temperature of the iron alloy that comprises the solid inner core. In the mantle we may integrate down from the temperature of the base of the lithosphere, another boundary layer, taking account of the phase changes within the mantle. The two procedures give independent estimates of the temperature of the CMB.

These calculations have been reviewed recently by *Poirier* [2000]. They give a mis-match at the CMB of about 1500 K, which is sometimes misleadingly referred to as a "temperature discontinuity". It is not a discontinuity, it is the departure from the adiabat in the mantle, which exists primarily in the boundary layer where radial convection and advection of heat are inhibited. 1500 K therefore represents the temperature difference down which all the core's heat flow, H, conducts through a boundary layer of thickness d and thermal conductivity k_{m}. Taking $k_{\mathrm{m}} = 4$ Wm^{-1}K^{-1} and $H = 4$ TW gives a boundary layer thickness of 250 km, about the average thickness of D''.

Lateral variations in heat flux will result in lateral variations in mean temperature of the boundary layer. D'' is similar to the lithosphere in many respects, and mantle convection calculations have shown the lateral variations can be large, easily 20% of the mean vertical heat flux. This would require a variation of a few hundred degrees in the boundary layer, consistent with the observed variations in seismic velocity [*Masters et al.*, 1996].

These rough calculations are consistent with the seismological layer D'' being a thermal lower boundary layer, but this is almost certainly not the whole story: estimates of CMB heat flux based on requirements of core convection and the geodynamo are several times larger than 4 TW [*Gubbins et al.*, 2001; *Labrosse, Poirier and LeMouël*, 1997] and the composition of D'' is unlikely to be uniform, as discussed in the next section. The thermal boundary layer may be very thick, or compositional variability may affect the numerical estimates. It remains, however, a reasonable and self-consistent starting point for discussing thermal interaction between the core and mantle.

Does the core influence mantle convection other than by imposing a uniform lower boundary temperature? Possibly. While core convection takes place on too short a time scale to affect mantle flow directly, the effects of rotation may impart a long term dependence on the geographic axis and therefore latitude. Studies of convection in a rotating spherical shell with homogeneous boundary conditions show that the surface heat flux varies with latitude. The precise form of the heat flux variation with latitude depends on the convective regime, which is largely unknown for the Earth's core,

but if such a variation exists it is likely to persist for very long periods of time. This could explain any observed dependence of mantle convection on rotation, such as the apparent clustering of the Euler poles of relative rotation of the plates near the geographic axes [*DeMets et al.*, 1990]. The core's influence would be destroyed by true polar wander following a change in the inertia tensor of the mantle.

Core Convection

The core responds passively to lateral variations in heat flux across its outer boundary by predominantly lateral fluid flow. These passive flows are called thermal winds; they arise in the atmosphere because the sun heats low latitudes more than polar regions. They carry heat from regions near the surface where heat flux out of the core is low to regions where it is high. The most efficient transport of heat is achieved by flow from places where the heat flux low to places where it is high, but the dominant Coriolis force deflects the flow and produces a quite different pattern, with flow across regions of high and low heat flux. The same Coriolis effect is seen on any weather chart, where winds blow parallel to isobars rather than, as one might expect were the charts not so familiar, blowing from high pressure to low pressure.

Thermal winds could be a significant contributor to geomagnetic secular variation if the flow speeds required to transport the necessary heat are comparable with those thought to drive the observed secular variation, say 10–30 km/yr, the value based on the westward drift. *Bloxham and Gubbins* [1987] estimated the heat transported by the following rough calculation. Suppose core fluid flows horizontally at speed v_{h} a distance L between two points on the CMB with temperature difference ΔT (Figure 1). The fluid cools by conduction against the solid boundary with the mantle as it moves. It travels the distance in time $\tau = L/v_{\mathrm{h}}$, during which it cools to a depth $\sqrt{\kappa\tau}$, where κ is the thermal diffusivity $\kappa = k/\rho C_p$ and C_p the specific heat. The heat lateral transfer is

$$H_L = \rho C_p (\kappa\tau)^{1/2} \frac{\Delta T}{\tau}. \qquad (2)$$

Bloxham and Gubbins [1987] use this formula to obtain a lateral heat flux of 1% of the radial heat flux for a temperature difference of 10^{-4} K forced by lateral temperature variations in the mantle for order 1 K. Lateral mantle temperatures of 1 K would give lateral variations in heat flux about 100 times less than the average radial heat flux.

Convection driven by heating from below is influenced by lateral variations in heat flux at the outer boundary [*Zhang and Gubbins*, 1993]. This is quite different from the passive thermal winds, which will also be present. A range of behaviour is possible, depending on the relative strengths of vertical and lateral heating, and the Ekman number $E = \nu/\Omega L^2$, where ν is the fluid viscosity, Ω the angular velocity, and L the size of the sphere. Convection driven by uniform internal heating and homogeneous boundary conditions takes the form of rolls aligned with the rotation axis that drift in azimuth [*Busse*, 1970], the number of rolls (or azimuthal wavenumber) being determined by E.

Lateral variations in heat flux on the boundary can lock the convection and prevent the rolls from drifting, provided the length scale of boundary anomalies is comparable with that of the convection [*Zhang and Gubbins*, 1993]. This case is illustrated in Figure 1. The convection is locked in a position favourable for lateral heat transport to match the boundary requirements. Upwelling fluid brings heat to the surface where the boundary condition requires heat flux to be high, and downwelling fluid takes heat away from the surface in places where the boundary condition requires heat flux to be low. This is different from the thermal wind, which tends to blow over the high and low heat flux regions.

Other effects of laterally varying boundary heat flux include secondary resonances, where stationary convection succeeds in meeting the demands of the outer boundary heat flux condition, and variable rates of drift, where upwelling flows dwell for longer beneath regions of high boundary heat flux [*Zhang and Gubbins*, 1993]. Locking is less effective at low Prandtl number because of the effects of inertia [*Zhang and Gubbins*, 1996]. If these effects hold in the Earth's core we should expect some correspondence between core convection and boundary heat flux: a similarity of length scales and downwelling where heat flux is high. Of course, these convection calculations are idealised and the core may not behave in the same way.

Apart from thermal interaction, two other mechanisms of long-term core-mantle interaction have been discussed: topography on the CMB and electrical conductivity in the lower mantle. Bumps on the CMB will have a thermal effect that will be very difficult to differentiate from the thermal effect of lateral temperature variations in the lower mantle. A growing protrusion into the core is like a cold finger, and will induce downwelling. The effect is transient and will only continue until the bump warms up to the ambient core temperature. Bumps also have a dynamical effect on the flow that has received relatively little quantitative study, the work of *Bell and Soward* [1996] being an exception.

Electrical conductivity in the lower mantle will influence the core flow and can even cause dynamo action [*Busse and Wicht*, 1992] provided the conductance, the product of conductivity and thickness of the conducting layer, is large enough. The dynamo effect needs high conductance: the whole of D″ having a conductivity within an order of magnitude of that of the core. Such a layer will shield the secular variation; *Herrero-Bervera and Runcorn* [1997] proposed a highly conducting layer beneath the Pacific to explain the low secular variation there. The main obstacle to this hypothesis is the lack of any mineralogical evidence for high conductivity in the region. A large iron component would raise the conductivity and is plausible because of the close proximity of D″ to the iron core, but *Poirier and LeMouël* [1992] argue against any percolation of iron into the lower mantle, favouring a very low electrical conductivity.

Buffett [1996] pointed out that aspherical mass anomalies in the mantle and inner core will lead to gravitational coupling between the two. The inner core is also strongly coupled magnetically to the outer core by its high electrical conductivity, thus providing another mechanism of core-mantle coupling.

Magnetic Fields and Dynamo Action

We do not observe convection in the core directly, we must infer it from the geomagnetic field. Frozen flux theory [*Backus*, 1968; *Roberts and Scott*, 1965], which assumes that diffusion of magnetic field by electrical resistance is negligible on timescales of decades to centuries, yields estimates of the fluid flow responsible for geomagnetic secular variation (e.g. *Bloxham and Jackson* [1991]), but this may not be the flow responsible for generating the magnetic field because dynamo action requires a balance with magnetic diffusion.

Kinematic dynamo models, in which a magnetic field is generated by a prescribed steady fluid flow, often have surface fields concentrated over regions of downwelling [*Gubbins et al.*, 2000]. The physical mechanism is shown in Figure 1; magnetic field lines ride on the flow as on a conveyor belt and are carried away from regions of upwelling, leaving a patch of weak radial magnetic field, and are concentrated over the region of downwelling. This effect will only be significant if the convection is relatively steady, or is locked to the boundary by, for example, thermal anomalies. In this case radial magnetic field would be concentrated over regions of cold lower mantle and high CMB heat flux.

This suggests another way to determine core flow, an alternative to frozen flux theory. Places where the magnetic field remains concentrated for long periods of time (a few centuries), such as the four main lobes over Siberia, North America, and in the Southern Hemisphere on the same longitudes [*Bloxham and Gubbins,* 1985], may be sites of downwelling, while places where the radial field has been persistently weak, such as over the North Pole and in the Pacific, may be sites of upwelling. The correspondence between magnetic field and fluid flow is made more complicated by the hemispheric nature of rotating convection: for example, if one end of a convection roll is locked by a boundary anomaly in the Northern Hemisphere, the other end will be locked on the same longitude in the Southern Hemisphere, even though no boundary anomaly exists at that point.

One effect of boundary thermal anomalies is therefore to promote steady flows. Core convection is unlikely to be locked completely to the CMB, but some correlation may appear in a time average. This effect has recently been observed in dynamo simulations [*Bloxham,* 2000; *Glatzmaier et al.,* 1999]. Another effect is to influence the length scales of convection and perhaps to give a field morphology that reflects the boundary anomalies. Lateral variations in heat flux that are large in spatial extent may, like the magnetic field, promote large scale convection. This influence on the length scales of convection in the core could have a dramatic effect on the geodynamo regime.

Non-magnetic convection at high rotation speeds like that of the core takes the form of small scale rolls with azimuthal order number m. The critical Rayleigh number for onset of convection is high because rotation inhibits flows that convect heat efficiently. An applied magnetic field tends to increase the length scale of convection and reduce the critical Rayleigh number by countering the rotational constraint. The length scale reaches the scale of the container when the magnetic field strength gives an Elsasser number, Λ, of order one:

$$\Lambda = \frac{B^2}{2\Omega\mu_0\rho\eta} \sim 1 \qquad (3)$$

where ρ is density and η is magnetic diffusivity.

The critical Rayleigh number, R_c, and azimuthal wavenumber of convection at onset have simple asymptotic forms as $E \to 0$ [*Jones, Soward and Mussa,* 1999; *Roberts,* 1968; *Zhang and Gubbins,* 2000a]:

$$\Lambda = 0: \quad R_c = O(E^{-4/3}), \quad m = O(E^{-1/3}) \qquad (4)$$

$$\Lambda = 1: \quad R_c = O(E^{-1}), \quad m = O(1) \qquad (5)$$

In the Earth's core a toroidal magnetic field of about 10 mT would give an Elsasser number of order one. The Ekman number is very small; about 10^{-15} for molecular viscosity and 10^{-9} for turbulent viscosity. Even for

turbulent diffusivities R_c is smaller by a factor of 10^3 for magnetoconvection than it is for non-magnetic convection, reflecting the greater efficiency of large scale convection in transporting heat.

The precise form of the magnetic field also affects the critical Rayleigh number drastically. Early calculations were done with force-free fields, magnetic fields that exert zero Lorentz force on the flow [*Chandrasekhar,* 1961; *Eltayeb and Kumar,* 1977]. Force-free fields simplify the stability analysis because the system is at rest until the critical Rayleigh number is exceeded. This early work showed that, while rotation and magnetic field applied individually would inhibit convection, the combination of the two did not inhibit convection as much. A plot of R_c against B, shown schematically in Figure 2 for fixed rapid rotation, contained a minimum where convection was easiest to excite [*Chandrasekhar,* 1961]. These results are for magnetoconvection, where the magnetic field is imposed by a fixed current source. If the magnetic field were self-generated by dynamo ac-

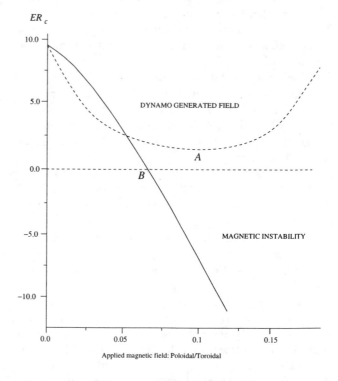

Figure 2. Variation of critical Rayleigh number with magnetic field strength for magnetoconvection. Dashed line is schematic for a force-free field, the minimum at A occurring when the separate stabilising effects of magnetic and Coriolis forces partially cancel. Solid line shows results from a calculation with applied poloidal and toroidal fields, with magnetic instabilities setting in at point B. Both A and B have been suggested as equilibration points for a self-generated magnetic field

tion, the minimum in $R_c(B)$ might indicate the strength at which the dynamo would equilibrate; the magnetic field would not grow stronger than this value.

These ideas have been changed somewhat by the study of *Zhang and Jones* [1996]. They used magnetic fields that exerted a force on the fluid and found no minimum in the $R_c(B)$ curve (Figure 2): the critical Rayleigh number decreases monotonically with increasing magnetic field strength, eventually reaching negative values where energy for the flow is provided by the applied magnetic field rather than by the convection. This is the regime of magnetic instability. It is not sustainable if the magnetic field is self-generated because there would be no external supply of energy from the magnetic field. The point where $R_c(B)$ passes through zero might therefore be an upper limit to the field strength that could be generated by this form of convection.

R_c depends critically on the form of the applied magnetic field: its dependence is rather weak for force-free fields, and is expected to be rather strong for applied fields that give strong magnetic instabilities. *Zhang and Gubbins* [2000a] give an example of extreme variation of R_c with applied field strength by applying a combination of toroidal and poloidal magnetic fields. This choice of applied field may be criticised as being particularly prone to magnetic instability, but the example serves to illustrate a serious difficulty when trying to understand magnetoconvection or dynamo action at low Ekman number. Equation (5) shows that the product $R_c E$ should become independent of E as $E \rightarrow 0$. Figure 2 shows this product as a function of the strength of the poloidal component of the applied field. It changes by a factor of 10 for a few per cent change in applied magnetic field. This is a change by a factor 10^{10} in R for the Ekman number relevant to the Earth's core. It is therefore likely that the vigour of convection in the core, which depends on the ratio of the Rayleigh number to its critical value, will vary dramatically with strength of the magnetic field.

This extreme variation in critical Rayleigh number can explain the failure to find a solution to the dynamo equations at zero Ekman number [*Fearn and Proctor*, 1987; *Walker, Barenghi and Jones*, 1998]. Attempts to integrate the equations in time from an initial solution invariably result in the magnetic field decaying away, often after a promising start. Fluctuations in the magnetic field lead to large changes in critical Rayleigh number and therefore large changes in the ratio R_a/R_a^c, which determines the strength, time dependence, and scale of the convection. Weaker magnetic fields lead to weaker, smaller scale, less chaotic convection. This re-

stricts the dynamo action and leads to an even weaker magnetic field. *Zhang and Gubbins* [2000a] suggest that this constitutes structural instability in the dynamo when $E = 0$, and that no physical solution exists. The demise of the generated magnetic field after integrating for a considerable period of time was called the dynamo catastrophe by *Jones* [2000].

Numerical Simulations of the Geodynamo

There are now several numerical integrations of the full three-dimensional nonlinear dynamo equations [*Bloxham*, 2000; *Christensen, Olson and Glatzmaier*, 1999; *Glatzmaier et al.*, 1999; *Glatzmaier and Roberts*, 1995; *Kuang and Bloxham*, 1997; *Olson, Christensen and Glatzmaier*, 1999], some of which include laterally varying boundary heat flux [*Bloxham*, 2000; *Glatzmaier et al.*, 1999]. These calculations show weak or no evidence of locking of the convection or magnetic field to boundary features, although that of Bloxham yields a time-averaged field that shows remarkable similarities to the time-averaged paleomagnetic field obtained by *Gubbins and Kelly* [1995]. This is taken by some as evidence against strong locking of the geodynamo to the lower mantle. They may well be right because core convection is unlikely to resemble the very simple convection models described in previous sections. However, even the most ambitious of geodynamo simulations fall far short of realistic geophysical parameters, and we should be cautious when interpreting the results of a small handful of parameter choices.

The major difficulties with simulating the geodynamo numerically are the very small Ekman number, which produces highly organised (non-chaotic) but small scale flow, the high Rayleigh number, which produces chaotic time dependence and small scale flow, and the very long time needed to establish a regime independent of the starting conditions. There is also the intractable problem of turbulence in the core, which forces everyone to increase the viscosity and thermal diffusivity by many orders of magnitude to be at least comparable with the electrical conductivity. These difficulties conspire to reduce the choice of parameters studied to just a few per scientific paper and runs that are rather short in time.

When extrapolating these results to the Earth it is essential to remain in the right regime. For example, E must be drastically reduced but at the same time R_a must be increased or convection (and dynamo action) will cease when it falls below the critical value, which rises as E falls. This much is obvious, but it is equally important not to have too high an R_a in the simulations. High R_a breaks the rotational constraints and related effects like boundary locking. The first geodynamo simulations have been run at relatively high R_a to ensure dynamo action, and we do not know the critical value below which dynamo action fails. They therefore lie in a chaotic regime. More recent models are closer to the critical value [*Christensen, Olson and Glatzmaier*, 1999].

The Rayleigh number in the core is very difficult to estimate, but ultimately it is controlled by the heat flux from the core and controls the average flow speed of the convection. Two independent recent estimates of the Rayleigh number in the core make it rather low. *Jones* [2000] bases his estimate on the convective flow speed and *Gubbins* [2001] on the heat flux. Both arrive at R_a about 1000 times the critical value for magnetoconvection and turbulent values of the diffusivities, or about equal to the critical value for non-magnetic convection. These are rather low values, and the heat flux estimate is an upper bound. R_a in the core could, therefore, be quite low. Boundary-locking and resonance effects in such a regime could be very important, particularly if lateral variations were large.

Most geodynamo simulations use hyperviscosity, a scale-dependent viscosity. This makes it difficult to identify the correct value of E, which is length-scale dependent. The small value of E associated with large length scales is the one usually quoted, as is appropriate when small scale effects are unimportant. The much larger value of E corresponding to small scales is relevant when small scale effects are significant. This is the case with the instability proposed by *Zhang and Gubbins* [2000b], which arises from collapse to small scales. This collapse is prevented by high E applied to the smallest length scales in the simulation. *Zhang and Jones* [1997] have stressed the need to use the viscosity relevant to the smallest scales, not the headline value applicable only to the large scale flow.

The real Earth is not restricted by numerical considerations. The Rayleigh number could be large enough to sustain non-magnetic convection. When the geomagnetic field is weak, such as during reversals or transitions, we could have a weak-field regime with small scale convection; we have no observations to contradict such a scenario. One possible description of the reversal process is a collapse of the geodynamo from a strong field regime, in which convection is large scale and vigorous, to a weak field regime in which the convection is small scale and rather feeble [*Zhang and Gubbins*, 2000a].

Another possibility I suggest here is that lateral heat flux variations on the CMB prevent the collapse to smale scale convection when the magnetic field is weak. Boundary anomalies promote relatively steady flows with the same length scale as the anomalies themselves;

they may therefore be a controlling factor of the stability of the geodynamo. The geodynamo may be operating well away from the simple locking regime studied by *Zhang and Gubbins* [1996], but we have virtually no understanding yet of the role played by boundary anomalies in the low E dynamo.

Can we drive the geodynamo entirely by lateral heating at the CMB? A simple thermodynamic argument shows this to be impossible, even if we allow enough radial heat flow to maintain the adiabatic gradient in the core. The balance of entropy (e.g. *Backus* [1975], *Gubbins* [1977], *Hewitt, McKenzie and Weiss* [1975]) shows that the ohmic heating cannot exceed the total heat transfer by more than an "efficiency" factor of $(T_{\max} - T_{\min})/T_{\min}$, which for radial heating is about 0.1 [*Backus*, 1975]. Core convection mixes the entropy so well that temperature changes at different places around the CMB will not exceed a fraction of a degree. The thermodynamic efficiency factor would therefore be tiny, perhaps 10^{-4}, placing quite impossible demands lateral heat flow at the CMB.

OBSERVATION

Seismology

Seismic tomography and splitting of normal mode frequencies both indicate a doughnut of fast seismic velocities in the lowermost mantle [*Dziewonski*, 1984; *Dziewonski and Woodhouse*, 1987; *Masters et al.*, 1996; *Masters, Laske and Dziewonski*, 2000]. The resolution of both the tomography of body waves reaching the lowermost mantle and the normal mode frequencies is about 150 km, so the results may be roughly interpreted as an average of the properties of the D'' region. The lower mantle doughnut lies approximately below the ring-of-fire subduction zones around the Pacific, suggesting that it results either from lithosphere that has been subducted all the way to the core mantle boundary or that the lower mantle has cooled by persistent subduction in the upper mantle above it. Subduction around the Pacific has retained roughly the same geometry for over 50 million years, enough time to cool to the base of the mantle. A map of shear wave velocity, which is sensitive to temperature, is shown in Figure 3.

The seismic anomalies at the base of the mantle could be caused by variations of temperature in a thermal boundary layer, or they could be caused by compositional variations. P- and S-velocities are likely to be affected in a similar way by temperature. The bulk sound speed v_c is defined by:

$$v_c^2 = \frac{K_S}{\rho} = v_P^2 - \frac{4}{3}v_S^2 \qquad (6)$$

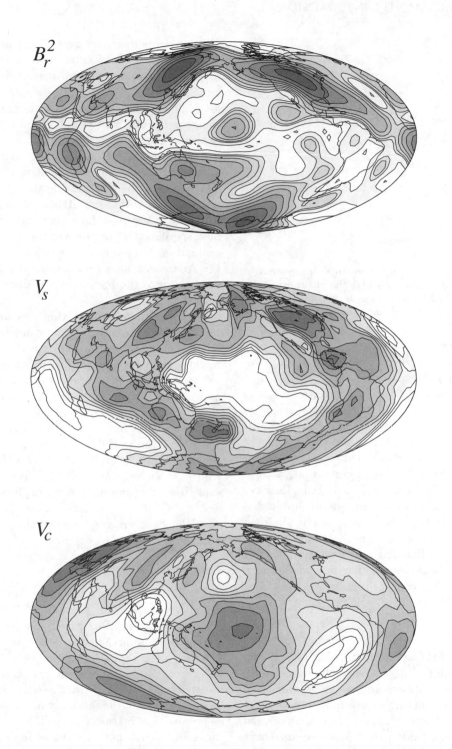

Figure 3. Maps of (top) B_r^2 at the CMB, arbitrary scale (middle) v_s variations in lowermost 150 km of the mantle, scale 1% (bottom) v_c, scale 1%, after *Masters, Laske and Dziewonski* [2000].

where K_S is the adiabatic bulk modulus. This should correlate with v_S if temperature is the cause of the anomalies. v_c is shown for the Scripps Model SB10L18 of *Masters et al.* [1996] model in Figure 3; regions of low velocity are anticorrelated with v_S, whereas regions of high velocity show weak correlation. Undoubtedly temperature makes a strong contribution to these anomalies, but it appears that low velocities contain effects of compositional variation. Variations in K_S that anticorrelate with v_S could be caused by elevated density, for

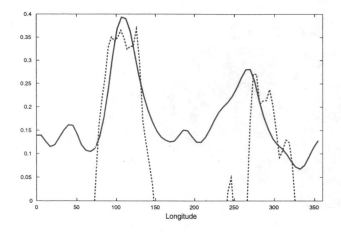

Figure 4. Latitude-averages of the vertical component of magnetic field squared (solid line) and the shear wave velocity near the CMB (dashed line, only the positive part shown). Note the coincidence of the maxima in longitude.

example by variations in the iron content of the perovskite structures in D″, which lowers the seismic velocities but leaves K_S the same.

Further evidence of compositional variations comes from reflection of short period body waves, such as PdP, from the top of D″ (e.g. *Weber and Davis* [1990]). Waves with periods of a few seconds and wavelengths of 10 km need a sharp boundary to be coherently reflected, too sharp to be sustained by temperature anomalies. *Weber and Davis* [1990] attribute their reflection beneath the Kurile Islands to a remnant of subducted plate. It seems likely that v_S gives a rough guide to the pattern of temperature in the lowermost mantle: the region beneath the Pacific rim is cold and the two zones of low velocity in the central Pacific and southern Africa are hot with some compositional variations from their proximity to the iron core.

Geomagnetism

The geomagnetic field at the CMB also reflects the lower mantle doughnut of high shear wave velocity, in keeping with the observation made in the previous section that downwelling core flow may concentrate magnetic flux. Figure 3 shows B_r^2, the square of the radial component of magnetic field at the CMB, averaged over the time interval 1840–1990 using the historical model of *Jackson, Jonkers and Walker* [2000]. The square is plotted rather than B_r because flux is concentrated without regard to sign.

The dominant axial dipole structure precludes any direct correspondence between magnetic field and lower mantle temperature and its proxy v_S. Removing the

latitude dependence by averaging (multiplying by $\sin\theta$ and integrating over θ) gives functions of longitude that may be compared directly. Figure 4 shows $\int_0^\pi B_r^2 \sin\theta d\theta$ and the positive part of $\int_0^\pi v_S \sin\theta d\theta$ as functions of longitude, ϕ. The correlation is remarkable, suggesting the geomagnetic field is influenced by the temperature of the lower mantle boundary layer.

A further suggestion of mantle influence on the geomagnetic field is the asymmetry in secular variation. The region of the CMB inside the lower mantle doughnut is quiet, with very little change in magnetic field throughout the past four centuries, while the SV has been significantly greater over the rest of the CMB. Four centuries is long enough for the westward drift seen in the Atlantic, which amounts to 0.1–0.2° per year, to pass well into the Pacific, but it does not appear to do so.

It is difficult to explain this east-west asymmetry in terms of lower mantle temperature anomalies. D″ beneath the central Pacific is dominated by low v_S, which might cause upwelling and consequent removal of flux to the rim. This would remove all radial magnetic field and leave no tracers for any flow that might exist at the top of the core. By contrast, the other large region of low v_S, that beneath southern Africa (Figure 3) is the site of the world's most intense SV. It seems that here toroidal flux might be expelled by fluid upwelling [*Bloxham and Gubbins*, 1985]. We seem to need the same upwelling mechanism to explain both intense SV and no SV at all! Perhaps there is no toroidal field near the core surface beneath the Pacific at present, and therefore none to expel.

Paleomagnetism

The historical record is too short to provide any definitive evidence of permanent features in the geomagnetic field imprinted from the solid mantle. Paleomagnetism and archeomagnetism are needed to extend the record further back into the past. Archeomagnetism has the advantage (sometimes) of providing very accurate dates but the disadvantage that artefacts are rarely oriented, so only the intensity of the magnetic field can be recovered. Intensity is a difficult quantity to measure, and we are here interested in departures from the axial dipole moment, which is also difficult to determine. *Constable, Johnson and Lund* [2000] have derived a model for the last 3,000 years; the most determined effort yet to study this difficult period.

A time average of the geomagnetic field may be obtained using paleomagnetic data from the last 5 Myr, during which plate motions may be neglected [*Gubbins* and Kelly, 1993; Johnson and Constable, 1995; Kelly and Gubbins, 1997]. Paleomagnetic measurements are taken from a set of sites within a rock unit. The sites are of different ages and the average of magnetic directions from all sites is intended to average out the secular variation [Butler, 1992]. This usually proves successful, and the average direction corresponds more closely to the geographic pole than individual determinations. The position of the magnetic pole derived from a single site, including the secular variation, is called a virtual geomagnetic pole (VGP); that derived from an average is called a paleomagnetic pole.

In plate reconstruction the paleomagnetic pole is assumed to give the geographic pole. Determining the time-average of the geomagnetic field is a much more difficult problem because one is looking for departures from the geocentric axial dipole, a signal amounting to $5°$ or less in the paleomagnetic direction. Nevertheless, Gubbins and Kelly [1993] obtained a map of B_r at the CMB that had similar concentrations in the northern hemisphere, under Siberia and Canada, as the modern field. Nothing appeared in the southern hemisphere, but little data were available from southern latitudes. Subsequent work [Johnson and Constable, 1995] also indicates some departures from axial symmetry reminiscent of the modern field. The non-axisymmetric part of the field depends mainly on declination, which may be in error because of tectonic block rotation. The debate was recently reviewed by Gubbins [1998].

While the mean direction for a rock unit gives an estimate of the time-averaged field, the scatter of directions within a rock unit gives an estimate of the SV. In a spherical harmonic analysis of paleomagnetic data, the SV contributes to the standard deviation of the spherical harmonic coefficients. The modern field gives a scatter in directions up to $12°$ in some places, while in the Pacific it is as low as $1°$ [Kelly and Gubbins, 1997]. They examined the geographical distribution of VGP scatter and found no evidence that it was lower in the Pacific region than elsewhere, the opposite to what is found in the historical record. Other recent studies [Khokhlov, Hulot and Carlut, 2001; Kono and Tanaka, 1995] have found significant geographical variations in the paleosecular variation in the form of anomalous standard deviations for the spherical harmonic coefficients g_2^1 and h_2^1. Gubbins and Kelly [1995] had pointed out earlier that these harmonics were important contributors to the VGP scatter when the modern field was analysed in the same way. These results suggest departure from axial symmetry in the paleosecular variation as well as the time averaged magnetic field. Dominant $m = 1$ and $m = 0$ harmonics may result from low secular variation in the Pacific.

Reversals

Geomagnetic reversals have occurred more frequently since the Cretaceous Normal Superchron. This change has take place over a longer time scale (tens of millions of years) than any known time scale for core dynamics and is attributed to changes in the lower mantle. The cause may be a change in heat flux from the core and consequent change in Rayleigh number and dynamical regime in the core (e.g. Larson and Olson [1991]), or it may be a change in the geographical pattern of lateral variations of heat flow across the CMB [Gubbins, 1987]. Several studies are currently underway designed to explore the dependence of reversal behaviour in dynamo models for different heating regimes.

Another apparent effect of the lower mantle on geomagnetic reversals is the tendency of VGPs to follow a great circle through Asia and the Americas during a polarity transition or even excursions [Laj et al., 1991]. This result has provoked some controversy; the observational evidence is not definitive—indeed it would be surprising if it were, as the geomagnetic field is unlikely to reverse in such a simple manner, but the coincidence of these paths with the preferred longitudes where the modern and time-averaged fields are concentrated (Figure 4) suggests a common explanation.

The early results of persistent transition paths were based on sediment records, which are thought to be rather poor magnetic recorders subject to over-printing and deformation after magnetisation. Prévot and Camps [1993] examined lava data and found no indication of preferred paths. Love and Mazaud [1997] performed a careful analysis of records from the last reversal, Matuyama-Brunhes, including both lava and sediment data, and found evidence for preferred paths (Figure 5). Love [1998] then reassessed the Prévot and Camps [1993] analysis. They had rejected data showing similar directions from the same region on the belief that they were obtained from lava flows that erupted almost simultaneously on the grounds that they would bias the dataset in favour of certain directions. Love [1998] discovered that some of the rejected data were from different volcanoes, and were therefore very unlikely to represent sampling of the same time interval. A stratigraphy was available for some flows; if two adjacent flows gave similar directions he rejected one, otherwise he kept them. With this new data compilation the preferred paths appear once more (Figure 6). In a later statistical analysis, Love [2000] concludes that the VGPs cluster into preferred

VGPs of MBD97

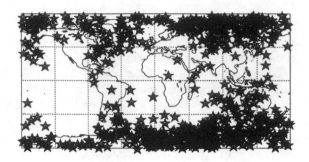

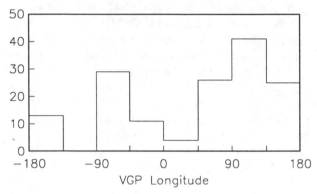

Figure 5. VGPS for the last reversal, Matuyama-Brunhes. From *Love and Mazaud* [1997].

longitudes with 95% probability. It seems inevitable that the *Prévot and Camps* [1993] analysis would eliminate the preferred paths because they removed a great deal of the very information that would produce them: similar directions at the same site. Love's procedure may well allow some repeat measurements through, but selection of the same longitudes as the sediment data, the modern field, and the time-averaged field, seems too much to be a coincidence.

It is very surprising to find systematics in the VGP paths because the field is not expected to remain dipolar during transition. Indeed, Figure 5 shows two different paths for the same reversal, so the field could not be dipolar. Symmetry properties of the dynamo can help in the interpretation. The geodynamo equations are invariant under change of sign of magnetic field, rotation about the spin axis, and reflection through the equator, as is the geometry. This has a paradoxical consequence for reversal paths: any mechanism that takes the north VGP along a specific path from the north geographic pole to the south geographic pole for a R-N transition must take the south VGP along the same path for a N-R transition. The result is paradoxical because, if the sign of the magnetic field is irrelevant one would perhaps expect the path to be the same each time. But

the the south magnetic pole (the one presently at the north geographic pole) is plotted by convention, and this pole starts at the south magnetic pole for an R-N transition. The same magnetic field configuration that brought the VGP through one particular path for the N-R transition will therefore follow the antipodal path for the R-N transition. If the N-R transition has a VGP path through the Americas, for example, the R-N path measured at the same site will have an Asian path.

[*Gubbins and Sarson*, 1994] published a kinematic dynamo model that oscillated with a dynamo wave mechanism [*Parker*, 1979], in which magnetic flux migrated from lobes concentrated at high latitudes towards the equator. The flux remained concentrated on the same longitudes, locations defined by downwelling limbs of the prescribed fluid flow. This dynamo produced VGP paths that were concentrated on the two longitudes where the flux was concentrated. The reason for this is simple: the VGP for each site is governed by the

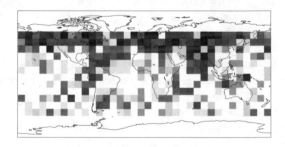

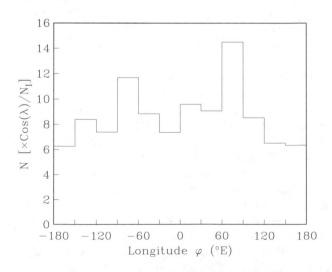

Figure 6. VGP concentrations for recent reversals and transitions, and histogram showing the weighted density of VGPs with longitudes. Note the peaks near ±90°. From *Love* [1998].

closest flux concentration on the CMB, reinforced by more distant flux of opposite sign. The flux patches migrate along lines of longitude, and so do the VGPs. The effect is site dependent, the path being determined by the sign and longitude of the nearest flux patch at the beginning of the transition.

The original dynamo model of *Gubbins and Sarson* [1994] is very simple and in no way geophysically realistic: being kinematic with a steady fluid flow the magnetic field can only grow, decay, oscillate, or remain stationary. What it does show is how the magnetic field configuration relates to the underlying fluid flow, which has two downwelling limbs on a great circle path passing through the poles (as the Earth might have if the Pacific rim were to induce downwelling). *Gubbins and Gibbons* [2002] have shown that the oscillatory behaviour is rare and depends critically on the meridian circulation; a time-dependent flow might therefore sustain a steady field most of the time, with occasional reversals reflecting the oscillatory solutions when the meridian circulation was exactly right. *Sarson and Jones* [1999] observed a reversal in a 2.5D dynamo model accompanied by a parcel of fluid rising along the axis, indicating a change in meridian circulation.

Sites in the northern hemisphere closest to the eastern (western) flux patch will follow an eastern (western) path, those in the southern hemisphere closest to the eastern (western) flux patch will follow an western (eastern) path. One might expect the path to change gradually as the site is moved across the equator or across some line delineating proximity to one flux patch or another, but this does not happen. In the model, the site on the equator measures an instantaeous reversal: the field strength goes to zero and grows in the opposite direction. This is a result of reflection symmetry about the equator. Something similar happens along the line midway between the two flux lobes, but the reversal is not instantaneous because the symmetry is rotational rather than reflectional.

To apply these ideas to thé Earth we must generalise away from the specific dynamo model, which at this stage only served to illustrate how VGP paths relate to the generated field. In the Earth we require the flux to remain concentrated on the cold longitudes of the Pacific rim, and to start from latitudes near the inner core tangent cylinder. The flux could migrate poleward or equatorward. Dynamo models exhibit both types of behaviour [*Parker*, 1979]. Path selection is then made according to whether the path is R-N or N-R and whether the flux migrates poleward or equatorward.

Gubbins and Love [1998] applied these ideas to the last reversal, using the dataset of *Love and Mazaud*

[1997]. They examined the VGP transition paths for each of the 11 available sites and denoted them east, west, or neither. They marked these on a global map divided into four quadrants by the equator and a great circle path defined approximately by the flux concentration in the modern field, close to the Pacific rim. If the hypothesis of flux concentration is correct, each quadrant should contain paths of one type and the sites showing neither path should lie near one of the nodal lines dividing the four quadrants. Figure 7 shows the results, an encouraging picture with only Tonging giving contradictory results, and Hawaii being rather far from a nodal line. The hypothesis therefore stands for the moment; more data are needed from other reversals before further tests can be done.

Excursions

Recent paleomagnetic results have confirmed the global nature of excursions, events in which the direction of the geomagnetic field becomes wayward or even reverses for a time before returning to its original configuration. Excursions seem to be quite common; since the last reversal 6 have been established globally by *Langereis et al.* [1997] and the total number might be 20 or more [*Lund et al.*, 1998]. New evidence shows some excursions in the Matuyama [*Channell et al.*, 2002]. This evidence of unstable geomagnetic behaviour is supported by continuous records of paleomagnetic intensity from sediments [*Guyodo and Valet*, 1999] and lavas from boreholes [*Garnier et al.*, 1996; *Laj et al.*, 2001; *Teanby*, 2001]: the dipole moment appears to be very irregular in time, occasionally dipping to very small values at times of excursions.

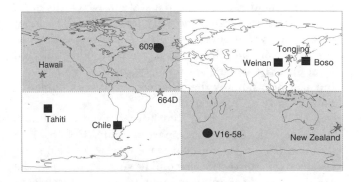

Figure 7. Site-dependence of the VGP paths for Matuyama-Brunhes. The 4 quadrants are separated by 2 great circles. No systematic VGP paths are expected from sites near these nodal lines, the remaining quadrants should see east and west paths alternately. Squares (circles) denote sites recording eastern (western) paths; stars sites with no discernible path. After *Gubbins and Love* [1998].

These two lines of evidence point to an inherently unstable dynamo, as outlined in Section 2.4, but one with the ability to recover from a collapse in the field and to maintain a strong field for 50–100 kyr, the interval between excursions, which is much longer than the overturn time. What imparts this degree of stability? I mention here three possibilities.

First, nothing is needed: we do not know enough about geodynamo behaviour at low Ekman number and core turbulence to be sure that a dynamo-generated field is not self-stabilising in some way. This is the view held by most geodynamo simulators, although their simulations are no proof of stability because the viscosity is too high. In this scenario an excursion may involve simply a transfer of magnetic flux from the dipole to smaller length scales. This is exactly what happens in the dynamo wave mechanism of *Gubbins and Sarson* [1994]; magnetic energy passes to spherical harmonic degree 4 or so as flux migrates towards the equator. This causes a drop in intensity at the Earth's surface by a factor of about 10 simply because of geometrical attenuation in the mantle. Geomagnetic excursions and reversals are accompanied by falls of intensity of factors of 2–10.

Secondly, the inner core may provide stability [*Hollerbach and Jones*, 1993]. While the magnetic field in the liquid outer core is changed by fluid motion, on a time scale of 500–1,000 years, it can only change in the solid inner core by electrical diffusion, a much slower process taking about 5,000 years. *Gubbins* [1998] pointed out that the ratio of time scales in the inner and outer cores is about 10, similar to the ratio of the number of excursions to that of reversals. He proposed that excursions represent flux reversal in the outer core. These relatively frequent events lead to a state in which the inner core retains its original polarity for a time. The inner core polarity will change slowly by diffusion as long as the outer core polarity remains reversed. This will tend to return the outer core polarity to its original state, but just occasionally the inner core reverses completely and we have a full reversal.

Finally, I suggest here that lateral variations of heat flux across the CMB can impart stability. Collapse of a dynamo from a strong field state to a weak field state involves the convection becoming very small scale, the scale set by the Ekman number in non-magnetic convection. Lateral variations in boundary conditions can impart a large scale to the flow, under the right conditions, and may help sustain the large scale component of the flow for sufficient time to allow the field strength to recover. We do not understand the convective regime in the core, and do not yet know whether the boundary anomalies have enough influence to provide this stability, but the observational evidence from reversal paths is compelling: the geomagnetic field in transition appears to reflect lower mantle temperatures, and the fluid flow is therefore similar, at least in the important large scales, to that holding in the core during stable polarity and at the present day.

CONCLUSIONS

We still understand so little about the convective regime in the Earth's core that theoretical studies must be guided by observations. The evidence for thermal core-mantle interactions comes from several rather different sources, none of which alone are completely convincing. The key is the existence of critical longitudes that coincide with high shear velocity in the lower mantle and subduction around the Pacific.

At the present day magnetic flux is concentrated near these two critical longitudes; it is also concentrated at latitudes just outside the inner core tangent cylinder, which tends to mask the longitude effect. These concentrations of flux, the main lobes that make up most of the Earth's dipole moment, have been relatively stationary throughout the historical period 1600–2000 AD; they have experienced some changes but not the regular westward drift we see in the Atlantic region, for example.

The time-average of paleomagnetic data over the last 5 Myr or so also show this concentration, although the effect is weaker. This smaller effect may be caused by genuine long term time variations, by poor data coverage (particularly in the southern hemisphere), or both. The coincidence of longitudes with the present day field is striking.

Paleosecular variation also leaves VGPs concentrated on these longitudes [*Constable*, 1992], as are reversal transition paths, suggesting the magnetic flux remains concentrated by downwelling induced by high heat flux at the CMB. The mechanism of flux concentration explains the tendency of VGPs to cluster on certain longitudes as well as the persistence of VGP paths during polarity transition. The argument that persistent VGP transition paths are an artefact of smoothing by sediments, averaging the direction between two non-antipodal stable directions [*Langereis, Hoof and Rochette*, 1992], is therefore spurious because both effects have the same cause: influence of the overlying mantle.

As is so often the case when dealing with the core and paleomagnetic data in particular, each one of the pieces of evidence listed above can be disputed, but two common features make the hypothesis of thermal core-mantle interaction compelling: a plausible unifying theory that has so far stood up to tests from new data and numerical experiments, and the same critical longitudes in each case.

REFERENCES

Backus, G. E., Kinematics of the secular variation, *Philos. Trans. R. Soc. London Ser. A*, *263*, 239–266, 1968.

Backus, G. E., Gross thermodynamic heat engines in deep interior of Earth, *Proc. Natl. Acad. Sci. U.S.A.*, *72*, 1555–1558, 1975.

Bell, P. I. and A. M. Soward, The influence of surface topography on rotating convection, *J. Fluid Mech.*, *313*, 147–180, 1996.

Bloxham, J., The effect of thermal core-mantle interactions on the paleomagnetic secular variation, *Philos. Trans. R. Soc. London Ser. A*, *358*, 1171–1179, 2000.

Bloxham, J. and D. Gubbins, The secular variation of the Earth's magnetic field, *Nature*, *317*, 777–781, 1985.

Bloxham, J. and D. Gubbins, Thermal core-mantle interactions, *Nature*, *325*, 511–513, 1987.

Bloxham, J. and A. Jackson, Fluid-flow near the surface of the Earth's outer core, *Rev. Geophys.*, *29*, 97–120, 1991.

Buffett, B. A., A mechanism for decade fluctuations in the length of day, *Geophys. Res. Lett.*, *23*, 3803–3806, 1996.

Busse, F. H., Thermal instabilities in rotating systems, *J. Fluid Mech.*, *44*, 444–460, 1970.

Busse, F. H. and J. Wicht, A simple dynamo caused by conductivity variations, *Geophys. Astrophys. Fluid Dyn.*, *64*, 135–144, 1992.

Butler, R. F., *Paleomagnetism*, Blackwell Scientific, Boston, MA, 1992.

Chandrasekhar, S., *Hydrodynamic and hydromagnetic stability*, Clarendon Press, Oxford, 1961.

Channell, J. E. T., A. Mazaud, P. Sullivan, S. Turner and M. E. Raymo, Geomagnetic excursions and paleointensities in the 0.9–2.15 Ma interval of the Matuyama Chron at ODP site 983 and 984 (Iceland Basin), *Earth Planet. Sci. Lett.*, , submitted, 2002.

Christensen, U., P. Olson and G. A. Glatzmaier, Numerical modelling of the geodynamo: a systematic parameter study, *Geophys. J. Int.*, *138*, 393–409, 1999.

Constable, C. G., Link between geomagnetic reversal paths and secular variation of the field over the past 5 My, *Nature*, *358*, 230–233, 1992.

Constable, C. G., C. L. Johnson and S. P. Lund, Global geomagnetic field models for the past 3000 years: transient or permanent flux lobes?, paper presented at Philos. Trans. R. Soc. London Ser. A, 2000.

DeMets, C., R. G. Gordon, D. F. Argus and S. Stein, Current plate motions, *Geophys. J. Int.*, *101*, 425–478, 1990.

Dziewonski, A. M., Mapping the lower mantle: determination of lateral heterogeneity in P velocity up to degree and order 6, *J. Geophys. Res.*, *89*, 5929–5952, 1984.

Dziewonski, A. M. and J. Woodhouse, Global images of the Earth's interior, *Science*, *236*, 37–48, 1987.

Eltayeb, I. A. and S. Kumar, Hydromagnetic convective instability of a rotating self-gravitating fluid sphere containing a uniform distribution of heat sources, *Proc. R. Soc.*, *A326*, 145–162, 1977.

Fearn, D. R. and M. R. E. Proctor, On the computation of steady, self-consistent spherical dynamos, *Geophys. Astrophys. Fluid Dyn.*, *38*, 293–325, 1987.

Garnero, E. J., J. Revenaugh, Q. Williams, T. Lay and L. H. Kellogg, Ultralow velocity zone at the core-mantle boundary, in *The core-mantle boundary region*, edited by M. Gurnis, B. Buffett, M. Wysession and E. Knittle, pp. 319–334, AGU Geophysical Monograph, AGU Geodynamics Series Volume 28, 1998.

Garnier, F., C. Laj, E. Herrero-Bervera, C. Kissel and D. M. Thomas, Preliminary determinations of Geomagnetic field intensity for the last 400 kyr from the Hawaii Scientific Drilling Project core, Big Island, Hawaii, *J. Geophys. Res.*, *101*, 11,665–11,673, 1996.

Glatzmaier, G. A., R. S. Coe, L. Hongre and P. H. Roberts, The role of the Earth's mantle in controlling the frequency of geomagnetic reversals, *Nature*, *401*, 885–890, 1999.

Glatzmaier, G. A. and P. H. Roberts, A three-dimensional convective dynamo solution with rotating and finitely conducting inner core and mantle, *Phys. Earth Planet. Int.*, *91*, 63–75, 1995.

Gubbins, D., Energetics of the Earth's core, *J. Geophys.*, *43*, 453–464, 1977.

Gubbins, D., Mechanism for geomagnetic polarity reversals, *Nature*, *326*, 167–169, 1987.

Gubbins, D., Interpreting the paleomagnetic field, in *The core-mantle boundary region*, edited by M. Gurnis, B. Buffett, M. Wysession and E. Knittle, pp. 167–182, AGU Geophysical Monograph, AGU Geodynamics Series Volume 28, 1998.

Gubbins, D., The Rayleigh number for convection in the Earth's core. , *Phys. Earth Planet. Int.*, *128*, 3–12, 2001.

Gubbins, D., D. Alfè, T. G. Masters, D. Price and M. J. Gillan, Gross thermodynamics of 2-component core convection, *Geophys. J. Int.*, , submitted, 2001.

Gubbins, D., C. N. Barber, S. Gibbons and J. J. Love, Kinematic dynamo action in a sphere: I Effects of differential rotation and meridional circulation on solutions with axial dipole symmetry, *Proc. R. Soc.*, *456*, 1333–1353, 2000.

Gubbins, D. and J. Bloxham, Morphology of the geomagnetic field and implications for the geodynamo, *Nature*, *325*, 509–511, 1987.

Gubbins, D. and S. Gibbons, Kinematic dynamo action in

a sphere III: Dynamo Waves, *Proc. R. Soc.*, , submitted, 2002.

Gubbins, D. and P. Kelly, Persistent patterns in the geomagnetic field during the last 2.5 Myr, *Nature*, *365*, 829–832, 1993.

Gubbins, D. and P. Kelly, On the analysis of paleomagnetic secular variation, *J. Geophys. Res.*, *100*, 14955–14964, 1995.

Gubbins, D. and J. Love, Geomagnetic reversal transition fields: a test of 4-fold symmetry, *Geophys. Res. Lett.*, *25* , 1079–1082, 1998.

Gubbins, D. and G. Sarson, Geomagnetic reversal transition paths from a kinematic dynamo model, *Nature*, *368*, 51–55, 1994.

Guyodo, Y. and J-P. Valet, Global changes in intensity of the Earth's magnetic field during the past 800 kyr, *Nature*, *399*, 249–252, 1999.

Haddon, R. A. W. and J. R. Cleary, Evidence for scattering of seismic PKP waves near the mantle-core boundary , paper presented at 8, 1974.

Herrero-Bervera, E. and S. K. Runcorn, Transition fields during geomagnetic reversals and their geodynamic significance, *Philos. Trans. R. Soc. London Ser. A*, *355*, 1713–1742, 1997.

Hewitt, J., D. P. McKenzie and N. O. Weiss, Dissipative heating in convective flows, *J. Fluid Mech.*, *68*, 721–738, 1975.

Hide, R., Motions of the earth's core and mantle, and variations of the main geomagnetic field, *Science*, *157*, 55–56, 1967.

Hilst, R. D. vander, S. Widiyantoro, K. C. Creager and T. J. McSweeney, Deep subduction and aspherical variationsin P-wavespeed at the base of Earth's mantle, in *The core-mantle boundary region*, edited by M. Gurnis, B. Buffett, M. Wysession and E. Knittle, pp. 5–20, AGU Geophysical Monograph, AGU Geodynamics Series Volume 28, 1998.

Hollerbach, R. and C. A. Jones, Influence of the Earth's inner core on geomagnetic fluctuations and reversals, *Nature*, *365*, 541–543, 1993.

Jackson, A., A. R. T. Jonkers and M. R. Walker, Four centuries of geomagnetic secular variation from historical records, *Philos. Trans. R. Soc. London Ser. A*, *358*, 957–990, 2000.

Johnson, C. and C. Constable, The time-averaged geomagnetic field as recorded by lava flows over the past 5Myr, *Geophys. J. Int.*, *122*, 489–519, 1995.

Jones, C. A., Convection-driven geodynamo models, *Proc. R. Soc.*, *873*, 873–897, 2000.

Jones, C. A., A. M. Soward and A. I. Mussa, The onset of thermal convection in a rapidly rotating sphere, *J. Fluid Mech.*, , submitted, 1999.

Kelly, P. and D. Gubbins, The geomagnetic field over the past 5 million years, *Geophys. J. Int.*, *128*, 315–330, 1997.

Kendall, J-M., The relative behavior of shear velocity, bulk sound speed, and compressional velocity in the mantle: implications for chemical and thermal structure, in *Earth's deep interior: Mineral physics and tomography from the atomic to the global scale*, edited by S. Karato, A. Forte, R. Lieberman, G. Masters and L. Stixrude, pp. 133–159, AGU Geophysical Monograph 117, 2000.

Kendall, J-M. and P. Shearer, Lateral variation in D″ thickness from long-period shear wave data, *J. Geophys. Res.*, *99*, 11,575–11,590, 1994.

Khokhlov, A., G. Hulot and J. Carlut, Towards a self-consistent approach to paleomagnetic field modelling, *Geophys. J. Int.*, *145*, 157–171, 2001.

Kono, M. and H. Tanaka, Mapping the Gauss coefficients to the pole and the models of paleosecular variation, *J. Geomagn. Geoelectr.*, *47*, 115–130, 1995.

Kuang, W. and J. Bloxham, An Earth-like numerical dynamo model, *Nature*, *389*, 371–374, 1997.

Labrosse, S., J-P. Poirier and J-L. LeMouël, On cooling of the Earth's core, *Phys. Earth Planet. Int.*, *99*, 1–17, 1997.

Laj, C., C. Kissel, V. Scao, J. Beer, R. Muscheler and G. Wagner, Geomagnetic Intensity Variations at Hawaii for the past 98 kyr from core SOH-4 (Big Island): New Results, *Phys. Earth Planet. Int.*, , in press, 2001.

Laj, C. A., A. Mazaud, M. Weeks, M. Fuller and E. Herrero-Bervera, Geomagnetic reversal paths, *Nature*, *351*, 447, 1991.

Langereis, C. G., M. J. Dekkers, G. J. de Lange, M. Paterne and P. J. M. van Santvoort, Magnetostratigraphy and astronomical calibration of the last 1.1 Myr from an eastern Mediterranean piston core and dating of short events in the Brunhes, *Geophys. J. Int.*, *129*, 75–94, 1997.

Langereis, C. G., A. A. M. van Hoof and P. Rochette, Longitudinal confinement of geomagnetic reversal paths as a possible sedimentary artefact, *Nature*, *358*, 226–230, 1992.

Larson, R. L. and P. Olson, Mantle plumes control magnetic reversal frequency, *Earth Planet. Sci. Lett.*, *107*, 437–447, 1991.

Lay, T. and D. V. Helmberger, A shear wave velocity in the lower mantle, *Geophys. Res. Lett.*, *10*, 63–66, 1983.

Lay, T., Q. Williams, E. J. Garnero, B. Romanowicz, L. Kellogg and M. E. Wysession, Seismic wave anisotropy in the D″ region and its implications, in *The core-mantle boundary region*, edited by M. Gurnis, B. Buffett, M. Wysession and E. Knittle, pp. 299–318, AGU Geophysical Monograph, AGU Geodynamics Series Volume 28, 1998.

Loper, D. E., Some thermal consequences of a gravitationally powered dynamo, *J. Geophys. Res.*, *83*, 5961–5970, 1978.

Love, J. and A. Mazaud, A database for the Matuyama-Brunhes magnetic reversal, *Phys. Earth Planet. Int.*, *103*, 207–245, 1997.

Love, J. J., Paleomagnetic volcanic data and geometric regularity of reversals and excursions, *J. Geophys. Res.*, *103*, 12,435–12,452, 1998.

Love, J. J., Statistical assessment of preferred transitional VGP longitudes based on paleomagnetic lava data , *Geophys. J. Int.*, *140*, 211–221, 2000.

Lowrie, W., A revised magnetic polarity timescale for the Cretaceous and Cainozoic, *Philos. Trans. R. Soc. London Ser. A*, *306*, 129–136, 1982.

Lund, S. P., G. Acton, B. Clement, M. Hastedt, M. Okada and R. Williams, Geomagnetic field excursions occurred often during the last million years., *EOS Trans. AGU*, *79*, 178–179, 1998.

Masters, T. G., S. Johnson, G. Laske and H. F. Bolton, A shear-velocity model of the mantle, *Philos. Trans. R. Soc. London Ser. A*, *354*, 1385–1411, 1996.

Masters, T. G., G. Laske and A. Dziewonski, The relative behavior of shear velocity, bulk sound speed, and compressional velocity in the mantle: implications for chemical and thermal structure, in *Earth's deep interior: Mineral physics and tomography from the atomic to the global scale*, edited by S. Karato, A. Forte, R. Lieberman, G. Masters and L. Stixrude, pp. 63–87, AGU Geophysical Monograph 117, 2000.

Olson, P., U. Christensen and G. Glatzmaier, Numerical modeling of the geodynamo: mechanisms of field generation and equilibration., *J. Geophys. Res.*, *104*, 10,383–10,404., 1999.

Parker, E. N., *Cosmical Magnetic Fields: Their Origin and Their Activity*, Clarendon Press, Oxford, 1979.

Poirier, J-P., *Introduction to the Physics of the Earth's Interior*, Cambridge Univ. Press, 2000.

Poirier, J-P. and J-L. LeMouël, Does infiltration of core material into the lower mantle affect the observed geomagnetic field?, *Phys. Earth Planet. Int.*, *73*, 29–37, 1992.

Prévot, M. and P. Camps, Absence of preferred longitude sectors for poles from volcanic records of geomagnetic reversals, *Nature*, *366*, 53–57, 1993.

Roberts, P. H., On the thermal instability of a self-gravitating fluid sphere containing heat sources, *Philos. Trans. R. Soc. London Ser. A*, *A263*, 93–117, 1968.

Roberts, P. H. and S. Scott, On the analysis of the secular variation. A hydromagnetic constraint: I. Theory, *J. Geomagn. Geoelectr.*, *17*, 137–151, 1965.

Sarson, G. R. and C. A. Jones, A convection driven geodynamo reversal model, *Phys. Earth Planet. Int.*, *111*, 3–20, 1999.

Teanby, N., Secular Variation on Hawaii, in *PhD Thesis*, University of Leeds, 2001.

Walker, M. R., C. F. Barenghi and C. A. Jones, A note on dynamo action at asymptotically small Ekman number, *Geophys. Astrophys. Fluid Dyn.*, *88*, 261–275, 1998.

Weber, M. and J. P. Davis, Evidence of a laterally variable lower mantle structure from P-waves and S-waves, *Geophys. J. Int.*, *102*, 231–255, 1990.

Zhang, K. and D. Gubbins, Convection in a rotating spherical fluid shell with an inhomogeneous temperature boundary condition at infinite Prandtl number, *J. Fluid Mech.*, *250*, 209–232, 1993.

Zhang, K. and D. Gubbins, Convection in a rotating spherical fluid shell with an inhomogeneous temperature boundary condition at finite Prandtl number, *Phys. Fluids*, *8*, 1141–1148, 1996.

Zhang, K. and D. Gubbins, Scale disparities and magnetohydrodynamics in the Earth's core, *Philos. Trans. R. Soc. London Ser. A*, *358*, 899–920, 2000a.

Zhang, K. and D. Gubbins, Is the geodynamo process intrinsically unstable?, *Geophys. J. Int.*, *140*, F1–F4, 2000b.

Zhang, K. and C. A. Jones, On small Roberts number magnetoconvection in rapidly rotating systems, *Proc. R. Soc.*, *452*, 981–995, 1996.

Zhang, K. and C. A. Jones, The effect of hyperviscosity on geodynamo models, *Geophys. Res. Lett.*, *24*, 2869–2872, 1997.

D. Gubbins, School of Earth Sciences, University of Leeds, Leeds LS2 9JT, UK.

The Range of Timescales on Which the Geodynamo Operates

Rainer Hollerbach[1]

Department of Geosciences, Princeton University, Princeton, New Jersey

We consider the range of timescales found in the Earth's core. We begin by reviewing the observational data, from geomagnetic jerks on annual timescales to variations in the average reversal rate on timescales of tens and hundreds of millions of years. We then turn to the theoretical origins of some of these timescales. By considering the relevant dispersion relations, we show why filtering out some of the shorter timescales is unlikely to succeed. We must simply accept that the geodynamo equations are intrinsically stiff. We therefore include a brief digression on stiff systems in general, and discuss why the standard methods for dealing with them are unfortunately extremely difficult to apply to the geodynamo system in particular. As a result, numerically attainable Rossby and/or Ekman numbers will continue to be many orders of magnitude too large. We conclude with a brief discussion of some of the implications of this, in terms of the interpretation, and even relevance, of the currently attainable numerical results.

1. INTRODUCTION

The internal structure of the Earth consists of a series of nested layers rather like the layers of an onion. Starting from the inside, there is the solid iron inner core, the liquid iron outer core, the mantle and crust, both consisting of rock of varying degrees of brittleness, and finally the oceans and the atmosphere. Unlike an onion though, all of these layers are in constant motion on a huge range of timescales, generating a great variety of geophysically interesting and important phenomena in the process. For example, the motions of the atmosphere and oceans, on timescales ranging from minutes to millenia, generate the familiar patterns of weather and climate. Similarly, the very slow motions of the mantle result in continental drift and all its associated phenomena,

whereas the occasional very rapid motions of the crust are better known as earthquakes. It is this vast range of phenomena occurring on all timescales from seconds to hundreds of millions of years that makes the Earth such an interesting planet to live on and to study.

As noted above, the inner/outer core system is also in motion. Because the only readily observable surface manifestation of these motions is the presence of the Earth's magnetic field, they do not command quite the same attention as the motions of the ocean/atmosphere and mantle/crust systems. Nevertheless, they are equally important in terms of our fundamental understanding of the dynamics of the Earth, and I will argue here that as challenging as the study of the ocean/atmosphere and mantle/crust systems may be, the dynamics of the core are more complicated still. In particular, we will consider the range of timescales present in the core, and discuss why it has proven virtually impossible so far to study subsets of this range in isolation — unlike in the mantle/crust system, for example, where it is relatively easy to isolate from one another the very slow motions of the mantle and the occasional very rapid motions of the crust. As a result, the core has probably the broadest range of inextricably intertwined timescales of any of these three systems that make up the Earth, and it is this feature more than anything else that makes the study of the core's dynamics so difficult, as we shall see.

[1] Permanently at: Department of Mathematics, University of Glasgow, Glasgow G12 8QW, United Kingdom (email: rh@maths.gla.ac.uk).

Earth's Core: Dynamics, Structure, Rotation
Geodynamics Series 31
10.1029/31GD12

2. OBSERVATIONAL BACKGROUND

Before presenting a theoretical analysis of why the underlying geodynamo equations might be expected to yield such a broad range of timescales, we begin with a very brief discussion of the timescales actually observed in the field. See also the far more detailed reviews by *Courtillot and LeMouël* [1988], *Courtillot and Valet* [1995], or *Dormy et al.* [2000]. Direct measurements of the field only date back at most three or four centuries, of course, but even in such relatively short periods of time the field can change by an easily measurable amount. For example, over three centuries ago it was noted at Greenwich observatory that the declination (the magnetic field's deviation from true north) varied slightly over the course of the years, and *Halley* [1692] even went on to speculate that motions in the Earth's interior might be the cause, it having previously been suggested by *Gilbert* [1600] that the field was of internal origin. Since then numerous other geomagnetic observatories have been established around the world, continuously monitoring the field. The evidence from all these observatories is unequivocal: the field is constantly fluctuating, on timescales as short as minutes or even seconds. It is generally recognized that these extremely rapid fluctuations must be of external origin, and so will not be considered further here. However, the fluctuations on timescales of a year and longer, known generically as secular variation, are generally accepted to be of internal origin. See, for example, *Bloxham et al.* [1989] or *Jackson et al.* [2000] for compilations of the historical record of secular variation.

Within this historical range of years to centuries, the general secular variation can be subdivided into a number of reasonably distinct phenomena. For example, at the relatively long end, there is the so-called westward drift, where certain features of the field migrate west at rates on the order of a degree in longitude per year [*Jault et al.*, 1988]. In contrast, at the very shortest end, there are so-called geomagnetic jerks, in which the second time-derivative of the field changes abruptly within a single year. Around ten such events occurred in the past century, of which at least three were global, that is, observatories around the world recorded similar jerks in the same year [*Alexandrescu et al.*, 1995; *LeHuy et al.*, 1998]. Events such as these clearly demand an explanation. Furthermore, it is entirely conceivable that the corresponding jerk within the core occurs even faster, since the weakly conducting mantle screens out any variations in the field on timescales shorter than about a year. (It is for this reason also that the variations in the field on timescales of minutes and seconds must be of external origin.)

For timescales longer than centuries, where we no longer have direct measurements, we rely on the very convenient property that in sediments or lavas minute particles of magnetic material not only align themselves with the ambient field, when the sediment/lava hardens into rock, this alignment is frozen in. By studying the magnetization of ancient rocks, it is thus possible to deduce the direction, and with a bit more effort even the intensity [*Jacobs*, 1998] of the field as it was thousands or even millions of years ago. The quality of this data obviously can't compete with that of present-day direct measurements, and from our perspective here two important limitations that need to be mentioned are, first, unless the sedimentation rate was unusually high all variation on timescales shorter than several centuries will be effectively averaged out, and second, since volcanos erupt so sporadically, it is almost impossible to deduce any information at all on timescales of variation from lavas. Nevertheless, the record preserved in these two types of rocks is our only source of information about the past history of the field, and even with all its limitations in terms of lack of resolution and sparseness of coverage, it turns out to be a very rich source indeed.

In particular, the most important paleomagnetic discovery is the fact that the field reverses occasionally. That is, the dominant dipolar part of the field is almost always more or less aligned with the rotation axis, but not always with the same orientation. There are then a great number of timescales associated with these reversals. There is first the time of the reversal itself, which seems to be a remarkably consistent five to ten thousand years or so. In sharp contrast, the time between successive reversals is quite irregular; over the past few million years reversals have occurred on average once every few hundred thousand years, but with considerable variation about that average. Furthermore, on timescales of tens and hundreds of millions of years, even this average reversal rate varies significantly; for example, between 83 and 121 million years ago there were no reversals at all. See also *Merrill and McFadden* [1999], *Love* [1999], or *Constable* [2000] for recent reviews of the reversal record.

And finally, even though the observational evidence of secular variation on the one hand and reversals on the other is quite distinct, one should not think of the two as fundamentally distinct processes. In particular, virtually everything in between is also present at some point in the record. For example, in addition to the usual reversals, *Langereis et al.* [1997] also found a number of so-called excursions, events too large and long-lasting (also around five to ten thousand years) to be classified simply as secular variation, but not actually reversals either. These excursions also occur irregularly, but roughly five or six times as frequently as reversals. Similarly, *Valet et al.* [1986] and *Herrero-Bervera and Theyer* [1986] found what is presumably secular variation during a reversal, namely evidence of fluctuations on timescales of a century or so during the reversal (although for the reasons mentioned above this cannot be as clearly established as the basic reversal itself).

Indeed, in terms of the underlying dynamics of the core, it is probably not very meaningful to even attempt to make distinctions between secular variation and small excursions, or between large excursions and back-to-back reversals. We must simply accept that the geodynamo operates on virtually all timescales from years or even shorter to tens of millions of years or even longer. In the following sections we will explore the origin of some of these timescales, discuss why it is so difficult to isolate any of them, and finally consider some of the implications for geodynamo modelling.

3. BASIC TIMESCALES AND EQUATIONS

Let us begin by simply listing some of the more basic timescales that might be expected to be present in the core. There are,

light waves, with $\tau = l/c \approx 10^{-2}$ sec,

sound waves, with $\tau = l/c_s \approx 10$ min,

the rotational timescale $\Omega^{-1} \approx 1$ day,

and the three diffusive timescales

magnetic, with $\tau = l^2/\eta \approx 10^5$ years,

thermal, with $\tau = l^2/\kappa \approx 10^{11}$ years,

viscous, with $\tau = l^2/\nu \approx 10^{11}$ years,

where we've used the outer core radius $r_o = 3480$ km for the lengthscale l. There are of course a great many other timescales in the core as well, foremost being the advective timescale l/u. However, most of these depend at least in part on the details of whatever solution ultimately emerges from the equations, and so we defer discussion of them until later.

We then note that two of our basic timescales, the viscous and thermal diffusive timescales, are actually longer than the age of the Earth. We will discuss in a moment why one should not in fact expect these two timescales to be relevant after all. However, even discounting these two, the remaining timescales still span 15 orders of magnitude! Since no reasonable model could possibly cope with such a huge range, we must filter out at least some of the more extreme ones. Fortunately, some of them are indeed very easily filtered out. For example, the induction equation for the magnetic field is derived by making the so-called magnetohydrodynamic approximation and thereby filtering out light waves. Similarly, we can filter out sound waves by making the Boussinesq approximation. The same is also true, incidentally, in the mantle. However, the crucial difference is that even after filtering out light and sound waves in the core, one is still left with a very broad range of timescales, whereas after filtering out sound (and shear) waves in the mantle one is only left with a relatively narrow range. And unfortunately we will find that from now on it becomes progressively much harder to further restrict this very broad range of timescales still present in the core.

So, having made the magnetohydrodynamic and Boussinesq approximations, the suitably nondimensionalized governing equations become,

$$\frac{\partial}{\partial t} \mathbf{B} = \nabla^2 \mathbf{B} + \nabla \times (\mathbf{U} \times \mathbf{B}), \qquad (1)$$

$$Ro\left(\frac{\partial}{\partial t} + \mathbf{U} \cdot \nabla\right) \mathbf{U} + 2\hat{\mathbf{z}} \times \mathbf{U} = -\nabla p$$

$$+ E\nabla^2 \mathbf{U} + (\nabla \times \mathbf{B}) \times \mathbf{B} + q\,Ra\,\Theta\,\mathbf{r}, \qquad (2)$$

$$\left(\frac{\partial}{\partial t} + \mathbf{U} \cdot \nabla\right) \Theta = q\nabla^2\Theta, \qquad (3)$$

$$\nabla \cdot \mathbf{U} = \nabla \cdot \mathbf{B} = 0,$$

where for our present purposes all we need to know about the nondimensionalization is that time has been scaled by the magnetic diffusion time. See, for example, *Roberts and Gubbins* [1987] for the full details.

See also *Braginsky and Roberts* [1995] for a set of equations in which the less drastic (and hence more realistic) anelastic approximation is made rather than the Boussinesq approximation. Although these equations are obviously considerably more complicated than the above, they also filter out sound waves. In terms of the timescales allowed by the equations they are therefore no worse than the above anyway, and have indeed been used in numerical geodynamo modelling [*Glatzmaier and Roberts*, 1996a]. Because the timescales allowed by the two sets are much the same though, we will here restrict attention to the simpler set above.

From the point of view of understanding the range of timescales on which the geodynamo operates, the three nondimensional parameters we most want to focus attention on are the Rossby and Ekman numbers

$$Ro = \eta/\Omega r_o^2, \qquad E = \nu/\Omega r_o^2,$$

measuring, respectively, the ratios of the rotational timescale Ω^{-1} to the magnetic and viscous diffusive timescales, and the Roberts number

$$q = \kappa/\eta,$$

measuring the ratio of the magnetic and thermal diffusive timescales. Our disparity of timescales above thus translates into rather extreme values for these three parameters, namely

$$q = O(10^{-6}),$$

$$Ro = O(10^{-9}), \qquad E = O(10^{-15}).$$

So one of the things we will want to consider in the remainder of this review is why this extreme smallness of these three parameters should lead to such difficulties, and what — if anything — can be done about them.

4. VISCOUS AND THERMAL DIFFUSIVE TIMESCALES

As noted above, the viscous and thermal diffusive timescales r_o^2/ν and r_o^2/κ are longer than the entire lifespan of the Earth. If we really believed these extremely long timescales to be relevant to the operation of the geodynamo, we would have to conclude that it will never reach an equilibrium state, that instead it will always retain a memory of its specific initial conditions early on in the Earth's history. And indeed, for certain aspects of the core's dynamics, such as the presence or absence of a stably stratified layer at the top, that may well be true [e.g., *Stevenson*, 1981]. However, in the bulk of the core, where we know (see below) that the advective timescale is far, far shorter, it seems likely that advection will so dominate diffusion that these extremely long diffusive timescales simply do not enter.

(There is in fact another reason why the viscous diffusive timescale at least does not apply, namely the well-known feature [e.g., *Duck and Foster*, 2001] that in rapidly rotating systems the spin-up time is not given by r_o^2/ν, but by $(r_o^2/\nu\Omega)^{1/2}$, that is, intermediate between the very long r_o^2/ν and the very short Ω^{-1}. However, this argument cannot necessarily be used to dismiss the thermal diffusive timescale, as there is no equivalent of the Coriolis force in (3).)

If advection dominates diffusion, that does mean though that one will end up with extremely short lengthscales rather than extremely long timescales. For example, according to (3) the temperature will exhibit structures as fine as $O(q^{1/2})$. The extreme smallness of q thus leads to problems with spatial rather than temporal resolution. The traditional way of dealing with these problems is to invoke turbulent diffusivities, for which one would almost certainly have

$$\eta_t \approx \nu_t \approx \kappa_t,$$

and hence

$$q_t = O(1),$$

which is indeed the range used in most numerical geodynamo models. One should be aware, though, that one has not thereby really "solved" the problems due to the smallness of the true, molecular Roberts number, one has merely deferred them to a proper investigation of the small-scale motions generating this turbulence. And indeed, *Matsushima et al.* [1999] showed in one such study that the use of isotropic turbulent diffusivities is almost certainly a gross oversimplification; both the rotation as well as the large-scale field introduce anisotropies into the small-scale motions that should properly be taken into account in the turbulent diffusivities used in numerical models. To date no one has done so though.

And finally, even if one accepts that invoking turbulent diffusivities is a (more or less) legitimate way of dealing with the smallness of q, the turbulent Rossby and Ekman numbers continue to be extremely small. In particular, the geodynamo's energy budget is sufficiently tight [e.g., *Buffett et al.*, 1996] that one cannot increase η_t much beyond η; there simply cannot be that much small-scale structure in the field. But then Ro_t and E_t are still at most $O(10^{-7})$ or so, which is still sufficiently small that it leads to severe difficulties.

5. ADDITIONAL TIMESCALES

Before considering some of these problems associated with Ro and E, let's continue simply listing a few more timescales,

namely those that depend at least in part on the solution itself. We have,

5.1. Advective Timescale(s)

If we consider the ratio of the magnetic diffusive timescale r_o^2/η to the advective timescale r_o/u, this defines the magnetic Reynolds number

$$Rm = ur_o/\eta.$$

It is then known [e.g., *Roberts*, 1967; *Moffatt*, 1978; *Roberts and Gubbins*, 1987] that to have any chance of obtaining dynamo action at all, Rm must be at least $O(100)$ or so. (The formal result is that $Rm > \pi^2$ is a necessary condition for dynamo action. Unfortunately this is not a sufficient condition though, and in practice no dynamos are known that operate at $Rm = O(10)$, so $Rm = O(100)$ is a more realistic requirement.) Intuitively this requirement is quite easy to understand; it simply means that the flow must be stretching and thereby amplifying the field more rapidly than diffusion is damping it. Using our above value of $r_o^2/\eta = O(10^5)$ years, $Rm \gtrsim O(100)$ then implies that $r_o/u \lesssim O(10^3)$ years, which fits quite nicely with the range of the observed secular variation.

In particular, one is tempted to interpret this previously mentioned westward drift [*Jault et al.*, 1988] simply as a case of advection of the field by the flow, and hence as direct evidence of the advective timescales. One must be careful with such an interpretation though, as it has also been suggested (see below) that a wave rather than a bulk advective motion may be the cause. Nevertheless, by observing the time-variation of the field one can certainly make some inferences about the fluid flow, particularly on relatively short (decadal) timescales. See, for example, *Bloxham and Jackson* [1991] for a review of the commonly used methods, and some of their limitations.

There may in fact be direct evidence of an advective timescale in the core, namely the rotation of the inner core relative to the mantle. Because the fluid directly above the inner core boundary is magnetically strongly coupled to it, it too must rotate at much the same rate. The inner core's rotation rate, variously determined to be 0.15°/year [*Vidale et al.*, 2000], 0.2 – 0.3°/year [*Creager*, 1997], 1°/year [*Song and Richards*, 1996], and even 3°/year [*Su et al.*, 1996], thus sets a definite lower bound on the differential rotation throughout the outer core [e.g., *Hollerbach*, 1998]. This uncertainty as to precisely what its rotation rate is — or indeed whether it has been firmly established that it rotates at all [*Laske and Masters*, 1999] — is therefore unfortunate. However, if a definite non-zero rotation rate is ever established, this would provide valuable evidence of at least one advective timescale, and not just at the surface, but deep within the core.

5.2. Waves

As noted above, it has also been suggested [e.g., *Hide*, 1966; *Braginsky*, 1967] that features like the westward drift are caused by waves in the core, so-called *MAC* waves. As the name implies, these waves are influenced by *M*agnetic (Lorentz), *A*rchimedean (buoyancy), and *C*oriolis forces, that is, by all three of the dominant forces in the Navier-Stokes equation. With periods on the order of centuries, they could indeed account for the relatively long-term secular variation. Note though that if these waves are confined to a thin, stably stratified layer at the top of the core, their periods are reduced to ~65 years [*Braginsky*, 1993], and would then contribute more to the intermediate secular variation.

Next in decreasing period are torsional oscillations, in which nested cylindrical shells parallel to the z-axis each oscillate in essentially solid-body rotation, that is, the flow is of the form $u(s)\hat{e}_\phi$, where (z, s, ϕ) are standard cylindrical coordinates. The balance of forces in these waves is between the Lorentz force (a large term) on the one hand, and inertia and viscosity (two small terms) on the other. Their periods could thus be extremely short, as we will see below. Under normal circumstances though, most of the magnetic torque will cancel itself, greatly increasing the period. Plausible assumptions about the internal structure of the field then lead to periods on the order of decades [*Braginsky*, 1970, 1984], or conversely observations of periodicities in length of day variations (which not surprisingly are affected by torsional oscillations in the core) allow inferences to be made about the internal structure of the field [*Zatman and Bloxham*, 1997, 1999; see also *Bloxham*, 1998]. One should also bear in mind though that because the periods identified by Zatman and Bloxham, 76 and 53 years, overlap with Braginsky's estimate of ~65 years for the MAC waves in this (conjectured) stably stratified layer, disentangling the effects of MAC waves and torsional oscillations could potentially be difficult in general.

Another particular torsional oscillation worth mentioning is that of the inner core itself. That is, the inner core's rotation rate could exhibit fluctuations about some long-term, possibly non-zero (as noted above) average, and the balance of forces governing these oscillations is exactly the same as that governing the above torsional oscillations in the outer core. Not surprisingly then, these inner core oscillations can also have extremely short periods [*Gubbins*, 1981], but once again most of the magnetic torque will normally cancel itself, leading to periods on the order of years [e.g., *Glatzmaier and Roberts*, 1996b; *Aurnou et al.*, 1998]. See also *Kuang* [1999] for a detailed study of some of the force balances involved in these oscillations.

Finally, there are a number of other waves potentially present in the core, such as inertial oscillations [*Aldridge and Lumb*, 1987], having periods on the order of days. The disparity between this timescale and the timescales observed in the field is sufficiently great that it is generally believed that these waves play no essential role in the dynamics of the main field, but merely ride along on top of it. However, one should remember that because the mantle is weakly conducting, we do not really know the shortest timescale on which the internally generated field fluctuates, and hence we do not

know whether this disparity is really all that great. Also, as we will discuss further below, the mere fact that the underlying equations also allow extremely short-period waves is usually enough to cause severe problems in any numerical model, even if these waves have nothing to do with the main phenomenon of interest (which of course is precisely why one filters them out if at all possible).

5.3. Reversals

We have noted above some of the timescales associated with reversals as they are actually observed in the paleomagnetic data; now we want to consider their possible theoretical origins. Concerning the five to ten thousand year timescale of the reversals themselves, it has been suggested [*Hollerbach and Jones*, 1993, 1995; *Gubbins*, 1999] that this is due to the magnetic diffusive timescale of the inner core; r_i^2/η is indeed around ten thousand years. On this view, reversals can't happen on shorter timescales because that simply wouldn't allow enough time to reverse the field in the inner core as well, where diffusion is the only process that can change the field. As to what triggers each individual reversal in the first place, and hence what determines the time between reversals, at this point we simply do not know. It is generally believed though that reversals are of entirely internal origin, and do not require an external triggering mechanism. See, for example, *Merrill and McFadden* [1988], who argue for purely internally triggered reversals based on an analysis of the actual data, or also *Sarson and Jones* [1999], who present one possible internal triggering mechanism. From a very general theoretical point of view it is certainly not difficult to imagine a chaotic dynamical system that intermittently switches from one state to another on timescales moderately long compared to the longest timescale "obviously" present in the equations (which we agreed is the magnetic diffusive timescale of $O(10^5)$ years). There is thus no difficulty in imagining reversals occurring on average once every few hundred thousand years to be of purely internal origin.

Somewhat more problematic are the variations in average reversal rate occurring on timescales of tens and hundreds of millions of year. The disparity between these timescales and the magnetic diffusive timescale is sufficiently great that these extremely long timescales are generally believed to be of external origin. This of course simply begs the question, what is that external origin? Given that tens and hundreds of millions of years is precisely the timescale of convection in the mantle, the most plausible answer is that changes in the pattern of mantle convection lead to changes in the temperature variations at the core-mantle boundary, which would quite likely lead to changes in the pattern of core convection, and hence possibly to changes in the average reversal rate (although each individual reversal continues to be purely internally triggered). This was certainly the reasoning adopted by *Glatzmaier et al.* [1999], who imposed different temperature inhomogeneities in their numerical model, and indeed found different stability properties when they ran each case for a few hundred thousand years.

However, one should be aware that this point of view is not universally shared; *McFadden and Merrill* [1995] for example argue that even these extremely long timescales are of internal origin. On their view, the geodynamo has not just two, but (at least) four quasi-stable states, namely two pairs of opposite polarity states. If one then hypothesizes that: (a) in the first pair polarity transitions occur relatively frequently, (b) in the second pair polarity transitions occur relatively infrequently, and (c) that a transition from one pair to another occurs extremely infrequently, then one obtains just such variations in average reversal rate on very long timescales. And while theirs is probably a minority point of view, it cannot be dismissed out of hand; dynamical systems can indeed exhibit intermittency on timescales very long compared to the longest timescale obviously present. Furthermore, the mere possibility that even these extremely long timescales might be of internal origin has (potentially disturbing) implications for numerical geodynamo modelling. For example, if Glatzmaier et al. wished to test this McFadden and Merrill hypothesis, instead of running a few cases for a few hundred thousand years each, they would have to run one case for a few hundred million years, and simply see if such extremely long timescales do or do not spontaneously emerge. Needless to say, such extremely long runs are currently not feasible.

5.4. Growth of the Inner Core

Even assuming that these variations in average reversal rate are of external rather than internal origin, there is still one process going on entirely within the core whose timescale is measured in hundreds of millions of years. This is the growth of the inner core [*Buffett et al.*, 1996], which could conceivably affect the operation of the geodynamo in a number of ways. Most obviously, the growth of the inner core gradually changes the geometry of the outer core from a very thick shell to an ever thinner one, in the process also shifting the location of the so-called tangent cylinder ever outward. Given the potential significance of such geometrical effects, it seems likely that these alone could have a considerable influence on the solution. For example, *Morrison and Fearn* [2000] systematically varied the size of the inner core in their model, and showed that the solution is indeed affected, sometimes in unexpected ways; they found that solutions with smaller inner cores were more stable than ones with larger inner cores, which is exactly the opposite of what one would have expected given the stabilizing role of the inner core mentioned above [*Hollerbach and Jones*, 1993, 1995]. The resolution to this particular paradox is presumably that when one varies the size of the inner core, one indirectly also varies the size of the region inside the inner core tangent cylinder, which in this case evidently more than outweighs the stabilizing role of the inner core itself.

Finally, it should be noted that even though it occurs on timescales of hundreds of millions of years, the growth of the inner core does not require one to do numerical runs for that long, unlike the test of the *McFadden and Merrill* [1995] hypothesis outlined above. The difference between the two is

that when testing their hypothesis, what one is trying to discover is precisely whether these extremely long timescales could emerge from within the core alone, and so one has no choice but to run for that long. In contrast, when considering varying inner core sizes, one is not trying to determine the growth rate — that after all has already been established (from energetic/thermodynamic considerations). What one is trying instead is to explore the effect of different sizes fixed for the duration of each run, which then need only be for the usual few hundred thousand years.

6. THE SMALLNESS OF Ro AND E

In this section we want to return to the geodynamo equations (1-3), and consider why the extremely small values of the Rossby and Ekman numbers lead to such difficulties. That is, why can't we simply filter out the short timescales caused by the smallness of these two parameters, for example by setting one or both of them identically equal to zero? It turns out that attempting to do so leads to both local and global difficulties, which we address in turn:

6.1. Local Analysis

Following *Walker et al.* [1998], let us consider the dispersion relation governing waves in the core. If we linearize (1) and (2) (with $Ra = 0$ for simplicity) about some large-scale field $\mathbf{B}_0$, and look for perturbations $\mathbf{b}$ and $\mathbf{u}$ proportional to $\exp[i(\mathbf{k} \cdot \mathbf{r} - \omega t)]$, the dispersion relation we ultimately obtain is

$$(Ro\,\omega + iEk^2)(\omega + ik^2) - \omega_M^2 = \omega_C(\omega + ik^2), \quad (4)$$

where $k \equiv |\mathbf{k}|$, and

$$\omega_M \equiv \mathbf{k} \cdot \mathbf{B}_0 \qquad \text{and} \qquad \omega_C \equiv \pm 2\,\mathbf{k} \cdot \hat{\mathbf{z}}/k \quad (5)$$

will turn out to be the dispersion relations (to within various factors of Ro) for pure Alfvén waves and pure inertial oscillations, respectively.

Being a simple quadratic, we can solve (4) to obtain

$$\omega = \frac{1}{2} Ro^{-1} \left[\omega_C - i(Ro + E)k^2 \pm D^{1/2} \right], \quad (6)$$

where

$$D = \omega_C^2 + 4Ro\,\omega_M^2 - (Ro - E)^2 k^4 + 2i(Ro - E)k^2\omega_C.$$

Since this is unfortunately rather messy, we look at the magnitudes of the various terms, and see if we can make any simplifying approximations. We begin by noting that if we've done our nondimensionalization correctly, then k, ω_M and ω_C will in general all be order unity (in fact k becomes arbitrarily large for sufficiently short wavelengths, but we're only concerned with very rough order of magnitude estimates at the moment). Therefore, of the four terms that make up D, one — $(Ro - E)^2 k^4$ — is so small that we will neglect it entirely, and another two — $4Ro\,\omega_M^2$ and $2i(Ro - E)k^2\omega_C$ — are also small (but not negligible). How we simplify $D^{1/2}$

therefore depends entirely on the size of the final term, ω_C^2. And there are then two cases to consider:

In general the orientations of $\mathbf{k}$ and $\hat{\mathbf{z}}$ will be such that $\omega_C = O(1)$, so we can Taylor-expand the square-root to obtain

$$\omega \approx \frac{1}{2} Ro^{-1} \left[\omega_C - i(Ro + E)k^2 \right. \tag{7}$$

$$\left. \pm \omega_C \left(1 + \frac{2Ro\,\omega_M^2}{\omega_C^2} + \frac{i(Ro - E)k^2}{\omega_C} \right) \right],$$

so the two roots we finally obtain are

$$\omega_1 = Ro^{-1}\omega_C \left(1 + Ro\,\frac{\omega_M^2}{\omega_C^2} \right) - i\,\frac{E}{Ro}\,k^2, \tag{8a}$$

$$\omega_2 = -\frac{\omega_M^2}{\omega_C} - ik^2. \tag{8b}$$

If, however, the orientations of $\mathbf{k}$ and $\hat{\mathbf{z}}$ are such that $\omega_C \lesssim O(Ro^{1/2})$, that is, if $\mathbf{k}$ is almost perpendicular to $\hat{\mathbf{z}}$, then (7) is not valid after all. Instead, we return to (6), and simply consider the special case $\omega_C \equiv 0$, yielding

$$\omega_3 = \pm Ro^{-1/2}\omega_M - \frac{i}{2}\left(1 + \frac{E}{Ro}\right)k^2. \tag{8c}$$

We have shown therefore that there are (at least) these three types of waves given by these three dispersion relations (8). So what we wish to consider next is what sort of timescales these waves occur on, and what sort of physics they correspond to. Addressing the timescales question first, it is clear that ω_1 is the fastest, occurring on $O(Ro)$ timescales, ω_3 is next, occurring on $O(Ro^{1/2})$ timescales, and ω_2 is the slowest, on $O(1)$ timescales.

As to what physics to associate with each mode, ω_1 is basically just an inertial oscillation, for which the dimensional, non-dissipative dispersion relation would be $\omega = \pm 2\mathbf{k} \cdot \mathbf{\Omega}/k$, so measuring the daily timescale of these waves on the magnetic diffusive timescale of our nondimensionalization will indeed introduce precisely this factor of Ro^{-1} in (8a). The factor of $(1 + Ro\,\omega_M^2/\omega_C^2)$ then shows that the background magnetic field $\mathbf{B}_0$ has very little effect on these modes, since (7) is only valid for $Ro\,\omega_M^2/\omega_C^2 \ll 1$. And finally, the reason the dissipative term contains a factor of E/Ro is that these modes decay on the viscous diffusive timescale, but once again we're measuring things on the magnetic diffusive timescale.

Similarly, ω_3 is just like an Alfvén wave in a non-rotating system, for which the dimensional, non-dissipative dispersion relation would be $\omega = \pm \mathbf{k} \cdot \mathbf{B}_0/\sqrt{\mu \rho_0}$, so remembering the nondimensionalizations of length, time, and $\mathbf{B}_d = \sqrt{\Omega \mu \rho_0 \eta}\, \mathbf{B}_{nd}$, we do indeed recover precisely the factor of $Ro^{-1/2}$ in (8c). The existence of these modes thus shows that even though we are in a system where rotation is ordinarily dominant, it is possible to have waves that behave as if they were in a non-rotating system, provided only the wave-vector

$\mathbf{k}$ is perpendicular to the rotation vector $\hat{\mathbf{z}}$. And we note in passing that if the field contains a predominantly toroidal component B_ϕ, then Alfvén waves running along these field lines will indeed have $\mathbf{k}$ perpendicular to $\hat{\mathbf{z}}$. However, (8c) also shows that if our nondimensionalization is correct, so that the field strength $\mathbf{B}_0 = O(1)$, then these Alfvén waves will be much slower than the inertial oscillations (8a), by a factor of $Ro^{-1/2}$. That is, the timescale of Alfvén waves in the core is on the order of decades instead of days. And again, the reason the dissipative term now contains a factor of $(1 + E/Ro)$ is that these modes are damped both magnetically (yielding the 1) and viscously (yielding the E/Ro).

Finally, ω_2 is a fundamentally new mode that has no analog in either rotating, non-magnetic or non-rotating, magnetic systems, but instead depends crucially on both ω_C and ω_M. Because its timescale is precisely our basic magnetic diffusive timescale though, we do not expect the existence of these so-called slow magnetohydrodynamic waves to pose any problems.

To summarize then, we have shown that if $\omega_C = O(1)$ we obtain modes with the two very different timescales $O(Ro)$ and $O(1)$, whereas if $\omega_C = 0$ we obtain modes with the single intermediate timescale $O(Ro^{1/2})$. And not surprisingly then, since the timescales obtained must depend continuously on ω_C, if we considered ω_C small but non-zero, we could obtain all timescales between $O(Ro)$ and $O(1)$ as well. We find therefore that the governing geodynamo equations do indeed support phenomena occurring on a vast range of timescales.

So, just as before with light and sound waves, we would like to filter out some of these timescales, by simplifying the equations in some way. Very much unlike light and sound waves though, we will find that in the process we introduce new complications that end up being worse than the original problem. In particular, we can easily enough filter out inertial oscillations simply by neglecting inertia, that is, setting $Ro \equiv 0$. (Note though that this is a singular limit, that is, $Ro \equiv 0$ and $Ro \to 0$ are not the same. See also *Kuang and Bloxham* [1999] on this point.) Our original dispersion relation (4) then yields just

$$\omega = -\frac{\omega_M^2}{\omega_C - iEk^2} - ik^2. \tag{9}$$

We have therefore succeeded in filtering out not just the inertial oscillations, but the Alfvén waves as well, with the only mode remaining being the analog of the slow magnetohydrodynamic mode (8b). However, before congratulating ourselves on having solved the problem, we need to realize that while (9) and (8b) may look superficially similar, there is in fact one crucial difference, namely that whereas (8b) is valid only for $\omega_C \gtrsim O(Ro^{1/2})$, and can therefore never yield timescales more rapid than $O(Ro^{1/2})$, (9) is valid for all ω_C, and can therefore yield timescales as rapid as $O(E)$. We thus realize that neglecting inertia accomplishes nothing; (9) may support fewer types of waves than (8) did, but if anything they can occur on even shorter timescales, namely $O(E)$ versus $O(Ro)$.

6.2. Global Analysis

As if these local difficulties weren't enough, it turns out there are global ones as well, having to do with the well-known *Taylor's* [1963] constraint, stating that in the $Ro = E = 0$ limit (2) has no solution at all unless the magnetic field satisfies

$$\int_{C(s)} \left[(\nabla \times \mathbf{B}) \times \mathbf{B} \right] \cdot \hat{\mathbf{e}}_\phi \, dS = 0, \qquad (10)$$

where the so-called geostrophic contours $C(s)$ consist of nested cylindrical shells parallel to the axis of rotation, and (10) must be satisfied on each such contour, that is, at each value of the cylindrical radius s.

The problem is that it's not obvious that the field will always satisfy (10); indeed, one can certainly imagine starting off with some initial condition that doesn't satisfy it. So what happens then? For Ro and E small but non-zero, an asymptotic analysis of the Ekman boundary layer then yields not (10), but instead

$$\frac{(1-s^2)^{1/4}}{4\pi s} \int_C \left[(\nabla \times \mathbf{B}) \times \mathbf{B} \right]_\phi dS$$

$$= Ro \, (1-s^2)^{3/4} \frac{\partial U_g(s)}{\partial t} + E^{1/2} U_g(s), \qquad (11)$$

where $U_g(s)\, \hat{\mathbf{e}}_\phi$ is the so-called geostrophic flow, consisting of these geostrophic contours each undergoing solid-body rotation. See, for example, *Roberts and Soward* [1992] for the details of this analysis.

According to (11) it is clear that if (10) breaks down completely, so that this Taylor integral is order one rather than almost zero, then U_g will either oscillate on timescales as short as $O(Ro)$ if inertia is more important, or it will be as large as $O(E^{-1/2})$ if viscosity is more important. In either case though one will obtain extremely short timescales, either the $O(Ro)$ timescale of the torsional oscillation itself, or else the $O(E^{1/2})$ timescale on which this enormously large flow advects the field, for example. See also the discussion by *Kuang* [1999].

In fact, in neither case will the timescales be quite as extreme as $O(Ro)$ or $O(E^{1/2})$, since the flow determined by (11) will always tend to push the Taylor integral back toward the "proper" balance (10), which should therefore never break down so completely that the Taylor integral is order one. Indeed, Taylor's original idea was that even if the field initially did not satisfy (10), after a short period of very rapid adjustment governed by (11) it would satisfy it, and thereafter everything would evolve only on much slower timescales. Indeed, he even gave an alternative prescription for solving for U_g, which is valid when $Ro = E = 0$ and (10) is satisfied identically, and therefore explicitly filters out these potentially very rapid adjustments governed by (11).

However, no one has yet succeeded in following Taylor's prescription (possibly because of the local difficulties that arise when setting $Ro = E = 0$?). And indeed, the more commonly accepted view today is that the field probably does not evolve ever closer to the constraint (10), but may instead almost satisfy it for long periods of time, resulting in evolution on moderate timescales, but may then suddenly fail to satisfy it, resulting in evolution on very short timescales. The existence of phenomena such as the geomagnetic jerks mentioned above certainly shows that the core is capable of generating global modes on very short timescales. So, we simply have to accept that in the geophysically relevant limit of small Ro and E the geodynamo equations permit both local and global structures on a very broad range of timescales, and that it is impossible to reduce this range by setting one or both of them identically equal to zero.

7. STIFF SYSTEMS

In the previous section we have seen some of the reasons why the geodynamo equations permit such a broad range of timescales, and why it does not appear to be possible to reduce this range by simplifying or approximating the equations in any way. So in this and subsequent sections we want to consider some of the implications for numerical geodynamo modelling. In particular, we begin by noting that there is nothing unusual about a system of differential equations supporting a broad range of timescales; such systems are sufficiently common that there is a name for them, namely stiff, and even a well-developed theory for how to deal with them. So what we want to consider in this section is the extent to which we can apply this theory to our particular stiff system.

Let's begin by considering the simplest possible stiff 'system', namely the single ordinary differential equation

$$\frac{dy}{dt} = -\epsilon^{-1} y + \epsilon^{-1} f(t) + f'(t), \quad y(0) = y_0, \qquad (12)$$

where $0 < \epsilon \ll 1$, and $f(t)$ is a given function varying on order one timescales. The analytic solution is simply

$$y = f(t) + C \exp(-\epsilon^{-1} t), \quad C = y_0 - f(0), \qquad (13)$$

which we note consists of two parts varying on very different timescales, namely $O(1)$ for $f(t)$, and $O(\epsilon)$ for $C \exp(-\epsilon^{-1}t)$. As simple as it is, (12) thus satisfies the definition of a stiff system. Furthermore, we note that regardless of the initial condition y_0 the second part will decay away extremely rapidly; indeed, this rapid $O(\epsilon)$ timescale is precisely the timescale on which it decays. Once this initial adjustment has taken place, the final solution therefore will vary only on the $O(1)$ timescale of the first part, $f(t)$. As contrived as it may seem, in this regard at least (12) is therefore exactly analogous to the very rapid initial adjustment but subsequent slow evolution envisioned by Taylor, for example.

So, suppose now that we wanted to solve (12) numerically, for example by the simplest possible scheme, namely the explicit Euler method

$$\frac{y_{n+1} - y_n}{\Delta t} = -\epsilon^{-1}y_n + \epsilon^{-1}f(t_n) + f'(t_n), \qquad (14)$$

that is,

$$y_{n+1} = (1 - \epsilon^{-1}\Delta t)y_n + \Delta t\left(\epsilon^{-1}f(t_n) + f'(t_n)\right).$$

We would soon discover that unless we reduce the timestep Δt in line with ϵ this scheme becomes violently unstable. That is, the mere fact that the underlying equation (12) allows solutions varying on an $O(\epsilon)$ timescale is enough to condemn this method to advance at that same very short timescale, even though we know that the eventual solution (13) varies only on the $O(1)$ timescale of $f(t)$. A stiff system is therefore not necessarily one where the solution exhibits a broad range of timescales, but merely where the equations allow a broad range of timescales, even if most of them do not appear in the final solution.

And this is a point that cannot be emphasized enough: standard, explicit methods such as (14) will invariably be restricted to timesteps no greater than the shortest timescale allowed by the equations even if — as is the case here — that shortest timescale consists of an extremely rapidly damped mode, with the subsequent evolution entirely on much slower timescales. It is therefore not sufficient to point out, for example, that the $O(Ro)$ and/or $O(E^{1/2})$ timescales allowed by (11) consist merely of an extremely rapid damping of any departures from (10), and that therefore once this adjustment to (10) has come about one should be able to timestep (11) explicitly using $O(1)$ timesteps. It simply doesn't work, just like timestepping (14) using $O(1)$ timesteps doesn't work, even after this adjustment to $y = f(t)$ has come about.

So, what does work? That is, is there any way of advancing (12) using $O(1)$ timesteps, and if so, can that method be extended to deal with (11) or indeed ultimately the full set of geodynamo equations (1-3)? Well, there certainly are ways of advancing (12) using $O(1)$ timesteps, for example the implicit Euler method

$$\frac{y_{n+1} - y_n}{\Delta t} = -\epsilon^{-1}y_{n+1} + \epsilon^{-1}f(t_{n+1}) + f'(t_{n+1}), \quad (15)$$

yielding

$$y_{n+1} = \left[y_n + \Delta t\left(\epsilon^{-1}f(t_{n+1}) + f'(t_{n+1})\right)\right] / (1 + \epsilon^{-1}\Delta t).$$

So first of all, why does (15) supposedly work where (14) didn't? To answer that, we need only consider the even simpler problem with $f(t) = 0$, so (14) and (15) reduce to

$$y_{n+1} = y_n(1 - \epsilon^{-1}\Delta t), \qquad (16a)$$

$$y_{n+1} = y_n/(1 + \epsilon^{-1}\Delta t), \qquad (16b)$$

respectively. We then note that for $\Delta t \ll \epsilon$, both of (16) are first-order accurate approximations to the analytic solution

$$y_{n+1} = y_n \exp[-\epsilon^{-1}\Delta t]. \qquad (17)$$

If we're willing to take such small timesteps, therefore, we can indeed use either of (14) or (15). Now let's see what happens if instead we try to take $\Delta t = O(1)$ though. We note first that then neither of (16) is even remotely close to approximating (17). However, (16b) is at least still strongly damping this mode, by essentially a factor of ϵ at every timestep, whereas (16a) is amplifying it, by a factor of $-\epsilon^{-1}$ at every timestep! That is, the implicit scheme (15) is getting the precise damping rate wrong, but at least it is still very strongly damping this mode, which is all we really care about. In contrast, the explicit scheme (14) is getting the 'damping' rate so completely wrong that in fact it's very strongly amplifying this mode, and so it can't possibly be stable for $\Delta t \gtrsim O(\epsilon)$.

Furthermore, this result, that explicit methods are restricted to $O(\epsilon)$ timesteps whereas implicit methods can use $O(1)$ timesteps, remains generally valid even for more sophisticated, higher-order methods, and also for more complicated systems than (12). The conclusion therefore seems quite clear: in order to cope with the stiffness inherent in the geodynamo equations we should solve them implicitly rather than explicitly. Unfortunately, this is easier said than done. In particular — as indeed the names suggest — with explicit methods such as (14) we are guaranteed to obtain an explicit formula for the updated variables in terms of the old ones, whereas with implicit methods such as (15) we must still solve the resulting system of equations. And whereas (15) was trivial to solve, being just a single equation, and linear at that, that is unfortunately not the case in general. In general this system of equations to be solved will be nonlinear, and very large, with however many equations as one has variables. So for example, if one is doing even relatively modest computations, with perhaps $\sim$ 30 modes in all three directions, that already adds up to $O(10^5)$ variables, and hence equations! Blindly attempting to apply standard methods such as Newton-Raphson iteration is therefore unlikely to be successful.

Fortunately, it is possible (using conjugate-gradient methods, for example) to apply Newton-Raphson without ever directly inverting, or even evaluating, the full 10^5 by 10^5 Jacobian matrix. Furthermore, the distinction between implicit versus explicit methods is not quite as 'either/or' as we've made it sound like so far. That is, it is possible to treat implicitly certain terms but not others — thereby gaining some but not all of the advantages, of course. For example, because it is still linear, treating the Coriolis force implicitly turns out to be relatively straightforward [e.g., *Hollerbach*, 2000], and allows one to take timesteps considerably larger than the $O(Ro)$ inertial oscillation timescale one would otherwise be restricted to. However, it does very little to help with some of the other difficulties, such as those associated with (11), which doesn't involve the Coriolis force at all. The only way to deal with these sources of stiffness would be to treat implicitly not only the Lorentz force $(\nabla \times \mathbf{B}) \times \mathbf{B}$ in (2), but also the advective terms $\nabla \times (\mathbf{U} \times \mathbf{B})$ in (1) and probably also $\mathbf{U} \cdot \nabla\Theta$ in (3), since it is the combination of an unbalanced Lorentz force generating a very large geostrophic flow which then acts back on $\mathbf{B}$ and Θ that causes the stiffness in this case. And these terms all being nonlinear, treating them implicitly is unfortunately considerably more difficult than treating the Coriolis force was. However, it might be possible to project out certain parts of these terms and treat only them implicitly, the idea obviously being to choose precisely those parts that help the most regarding the stiffness, but cost the least regarding the implicitness. Once again, this is clearly (a lot!) easier said than done, but I believe that some such approach will eventually be required if we are ever to get anywhere near realistic values for Ro and E.

And finally, one might just note that even if one did succeed in treating these terms implicitly, one would thereby still only have dealt with some of the many sources of stiffness in the geodynamo equations. Even treating all of these terms implicitly though might still not help much with some of the other sources, such as the existence of some of these waves (8) or (9). If the physics of the core are such that some of these extremely rapid oscillations really are excited to a significant extent, then one would have no choice but to resolve them numerically as well, that is, to take such small timesteps. That is precisely why it would have been so nice to be able to simplify the equations in such a way as to ensure that these modes do not exist at all, but as we saw in section 6.1, that does not appear to be possible.

8. IMPLICATIONS

Barring any such fundamental breakthroughs in the derivation of new equations and/or the development of new numerical methods, it seems unlikely that we will be able to reduce Ro and/or E much beyond $O(10^{-5})$ or so any time soon. Advances in hardware will certainly help some, but not nearly enough. As we've seen here, the shortest allowed timescales — and similarly the shortest allowed lengthscales, and in all three dimensions — simply decrease so quickly as one attempts to reduce Ro and E that reducing them by a factor of 10 might require 100, 1000, or even more times the computational power. Even allowing for Moore's Law, that computational speed doubles every 18 months, one would therefore have to wait quite some time before being able to systematically explore the $O(10^{-6})$ range.

So, what implications do these limitations on the numerically achievable Rossby and Ekman numbers have for geodynamo modelling? First and foremost, there is the very real danger that the results we obtain aren't even qualitatively in the right regime. In particular, there is the well-known distinction between so-called weak and strong field regimes that has been conjectured to exist at sufficiently small values, but has thus far never been unambiguously observed. That suggests that the $O(10^{-4})$ to at best $O(10^{-5})$ values currently being used are perhaps still too big for these two states to constitute distinct regimes at all. And given the general theoretical importance traditionally attached to this distinction [e.g., *Hollerbach*, 1996; *Fearn*, 1998; *Roberts and Glatzmaier*, 2000], that in turn suggests that maybe the results to

date aren't even qualitatively applicable to the real Earth, and that the surprisingly good agreement obtained already [e.g., *Glatzmaier and Roberts*, 1997; *Kuang and Bloxham*, 1997] is just a coincidence, and doesn't really tell us about the true workings of the geodynamo.

Even leaving aside this rather gloomy thought, computing at Rossby and Ekman numbers orders of magnitude too large has an additional implication in terms of the interpretation of the results. In particular, in the real Earth we certainly know exactly what we mean by a day, a year, etc., but what about in the computational model? If we increase Ro and E by several orders of magnitude, should we think of that as an increase in η and ν, in which case our magnetic diffusion time will suddenly be on the order of years rather than hundreds of thousands of years, or should we think of that as a decrease in Ω, in which case our 'days' will be on the order of decades. Or we could also think of it as a decrease in r_o, corresponding to a really small Earth! The point is that all of these, even the seemingly silly decrease in size, are equally valid ways of translating the nondimensional numbers back into dimensional quantities, given that one can't get all the dimensional quantities right. And given the arbitrary nature of this translation back to dimensional quantities, and the inevitable confusion as to what is then meant by 'a day' or 'a year', it is probably best if one avoids it entirely, and presents one's results nondimensionally, with any subsequent translation explicitly separated from the basic calculation.

9. CONCLUSION

In this review we have considered the timescales that exist in the Earth's core, and have seen how these range from the 10^{-2} second timescale of light waves to potentially longer than the age of the Earth. And even after filtering out some of the shortest, and discounting some of the longest, we found that we were still left with everything from inertial oscillations, occurring on timescales of days, to the average time between reversals, on timescales of hundreds of thousands of years, and possibly even variations in this average reversal rate on timescales of tens or hundreds of millions of years. Even just this restricted range of timescales therefore still spans 8 to 10 orders of magnitude, by far the broadest range of any geophysical system.

Furthermore, we saw why it does not appear to be possible to make any further progress in reducing this range by filtering out additional waves such as these inertial oscillations. Attempting to do so simply introduces other waves on comparable (or even shorter) timescales. The only way to reduce this range to something numerically feasible therefore is to arbitrarily increase the Rossby and/or Ekman numbers by several orders of magnitude, despite the potential dangers that entails in terms of eliminating the distinction between weak and strong field regimes.

Indeed, pinpointing precisely how small Ro and E must be to obtain this distinction is almost certainly the single most pressing issue in numerical geodynamo modelling: Until we know that $O(10^{-5})$ or so is small enough we cannot be certain that any of the results obtained to date — as impressive as they undoubtedly are — are really applicable to the Earth. Conversely though, if we could establish that $O(10^{-5})$ say is definitely small enough, then we could be reasonably confident not only that the results are applicable to the Earth, but also that we should not invest vast additional resources into reducing them yet further, but rather into things like simply doing longer runs, to begin to address some of the questions mentioned earlier.

Regardless of what ultimately turns out to be small enough though, we will simply have to accept that the geodynamo equations are an extremely stiff system, and will provide challenges for years to come. Indeed, given the difficulty of the problem, the progress to date is nothing short of remarkable, and one can only hope that future progress will be equally impressive!

Acknowledgments. This review was written while the author was on sabbatical at Princeton University. It is a pleasure to acknowledge the hospitality of the entire Department of Geosciences, and especially of Prof. H.-P. Bunge, without whose support this visit would not have been possible at all, let alone as pleasant and productive as it was.

REFERENCES

Aldridge, K. D., and L. I. Lumb, Inertial waves identified in the Earth's fluid outer core, *Nature, 325,* 421–423, 1987.

Alexandrescu, M., D. Gibert, G. Hulot, J. L. LeMouël, and G. Saracco, Detection of geomagnetic jerks using wavelet analysis, *J. Geophys. Res., 100,* 12,557–12,572, 1995.

Aurnou, J., D. Brito, and P. Olson, Anomalous rotation of the inner core and the toroidal magnetic field, *J. Geophys. Res., 103,* 9721–9738, 1998.

Bloxham, J., Dynamics of angular momentum in the Earth's core, *Ann. Rev. Earth Planet. Sci., 26,* 501–517, 1998.

Bloxham, J., D. Gubbins, and A. Jackson, Geomagnetic secular variation, *Phil. Trans. Roy. Soc. Lond. A, 329,* 415–502, 1989.

Bloxham, J., and A. Jackson, Fluid flow near the surface of Earth's outer core, *Rev. Geophys., 29,* 97–120, 1991.

Braginsky, S. I., Magnetic waves in the Earth's core, *Geomagn. Aeron., 7,* 851–859, 1967.

Braginsky, S. I., Torsional magnetohydrodynamic vibrations in the Earth's core and variations in day length, *Geomagn. Aeron., 10,* 1–8, 1970.

Braginsky, S. I., Short-period geomagnetic secular variation, *Geophys. Astrophys. Fluid Dynam., 30,* 1–78, 1984.

Braginsky, S. I., MAC-oscillations of the hidden ocean of the core, *J. Geomagn. Geoelec., 45,* 1517–1538, 1993.

Braginsky, S. I., and P. H. Roberts, Equations governing convection in Earth's core and the geodynamo, *Geophys. Astrophys. Fluid Dynam., 79,* 1–97, 1995.

Buffett, B. A., H. E. Huppert, J. R. Lister, and A. W. Woods, On the thermal evolution of the Earth's core, *J. Geophys. Res., 101,* 7989–8006, 1996.

Constable, C., On rates of occurrence of geomagnetic reversals, *Phys. Earth Planet. Inter., 118,* 181–193, 2000.

Courtillot, V., and J. L. LeMouël, Time variations of the Earth's magnetic field: from daily to secular, *Ann. Rev. Earth Planet. Sci., 16,* 389–476, 1988.

Courtillot, V., and J. P. Valet, Secular variation of the Earth's magnetic field: from jerks to reversals, *C. R. Acad. Sci. Paris II, 320,* 903–922, 1995.

Creager, K. C., Inner core rotation rate from small-scale heterogeneity and time-varying travel times, *Science, 278,* 1284–1288, 1997.

Dormy, E., J. P. Valet, and V. Courtillot, Numerical models of the geodynamo and observational constraints, *Geochem. Geophys. Geosys., 1,* 62–103, 2000.

Duck, P. W., and M. R. Foster, Spin-up of homogeneous and stratified fluids, *Ann. Rev. Fluid Mech., 33,* 231–263, 2001.

Fearn, D. R., Hydromagnetic flow in planetary cores, *Rep. Prog. Phys., 61,* 175–235, 1998.

Gilbert, W., *De Magnete,* 1600 (Reprinted by Dover, New York, 1991).

Glatzmaier, G. A., R. S. Coe, L. Hongre, and P. H. Roberts, The role of the Earth's mantle in controlling the frequency of geomagnetic reversals, *Nature, 401,* 885–890, 1999.

Glatzmaier, G. A., and P. H. Roberts, An anelastic evolutionary geodynamo simulation driven by compositional and thermal convection, *Physica D, 97,* 81–94, 1996a.

Glatzmaier, G. A., and P. H. Roberts, Rotation and magnetism of Earth's inner core, *Science, 274,* 1887–1891, 1996b.

Glatzmaier, G. A., and P. H. Roberts, Simulating the geodynamo, *Contemp. Phys., 38,* 269–288, 1997.

Gubbins, D., Rotation of the inner core, *J. Geophys. Res., 86,* 1695–1699, 1981.

Gubbins, D., The distinction between geomagnetic excursions and reversals, *Geophys. J. Int., 137,* F1–3, 1999.

Halley, E., An account of the cause of the change of the variation of the magnetical needle, with an hypothesis of the structure of the internal parts of the Earth, *Phil. Trans. Roy. Soc. Lond., 16,* 563–578, 1692.

Herrero-Bervera, E., and F. Theyer, Non-axisymmetrical behavior of Olduvai and Jaramillo polarity transitions recorded in north-central Pacific deep-sea sediments, *Nature, 322,* 159–162, 1986.

Hide, R., Free hydromagnetic oscillations of the Earth's core and the theory of the geomagnetic secular variation, *Phil. Trans. Roy. Soc. Lond. A, 259,* 615–647, 1966.

Hollerbach, R., On the theory of the geodynamo, *Phys. Earth Planet. Inter., 98,* 163–185, 1996.

Hollerbach, R., What can the observed rotation of the Earth's inner core reveal about the state of the outer core?, *Geophys. J. Int., 135,* 564–572, 1998.

Hollerbach, R., A spectral solution of the magneto-convection equations in spherical geometry, *Int. J. Num. Meth. Fluids, 32,* 773–797, 2000.

Hollerbach, R., and C. A. Jones, Influence of the Earth's inner core on geomagnetic fluctuations and reversals, *Nature, 365*, 541–543, 1993.

Hollerbach, R., and C. A. Jones, On the magnetically stabilizing role of the Earth's inner core, *Phys. Earth Planet. Inter., 87*, 171–181, 1995.

Jackson, A., A. R. T. Jonkers, and M. R. Walker, Four centuries of geomagnetic secular variation from historical records, *Phil. Trans. Roy. Soc. Lond. A, 358*, 957–990, 2000.

Jacobs, J. A., Variations in the intensity of the Earth's magnetic field, *Surv. Geophys., 19*, 139–187, 1998.

Jault, D., G. Gire, and J. L. LeMouël, Westward drift, core motions and exchanges of angular momentum between core and mantle, *Nature, 333*, 353–356, 1988.

Kuang, W. J., Force balances and convective state in the Earth's core, *Phys. Earth Planet. Inter., 116*, 65–79, 1999.

Kuang, W., and J. Bloxham, An Earth-like numerical dynamo model, *Nature, 389*, 371–374, 1997.

Kuang, W. J., and J. Bloxham, Numerical modeling of magnetohydrodynamic convection in a rapidly rotating spherical shell: weak and strong field dynamo action, *J. Comp. Phys., 153*, 51–81, 1999.

Langereis, C. G., M. J. Dekkers, G. J. de Lange, M. Paterne, and P. J. M. van Santvoort, Magnetostratigraphy and astronomical calibration of the last 1.1 Myr from an eastern Mediterranean piston core and dating of short events in the Brunhes, *Geophys. J. Int., 129*, 75–94, 1997.

Laske, G., and G. Masters, Limits on differential rotation of the inner core from an analysis of the Earth's free oscillations, *Nature, 402*, 66–69, 1999.

LeHuy, M., M. Alexandrescu, G. Hulot, and J. L. LeMouël, On the characteristics of successive geomagnetic jerks, *Earth Planets Space, 50*, 723–732, 1998.

Love, J., Reversals and excursions of the geodynamo, *Astron. Geophys., 40*, 14–19, 1999.

Matsushima, M., T. Nakajima, and P. H. Roberts, The anisotropy of local turbulence in the Earth's core, *Earth Planets Space, 51*, 277–286, 1999.

McFadden, P. L., and R. T. Merrill, Fundamental transitions in the geodynamo as suggested by paleomagnetic data, *Phys. Earth Planet. Inter., 91*, 253–260, 1995.

Merrill, R. T., and P. L. McFadden, Secular variation and the origin of geomagnetic field reversals, *J. Geophys. Res., 93*, 11,589–11,597, 1988.

Merrill, R. T., and P. L. McFadden, Geomagnetic polarity transitions, *Rev. Geophys., 37*, 201–226, 1999.

Moffatt, H. K., *Magnetic Field Generation in Electrically Conducting Fluids*, Cambridge Univ. Press, Cambridge, 1978.

Morrison, G., and D. R. Fearn, The influence of Rayleigh number, azimuthal wave number and inner core radius on 2-1/2 D hydromagnetic dynamos, *Phys. Earth Planet. Inter., 117*, 237–258, 2000.

Roberts, P. H., *An Introduction to Magnetohydrodynamics*, Elsevier, New York, 1967.

Roberts, P. H., and G. A. Glatzmaier, Geodynamo theory and simulations, *Rev. Mod. Phys., 72*, 1081–1123, 2000.

Roberts, P. H., and D. Gubbins, Origin of the main field: dynamics, *in Geomagnetism, vol 2*, edited by J. A. Jacobs, pp. 185–249, Academic Press, London, 1987.

Roberts, P. H., and A. M. Soward, Dynamo theory, *Ann. Rev. Fluid Mech., 24*, 459–512, 1992.

Sarson, G. R., and C. A. Jones, A convection driven geodynamo reversal model, *Phys. Earth Planet. Inter., 111*, 3–20, 1999.

Song, X., and P. G. Richards, Seismological evidence for differential rotation of the Earth's inner core, *Nature, 382*, 221–224, 1996.

Stevenson, D. J., Models of the Earth's core, *Science, 214*, 611–619, 1981.

Su, W., A. M. Dziewonski, and R. Jeanloz, Planet within a planet: rotation of the inner core of Earth, *Science, 274*, 1883–1887, 1996.

Taylor, J. B., The magnetohydrodynamics of a rotating fluid and the Earth's dynamo problem, *Proc. Roy. Soc. Lond. A, 274*, 274–283, 1963.

Valet, J. P., C. Laj, and P. Tucholka, High-resolution sedimentary record of a geomagnetic field reversal, *Nature, 322*, 27–32, 1986.

Vidale, J. E., D. A. Dodge, and P. S. Earle, Slow differential rotation of the Earth's inner core indicated by temporal changes in scattering, *Nature, 405*, 445–448, 2000.

Walker, M. R., C. F. Barenghi, and C. A. Jones, A note on dynamo action at asymptotically small Ekman number, *Geophys. Astrophys. Fluid Dynam., 88*, 261–275, 1998.

Zatman, S., and J. Bloxham, Torsional oscillations and the magnetic field within the Earth's core, *Nature, 388*, 760–763, 1997.

Zatman, S., and J. Bloxham, On the dynamical implications of models of B_s in the Earth's core, *Geophys. J. Int., 138*, 679–686, 1999.

Geodynamo Modeling and Core-Mantle Interactions

Weijia Kuang,[1]

Joint Center for Earth Systems Technology, UMBC, Baltimore, Maryland

Benjamin F. Chao

NASA Goddard Space Flight Center, Greenbelt, Maryland

We discuss possible mechanisms for the transfer of axial angular momentum between the Earth's solid mantle and fluid outer core, and their impacts on the observed length of day (LOD) variation on time scales of decades and longer. We use our modular, scalable, self-consistent, three-dimensional (MOSST) numerical geodynamo model to estimate the coupling torques arising from the magnetic stress on the core-mantle boundary (CMB), and that due to the CMB topography under the (non-hydrostatic) pressure that are dynamically consistent with the convective processes in the outer core. Our results show that, for the axial angular momentum exchange, a significant torque can be provided: (i) by the magnetic coupling depending on the electrical conductivity of the lower mantle; and (ii) by the topographic coupling if the amplitude of the CMB topography is larger than about 3 km. This implies that the core-mantle interactions are far more complex than have been projected, and that there is unlikely a single dominant coupling mechanism for the decadal LOD variation.

1. INTRODUCTION

It has long been known that the Earth's rotation is not constant. Its rotation rate, as well as the direction of the rotation axis vary slightly on time scales ranging from sub-daily to geological time scales. In addition to the (external) interaction in the Sun-Earth-Moon system, the Earth's rotation varies as a result of a host of (internal) geophysical processes involving large mass transports on or within the Earth. The Earth rotation can be measured very precisely by modern space geodetic techniques [e.g. *Chao et al.*, 2000], and hence studying the Earth's rotation variation could further our understanding of the geophysical processes from the surface all the way to the deep interior of the Earth.

The dynamical quantity governing the Earth rotation is the angular momentum M:

$$M = M_m + M_{core} + M_{sf}, \qquad (1)$$

where M_m is the angular momentum of the solid mantle, M_{core} the angular momentum of the core and M_{sf} the angular momentum of the surface geophysical fluid systems (e.g. atmosphere, ocean and hydrosphere). De-

---

[1] Also at NASA Goddard Space Flight Center, Greenbelt, Maryland

Earth's Core: Dynamics, Structure, Rotation
Geodynamics Series 31
10.1029/31GD13

noting by I_m the moment of inertia tensor and by $\boldsymbol{\Omega}$ the angular velocity vector of the mantle, we can write

$$\boldsymbol{M}_m \equiv \boldsymbol{I}_m \cdot \boldsymbol{\Omega}. \qquad (2)$$

Ignoring the external interactions, the total angular momentum of the Earth should be conserved, i.e. $d\boldsymbol{M}/dt = 0$. By (1) and (2), we have,

$$\boldsymbol{I}_m \cdot \frac{d\boldsymbol{\Omega}}{dt} = -\frac{d\boldsymbol{M}_{sf}}{dt} - \frac{d\boldsymbol{I}_m}{dt} \cdot \boldsymbol{\Omega} - \frac{d\boldsymbol{M}_{core}}{dt}, \qquad (3)$$

Thus, any change in either the angular momentum $\boldsymbol{M}_{sf}$ of the surface geophysical fluid systems, or that $\boldsymbol{M}_{core}$ of the core, or in the moment of inertia $\boldsymbol{I}_m$ of the mantle, or any combination thereof, will be compensated by an equal but opposite quantity on the left side of (3), manifesting itself as variation in the rotation of the solid Earth. In this paper, we consider only the variation of the Earth's rotation rate Ω, or the length of day (LOD)

$$\mathrm{LOD} \equiv \frac{2\pi}{\Omega}. \qquad (4)$$

This variation component is given along the direction of the rotation axis, or the z-axis. The rotation variation perpendicular to the z-axis, known as the polar motion of the rotation axis, involves more complex dynamics and is beyond our present scope.

Perhaps the most significant difference among the contributions to the LOD variation is the time scale on which one particular angular momentum variation dominates the others. For example, LOD variations on time scales from several days to several years have been demonstrated to result from the axial angular momentum exchange between the solid mantle and its fluid envelope that includes the atmosphere and hydrosphere [see reviews of, e.g., *Rosen*, 1993; *Eubank*, 1993; *Dickey*, 1993; *Marcus et al.*, 1998]. In the other extreme, the secular LOD variation is mainly caused by tidal breaking of the Earth in the Sun-Earth-Moon system (an external torque), and in part by the post-glacial rebound (that alters the moment of inertia $\boldsymbol{I}_m$ of the solid Earth) [e.g. *Munk and MacDonald*, 1960; *Stephenson and Morrison*, 1995]. In between, the observed decadal fluctuation of LOD, on the order of ± 1 ms (millisecond) in amplitude, is believed to arise from the axial angular momentum exchange betweenthe mantle and the core [*Hide*, 1969; *Lambeck*, 1980]. It is this part of the core-mantle angular momentum exchange that we shall focus on in this paper. We disregard any decadal meteorological influences that may be present but are smaller by more than an order of magnitude.

Jault et al. [1988] first demonstrated the correlation between the axial angular momentum of the fluid core

and the decadal LOD variation. Their work can be summarized as follows (more details can be found in the following sections of this paper and in their original paper). First the poloidal part of the geomagnetic field at the core-mantle boundary (CMB) is downward continued from the observed geomagnetic field at the surface of the Earth. Then the core flow beneath the CMB is determined from the time variation of the field, with the "frozen flux" approximation (i.e. the dissipation of the geomagnetic field in the core is negligible compared to the advection of the field) and the assumption that the flow is tangentially geostrophic. Assuming further that the part of the zonal flow (i.e. the flow in the longitude direction) symmetric about the equator is invariant along the rotation axis, the time-varying axial angular momentum of the core can then be evaluated. With this recipe, *Jault et al.* [1988] found that the variation of the core's axial angular momentum matches well with the decadal LOD variation for the period of 1969 to 1985. Using an independent geomagnetic field model, *Jackson et al.* [1993] extended the result to encompass the period 1840 – 1990 .

These kinematic studies provide observational evidence on one-to-one correspondence between the axial angular momentum variation of the fluid outer core and that of the solid mantle, thus suggesting strongly that the decadal LOD variation arises from the core-mantle angular momentum exchange. However, they do not identify the actual coupling mechanisms, i.e. the torques between the core and the mantle, that are responsible for the angular momentum exchange. To do so we need to find first the magnitude of the required coupling torque between the mantle and the core. Then we need to estimate the contributions of all possible coupling mechanisms. Note that here we are only concerned about the axial component, even though our formulas derived below are in the general 3-dimensional vector form.

The required coupling torque between the mantle and the core can be readily estimated as follows. Observations [e.g. *Chao and Ray*, 1997] show that the decadal LOD variation is about 2 ms. With the mean rotation rate $\Omega \approx 7.29 \times 10^{-5} \mathrm{rad\,s}^{-1}$ and the moment of inertia $C \approx 8 \times 10^{37} \mathrm{kg\,m^2}$ of the Earth, one could yield the required torque

$$\Gamma \approx 5 \times 10^{17} \mathrm{\,Nm}. \qquad (5)$$

To identify the coupling mechanisms that could provide torques comparable to that required (5) proves to be much more difficult. This problem has been investigated in the past decades with limited success, because there are multiple coupling mechanisms across the CMB, and more importantly, because the estimation of the coupling torques depends on detail structure of the core dynamical states that have remained mostly unexplored until very recently.

Most of the previous studies have been concentrated on evaluating the coupling torques arising from aspherical CMB topography, and from a finitely electrically conducting D''-layer at the top of the CMB. However, these studies are either kinematic [e.g. *Stix and Roberts,* 1984; *Jault and Lemouël,* 1989] or quasi-dynamic [e.g. *Anufriev and Braginsky,* 1975, 1977; *Kuang and Bloxham,* 1993]. Therefore the results may be prone to dynamical inconsistencies in the studies.

In the kinematic studies, the core flow are derived with several approximations that may not be consistent with the force balances in the core. In the quasi dynamic studies, the ambient magnetic field and the velocity field are not consistent with the geodynamo processes in the core. These problems could potentially prohibit proper interpretations of the core-mantle interactions. Furthermore, they can not be improved with observations.

In the past few years, great progress has been made in numerical modeling of core dynamics [e.g. *Glatzmaier and Rroberts,* 1995; *Kuang and Bloxham,* 1997b]. The numerical models provide a new, powerful tool to study the dynamics of the core-mantle interactions. For example, *Kuang and Chao* [2001] showed that the dynamo modeling can be used to study core-mantle interactions in a way consistent with dynamical processes in the outer core. Here we shall review our understanding of the dynamics of various coupling mechanisms, and present further numerical results from our improved geodynamo model on the electromagnetic and topographic couplings. Their geodynamo model, called MOSST (Modular, Scalable, Self consistent, Three dimensional) model in the rest of this paper, is based upon the model originally developed by *Kuang and Bloxham* [1999], but several key modifications are implemented in order to investigate the core-mantle interactions.

In this paper, we shall review our understanding of the dynamics of various core-mantle coupling mechanisms, and present further results from MOSST dynamo model on the electromagnetic and topographic couplings. We shall compare our numerical results (that are dynamically consistent with the convective flow in the core) with those from previous studies, and assess quantitatively the importance of the couplings against the torque (5) required for the decadal LOD variation.

The paper is organized as follows: the force balances in the core and relevant modifications to the original

Kuang and Bloxham [1999] model are described in Section 2. The viscous coupling and the gravitational coupling are briefly summarized in Sections 3 and 4, respectively. The electromagnetic coupling is discussed in Section 5, followed by the discussion of the topographic coupling in Section 6. Conclusions are given in Section 7.

2.   CORE DYNAMICS AND GEODYNAMO MODELING

2.1.   *Force Balances in the Core*

It is now widely accepted that the Earth's fluid outer core is in vigorous convection, presumably driven by gravitational energy released from the thermal evolution of the Earth. This gravitational energy could be described in the form of density distribution slightly different from the (axi-symmetric) adiabatic density distribution. The convective flow generates and maintains a strong magnetic field (the geodynamo), resulting in geomagnetic signals observable at the surface of the Earth. However, the magnetic field is not passive to the flow: it interacts with the core flow via the Lorentz force. The core dynamics is not independent of the mantle. On very long time scales ($\sim 10^8$ years), the thermal convective process in the mantle controls the core convection via thermal boundary conditions across the CMB. The core flow, in turn, affects the rotation of the mantle on much shorter time scales ($\sim 10^2$ years) through couplings across the CMB.

To understand the dynamics of the core-mantle interaction, we start with the force balances in the Earth's fluid outer core. The principal force balance is described by the following momentum equation defined in the reference frame co-rotating with the mantle,

$$\rho \left[\frac{\partial v}{\partial t} + (v \cdot \nabla)v \right] + 2\rho\,\Omega \times v = -\nabla p + J \times B$$
$$+ \rho g + \rho\nu\nabla^2 v, \qquad (6)$$

where v is the fluid velocity, $J \equiv (\nabla \times B)/\mu$ (μ is the magnetic permeability of the core) is the electrical current density, ν is the kinematic viscosity of the fluid and g is the gravitational acceleration. The Lorentz force $J \times B$ in the equation describes the effect of magnetic field on flow.

In theoretical studies, the momentum balance (6) is often nondimensionalized. Using the mean radius r_o of the CMB as the length scale and the magnetic diffusive time $\tau_\eta \equiv r_0^2/\eta$ (where η is the magnetic diffusivity of the outer core) as the time scale, (6) can be written as

$$R_o \left[\frac{\partial \boldsymbol{v}}{\partial t} + (B - V \cdot \nabla) \boldsymbol{v} \right] + \hat{\boldsymbol{z}} \times \boldsymbol{v} = -\nabla p$$
$$+ (\nabla \times \boldsymbol{B}) \times \boldsymbol{B} - R_{th} \Theta \hat{\boldsymbol{g}} + E \nabla^2 \boldsymbol{v}, \quad (7)$$

where R_{th} is the Rayleigh number that measures the strength of the buoyancy force, Θ denotes the non-dimensional density perturbation, $\hat{\boldsymbol{z}}$ is the unit vector along the mean rotation axis of the Earth and $\hat{\boldsymbol{g}}$ is the unit vector parallel to the gravitational acceleration [for more details, see *Kuang and Bloxham, 1999*]. In (7), the fluid inertia and the fluid viscosity are measured relative to the Coriolis force and are described by the (non-dimensional) Rossby number R_o and the Ekman number E, respectively:

$$R_o = \frac{\eta}{2\Omega r_0^2}, \quad E = \frac{\nu}{2\Omega r_o^2}. \quad (8)$$

In the Earth's core, the fluid inertia and the viscous force are very small compared to the Coriolis force,

$$R_o \approx 10^{-9}, \quad E \approx 10^{-15}. \quad (9)$$

(Here the molecular viscosity of the core fluid is used for estimation. $E \approx 10^{-9}$ even if the core flow is assumed to be turbulent.)

The leading force balance in a rapidly rotating fluid is the geostrophic balance

$$\hat{\boldsymbol{z}} \times \boldsymbol{v}_G = -\nabla p, \quad (10)$$

where $\boldsymbol{v}_G$ represents the geostrophic flow. However, there are two other force balances in the core that are important to geodynamo. The first is the magnetostrophic balance

$$\hat{\boldsymbol{z}} \times \boldsymbol{v} \approx -\nabla p + \boldsymbol{J} \times \boldsymbol{B} - R_{th} \Theta \hat{\boldsymbol{g}}. \quad (11)$$

The corresponding flow, the so-called MAC waves [*Braginsky, 1967*], varies on the time scales of several hundred years and longer.

The other force balance arises when one integrates the ϕ-component of (7) over the co-axial cylindrical surfaces A across the core (the "Taylor cylinders") :

$$R_o \int_A \left(\frac{D\boldsymbol{v}}{Dt} \right)_\phi dS = \int_A (\boldsymbol{J} \times \boldsymbol{B})_\phi dS +$$
$$E \int_A (\nabla^2 \boldsymbol{v})_\phi dS - R_{th} \int_A \Theta(\hat{\boldsymbol{g}} \cdot \hat{\phi}) dS. \quad (12)$$

The last term vanishes if the gravitational acceleration is axi-symmetric. This force balance actually describes the torque balance on the Taylor cylinders [*Kuang and Bloxham, 1999*] (referred to simply as the torque balance in this paper).

Compared to the leading order force balances, (12) is of higher order effect, due to the small R_o and E in the core and the nearly axi-symmetric mass distribution in the Earth. However, it is critical for understanding the transfer of the axial angular momentum between the core and mantle. The flow associated with this force balance can be characterized as co-axial cylindrical fluid parcels oscillating about the Earth's rotation axis on decadal time scales, and thus called torsional oscillations [*Braginsky, 1976*]. In particular, they carry axial angular momentum and are proposed by *Braginsky* [1984] to be important for the variation in the core axial angular momentum, and hence decadal LOD variation, and are further emphasized in e.g. *Jault et al.* [1988], *Zatman and Bloxham* [1997].

To analyze the variation of the mantle angular momentum $\boldsymbol{M}_m$ due to the core dynamical processes, we integrate $\boldsymbol{r} \times (6)$ in the outer core, add to it the angular momentum variation of the inner core, and obtain (see Appendix A)

$$\frac{d\boldsymbol{M}_m}{dt} + \boldsymbol{\Omega} \times \boldsymbol{M}_m = \frac{1}{\mu} \int_{CMB} dS \, (\hat{\boldsymbol{n}} \cdot \boldsymbol{B}) \boldsymbol{r} \times \boldsymbol{B} -$$
$$\int_{CMB} dS \, (\boldsymbol{r} \times \hat{\boldsymbol{n}}) p - \int_{core} d\boldsymbol{r} \, \boldsymbol{r} \times \rho \boldsymbol{g} +$$
$$\rho \nu \int_{CMB} dS \, [\boldsymbol{r} \times (\hat{\boldsymbol{n}} \cdot \nabla) \boldsymbol{v} - \hat{\boldsymbol{n}} \times \boldsymbol{v}], \quad (13)$$

where $\hat{\boldsymbol{n}}$ is the unit vector normal to the CMB, and the subscript "core" includes both the outer and inner cores.

Equation (13) shows that there are four kinds of core-mantle interactions. The first term on the right hand side (r.h.s.) is the electromagnetic torque arising from a finite Lorentz stress on the CMB, known as the electromagnetic coupling. The second term is the pressure torque arising from the non-hydrostatic pressure p acting on the CMB topography, and is called the topographic coupling. The third term is the gravitational torque arising from aspherical density distributions and aspherical gravitational acceleration. The last term in (13) is the viscous torque resulting from fluid friction on the CMB.

2.2. Numerical Geodynamo Modeling

To evaluate these coupling torques, one must first have sufficient knowledge about the velocity field $\boldsymbol{v}$, the magnetic field $\boldsymbol{B}$, and the density perturbation Θ in the core. The most effective means to obtain such knowledge at present is to rely on numerical modeling. The numerical model used here, the MOSST geodynamo model, is based on the dynamo model developed by *Kuang and Bloxham* [1999], but extensively modified and strengthened for the study of the core-mantle interactions.

In addition to the momentum balance (7), the model also solves the (non-dimensional) induction equation

$$\frac{\partial \boldsymbol{B}}{\partial t} = \nabla \times (\boldsymbol{v} \times \boldsymbol{B}) + \nabla^2 \boldsymbol{B} \qquad (14)$$

that describes the magnetic field in the core, and the (non-dimensional) thermal equation

$$\frac{\partial \Theta}{\partial t} = -\boldsymbol{v} \cdot \nabla T_0 - \boldsymbol{v} \cdot \nabla \Theta + q_\kappa \nabla^2 \Theta, \qquad (15)$$

that describes the density perturbation in the core. In (15), $q_\kappa \equiv \kappa/\eta$ is the ratio of the thermal conductivity κ and the magnetic diffusivity η of the core fluid, and T_0 is the primary conducting temperature profile for given heat flux (or temperature) boundary conditions. Equation (15) implies that the density perturbation in the core can be described by temperature perturbations.

The electrical conductivity of the lower mantle is modeled by a thin D''-layer at the top of the CMB with a thickness d_m and an electrical conductivity σ_m (the corresponding magnetic diffusivity is $\eta_m \equiv 1/\mu_0 \sigma_m$). The inner core is assumed to have the same electrical conductivity as the outer core. The equations (7), (14) and (15) are solved with appropriate boundary conditions at the CMB and at the inner core boundary (ICB).

As in the original model [*Kuang and Bloxham*, 1999], the velocity field $\boldsymbol{v}$ and the magnetic field $\boldsymbol{B}$ are decomposed into poloidal P and toroidal T potentials:

$$\boldsymbol{v} = \nabla \times (T_v \hat{\boldsymbol{r}}) + \nabla \times \nabla \times (P_v \hat{\boldsymbol{r}}), \qquad (16)$$
$$\boldsymbol{B} = \nabla \times (T_b \hat{\boldsymbol{r}}) + \nabla \times \nabla \times (P_b \hat{\boldsymbol{r}}). \qquad (17)$$

These potentials and the density perturbation Θ are expanded in spherical harmonic functions, e.g.

$$\begin{bmatrix} P_v \\ T_v \\ P_b \\ T_b \end{bmatrix} = \sum_{m=0,M}^{l=m,L} \begin{bmatrix} v_l^m(r,t) \\ \omega_l^m(r,t) \\ b_l^m(r,t) \\ j_l^m(r,t) \end{bmatrix} Y_l^m(\theta,\phi) + C.C.$$
$$(18)$$

where $\{Y_l^m\}$ are the orthonomal spherical harmonic functions and $C.C.$ denotes the complex conjugate.

One major modification in the MOSST model is to redefine the above system in the reference frame fixed with the mantle, so that we could better incorporate lower mantle heterogeneity in modeling core-mantle interactions and solid-body rotation of the mantle. In

this new reference frame, the momentum balance (6) is modified by adding to the left side the Poincaré term $\rho \dot{\boldsymbol{\Omega}} \times \boldsymbol{r}$ ($\dot{\boldsymbol{\Omega}}$ is the time derivative of the mantle angular velocity $\boldsymbol{\Omega}$).

Another modification in MOSST is the addition of the horizontal fluid inertia ($m = 1$) into the new model. Therefore, the total (3−dimensional) angular momentum is conserved in the numerical modeling. With this addition, we are able to study not only the LOD variation, but also the equatorial angular momentum that affects the polar motion of the mantle.

With these modifications in MOSST, we are able to evaluate the angular momentum variation of the mantle and of the inner core through integrating the solid body rotations of the mantle and of the inner core, which are described by the equations

$$\boldsymbol{I}_m \cdot \dot{\boldsymbol{\Omega}} + \boldsymbol{\Omega} \times \boldsymbol{I}_m \cdot \boldsymbol{\Omega} = \tilde{\boldsymbol{\Gamma}}_m, \qquad (19)$$
$$\boldsymbol{I}_i \cdot \dot{\boldsymbol{\omega}} + (\boldsymbol{\Omega} + \boldsymbol{\omega}) \times \boldsymbol{I}_i \cdot (\boldsymbol{\Omega} + \boldsymbol{\omega})$$
$$+ \boldsymbol{I}_i \cdot (\boldsymbol{\Omega} \times \boldsymbol{\omega}) = \tilde{\boldsymbol{\Gamma}}_i - \boldsymbol{I}_i \cdot \dot{\boldsymbol{\Omega}}, \quad (20)$$

where $\boldsymbol{I}_i$ is the moment of inertia of the inner core, $\boldsymbol{\omega}$ is the angular velocity of the inner core relative to the mantle, $\tilde{\boldsymbol{\Gamma}}_m$ and $\tilde{\boldsymbol{\Gamma}}_i$ are the total torque on the mantle and on the inner core, respectively. It should be noted that $\dot{\boldsymbol{\omega}}$ in (20) is the time derivative in the mantle reference frame. We do not consider any external torque in our numerical model. Therefore, the total angular momentum of the system (i.e. the inner core, the outer core and the mantle) is conserved (within numerical accuracy) [see *Kuang and Bloxham*, 1999]. We select the initial conditions from our previous (well developed) numerical dynamo solutions, so as to minimize the transient periods in numerical calculations.

One constraint in geodynamo modeling is that the parameters (9) for the Earth's core are too small to be resolved numerically. In fact, the numerical parameters used in all geodynamo modeling are at least three orders of magnitude larger than those in (9). For example, in our modeling,

$$R_o = E = 1.25 \times 10^{-6} \qquad (21)$$

(The Prandtl number $q_\kappa = 1$ in our simulation). Therefore it is necessary to maintain appropriate force balances in numerical modeling. In particular, maintaining the torque balance (12) is important in studying the core-mantle interaction. *Kuang and Bloxham* [1997b] and *Kuang* [1999] demonstrate that by applying the "friction-free" boundary conditions, (12) can be appropriately maintained in numerical modeling. It should be pointed out here that the numerical solutions for the parameters (21) are strong-field dynamo solutions, with the effective Elsasser number

$$\tilde{\Lambda} \equiv \frac{\int_0^{2\pi} |\nabla \times (\boldsymbol{J} \times \boldsymbol{B})| d\phi}{\int_0^{2\pi} |\partial_z \boldsymbol{v}| d\phi} \approx 1 \qquad (22)$$

in the bulk of the outer core [*Kuang, 1999*]. This is consistent with the strong-field geodynamo.

To examine the effect of viscous coupling in time scales longer than several decades, we introduce in the new model "partial-slippery" boundary conditions to retain a weak viscous torque at the boundaries. These (non-dimensional) asymptotic boundary conditions are

$$-\hat{\mathbf{n}} \times (\hat{\mathbf{n}} \cdot \nabla) \boldsymbol{u} = \hat{\mathbf{n}} \times (\boldsymbol{u} + \hat{\boldsymbol{z}} \times \boldsymbol{u}), \qquad (23)$$

where $\boldsymbol{u}$ is the difference between the core flow $\boldsymbol{v}$ and the velocity $\boldsymbol{v}_B$ of the boundaries:

$$\boldsymbol{u} = \boldsymbol{v} - \boldsymbol{v}_B. \qquad (24)$$

With these boundary conditions, the viscous torque at the boundaries are proportional to the Ekman number E (see Appendix B).

To examine the electromagnetic coupling across the CMB, we select different profiles for the uniform, electrically conducting layer (D''-layer) at the top of the CMB, in which the layer thickness d_m and the layer conductivity σ_m have different values while the total conductance $d_m \sigma_m$ is kept constant,

$$\begin{array}{llll} \text{Profile I:} & \sigma_m = \sigma/200, & d_m = 200\,\text{km}, \\ \text{Profile II:} & \sigma_m = \sigma/20, & d_m = 20\,\text{km}, \end{array} \qquad (25)$$

where σ is the electrical conductivity of the outer core. In this paper, Profile I is also called the thick layer, while Profile II is called the thin layer.

To examine topographic core-mantle coupling, we introduce in MOSST an aspherical CMB, which is defined as

$$r_{\text{CMB}} = r_o [1 + \varepsilon H(\theta, \phi)], \qquad (26)$$

where ε is the relative amplitude of the boundary topography H. In our study, we adopt H from the seismic results of *Morelli and Dziewonski* [1987].

Since the amplitude ε is in general very small, of order $\mathcal{O}(10^{-3})$, the CMB topography is ignored in volume calculations (i.e. it is assumed spherical in solving convective states inside the core). It is considered only in the boundary conditions. For example, the (non-dimensional) boundary condition

$$[\hat{\mathbf{n}} \cdot \boldsymbol{B}] = 0$$

is approximated as

$$[B_r] = \varepsilon (\nabla H) \cdot [\boldsymbol{B}], \qquad (27)$$

where $[\cdot]$ denotes the difference across the CMB. Similar expansions also apply to other boundary conditions for the velocity field $\boldsymbol{v}$ and the temperature perturbation Θ. Since it is difficult to solve the boundary conditions (27) implicitly, we introduce in MOSST an iterative method to solve the variables $\boldsymbol{v}$, $\boldsymbol{B}$ and Θ. The iteration can be described as

$$\begin{aligned} [B_r^{(n)}(t + \Delta t)] &= \varepsilon (\nabla H) \cdot [\boldsymbol{B}^{(n-1)}(t + \Delta t)], \\ \boldsymbol{B}^{(0)}(t + \Delta t) &= \boldsymbol{B}(t), \end{aligned}$$

throughout the core (and similarly for the other variables). This method converges very quickly. It usually takes $n = 3$ for a relative error of order $\mathcal{O}(10^{-4})$. Consequently, the total CPU time is approximately tripled.

As in the previous modeling [*Kuang and Bloxham, 1999*], the heat fluxes are assumed to be fixed at the CMB and at the ICB.

One important concern in modeling the core-mantle interactions is the accuracy of numerical solutions, because the Earth's rotation variation is very small compared to the mean rotation rate. In our model, the typical non-dimensional rotation variation is $\Delta\omega \approx 20$. On the other hand, the typical non-dimensional velocity field is $|\boldsymbol{v}| \approx 1000$. Therefore the rotation rate variation can well be resolved within the numerical accuracy (10^{-4} relative errors with the above truncation orders) [*Kuang and Bloxham, 1999*].

Quantitatively, one should also consider the time scales of the solutions. The small parameters R_o and E also determine the time scale of the fast modes, such as the torsional oscillations. Therefore, appropriate asymptotic scaling rules are necessary to extrapolate numerical solutions to Earth-like parameter regimes (9), which may be obtained by examining variation of numerical solutions with R_o and E.

We select the truncation levels $L = 33$ in latitude and $N = 40$ in radius in the outer core. The radial grid points in the D''-layer are defined as

$$\begin{aligned} r_i &= 1.0 + (r_d - 1.0)x_i^2, \\ x_i &= i/N_d, \end{aligned} \quad \text{for} \quad i = 0, 1, \cdots, N_d, \qquad (28)$$

where r_d is the non-dimensional mean radius at the top of the D''-layer and N_d is the truncation order for which we select $N_d = 10$. The purpose of an unevenly distributed grid points (28) is to resolve possible boundary layer (e.g. skin-depth) at the top of the CMB. We apply a hyper-dissipation to ensure convergence, but only to the terms of the order $l \geq 10$, and the coefficient $\epsilon = 10^{-2}$ (see *Kuang and Bloxham* [1999] for detailed expressions). Consequently, the effective dissipation doubles for small scale solutions $l \geq 20$. However, as we shall discuss in more detail below, the part of the solutions contributing most to the core-mantle interaction are for $l \leq 6$. Thus the effect of the hyperdissipation is minimized.

3. VISCOUS COUPLING

The viscous coupling torque arises from a finite friction stress on the CMB. This torque is basically ignored in the study of the decadal LOD variation because the the Earth rotates very rapidly, thus the viscous effect of the core flow is very small, and the resultant torque negligible compared to the required value (5).

To demonstrate this in more detail, we first find from (13) that the viscous torque $\widetilde{\boldsymbol{\Gamma}}_\nu$ acting on the CMB is

$$
\begin{aligned}
\widetilde{\boldsymbol{\Gamma}}_\nu &= \rho\nu \int_{\text{CMB}} dS \left[\boldsymbol{r} \times (\hat{\boldsymbol{n}} \cdot \nabla) \boldsymbol{v} - \hat{\boldsymbol{n}} \times \boldsymbol{v} \right] \\
&= \Gamma_0 E \int_{r=1} dS \, \hat{\boldsymbol{r}} \times (\partial_r \boldsymbol{v} - \boldsymbol{v}) \\
&\equiv \Gamma_0 \, \boldsymbol{\Gamma}_\nu
\end{aligned}
\tag{29}
$$

where

$$
\Gamma_0 \equiv 2\rho\Omega r_0^5 / \tau_\eta
\tag{30}
$$

is the torque scaling factor. The axial component of the (non-dimensional) torque $\boldsymbol{\Gamma}_\nu$ is therefore

$$
\Gamma_\nu^z \equiv \hat{\boldsymbol{z}} \cdot \boldsymbol{\Gamma}_\nu = E \int_{r=1} dS \sin\theta \, (\partial_r v_\phi - v_\phi) \, ,
\tag{31}
$$

where θ is the co-latitude. Because the Ekman number E is very small in the Earth's core, a thin boundary layer of a thickness $\mathcal{O}(\sqrt{E})$ (the Ekman layer) develops at the CMB. Asymptotic analysis of the boundary layer demonstrates that [e.g. *Greenspan*, 1968]

$$
|\partial_r B - V_H| \approx E^{-1/2} |B - V_H| \, .
\tag{32}
$$

Therefore, we may obtain from (31) that

$$
|\Gamma_\nu^z| \approx \sqrt{E} |v_\phi| \, .
\tag{33}
$$

The corresponding dimensional viscous torque is

$$
\widetilde{\Gamma}_\nu^z \approx \sqrt{E} \rho_0 2\Omega r_0^4 u_\phi \, ,
\tag{34}
$$

where u_ϕ is the typical dimensional zonal flow beneath the CMB. With the value in (9), we have $\widetilde{\Gamma}_\nu^z \approx 5 \times 10^{15}$ Nm, about two orders of magnitude smaller than the required torque (5).

The inefficiency of the viscous coupling on the angular momentum transfer across the CMB can also be explained by the viscous spin-up time of the fluid through the Ekman layers (Ekman "pumping"). For any velocity variation at the boundary, asymptotic analysis shows that the velocity difference will propagate into the interior through the viscous layer over the time [e.g. *Greenspan*, 1968]

$$
\tau_E \approx E^{-1/2} \frac{1}{2\Omega} \approx 10^3 \, \text{years} \, ,
\tag{35}
$$

which is two orders of magnitude longer than the decadal time scales of interest here.

Nevertheless, it has been argued that the convective core flow may be highly turbulent. Turbulent mixing processes can enhance the viscous effect throughout the core. With a simple argument that all dissipative processes in turbulent flow are comparable, one would expect

$$
E \approx R_o \approx 10^{-9} \, .
\tag{36}
$$

This could yield a turbulent viscous coupling two orders of magnitude stronger than the molecular viscous coupling, thus not necessarily negligible in the core-mantle angular momentum exchange. However, since the turbulent flow can be anisotropic in the core [e.g. *Matsushima et al.*, 1999], and since advection does not help transferring axial angular momentum globally, the jury is still out as to possible importance of the turbulent viscous coupling.

4. GRAVITATIONAL COUPLING

The gravitational coupling arises from aspherical density anomalies in both the mantle and the core. It was first suggested by *Jault and LeMouël* [1989], emphasizing on the contribution from the outer core/mantle density anomalies. More recently, *Buffett* [1996a,b] proposed strong gravitational coupling arising from misalignment of a non-axisymmetric inner core to a non-axisymmetric mantle.

From (13) we find that the gravitational torque $\widetilde{\boldsymbol{\Gamma}}_g$ is

$$
\widetilde{\boldsymbol{\Gamma}}_g = - \int_{\text{core}} d\boldsymbol{r} \, \boldsymbol{r} \times \rho \boldsymbol{g} \, .
\tag{37}
$$

The axial component of the coupling torque is then given by

$$
\widetilde{\Gamma}_g^z = - \int_{\text{core}} d\boldsymbol{r} \, r\rho g_\phi \sin\theta \, ,
\tag{38}
$$

where θ is the colatitude and g_ϕ is the longitudinal component of the gravitational acceleration. Clearly, only

Figure 1. The density anomalies $R_{th}\Theta$ in the fluid outer core. The horizontal axis is the time scaled by the magnetic free-decay time τ_d ($\tau_d \equiv \tau_\eta/\pi^2 \approx 20000$ years).

the (small) non-axisymmetric part of the density distributions contributes to the torque. If we denote by $\Delta\rho$ and Δg the non-axisymmetric density and the non-axisymmetric gravitational acceleration, i.e.

$$\rho \equiv \rho_0 (1 + \Delta\rho), \qquad g \equiv g_0 (-r + \Delta g), \quad (39)$$

(where ρ_0 is the mean density in the outer core and g_0 is the magnitude of the axisymmetric gravitational acceleration at the CMB), the coupling torque (38) can then be written as

$$\widetilde{\Gamma}_g^z = -\rho_0 g_0 \int_{\text{core}} dr\, r\Delta\rho\Delta g_\phi \sin\theta, \quad (40)$$

Two kinds of contributions to the torque (40) can be considered: one arising from the density anomalies $\Delta\rho$ in the outer core [*Jault and Lemouël*, 1989], and the other one arising from the relative motion between the aspherical mantle and the aspherical inner core [*Buffett*, 1996b].

Due to thermal processes in the Earth, the density distribution in the outer core is slightly different from the (axisymmetric) adiabatic density distribution. This departure provides the driving force for the geodynamo in the core. In our MOSST model, as well as in *Kuang and Bloxham* [1999] model, we have

$$\Delta\rho = \frac{2\eta\Omega R_{th}}{gr_0}\Theta$$
$$\approx 8\times10^{-12} R_{th}\Theta. \quad (41)$$

Figure 1 shows numerical rms density anomalies $R_{th}\Theta$ in (non-dimensional) time (scaled by the magnetic free decay time $\tau_d \equiv \tau_\eta/\pi^2 \approx 20000$ years for the Earth) for $R_o = E = 1.25 \times10^{-6}$ and $R_{th} = 15000$. From the figure we find that $R_{th}\Theta \approx 3000$, resulting in

$$\Delta\rho \approx 2\times10^{-8}. \quad (42)$$

The density anomalies under different choices of the parameters (e.g. $R_o = E = 2\times10^{-5}$) in our modeling are of the same order. This insensitivity can be well explained with the force balance argument: with small R_o and E, the numerical solutions are the strong-field dynamo solutions, which imply that to leading order the magnetostrophic force balance (11) is established in the core, i.e.

$$|R_{th}\Theta| \approx |v|. \quad (43)$$

Since the numerical velocity v in the core does not vary much with different R_o and E, the density anomalies are similar in our numerical parameter regime.

Regardless the unrealistic values of R_o and E in our modeling and the uncertainty of the Rayleigh number R_{th} in the Earth's core, the numerical estimation (42) is very close to the real Earth value, provided that the geodynamo is a strong-field dynamo. This can be seen as follows: with the magnetostrophic balance (11), we have

$$\Delta\rho \approx \frac{2\Omega\, v}{g_0}, \quad (44)$$

where v is the typical velocity in the core. Using the core flow $v \approx 5\times10^{-4}\,\text{ms}^{-1}$ inverted from geomagnetic studies [e.g. *Bloxham and Jackson*, 1991], we obtain

$$\Delta\rho \approx 10^{-8}, \quad (45)$$

comparable to (42).

With this estimation, we have

$$|\widetilde{\Gamma}_g| \leq \rho_0\, g_0 \int_{\text{outer core}} dr\, r \sin\theta|\Delta\rho\Delta g_\phi|$$
$$\approx 5\times 10^{22}\,|\Delta g|\,\text{Nm}. \quad (46)$$

For this coupling torque comparable to the torque (5) necessary for the decadal LOD variation, it is required that

$$\Delta g \approx 10^{-5}. \tag{47}$$

If this gravitational acceleration anomaly arises from mantle density anomalies, then by (39), $\Delta \rho_{\text{mantle}} \approx 10^{-5}$. Considering possible cancellation (40), Δg and hence $\Delta \rho_{\text{mantle}}$ are likely larger, but still not unrealistic for the mantle.

Buffett [1996a, b] suggested that another possible source of gravitational coupling arises from the interaction between the non-axisymmetric inner core and the non-axisymmetric mantle. We reformulate this idea as follows. Neglecting the density anomalies in the fluid outer core, the gravitational torque may then be approximated as

$$\widetilde{\mathbf{\Gamma}}_g = \int_{\text{CMB}} dS \, (\mathbf{r} \times \hat{\mathbf{n}}) \rho \Phi, \tag{48}$$

where Φ is the gravitational potential and ρ is now the mean density in the core. The axial component of the torque (48) is

$$\widetilde{\Gamma}_g^z = \int_{\text{CMB}} dS \, \rho \Phi r \sin \theta \, \hat{n}_\phi, \tag{49}$$

where $\hat{n}_\phi$ is the longitudinal component of the normal vector $\hat{\mathbf{n}}$ of the CMB.

In the (equilibrium) hydrostatic balance, both the ICB and the CMB are equal potential surfaces, and are non-axisymmetric, because of non-axisymmetric density distribution in the mantle, i.e. $\hat{n}_\phi$ in (49) is finite (and is a function of ϕ). By virtue of dynamical processes in the outer core, there is a strong electromagnetic torque (arising from the finite electrical conductivity of the inner core) on the ICB that drives the inner core moving relative to the mantle. Thus, both CMB and ICB are no longer in the equal potential positions. Therefore the potential Φ in (49) is not axisymmetric, which yields a finite gravitational coupling torque between the inner core and the mantle to restore the relative orientation between the inner core and the mantle to the equilibrium position.

Based on the currently estimated density anomalies in the mantle, *Buffett* [1996b] concluded that the gravitational restoring torque could be too strong for the decadal LOD variation. To resolve this difficulty, he assumes that the inner core is visco-elastic, i.e. the ICB can deform and restore to be equal potential when the inner core rotates relative to the mantle. Using numerical modeling, *Buffett and Glatzmaier* [2000] demonstrate that by selecting appropriate relaxing time scales and the viscosity of the inner core, the gravitational restoring torque could explain the decadal LOD variations.

Nevertheless, certain problems need to be addressed with respect to this coupling mechanism. First of all, there are large uncertainties in the estimation of mantle density anomalies. This could be the ultimate constraint on the coupling mechanism. Another concern is on the inner core viscosity and the relaxation time scales. They should be consistent with the differential rotation rate of the inner core [*Buffett and Creager*, 1999], which can be difficult because neither geodynamo modeling nor seismic observation could at present produce very consistent results. Also it is necessary to examine the combined gravitational torque arising from the both sources (46) and (49).

5. ELECTROMAGNETIC COUPLING

The electromagnetic coupling arises when the D''-layer is electrically conducting. This coupling mechanism was proposed more than a half century ago [e.g. *Bullard et al.*, 1950; *Munk and Revelle*, 1952]. The electrical current in the layer interacts with the magnetic field (in the form of the Lorentz force) and produce a finite coupling torque on the CMB.

By (13), the coupling torque is

$$\widetilde{\mathbf{\Gamma}}_b = \frac{1}{\mu} \int_{\text{CMB}} dS \, (\hat{\mathbf{n}} \cdot \mathbf{B}) \mathbf{r} \times \mathbf{B}. \tag{50}$$

Assuming a spherical CMB, and applying similar scaling (30), one gets

$$\widetilde{\mathbf{\Gamma}}_b = \Gamma_0 \int_{r=1} dS \, B_r (\mathbf{r} \times \mathbf{B}) \equiv \Gamma_0 \mathbf{\Gamma}_b, \tag{51}$$

where $\mathbf{B}$ is the (scaled) non-dimensional magnetic field. The axial component of the torque that contributes to the LOD variation of the Earth is

$$\Gamma_b^z = \hat{\mathbf{z}} \cdot \mathbf{\Gamma}_b = \int_{r=1} dS \, r \sin \theta \, B_r B_\phi, \tag{52}$$

where B_ϕ is the longitudinal component of the magnetic field.

It has long been speculated in the past decades that the electromagnetic coupling torque (52) is sufficient for the core-mantle angular momentum exchange on decadal time scales. However, most of the previous studies are kinematic: the coupling torque (52) is evaluated from the geomagnetic secular variation observed at the surface of the Earth. The technique for the torque evaluation was first presented systematically by *Stix and Roberts* [1984]. It has since become more or less a standard procedure in the subsequent studies [e.g. *Gubbins and Roberts*, 1987; *Love and Bloxham*, 1994; *Holme*, 1998a]. There are several important assumptions made in the procedure and are summarized as follows for a spherically symmetric D''-layer with a constant electrical conductivity σ_m.

The first assumption is that the poloidal field is slowly varying in the D''-layer on the decadal time scales (low-frequency assumption). Following *Roberts* [1972], this approximation can be expressed as

$$b_l^m = b_l^{m(0)} + b_l^{m(1)} + \cdots, \qquad (53)$$

$$\partial_t b_l^{m(0)} = -\eta_{mc}(\partial_r^2 + \hat{L}/r^2) b_l^{m(1)}, \qquad (54)$$

where $b_l^{m(0)}$ are the coefficients for the potential field (that can be derived from observations), $\eta_{mc} \equiv \eta_m/\eta$ (η_m is the magnetic diffusivity of the D''-layer), $\hat{L}$ is the angular momentum operator [*Kuang and Bloxham*, 1999]. With this assumption, one could infer from surface observations the time-varying radial component B_r of the magnetic field at the CMB.

The second assumption is the well-known "frozen-flux" approximation in the fluid outer core,

$$\partial_t \boldsymbol{B} \approx \nabla \times (\boldsymbol{v} \times \boldsymbol{B}), \qquad (55)$$

i.e. on the decadal time scales, the dissipation of the magnetic field is negligible compared to the advection of the field by the core flow (so that the field lines are "frozen" into the flow). The horizontal core flow $\boldsymbol{v}_h$, i.e. the flow tangential to the CMB can then be obtained by taking the radial component of (55). However, the velocity field can not be uniquely determined by (55) [*Backus*, 1968], unless an additional constraint is introduced to remove the non-uniqueness, such as that the flow is steady or is tangentially geostrophic [e.g. *Bloxham and Jackson*, 1991].

The low-frequency approximation is also applied to obtain the toroidal field $\{j_l^m\}$ in the D''-layer,

$$\left(\partial_r^2 + \hat{L}/r^2\right) j_l^m \approx 0. \qquad (56)$$

This equation is solved subject to the boundary conditions at the top of the D''-layer and at the CMB. The toroidal field vanishes at the top of the layer due to the insulating mantle,

$$j_l^m = 0.$$

However, the boundary conditions at the CMB depends on the toroidal field inside the core. Because of the continuity of the electrical field $\mathbf{E}$ across the CMB, we have [*Kuang and Bloxham*, 1999]

$$\left[\eta_{mc}\frac{\partial j_l^m}{\partial r}\right]_{D''-\text{layer}} = \left[\frac{1}{l(l+1)}(\hat{\mathbf{r}} \cdot \nabla \times B_r \boldsymbol{v})_l^m\right]_{\text{core}}$$
$$+ \left[\frac{\partial j_l^m}{\partial r}\right]_{\text{core}} \equiv j_l^{m(a)} + j_l^{m(d)}, \qquad (57)$$

where $j_l^{m(a)}$ represents the contribution from advection of the poloidal field by the core flow and is called the advective term, $j_l^{m(d)}$ represents the effect of the core toroidal field diffusing into the D''-layer and is called the leakage term. This leakage term has been neglected in all kinematic studies due to the lack of knowledge of the core toroidal field. With these approximations, the toroidal field coefficients j_l^m in the D''-layer can be solved with the equation (56) and the two boundary conditions.

By the above procedure, both B_r and B_ϕ at the CMB can be evaluated, the magnetic coupling torque (52) therefore follows.

Love and Bloxham [1994] studied extensively the magnetic torque generated by different core flow models with a spherically symmetric, weakly conducting D''-layer. They demonstrated that the advective term is dominant in providing the coupling torque. However, while the magnitude of the torque could be sufficient, their results seem not to favor the electromagnetic coupling [*Bloxham*, 1998]: the coupling torque is difficult to account for the temporal variability of LOD.

Holme [1998a,b] revisited the electromagnetic coupling through a careful examination of the core flow models. He demonstrated that, while the steady part of the core flow could explain most of the observed geomagnetic secular variation, the time-varying part of the core flow could result a coupling torque capable of explaining the observed decadal LOD variation. The coupling torque is stronger with a better electrically conducting and thinner D''-layer (the total conductance $D = \sigma_m d_m$ of the D''-layer is unchanged). He further considered the effect of laterally varying conductivity of the D''-layer on the electromagnetic coupling. He argued that large-scale lateral variation could affect the long term trends of the torque; it does not on average enhance the magnitude and the decadal variation of the torque [*Holme*, 2000].

However, the toroidal field and the core flow derived in these studies are not necessarily consistent with the core geodynamo processes. For example, the Lorentz force is stronger near the CMB with a better electrically conducting D''-layer. Therefore the tangential geostrophic core flow beneath the CMB may not be a good approximation with a thin, highly conducting D''-layer. Also, using the low-frequency approximation indiscriminately regardless the properties of the D''-layer could potentially alter the solutions of the kinematic studies.

Most importantly, the kinematic studies can not address the "leakage" of the core toroidal field in the D''-layer. To assess this part of the electromagnetic core-mantle coupling is by no means straightforward and in principle impossible from surface observations.

Therefore we use our MOSST dynamo model to study the electromagnetic coupling. In particular, we focus on the dependence of the torque upon the properties of a uniform D''-layer and the importance of the leakage toroidal field. For our study, we implement in our model two different profiles of the D''-layer specified in (25). The total conductance of the two profiles are the same, where a thicker layer has a lower electrical conductivity. All simulations start from the same given initial state. The solutions stabilized after a short transient period. Then we examine the coupling torque with the well developed numerical solutions.

The resultant magnetic torques are shown in Figure 2. Our results demonstrate clearly that the coupling torques with the two D''-layer profiles (25) are very different. While the torque is weaker and of DC type with the thick layer (the dashed line in Figure 2), the coupling torque with the thin layer (the solid line in Figure 2) is stronger and also varies significantly with time (AC type).

To examine the importance of the core toroidal field leaking to the D''-layer, we plot in Figure 3 the following L_2-norms

$$T_{br}^{(t)} \equiv \eta_{mc} \left[\sum_{\substack{0 \le m \le l}}^{l \le L} \left| \frac{\partial j_l^m}{\partial r} \right|_{D''-\text{layer}}^2 \right]^{1/2}, \quad (58)$$

$$T_{br}^{(a)} \equiv \eta_{mc} \left[\sum_{\substack{0 \le m \le l}}^{l \le L} \left| j_l^{m(a)} \right|^2 \right]^{1/2}, \quad (59)$$

$$T_{br}^{(d)} \equiv \eta_{mc} \left[\sum_{\substack{0 \le m \le l}}^{l \le L} \left| j_l^{m(d)} \right|^2 \right]^{1/2}, \quad (60)$$

for the thin layer profile. Clearly, when the electrical conductivity of the D''-layer is closer to that of the core, the contribution of the leakage term $T_{br}^{(d)}$ (the dotted line in the figure) is comparable to that of the advective term $T_{br}^{(a)}$ (the dashed line in the figure). Furthermore, they are approximately in phase most of the time. This implies that the toroidal field in the D''-layer and

hence the electromagnetic torque could be significantly underestimated in the kinematic studies.

From our numerical results we find that the toroidal field becomes stronger with the thin D''-layer profile. Therefore, the increase in the magnetic torque results from the increase of the toroidal field strength. Furthermore, the importance of the "leakage" toroidal field with the thin D''-layer (see Figure 3) suggests that the "frozen-flux" approximation (55) does not apply to the toroidal field. On the other hand, with our numerical parameters (21), (55) is a good approximation for the poloidal field.

Since the frequencies of the torsional oscillations depend on the magnetic Rossby number R_o, as shown in (12), the frequencies of the torsional oscillations in our numerical solutions are approximately one and a half orders of magnitude smaller than those in the Earth's core (*Braginsky* [1976] showed that the frequencies are proportional to $R_o^{1/2}$). Therefore we are limited to examine the low-frequency approximations (53) and (56). However, considering that the toroidal field varies faster with the thin D''-layer, we would not be surprised to find that (56) is not a good approximation.

It is interesting to examine the "skin-depth" in our numerical modeling. From Figure 2 we find that the frequency ω_t of the torque is approximately

$$\omega_t \approx \frac{2\pi}{0.0025\tau_\eta}. \quad (61)$$

Therefore, the non-dimensional skin-depth d_s is approximately

$$d_s = \left(\frac{2\eta_m}{\omega_t r_0^2} \right)^{1/2} = \left(\frac{40}{\omega_t \tau_\eta} \right)^{1/2} \approx 0.1, \quad (62)$$

which implies that $r_0 d_s \approx 300$km, thicker than the D''-layer in our modeling. It should be pointed out that even if the frequency ω_t decreases by two orders of mag-

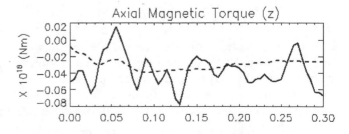

Figure 2. The electromagnetic coupling torques $\widetilde{\Gamma}_b^z \equiv \Gamma_0 \Gamma_b^z$ for the thick (weakly conducting) D''-layer (dashed line) and for the thin (highly conducting) D''-layer (the solid line). The horizontal axis is the time scaled by τ_d.

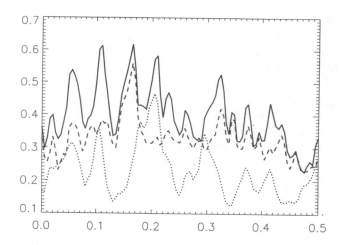

Figure 3. The non-dimensional L_2-norms $T_{br}^{(t)}$ of the total radial derivative of the toroidal scalar at the bottom of the D''-layer (the solid line), $T_{br}^{(a)}$ of the advective term (the dashed line) and $T_{br}^{(d)}$ of the leakage term (the dotted line) at the top of the outer core for the thin-layer profile. The definitions of the norms are given by (58-60). The horizontal axis is the time is scaled by τ_d.

nitude, $r_0 d_s$ is still larger than the D''-layer thickness, i.e. the time-varying field should be observable at the surface of the Earth.

6. TOPOGRAPHIC COUPLING

The topographic coupling arises from non-hydrostatic pressure p acting on the CMB topography H (26). This coupling mechanism was first suggested by *Hide* [1969], analogous to the atmospheric winds acting on mountainous surface of the Earth.

By (13) and (26), we may obtain the coupling torque

$$\widetilde{\mathbf{\Gamma}}_p \;=\; \varepsilon\, r_0^2 \int_{\mathrm{CMB}} dS\, (\hat{\mathbf{r}} \times \nabla H)\, p. \qquad (63)$$

Again, using the scaling (30), we have

$$\widetilde{\mathbf{\Gamma}}_p \;=\; \Gamma_0\, \varepsilon \int_{r=1} dS\, (\hat{\mathbf{r}} \times \nabla H)\, p \;\equiv\; \Gamma_0\, \mathbf{\Gamma}_p, \qquad (64)$$

where p is the (scaled) non-dimensional pressure at the CMB. In deriving (64) we have assumed that the surface integration in (63) is approximated over the mean CMB. From (64), the axial component Γ_p^z of the coupling torque that contributes to the LOD variation of the Earth is

$$\Gamma_p^z \;=\; \varepsilon \int_{r=1} dS\, \frac{\partial H}{\partial \phi}\, p \;=\; -\varepsilon \int_{r=1} dS\, H\, \frac{\partial p}{\partial \phi}. \quad (65)$$

Hide [1986] proposed not only the coupling mechanism, but also an approach to evaluate the torque (65) from observed geomagnetic field at Earth's surface. This procedure is in principle kinematic and can be summarized as follows: using the "frozen-flux" approximation (55) and the "tangential geostrophy" approximation

$$2\rho\, (\mathbf{\Omega} \times \mathbf{v})_h \;\approx\; -\nabla_h p, \qquad (66)$$

the flow $\mathbf{v}$ and the non-hydrostatic pressure p at the CMB can be evaluated from the time varying poloidal field at the CMB inverted from the surface-observed geomagnetic secular variation. Then the topographic torque (65) can be evaluated by specifying CMB topography H. With this approach, the coupling torque Γ_p^z is proportional to the boundary topography amplitude:

$$\Gamma_p^z \;\approx\; \mathcal{O}(\varepsilon\, H\, v_0), \qquad (67)$$

where v_0 is the typical velocity beneath the CMB. Such coupling torque can be more than adequate to account for the observed decadal LOD variation: with an 1 km CMB topography amplitude, the coupling torque can be much larger than the required magnitude (5) within the geomagnetic data accuracy [*Jault and Lemouël*, 1989; *Hide et al.*, 1993].

While this analysis does provide a feasible way of evaluating the topographic torque, it does not lend itself to understanding the dynamics of the topographic coupling, i.e. how the core flow and the magnetic field adjusting to an CMB topography and thus causing the angular momentum variation in the core. In addition, there are potential problems in this approach. In addition to possible dynamical inconsistencies in the approximations (55) and (66), the evaluation of the core flow $\mathbf{v}$ and the pressure p may not be consistent with the CMB topography H used in the torque evaluation.

An alternative, quasi-dynamic approach has been developed through perturbations of boundary topography to core flow [e.g. *Anufriev and Braginsky*, 1975, 1977; *Kuang and Bloxham*, 1993]. This approach takes into partial consideration of the interactions between the core flow and the magnetic field, and their reaction to small boundary topographies. It can be summarized as follows. Given an ambient magnetohydrodynamic (MHD) state $(\mathbf{v}_a, \mathbf{B}_a)$, the perturbations due to the boundary topography (26) can be expressed in following asymptotic expansions,

$$\begin{bmatrix} \mathbf{v} \\ \mathbf{B} \end{bmatrix} = \begin{bmatrix} \mathbf{v}_a \\ \mathbf{B}_a \end{bmatrix} + \varepsilon \begin{bmatrix} \mathbf{v}_1 \\ \mathbf{B}_1 \end{bmatrix} + \mathcal{O}(\varepsilon^2). \quad (68)$$

The first order perturbation $(\boldsymbol{v}_1, \boldsymbol{B}_1)$ can then be solved by linearized equations of (7) and (14) with appropriate boundary conditions. The perturbed pressure p_1 at the boundary is obtained with the magnetostrophic approximation

$$\hat{\boldsymbol{z}} \times \boldsymbol{v}_1 = -\nabla p_1 + \boldsymbol{J}_a \times \boldsymbol{B}_1 + \boldsymbol{J}_1 \times \boldsymbol{B}_a, \qquad (69)$$

where $\boldsymbol{J}_a$ and $\boldsymbol{J}_1$ are the ambient and the perturbed electrical current densities, respectively. The coupling torque (65) can then be evaluated. In this approach, the pressure p_1 is proportional to ε, as shown in the expansion (68). Therefore,

$$\Gamma_p^z \approx \mathcal{O}(\varepsilon^2 H^2 v_0). \qquad (70)$$

This result is obviously much smaller than the kinematic result (67). Based on an planar layer system, *Kuang and Bloxham* [1993] suggested that the torque Γ_p^z could contribute significantly to the total coupling torque (5) only if the CMB topography amplitude is about 3 km or larger.

Although some dynamics are included in this approach, it is still not fully dynamically consistent: the ambient states $(\boldsymbol{v}_a, \boldsymbol{B}_a)$ are not determined dynamically. In fact, they do not even represent a dynamo solution. This leaves questions as to a full discrimination of the two results (67) and (70). For example, based on the result (70) and the fact that a geostrophic flow does not contribute to topographic coupling, *Kuang and Bloxham* [1997a] suggested that the result (67) is very unlikely and that the topographic coupling may not be the dominant coupling mechanism for the decadal LOD variation. But this is questioned by *Jault and Lemouël* [1999].

To conciliate the differences, one may have to understand whether there is a large scale cancellation occurring in the surface integration (65). The result (70) may indicate that the local effect of the pressure p acting on the CMB topography H cancels almost each other, resulting a much smaller net torque on the CMB. But we do not observe such cancellation in the result (67). One possible explanation is that the pressure p is not evaluated consistently with the CMB topography. It is reasonable to argue that the observed geomagnetic secular variation does include the effect of some CMB topography H_1. However, it is not at all sure that this H_1 is the same as the H used in the torque integration (65). The differences between the two topographies may be critical for the magnitude of the coupling torque. To resolve this scenario, we must know the flow that are

consistent with both the dynamo processes in the core and the boundary conditions at the CMB.

Thus, again, we resort to our MOSST dynamo model, so that the magnetic field and the velocity field (and hence the pressure field) in the core can be determined dynamically with a given CMB topography.

Kuang and Chao [2001] first used the MOSST model to examine the topographic coupling, where a CMB topography based on seismic tomography *Morelli and Dziewonski* [1987] was implemented with the amplitude varying from 1 km to 4 km, all within the resolution of the model. The CMB topography in their study is approximated by the normal vector (in non-dimensional expression)

$$\hat{\mathbf{n}} = \hat{\mathbf{r}} - \varepsilon \nabla H. \qquad (71)$$

The boundary conditions are approximated up to order $\mathcal{O}(\varepsilon)$. Their numerical results demonstrate clearly that the non-dimensional coupling torque Γ_p^z is

$$\Gamma_p^z = \alpha (\varepsilon H)^2 v_0, \qquad (72)$$

where α is an order one coefficient and does not vary significantly with the Rossby number R_o and the Ekman number E used in the modeling. This result corroborates the result (70) of the quasi-dynamic studies. Their numerical results also demonstrated that if the coupling torque Γ_p^z is scaled linearly with ε

$$\Gamma_p^z = \alpha_t \varepsilon H v_0, \qquad (73)$$

then the scaling coefficient α_t is in average very small [*Kuang and Chao*, 2001],

$$\alpha_t \leq 10^{-3} \qquad (74)$$

and varies significantly with ε. The above results (72) and (70) imply that the topographic coupling arises from part of the flow adjusting to small CMB topography. In other words, there is indeed a large-scale cancellation within the integration (65).

If the CMB topography introduced in (65) is different from that in evaluating the dynamo state, we would expect such cancellation does not exist. Therefore, the coupling torque Γ_p^z can be sufficiently large to be linearly proportional to the CMB topography amplitude ε, as is shown in the kinematic studies (67).

To verify this, we performed the following numerical test. We retain the same non-dimensional parameters (e.g. the Rayleigh number R_{th}, the Rossby number R_o, the Ekman number E and the Prandtl number q_κ) as in the *Kuang and Chao* [2001], and assume a spherical CMB topography in integrating numerically the core flow $\boldsymbol{v}$ and the magnetic field $\boldsymbol{B}$. The CMB topogra-

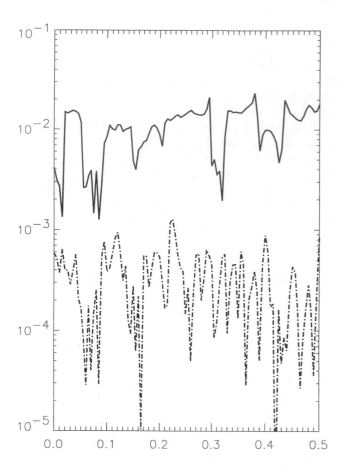

Figure 4. The scaling coefficients $\alpha_t^{(1)}$ of the testing topographic torque $\Gamma_p^{z(1)}$ (75) (the solid line) and α_t of the dynamically consistent topographic torque Γ_p^z (73) (the dashed line). The horizontal axis is again the time scaled by τ_d.

phy H is introduced only in the torque calculation (65). With this procedure, we obtain a torque

$$\Gamma_p^{z(1)} = \alpha_t^{(1)} \varepsilon H v_0 \qquad (75)$$

that mimics that of the kinematic studies. Then we compare the scaling coefficients $\alpha_t^{(1)}$ (75) of the testing torque with α_t (73) of the dynamically consistent torque. They are shown in Figure 4. We can now observe clearly that on average $\alpha_t^{(1)}$ (the solid line in the figure) is more than one and a half orders of magnitude larger than α_t (the dashed line). We also observe from the figure that α_t varies much faster than $\alpha_t^{(1)}$. This numerical test indeed shows the importance of the core flow and the magnetic field, and hence the pressure in the core, to be dynamically consistent with the CMB topography in estimating the coupling torque.

We should point out here that the large scale CMB topography H used in *Kuang and Chao* [2001] is expressed with an $l \leq 4$ spherical harmonic expansion [*Morelli and Dziewonski,* 1987]. Thus only the components of the pressure p with $l \leq 4$ will contribute to the torque (65). On the other hand, the hyper-dissipation is introduced for $l \geq 10$. Therefore, its effect on the torque calculation is minimal. However, if a more complicated, small-scale CMB topography H is used for future studies, then the numerical truncation order L must increase and the hyper-dissipation must be further reduced to eliminate possible numerical artifact. But we anticipate that the result (72) should hold for any small amplitude CMB topography with the length scales larger than those of the boundary layers (e.g. Ekman layer) at the CMB.

7. CONCLUSION

The problem of core-mantle interaction can generally be approached in two ways: kinematic and dynamic. The kinematic approach emphasizes on deriving the coupling torques at the CMB using surface geomagnetic observations. The dynamic approach seeks to understand the interaction of the flow and the magnetic field in the core with specified D''-layer above the CMB. Using the dynamic approach, we discussed the four possible core-mantle coupling mechanisms, while focusing on the electromagnetic and topographic couplings. We also presented some new results on the latter two coupling torques from the dynamo solutions obtained from our MOSST dynamo model .

In the kinematic electromagnetic coupling studies, the non-potential part of the magnetic field in the electrically conducting D''-layer above the CMB is derived using observed geomagnetic secular variation at the surface of the Earth, with a number of assumptions on the properties of the field and the flow in the core [e.g. *Roberts,* 1972]. The resultant field in the D''-layer can then be used to evaluate the magnetic torque Γ_b^z (52). The central purpose of the assumptions in the kinematic studies is to eliminate the effect of unobservable core field from the analysis. While some of the assumptions, such as the "low-frequency" assumption (53)-(54) and the "frozen-flux" assumption (55) are based on envelop analysis. Others, such as ignoring the toroidal field diffusing to the D''-layer(the leakage toroidal field), are simply introduced to eliminate unobservable. These assumptions can potentially introduce dynamic inconsistencies in evaluating the coupling torques. Though some of the assumptions have been questioned on that account [e.g. *Love and Bloxham,* 1994], they are still used in subsequent studies [e.g. *Holme,* 1998a]. At any rate, ignoring the leakage toroidal field in the D''-layer may ultimately result in an erroneous coupling torque.

Results from our MOSST model demonstrate that within the numerically small Rossby number R_o and the Ekman number E, the coupling torque is stronger and more time-variable with a highly electrically conducting thin D''-layer than that with a poorly electrically conducting but thicker D''-layer, as shown in Figure 2. We also used our solutions to examine the significance of the core toroidal field leaking into the D''-layer. Our findings, shown in Figure 3, indicate that with a thin, highly conducting D''-layer, the leakage toroidal field can be comparable to the advective toroidal field in the D''-layer. This suggests that the toroidal field in the D''-layer and hence the coupling torque Γ_b^z on the CMB can be much stronger if the leakage field is included in the kinematic studies. Our future work shall include examining the "low-frequency" and the "frozen-flux" assumptions with our model, and studying the electromagnetic coupling with a heterogeneous D''-layer.

We should point out here that the strength of the numerical electromagnetic torque Γ_b^z on the CMB, shown in Figure 2, is about $\mathcal{O}(6{\times}10^{16}\,\mathrm{Nm})$, almost an order of magnitude smaller than the required value (5) to explain the observed decadal LOD variation. Also the time scales of the torque variation are very long, about 5% of the magnetic free-decay time τ_d, equivalent to about 1000 years. Two considerations should be noted to interpret these results. First, the dimensional numerical magnetic field at the CMB in our modeling is only about 40% in magnitude of the field inverted from observations. Thus, the numerical torque $\Gamma_b^z \approx 4{\times}10^{17}\mathrm{NM}$ if the numerical field strength is comparable to that of observations. In this case, it is close to the required torque (5). Second, the Rossby numer $R_o = 1.25{\times}10^{-6}$ in our numerical modeling, which is about three orders of magnitude larger than the value appropriate for the Earth's core. Assuming that the asymptotic scaling rules of *Braginsky* [1967] hold for our numerical parameters, the frequencies of the numerical torsional oscillations are approximately one and a half orders of magnitude smaller than those of the torsional oscillations in the Earth's core. Therefore, the magnetic torque could vary on the time scales of 30 years and thus consistent with the decadal LOD variations.

Kinematic studies of the topographic core-mantle interaction depend also on geomagnetic observations. In these studies, the core pressure is obtained from tangentially geostrophic flow beneath the CMB that is inverted from observed geomagnetic secular variation. Then the coupling torque Γ_p^z can be evaluated via (65), provided that the CMB topography εH is specified [*Hide,* 1986]. Such studies obtain a coupling torque (67) proportional to the topography amplitude ε. Alternatively, in quasi-dynamic studies, the core flow reacting to an imposed CMB topography is analyzed [e.g. *Anufriev and Braginsky,* 1975; *Kuang and Bloxham,* 1993]. Such studies yield a coupling torque (70) proportional to ε^2, significantly smaller than that (67) of the kinematic studies. However, both approaches are prone to the problems of dynamical inconsistencies with core geodynamo processes.

In a fully dynamical approach, *Kuang and Chao* [2001] reported the results from their MOSST dynamo model on topographic core-mantle coupling. In their studies, the flow, the magnetic field and therefore the pressure in the core are solved dynamically, with an imposed CMB topography based on seismic tomography [*Morelli and Dziewonski,* 1987]. Their results support quadratic dependence of the coupling torque on boundary topography (72), corroborating the quasi-dynamic results. This implies that only the part of flow (of order ε) adjusting to the CMB topography contributes to the coupling. In other words, the local pressure torque on the CMB topography nearly cancels each other, leaving only a very small net torque over the CMB. Our studies in this paper demonstrate that this large-scale cancellation will not happen (and therefore result in a much larger net torque) if one had inconsistently used different CMB topographies in deriving the pressure and in the torque evaluation, as shown in Figure 4. This is very likely the cause of the over-estimated torque (67) from kinematic studies.

From numerical results, we may estimate the magnitude of the topographic coupling torque: it can contribute significantly, but not dominantly, to the decadal core-mantle angular momentum exchange if the topography amplitude is about 3 km or larger.

Recently, *Garcia and Souriau* [2000] reported that the CMB topography amplitude is unlikely larger than 2 km. With this estimate, the topographic coupling torque could at most contribute a small fraction to the total torque (5) required for the decadal LOD variation. However, our estimation may differ if the length scale of the CMB topography is very small, comparable to the core boundary layer thickness at the CMB (e.g. ≤ 1 km).

Buffett and Glatzmaier [2000] studied the gravitational coupling between the inner core and the mantle using the Glatzmaier-Roberts dynamo model. Their results suggest that the torque can be sufficient for the observed decadal LOD variation with a heterogeneity equivalent to an order $100\,\mathrm{m}$ non-axisymmetric CMB. With this value, the topographic coupling becomes certainly negligible.

The above two studies, however, may not accommodate each other. The gravitational coupling between the inner core and the mantle depends critically on the viscous relaxation of the inner core and on the density heterogeneity of the mantle. Assuming that the CMB topography represents an equal potential surface, then the larger the CMB topography, the greater the coupling torque. Considering the constraints imposed by the inner core differential rotation and by the decadal LOD variation, different inner core models must be introduced, thus new uncertainties may surface in the studies of core-mantle interactions. A real resolution may await improved estimates, seismological or otherwise, of the amplitude of the CMB topography.

Recent studies and ours [e.g. *Holme*, 1998a; *Buffett and Glatzmaier*, 2000; *Kuang and Chao*, 2001] suggest that except the viscous core-mantle coupling, other three core mantle coupling mechanisms (i.e. gravitational, electromagnetic and topographic) may all contribute significantly to the exchange of the axial angular momentum between the core and the mantle and hence to the decadal LOD variation. Their relative importance depend critically on many properties and various aspects of the dynamic behavior of the core and its boundaries. Our study demonstrates that geodynamo modeling is a powerful tool towards gaining insights in understanding the entire time-variable dynamic processes, and that, although past studies mostly focused on one particular coupling mechanism while ignoring others, incorporating all coupling mechanisms in geodynamo modeling is necessary.

APPENDIX A: ANGULAR MOMENTUM VARIATION OF THE CORE

To obtain the angular momentum variation of the outer core, we integrate $r \times (6)$ over the fluid outer core

$$\int_{\mathrm{OC}} dr\, \rho r \times \left(\frac{\partial v}{\partial t} + 2\Omega \times v \right) = \int_{\mathrm{OC}} dr\, r \times [-\nabla p$$
$$+ J \times B + \rho g + \rho \nu \nabla^2 v], \quad (A1)$$

where the subscript OC means the outer core. The advection $(v \cdot \nabla)v$ in (6) transfers angular momentum within the fluid core. It does not contribute to the total angular momentum variation of the outer core and vanishes in (A1). The terms on the right hand side of (A1) describe the torques acting on the fluid outer core.

The first one is the pressure torque

$$\widetilde{\Gamma}_p \equiv -\int_{\mathrm{outer\ core}} dr\, r \times \nabla p$$
$$= -\int_{\mathrm{CMB}} dS (r \times \hat{n}) p + \int_{\mathrm{ICB}} dS (r \times \hat{n}) p$$
$$\equiv \widetilde{\Gamma}_p^{\mathrm{C}} - \widetilde{\Gamma}_p^{\mathrm{I}}, \quad (A2)$$

where $\hat{n}$ denotes the normal of the boundaries. The Gaussian theorem is used to obtain (A2). Similar expressions can be obtained for the magnetic torque $\widetilde{\Gamma}_B$ and the viscous torque $\widetilde{\Gamma}_\nu$, with

$$\widetilde{\Gamma}_B^{\mathrm{C,I}} \equiv \frac{1}{\mu} \int_{\mathrm{CMB,ICB}} dS\, (\hat{n} \cdot B)(r \times B) \quad (A3)$$

$$\widetilde{\Gamma}_\nu^{\mathrm{C,I}} \equiv \int_{\mathrm{CMB,ICB}} dS\, \rho \nu\, [r \times (\hat{n} \cdot \nabla)v - \hat{n} \times v] \quad (A4)$$

In deriving (A4) we assume the mean density ρ of the outer core. The gravitational torque

$$\widetilde{\Gamma}_g \equiv \int_{\mathrm{outer\ core}} dr\, \rho r \times g \quad (A5)$$

can not be simplified to surface integral.

If we denote by M_{ic} the angular momentum of the inner core, we may obtain the following equation for the angular momentum variation of the inner core

$$\frac{dM_{ic}}{dt} + \Omega \times M_{ic} = \widetilde{\Gamma}_p^{\mathrm{I}} + \widetilde{\Gamma}_B^{\mathrm{I}} + \widetilde{\Gamma}_\nu^{\mathrm{I}}. \quad (A6)$$

Denote by M_{oc} and M_m the angular momentum of the outer core and the mantle, respectively. If we don not consider external couplings on the core-mantle system, the total angular momentum $M \equiv M_{ic} + M_{oc} + M_m$ must be conserved

$$\frac{dM}{dt} + \Omega \times M = 0, \quad (A7)$$

By (A1)-(A6), we obtain

$$\frac{dM_m}{dt} + \Omega \times M_m = -\widetilde{\Gamma}_p^{\mathrm{C}} - \widetilde{\Gamma}_B^{\mathrm{C}}$$
$$- \widetilde{\Gamma}_\nu^{\mathrm{C}} - \int_{\mathrm{core}} dr\, \rho r \times g, \quad (A8)$$

where the subscript "core" means the outer core and the inner core.

The gravitational torque in (A8) can be simplified if we could attribute all density anomalies in the core to boundary topographies. In this situation, the density could be replaced by its mean value and

$$\widetilde{\Gamma}_g \equiv -\int_{\mathrm{core}} dr\, \rho r \times g = \int_{\mathrm{CMB}} dS\, \rho(r \times \hat{n})\Phi, \quad (A9)$$

where Φ is the gravitatial potential field ($\nabla\Phi \equiv -g$).

APPENDIX B: PARTIAL SLIPPERY BOUNDARY CONDITIONS

By (29), the (non-dimensional) viscous torque $\mathbf{\Gamma}_n u$ on a boundary is

$$\mathbf{\Gamma}_\nu = E \int dS \left[\mathbf{r} \times (\hat{\mathbf{n}} \cdot \nabla) \mathbf{v} - \hat{\mathbf{n}} \times \mathbf{v} \right], \qquad (B1)$$

where $\hat{\mathbf{n}}$ is normal unit vector of the boundary. Assuming no-slip boundary conditions, boundary layer analysis [e.g. *Greenspan, 1968*] shows that the normal derivatives of the horizontal flow are of the order

$$\left| \mathbf{r} \times (\hat{\mathbf{n}} \cdot \nabla) \mathbf{v} \right| \approx E^{-1/2} \left| \mathbf{r} \times \mathbf{v} \right|. \qquad (B2)$$

Therefore, the torque is of order $\mathcal{O}(\sqrt{E})$. To reduce the viscous torque in our modeling to be comparable to that of the Earth's core, one approach is to introduce an asymptotic approximation to the normal derivatives of the flow near the boundaries.

To examine how boundary conditions affect fluid flow, we consider first a fluid layer rotating rapidly about an axis

$$\mathbf{1}_\Omega = \hat{\mathbf{z}} \cos \alpha + \mathbf{1}_\mathbf{x} \sin \alpha, \qquad (B3)$$

with an impenetrable boundary $z = 0$ and a mainstream flow $\mathbf{v}$ far away from the boundary:

$$\mathbf{v} = u \mathbf{1}_\mathbf{x} \qquad \text{as} \qquad z \gg 1, \qquad (B4)$$
$$\hat{\mathbf{z}} \cdot \mathbf{v} = 0, \qquad \text{at} \qquad z = 0, \qquad (B5)$$

where u is a time-varying, large scale flow

$$u = u_0 e^{i(\omega t + ky)} + c.c., \qquad (B6)$$

with $\omega, k = \mathcal{O}(1)$. The flow inside the system is determined to leading order by the equation

$$\mathbf{1}_\Omega \times \mathbf{v} = -\nabla p + E \nabla^2 \mathbf{v}, \qquad (B7)$$

where p is the modified pressure. By the matching condition (B4) we can write

$$\mathbf{v} = \mathbf{u}(z) e^{i(\omega t + ky)} + c.c.. \qquad (B8)$$

Taking the x and y components of $\nabla \times$(B7), we may obtain

$$f^2 \frac{du_x}{dz} = -E^2 \left(\frac{d^4}{dz^4} - 2k^2 \frac{d^2}{dz^2} + k^4 \right) \frac{du_x}{dz}, (B9)$$

$$f \frac{du_y}{dz} = -E \left(\frac{d^2}{dz^2} - k^2 \right) u_x, \qquad (B10)$$

where $f \equiv \cos \alpha$. The solution of the equations is

$$u_x = u_0 + e^{-\sqrt{\frac{f_+}{2E}} z} \left[c_1 \cos \sqrt{\frac{f_-}{2E}} z \right.$$
$$\left. + c_2 \sin \sqrt{\frac{f_-}{2E}} z \right], \qquad (B11)$$

$$u_y = \text{sgn}(f) e^{-\sqrt{\frac{f_+}{2E}} z} \left[c_2 \cos \sqrt{\frac{f_-}{2E}} z \right.$$
$$\left. - c_1 \sin \sqrt{\frac{f_-}{2E}} z \right], \qquad (B12)$$

where

$$f_\pm \equiv |f| \pm k^2 E, \qquad (B13)$$

and $c_{1,2}$ are the constants to be decided by the boundary conditions at $z = 0$. The solutions (B11) and (B12) show that a boundary layer with a thickness of $\mathcal{O}(E^{1/2})$ appears if c_1 and c_2 do not vanish simultaneously.

If the stress-free boundary conditions are used, i.e.

$$\frac{du_x}{dz} = \frac{du_y}{dz} = 0 \qquad \text{at} \qquad z = 0,$$

the solutions are trivial:

$$c_1 = c_2 = 0. \qquad (B14)$$

In this case, there is no boundary layer at $z = 0$. With the no-slip boundary conditions

$$u_x = u_y = 0 \qquad \text{at} \qquad z = 0,$$

we have

$$c_1 = -u_0, \qquad c_2 = 0. \qquad (B15)$$

Thus we observe the typical Ekman layer structure in the solutions: there is a weak vertical flow u_z of order $\mathcal{O}(E^{1/2})$ in the fluid layer, often called the Ekman pumping [*Greenspan, 1968*]. But we do not provide the details of the flow because it does not serve our discussions here. By (B11), (B12) and (B15), we can easily verify the Ekman spiral $\phi_E = \pi/4$ across the boundary layer:

$$\phi_E = \tan^{-1} \left(\frac{u_y}{u_x} \right) = \tan^{-1}[\text{sgn}(f)] \quad \text{at} \quad z = 0. \qquad (B16)$$

The viscous stress $\boldsymbol{\tau}_\nu$ at the boundary $z = 0$ is

$$\hat{\mathbf{n}} \cdot \boldsymbol{\tau}_\nu \sim u_0 \sqrt{\frac{E|f|}{2}} [1, \text{sgn}(f)]. \qquad (B17)$$

However, if we assume that the normal derivative of the horizontal flow v_H is proportional to itself,

$$-d\left(\hat{n}\cdot\nabla\right)v_H = v_H + (1_{\Omega}\times v)_H \quad \text{at} \quad z = 0, \quad \text{(B18)}$$

where d is a finite scaling factor (e.g. the mean radius r_0 of the CMB), we may obtain that

$$c_1 \approx -\frac{2 + [1 + \mathrm{sgn}(f)]d\sqrt{|f|/2E}}{(1 + d\sqrt{|f|/2E})^2 + [\mathrm{sgn}(f) + d\sqrt{|f|/2E}]^2}\,u_0, \tag{B19}$$

$$c_2 \approx \frac{[1 - \mathrm{sgn}(f)]d\sqrt{|f|/2E}}{(1 + d\sqrt{|f|/2E})^2 + [\mathrm{sgn}(f) + d\sqrt{|f|/2E}]^2}\,u_0. \tag{B20}$$

The two constants are proportional to $\sqrt{E}$ when $E \ll 1$. It is explicit from (B19) and (B20) that modification on the flow in the boundary layer is of higher order effect compared to the mainstream u_0. The viscous stress on the boundary $z = 0$

$$|\hat{n}\cdot\tau_\nu| \sim \mathcal{O}(E\,u_0), \tag{B21}$$

much weaker than the actual viscous stress (B17) with the no-slip boundary conditions.

It is interesting to point out that the spiral

$$\phi_E = \tan^{-1}\left(\frac{c_2}{u_0}\right) \approx \frac{1 - \mathrm{sgn}(f)}{2d}\sqrt{\frac{2E}{|f|}}, \tag{B22}$$

is small and is in the same direction as the actual spiral (B16) when $f > 0$. It vanishes if $f < 0$.

As an example, we use spherical coordinates to examine the validity of the asymptotic boundary condition (B18) in curvilinear coordinate systems. Near the spherical boundaries, the radial derivative dominates horizontal derivatives. Taking the horizontal components of $\nabla\times(7)$, we have to leading order that

$$\cos\theta\,\frac{\partial v_\theta}{\partial r} = E\,\frac{\partial^3 v_\phi}{\partial r^3}, \tag{B23}$$

$$\cos\theta\,\frac{\partial v_\phi}{\partial r} = -E\,\frac{\partial^3 v_\theta}{\partial r^3}. \tag{B24}$$

Introducing the stretching coordinates

$$\zeta = \delta^{-1}(1 - r), \qquad \delta = \sqrt{E/|f|} \tag{B25}$$

near the CMB, we may obtain the following solutions

$$v_\theta = v_\theta^0 + e^{-\zeta/\sqrt{2}}\left[c_1\cos\zeta/\sqrt{2} + c_2\sin\zeta/\sqrt{2}\right], \tag{B26}$$

$$v_\phi = v_\phi^0 + \mathrm{sgn}(f)\,e^{-\zeta/\sqrt{2}}\left[c_2\cos\zeta/\sqrt{2} - c_1\sin\zeta/\sqrt{2}\right], \tag{B27}$$

where (v_θ^0, v_ϕ^0) are the mean flow (i.e. independent of the stretched coordinate ζ). It is explicit that (B26)-(B27) are almost identical to the solutions (B11)-(B12) in the Cartesian coordinate systems. The partial slippery boundary conditions (B18) are

$$d\delta^{-1}\frac{\partial v_\theta}{\partial\zeta} = v_\theta - f\,v_\phi,$$

$$d\delta^{-1}\frac{\partial v_\phi}{\partial\zeta} = v_\phi + f\,v_\theta,$$

at $\zeta = 0$, which yield to leading order the coefficients

$$c_1 \approx -\frac{\delta}{d\sqrt{2}}\left[(1 + |f|)\,v_\theta^0 + \mathrm{sgn}(f)\,(1 - |f|)\,v_\phi^0\right], \tag{B28}$$

$$c_2 \approx \frac{\delta}{d\sqrt{2}}\left[(1 - |f|)\,v_\theta^0 - \mathrm{sgn}(f)\,(1 + |f|)\,v_\phi^0\right]. \tag{B29}$$

Again the coefficients (c_1, c_2) are of orde $\mathcal{O}(\sqrt{E})$. The resultant viscous stress at the boundaries is of order $\mathcal{O}(\sqrt{E})$. This conclusion can be also applied to the moving boundaries.

Based on the above analysis, we introduce the partial slippery boundary conditions for our geodynamo simulation:

$$-\hat{n}\times d\left(\hat{n}\cdot\nabla\right)u = \hat{n}\times(u + \hat{z}\times u) \quad \text{at} \quad r = r_{io}, 1, \tag{B30}$$

where u is the differences between the core flow v and the boundary velocity v_B:

$$u = v - v_B. \tag{B31}$$

By (16) and (18), we have

$$d\frac{\partial^2 v_l^m}{\partial r^2} - \left[\frac{d}{r}\pm 1 \mp \frac{im}{l(l+1)}\right]\frac{\partial v_l^m}{\partial r}$$
$$\mp\left[\frac{l-1}{l}c_l^m\omega_{l-1}^m + \frac{l+2}{l+1}c_{l+1}^m\omega_{l+1}^m\right] = a_l^m, \tag{B32}$$

$$d\frac{\partial\omega_l^m}{\partial r} - \left[\frac{d}{r}\pm 1 \mp \frac{im}{l(l+1)}\right]\omega_l^m$$
$$\pm\left[\frac{l-1}{l}c_l^m\frac{\partial v_{l-1}^m}{\partial r} + \frac{l+2}{l+1}c_{l+1}^m\frac{\partial v_{l+1}^m}{\partial r}\right] = b_l^m, \tag{B33}$$

at the ICB and at the CMB, respectively. In the equations (a_l^m, b_l^m) are the spectral coefficients of the inner core and the mantle rotations (In the reference frame rotating with the mantle, $a_l^m = b_l^m = 0$ at the CMB), and

$$c_l^m \equiv \left[\frac{(l-m)(l+m)}{(2l-1)(2l+1)}\right]^{1/2}.$$

Acknowledgments. This work is supported by NASA Solid Earth and Natural Hazards Program and by NSF CSEDI program under Grant EAR0079998. We also thank NCCS at NASA's Goddard Space Flight Center for supporting our numerical simulation.

REFERENCES

Anufriev, A. P. and S.I. Braginsky, Influence of irregularities of the boundary of the Earth's core on the velocity of the liquid and on the magnetic field, *Geomag. Aeron., 15,* 754-757, 1975.

Anufriev, A. P. and S.I. Braginsky, Influence of irregularities of the boundary of the Earth's core on the velocity of the liquid and on the magnetic field II, *Geomag. Aeron., 17,* 78-82, 1977.

Backus, G. E., Kinematics of secular variation in a perfectly conducting core, *Phil. Trans. Roy. Soc. Lond. A, 263,* 239-266, 1968.

Bloxham, J. , Dynamics of angular momentum in the Earth's core, *Ann. Rev. Planet. Sci., 26,* 501-517, 1998.

Bloxham, J. and A. Jackson, Fluid flow near the surface of the Earth's outer core, *Rev. Geophys., 29,* 97-120, 1991.

Braginsky, S. I., Magnetic waves in the Earth's core, *Geomag. Aeron., 7,* 851-859, 1967.

Braginsky, S. I., Torsional magnetohydrodynamic vibrations in the Earth's core and variations in day length, *Geomag. Aeron., 10,* 1-8, 1976.

Braginsky, S. I., Short-Period Geomagnetic Secular Variation, *Geophys. Astrophys. Fluid Dynam., 30,* 1-78, 1984.

Buffett, B. A. , Gravitational oscillations in the length of day, *Geophys. Res. Lett., 23,* 2279-2286, 1996a.

Buffett, B. A. , A mechanism for decadal fluctuations in the length of day, *Geophys. Res. Lett., 23,* 3803-3806, 1996b.

Buffett, B. A. and K. C. Creager, A comparison of geodetic and seismic estimates of inner-core rotation, *Geophys. Res. Lett., 26,* 1509-1512, 1999.

Buffett, B. A. and G. A. Glatzmaier, Gravitational braking of the inner-core rotation in geodynamo simulations, *Geophys. Res. Lett., 27,* 3125-3128, 2000.

Bullard, E. C., C. Freedman, H. Gellman and J. Nixon, The westward drift of the Earth's magnetic field, *Phil. Trans. Roy. Soc. Lond. A, 243,* 67-92, 1950.

Chao, B. F. and R. D. Ray, Oceanic tidal angular momentum and Earth's rotation variations, *Prog. Oceanog., 40,* 399-421, 1997.

Chao, B. F., V. Dehant, R. S. Gross, R. D. Ray, D. A. Salstein, M. M. Watkins and C. R. Wilson, Space geodesy monitors mass transports in global geophysical fluids, *EOS, Trans. Amer. Geophys. Union, 81,* 247-250, 2000.

Dickey, J. O., Atmospheric excitation of the Earth's rotation: Progress and prospect via space geodesy, in *Contributions of Space Geodesy to Geodynamics: Earth Dynamics,* edited by D. Smith and D. Turcotte, pp. 55-70, AGU, Washington, D.C., 1993.

Eubanks, T. M., Variations in the orientation of the Earth, in *Contributions of Space Geodesy to Geodynamics: Earth Dynamics,* edited by D. Smith and D. Turcotte, pp. 1-54, AGU, Washington, D.C., 1993.

Garcia, R. and A. Souriau, Amplitude of the core-mantle boundary topography estimated by stochastic analysis of core phases, *Phys. Earth Planet. Inter., 117,* 345-359, 2000.

Glatzmaier, G. A. and P. H. Roberts, A three-dimensional self-consistent computer simulation of a geomagnetic field reversal, *Nature, 377,* 203-209, 1995.

Greenspan, H. P., *The Theory of Rotating Fluids,* 325 pp., Cambridge University Press, Cambridge, 1968.

Gubbins, D. and P. H. Roberts, Magnetohydrodynamics of the Earth's core, in *Geomagnetism,* edited by J. A. Jacobs, Volume 2, Chapter 1, Academic, London, 1987.

Hide, R., Interaction between the Earth's liquid core and solid mantle, *Nature, 222,* 1055-1056, 1969.

Hide, R., The Earth's differential rotation, *Quart. J. Roy. Astr. Soc., 278,* 3-14, 1986.

Hide, R., R. W. Clayton, B. H. Hager, M. A. Spieth and C. V. Voorhies, Topographic core-mantle coupling and fluctuations in the Earth's rotation, in *Relating geophysical structure and processes, The Jeffreys volume,* edited by K. Aki and R. Dmowska, *Geophys. Monog., 76,* pp. 107-120, AGU, Washington, D.C., 1993.

Holme, R., Electromagnetic core-mantle coupling-I. Explaining decadal changes in the length of day, *Geophys. J. Int., 132,* 167-180, 1998.

Holme, R., Electromagnetic core-mantle coupling-II. Probing deep mantle conductance, in *The core-mantle boundary region,* edited by M. Gurnis, M. Wysession, E. Knittle and B. Buffett, *Geodynamics, 28,* pp. 139-151, AGU, Washington, D.C., 1998.

Holme, R., Electromagnetic core-mantle coupling-III. Laterally varying mantle condutance, *Phys. Earth Planet. Inter., 117,* 329-344, 2000.

Jackson, A., J. Bloxham and D. Gubbins, Time-dependent flow at the core surface and conservation of angular momentum in the coupled core-mantle system, in *Dynamics of the Earth's deep interior and Earth rotation,* edited by J.-L. LeMouël, D. Smylie and T. Herring, *Geophys. Monog., 72,* pp. 97-107, AGU, Washington, D.C., 1993.

Jault, D., C. Gire and J.-L. LeMouël, Westward drift, core motions and exchanges of angular momentum between core and mantle, *Nature, 333,* 353-356, 1988.

Jault, D. and J.-L. LeMouël, The topographic torque associated with a tangentially geostrophic motion at the core surface and inferences on the flow inside the core, *Geophys. Astrophys. Fluid Dynam., 48,* 273-296, 1989.

Jault, D. and J.-L. LeMouël, Comment on 'On the dynamics of topographical core-mantle coupling' by Weijia Kuang and Jeremy Bloxham, *Phys. Earth Planet. Inter., 114,* 211-215, 1999.

Kuang, W. , Force balances and convective state in the Earth's core, *Phys. Earth Planet. Inter., 116,* 65-79, 1999.

Kuang, W. and J. Bloxham, On the effect of boundary topography on flow in the Earth's core, *Geophys. Astrophys. Fluid Dynam., 72,* 161-195, 1993.

Kuang, W. and J. Bloxham, On the dynamics of the topographical core-mantle coupling, *Phys. Earth Planet. Inter., 99,* 289-294, 1997a.

Kuang, W. and J. Bloxham, An Earth-like numerical dynamo model, *Nature, 389,* 371-374, 1997b.

Kuang, W. and J. Bloxham, Numerical modeling of magnetohydrodyanmic convection in a rapidly rotating spherical shell: weak and strong field dynamo actions, *J. Comp. Phys., 153,* 51-81, 1999.

Kuang, W. and B. F. Chao, Topographic core-mantle coupling in geodynamo modeling, *Geophys. Res. Lett.*, *28*, 1871-1894, 2001.

Lambeck, K., *The Earth's Variable Rotation*, 449 pp., Cambridge University Press, Cambridge, 1980.

Love, J. J. and J. Bloxham, Electromagnetic coupling and the toroidal field at the core-mantle boundary, *Geophys. J. Int.*, *117*, 235-256, 1994.

Marcus, S. L., Y. Chao, J. O. Dickey and P. Gegout, Detection and modeling of nontidal oceanic effects on Earth's rotation rate, *Science*, *281*, 1656-1659, 1998.

Matsushima, M., T. Nakajima and P. H. Roberts, The anisotropy of local turbulence in the Earth's core, *Earth Planets Space*, *51*, 277-286, 1999.

Morelli, A. and A. M. Dziewonski, Topography of the core-mantle boundary and lateral homogeneity of the liquid core, *Nature*, *325*, 678-683, 1987.

Munk, W.H. and G. J. F. MacDonald, *The Rotation of the Earth*, 323 pp.,Cambridge University Press, Cambridge, 1960.

Munk, W. H. and R. Revelle, On the geophysical interpretation of irregularities in the rotation of the Earth, *Mon. Not. Roy. Astr. Soc.*, *6*, 331-347, 1952.

Roberts, P. H., Electromagnetic core-mantle coupling, *J. Geomag. Geoelectr.*, *24*, 231-259, 1972.

Rosen, R. D. , The axial momentum balance of Earth and its fluid envelope, *Surveys Geophys.*, *14*, 1-29, 1993.

Stephenson, F. R. and L. V. Morrison, Long-term fluctuations in the Earth's rotation: 700 BC to AD 1990, *Phil. Trans. Roy. Soc. Lond. A*, *353*, 165-202, 1995.

Stix, M. and P. H. Roberts, Time-dependent electromagnetic core-mantle coupling, *Phys. Earth Planet. Inter.*, *36*, 49-60, 1984.

Zatman, S. A. and J. Bloxham, Torsional oscillations and the magnetic field within the Earth's core, *Nature*, *388*, 760-763, 1997.

W. Kuang, Joint Center for Earth Systems Technology, University of Maryland, Baltimore County, 1000 Hilltop Circle, Baltimore, MD 21250. (e-mail: wkuang@umbc.edu, kuang@bowie.gsfc.nasa.gov)

B. F. Chao, Space Geodesy Branch, Code 926, NASA Goddard Space Flight Center, Greenbelt, MD 20771. (e-mail: chao@bowie.gsfc.nasa.gov)

Thermal Interactions Between the Mantle, Outer and Inner Cores, and the Resulting Structural Evolution of the Core

Ikuro Sumita

Department of Earth Sciences, Kanazawa University, Kanazawa, Japan

Shigeo Yoshida

Department of Earth and Planetary Sciences, Nagoya University, Nagoya, Japan

We review our current understanding of the thermal interactions between the mantle, outer and inner cores, and how they determine the present geomagnetic field pattern and the seismic structure of the core. First, we describe the evolution of the radial structure of the core. The formation of several structures is placed in the context of the Earth's history. We review the heat flow across the core-mantle boundary, and show a simple model of the inner core growth. We present a model of the initial chemical stable stratification of the core and its subsequent disruption. Model calculations show that viscous compaction efficiently expels liquid from the inner core, and also causes a crust-like structure to form beneath the inner-core boundary because the inner core growth rate gradually decreases. Next, we consider how the radial structure is modified as a result of the lateral variation of heat transfer in the core. The inner core grows in an anisotropic manner as a consequence of the columnar convection in the outer core, which results in a latitudinal variation of heat transfer. This anisotropic growth gives rise to the observed seismic anisotropy in the inner core. In addition to the latitudinal variation, a longitudinal variation of convective heat flux is likely in the outer core because the mantle is thermally heterogeneous. We show, from experimental and theoretical methods, how this modifies the pattern of the outer core flow and the inner core growth. Spatial and temporal characteristics of the geomagnetic field and the longitudinally heterogeneous seismic structure of the inner core can be interpreted in terms of this modification.

1. INTRODUCTION

There are two major solid-liquid interfaces in the Earth's interior, the core-mantle boundary (CMB) and

Earth's Core: Dynamics, Structure, Rotation
Geodynamics Series 31
Copyright 2003 by the American Geophysical Union
10.1029/31GD14

the inner-core boundary (ICB), at which the mantle, outer and inner cores are coupled. They can be coupled thermodynamically, by transferring heat (and composition) between these layers. Alternatively, they can be coupled mechanically, by the torques which act between them by transferring angular momentum from one layer to another. The two coupling mechanisms mentioned above have very different time scales. Mechanical coupling is involved in relatively short time scale phenomena. It can be instantaneous such as by gravitational locking, to less than 10^4 years which is the spin-up time of the outer core. On the other hand, thermal coupling has much longer time scales. The thermal core-mantle coupling can be considered as a quasi-steady process for the outer core; the outer core responds to the mantle instantaneously on the time scales of mantle convection, because the dynamical time scales of the outer core are much shorter than those of the mantle. In that sense, thermal (and compositional) coupling is a geological process, and its time scale is determined by the mantle overturn time of the order of 10^8 y. The thermal coupling between the outer and inner cores is also a geological process, whose time scale is the age of the inner core, of the order of 10^9 y.

The understanding of thermal coupling is important in interpreting the non-dipole features of the geomagnetic field, which can be the manifestation of the non-axisymmetric flow pattern controlled by the thermally heterogeneous upper and lower boundaries of the outer core (i.e., mantle and inner core). The thermal influence of the outer core on the inner core is important in interpreting the seismic structure of the inner core, because, in our view, the structure is developed through its growth, which is essentially controlled by how the heat is being transferred in the outer core.

In this paper, we review some models of thermal interactions between the mantle, outer and inner cores, and discuss how they are related to observational evidence. In section 2, we describe the radial structure of the core and its evolution in the Earth's history. We also show how the partially molten structure of the inner core is developed as a result of its growth. In section 3, we describe how this radial structure is modified due to lateral variation of heat transfer. We show that seismic anisotropy of the inner core can be considered as a consequence of anisotropic heat transfer in the outer core controlled by rotation. We also show how lateral thermal heterogeneities on the CMB can control outer core flow and hemispherically varying inner core seismic structure. In section 4, we describe future prospects which are needed to obtain better models for the core.

2. RADIAL CORE STRUCTURE AND ITS EVOLUTION

2.1. History of the Earth's Core

The structure of the Earth's core has been evolving throughout the long history of the Earth. Hence the understanding of the history of the Earth's core is necessary in order to understand various structures in the core.

The Earth formed as a result of the accretion of planetesimals in the primordial solar system. The bombardment of planetesimals released a substantial amount of energy, heating the Earth above the melting temperature of rocks. Molten rock formed the magma ocean, in which metallic components, mostly iron, sank to the center to form the Earth's core [e.g., *Stevenson*, 1981, 1990].

As the Earth gradually cools down after the core formation, the inner core solidifies (grows) from the Earth's center [*Jacobs*, 1953]. This process has been shown in a number of thermal history calculations [e.g., *Gubbins et al.*, 1979; *Stevenson et al.*, 1983; *Buffett et al.*, 1992, 1996; *Sumita et al.*, 1995; *Labrosse et al.*, 1997, 2001]. As the inner core solidifies, light elements, such as sulfur, oxygen, carbon, or hydrogen [e.g., *Jeanloz*, 1990; *Poirier*, 1994] are expelled from the inner core. The release of gravitational energy due to the release of light elements is considered to be essential in driving the Earth's dynamo [e.g., *Gubbins*, 1977; *Gubbins and Masters*, 1979].

In the remainder of this section, we first give estimates of the heat flow across the CMB, which is important in the thermal history of the Earth's core, and then discuss three consequent one-dimensional models of the structures of the core: the inner core growth, possible initial stable density stratifications in the outer core, and the partially molten structure in the inner core.

2.2. Heat Flow Across the CMB

There are various estimates of the heat flow across the CMB. They are obtained by inferring the cooling rate of the core, by inferring the heat transport by mantle plumes, which may be considered to originate from the CMB, or by inferring the dissipation due to dynamo action.

The CMB heat flow can be estimated if we know the cooling rate of the core as

$$Q_{\mathrm{CMB}} = -C_{\mathrm{eff}} \frac{d\Theta}{dt}, \qquad (1)$$

where Q_{CMB} is the heat flow across the CMB, Θ is the potential temperature of the core, and C_{eff} is the effective heat capacity, expressed as

$$C_{\mathrm{eff}} = C_{\mathrm{cool}} + C_{\mathrm{latent}} + C_{\mathrm{grav}}. \qquad (2)$$

Here C_{cool} represents the cooling of the core, C_{latent} represents the effect of the latent heat due to the inner core solidification, and C_{grav} represents the gravitational energy release due to the light elements [e.g., *Gubbins et al.*, 1979]. There are other effects, such as effects from thermal contraction [*Stacey*, 1992], but the main contribution comes from the three effects above. Since the three heat capacities have similar values of about 2×10^{27} J/K, the effective heat capacity C_{eff} is about 6×10^{27} J/K. Note that the values of C_{latent} and C_{grav} depend on the difference between the liquidus temperature gradient and the adiabatic gradient $dT_L/dp - dT_{\text{ad}}/dp$, which is highly uncertain. We use the value 2.5×10^{-9} K/Pa for the estimate above. Note also that the values of C_{latent} and C_{grav} change with time. They are approximately proportional to the inner core radius.

An estimate of the cooling rate is given by assuming that the mantle was completely molten at the formation of the Earth. The mantle solidified very rapidly after the period of heavy bombardment because the heat transport of the molten mantle was very efficient [e.g., *Abe*, 1997]. The solidification of the mantle dramatically decreased the efficiency of heat transport, and after that the cooling became gradual. If we assume that the temperature at the CMB 4.5 Ga ago was about 4300 K, the extrapolated solidus of pyrolite [*Boehler*, 2000], and the present CMB temperature is about 4000 K, which is deduced from melting temperature of Fe-O-S system [*Boehler*, 1996, 2000], the temperature drop in 4.5 Ga is about 300 K. The mean cooling rate thus becomes 7×10^{-8} K/y, which gives the heat flow of about 8×10^{12} W if we take the value of 4×10^{27} J/K as the mean effective heat capacity.

Another clue can be found from the cooling rate of the mantle. The cooling rate of the core becomes similar to that of the mantle if the CMB heat flow strongly depends on the temperature difference of the core and mantle. Estimates of the cooling rate of the mantle are given by calculations of thermal history [e.g., *Sleep and Langan*, 1981; *Christensen*, 1985], and have a range of 140 - 350 K / 3.5 Ga. Petrological evidence suggest cooling rates of < 50K/2.7Ga [*Campbell and Griffiths*, 1992], 120K/3.2Ga [*Ohta et al.*, 1996], ∼ 160K/2.7Ga [*Abbott et al.*, 1994], and 300K/3.5Ga [*Green*, 1981]. These estimates are about the same as those of the cooling rate of the core given above, suggesting that the CMB heat flow is controlled by the temperature difference between the mantle and the core.

A few thermal history calculations which include both the mantle and the core have been carried out [*Stevenson et al.*, 1983; *Mollett*, 1984; *Stacey and Loper*, 1984]. *Stacey and Loper* [1984] used the constraint that the CMB was at the solidus of the mantle 4.5 Ga ago, and that the inner core has grown to its present size. The parameterizations of *Stevenson et al.* [1983] and *Mollett* [1984] make the temperature decrease of the core

follow that of the mantle. They therefore arrived at the cooling rates similar to the two simple estimates given above, despite the differences in the formulations of the CMB heat flow. The obtained CMB heat flows are in the range of 2×10^{12} W [*Stacey and Loper*, 1984] to 9×10^{12} W [*Mollett*, 1984].

Another type of estimate of the CMB heat flow can be given by assuming that mantle plumes originate from CMB, and that they carry most of the heat flow from the core. *Sleep* [1990] estimated the heat flow of about 3×10^{12}W, whereas *Davies* [1988] estimated it to be 2.5×10^{12}W. This value gives a lower bound of the heat flow because other contributions to the heat flow are neglected. Mantle convection calculations such as by *Tackley et al.* [1994] give a CMB heat flow of about 7×10^{12}W, although the effect of secular cooling is neglected in their calculations.

Yet another estimate is given by *Gubbins et al.* [1979], who calculated the heat flow necessary to maintain dynamo action to be about $(2-5) \times 10^{12}$ W, although this value depends strongly on the type of dynamo model assumed. Similarly, this would give the lower bound on the heat flow, since not all of the heat is dissipated through dynamo action.

To summarize, all the estimates of the CMB heat flow are of the order of several 10^{12} W. We use a core heat flow of $(3-5) \times 10^{12}$W as a typical estimate in the discussion below.

2.3. Inner Core Growth

The change of the inner core radius as a function of time is shown in various thermal history calculations. Here we show by a simple argument, that the inner core radius grows approximately proportional to $\sqrt{t}$ if the heat flow is constant [*Gubbins et al.*, 1979; *Buffett et al.*, 1992, 1996; *Sumita et al.*, 1995; *Lister and Buffett*, 1998].

When the inner core is very small, we can neglect the effect of latent heat and gravitational energy on the energy budget, which becomes

$$Q_{\text{CMB}} = -C_{\text{OC}} \frac{d\Theta}{dt}, \qquad (3)$$

where C_{OC} is the heat capacity of the outer core.

Since the temperature of the ICB is at the liquidus, we can relate the cooling rate of the ICB and the change of the inner core radius as

$$\frac{dT_{ICB}}{dt} = \frac{dT_{ICB}}{dp}\frac{dp}{dr}\frac{dr}{dt} = -\rho g \frac{dR_{IC}}{dt}\frac{dT_L}{dp}, \qquad (4)$$

where T_{ICB} is the ICB temperature, T_L is the liquids temperature, p is the pressure, r is the radical coordinate, R_{IC} is the radius of the inner core, ρ is the density, and g is the gravitational acceleration. Here we omit the effect of composition on the liquidus, because the composition of the outer core does not change much when the inner core is small. On the other hand, if the temperature gradient of the core is adiabatic, the ICB temperature decrease may be related to the potential temperature change $d\Theta/dt$ as

$$\begin{aligned}
\frac{dT_{ICB}}{dt} &= \frac{d\Theta}{dt} + \frac{dT_{ad}}{dp}\frac{dp}{dr}\frac{dr}{dt} \\
&= \frac{d\Theta}{dt} - \rho g \frac{dR_{IC}}{dt}\frac{dT_{ad}}{dp},
\end{aligned} \qquad (5)$$

where dT_{ad}/dp is the adiabatic temperature gradient. Combining these two equations, we get

$$\frac{d\Theta}{dt} = -\rho g \frac{dR_{IC}}{dt}\left(\frac{dT_L}{dp} - \frac{dT_{ad}}{dp}\right). \qquad (6)$$

Combining equations (3) and (6), and expressing the radial gravity dependence as

$$g = \gamma r \qquad (7)$$

where $\gamma = 4\pi G\rho/3$, we obtain

$$Q_{CMB} = C_{OC}\rho\gamma\left(\frac{dT_L}{dp} - \frac{dT_{ad}}{dp}\right)\frac{d}{dt}\frac{R_{IC}^2}{2}. \qquad (8)$$

When the CMB heat flow is constant, this equation be easily integrated to give an explicit expression of the inner core radius as a function of time as

$$R_{IC} = \left[\frac{2Q_{CMB}}{C_{OC}\rho\gamma\left(dT_L/dp - dT_{ad}/dp\right)}t\right]^{\frac{1}{2}}. \qquad (9)$$

This equation gives a rough estimate of the age of the inner core as

$$\begin{aligned}
\tau_{IC-growth} &= \left(1 \times 10^9 \text{y}\right) \times \left(\frac{Q_{CMB}}{4 \times 10^{12}\text{W}}\right)^{-1} \\
&\times \left(\frac{dT_L/dp - dT_{ad}/dp}{2 \times 10^{-9}\text{K/Pa}}\right).
\end{aligned} \qquad (10)$$

This estimate indicates that the inner core has grown to its present size in a time scale of the age of the Earth. The corresponding growth rate of the inner core at present is about 0.1 mm/y, which is a rate comparable to or larger than the sedimentation rate of the deep sea sediment.

If we consider a partially molten inner core, the inner core would grow faster, but this effect is small because the melt is efficiently expelled by compaction (section 2.5). The effects which act as additional heat sources (latent heat and gravitational energy) tend to increase the age of the inner core, and more detailed calculations will give an older age. Indeed, with the CMB heat flow of 4×10^{12} W, *Lister and Buffett* [1998] and *Buffett* [2000], for example, obtained the age of 4.5×10^9 y and 2.2×10^9 y, respectively, with the difference between the two being due mainly to the uncertainty in the estimate of $(dT_L/dp - dT_{ad}/dp)$. *Lister and Buffett* [1998] also showed that the inner core radius increases as $t^{5/6}$ when the outer core is stably stratified initially. The problem of the initial stratification is discussed below.

2.4. Initial Density Stratification of the Outer Core and its Destruction

In this section we discuss the stable density stratification of the whole outer core which may have existed at the beginning of the Earth's history [*Muromachi*, 1991; *Stevenson*, 1992; *Kumazawa et al.*, 1994]. When the Earth formed, the core grew as iron sank in the magma ocean [e.g., *Stevenson*, 1981, 1990]. The iron should have been in equilibrium with the lowermost mantle, whose condition changes with time. The core grew simultaneously as the size of the Earth increased through the accretion of planetesimals. The pressure and temperature of the lowermost mantle increased with time accordingly. Light elements such as oxygen [e.g., *Ohtani and Ringwood*, 1984; *Ohtani et al.*, 1984; *Kato and Ringwood*, 1989] and hydrogen [e.g., *Stevenson*, 1977; *Fukai*, 1984, 1992; *Yagi and Hishinuma*, 1995; *Kuramoto and Matsui*, 1996; *Okuchi*, 1997, 1998] are known to become more soluble in iron as the temperature and pressure increase. It follows that the iron accreted later in time may have contained more light materials, causing the core to become stably stratified.

This stratification could have been sufficiently large as to prevent dynamo action. However, the evidence of the magnetic field at 3.5 Ga [*McElhinny and Senanayake*, 1980; *Hale and Dunlop*, 1984; *Yoshihara and Hamano*, 2000] and the magnetic field reversal at 3.2 Ga [*Layer et al.*, 1996] signify that most of the outer core was already vigorously convecting at 3.5 Ga. The initial stable stratification, therefore, should have been disrupted by some mechanism during the first 1 Ga of the Earth's history.

The mechanisms of the disruption of the stable stratification can be gradual or catastrophic. Gradual mechanisms include encroachment or entrainment [*Lister and Buffett*, 1998]. Catastrophic mechanisms include tidal resonance of inertial-gravity waves [*Kumazawa et*

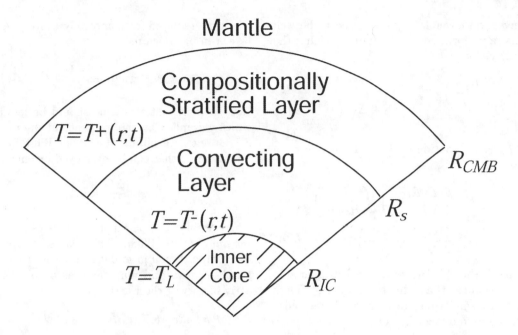

Figure 1. A sketch showing a model where a chemically stratified layer is gradually destroyed by the release of light elements by inner core growth. The interface between the convecting layer and the stratified layer is denoted by R_s.

al., 1994] or the 1/3-annual forced nutation [*Williams*, 1994].

Here we investigate the encroachment mechanism based on the theory of *Lister and Buffett* [1998]. The situation is shown in Fig. 1. Encroachment is a mechanism in which stratification is gradually disrupted by the convection developing under the stratified layer. The position of the interface between the two layers is determined by the condition that the buoyancies just above and below the interface are the same [*Turner*, 1973]. *Lister and Buffett* [1998] examined the movement of the interface between a thermally stratified layer beneath the CMB and a compositionally convecting outer core, but their theory can easily be extended for the case of initial chemical stratification. In the present case, the thickness of the stratified layer decreases as the light element content in the convecting region increases with inner core growth.

Let us assume that the initial radial profile of light element concentration takes the form of

$$C = C_0 + \Delta C \left(\frac{r}{R_{CMB}}\right)^2, \quad (11)$$

where C is the concentration, R_{CMB} is the CMB radius, and C_0 and ΔC are constants. The functional form is derived as follows. The pressure at the CMB will increase as R_{CMB}^2 as the Earth grows. If the solu-

bility is proportional to the pressure, the concentration will vary as r^2 accordingly. We can estimate that the temperature effect on solubility would work in a similar way. If the mantle convection is vigorous in the early Earth, the temperature distribution would follow the adiabatic gradient. Then the temperature would be proportional to the pressure. Assuming that the solubility is proportional to temperature, the concentration would also vary as r^2.

Let us denote the radius at the bottom of the stable layer as R_s as shown in Fig.1. If the compositional buoyancy is much larger than the thermal one, R_s is determined by the continuity of the concentration given by

$$C(R_s^+) = C(R_s^-). \quad (12)$$

The concentration of the stable layer is given by Eq (11), whereas that of the convecting lower layer is given by the conservation of light material as

$$\int_{R_{IC}}^{R_s} C^- 4\pi r^2 dr = \int_0^{R_s} \left[C_0 + \Delta C \left(\frac{r}{R_{CMB}}\right)^2\right] 4\pi r^2 dr, \quad (13)$$

which becomes,

$$C^-(R_s^3 - R_{IC}^3) = C_0 R_s^3 + \frac{3}{5}\Delta C \left(\frac{R_s}{R_{CMB}}\right)^2 R_s^3. \quad (14)$$

Here the convecting lower layer is assumed to be well mixed, and has a uniform composition of C^-. Thus the continuity (12) yields

$$\frac{C_0 R_s^3 + \frac{3}{5}\Delta C \left(\frac{R_s}{R_{CMB}}\right)^2 R_s^3}{R_s^3 - R_{IC}^3} = C_0 + \Delta C \left(\frac{R_s}{R_{CMB}}\right)^2, \tag{15}$$

which gives R_s as a function of R_{IC}. Eq.(15) can be rearranged as

$$\left(\frac{R_s}{R_{CMB}}\right)^5 - \frac{5}{2}\left(\frac{R_{IC}}{R_{CMB}}\right)^3\left(\frac{R_s}{R_{CMB}}\right)^2$$
$$- \frac{5}{2}\frac{C_0}{\Delta C}\left(\frac{R_{IC}}{R_{CMB}}\right)^3 = 0. \tag{16}$$

When $C_0 \ll \Delta C$, the third term of the left-hand side of Eq. (16) is smaller than the second term unless R_s/R_{CMB} is very small (i.e., beginning of inner core growth), and can be omitted. This leads to

$$\frac{R_s}{R_{CMB}} = \left(\frac{5}{2}\right)^{\frac{1}{3}} \frac{R_{IC}}{R_{CMB}}, \tag{17}$$

from which we obtain

$$\frac{R_s}{R_{IC}} = \left(\frac{5}{2}\right)^{\frac{1}{3}}. \tag{18}$$

This equation shows that a thick chemically stable layer still remains even now, which is probably not the case. Hence a strongly stratified layer, $C_0 \ll \Delta C$, cannot be disrupted by encroachment.

On the other hand, when $C_0 \gg \Delta C$, the second term of the left-hand side of Eq. (16) is smaller than the third term, and can be omitted. This leads to

$$\frac{R_s}{R_{CMB}} = \left(\frac{5}{2}\frac{C_0}{\Delta C}\right)^{\frac{1}{5}}\left(\frac{R_{IC}}{R_{CMB}}\right)^{\frac{3}{5}}. \tag{19}$$

We can use the theory of *Lister and Buffett* [1998] to obtain the initial change of the depth of the stably stratified layer as

$$\frac{R_s}{R_{CMB}} = \sqrt{15\frac{C_0}{\Delta C}\left(\mathcal{L} - \frac{5}{2}\mathcal{C}\right)^{-1}\frac{t}{\tau_{\text{therm}}}}, \tag{20}$$

where $\mathcal{L}$ and $\mathcal{C}$ are non-dimensional constants and τ_{therm} is the thermal diffusion time of the core of about 10^{11} y (see Appendix for derivation and the definition of the parameters). Here we asumed that the CMB heat flow is equal to the heat flow conducted down the adiabat.

The time required for completely destroying the stable layer can be estimated as

$$\tau_{\text{encroach}} = \frac{1}{15}\tau_{\text{therm}}\frac{\Delta C}{C_0}\left(\mathcal{L} - \frac{5}{2}\mathcal{C}\right). \tag{21}$$

If we require that this time should be less than 1 Ga, which is implied by the existence of the magnetic field 3.5 Ga ago, we obtain the upper limit for the initial stratification that can be destroyed by encroachment as

$$\frac{\Delta C}{C_0} < 0.2, \tag{22}$$

where we use the parameter values estimated by *Lister and Buffett* [1998]. This places an upper limit to the compositional gradient in the outer core when the Earth formed, and may serve as a constraint for the early history of the Earth.

2.5. Partially Molten Structure of the Inner Core

The Earth's core contains some light elements such as oxygen, sulfur etc. They depress the melting point of the solid phase and can give rise to the formation of a partial melt between the solidus and liquidus temperatures. *Fearn et al.* [1981] argued that a thick partially molten region exists in the inner core because of compositional super-cooling, i.e., inhibition of solidification from excess light element content. If the ICB were flat, the liquid-side of the growing solid-liquid interface would become supercooled because light elements diffuse slowly and depresses melting point. As a result, the growing interface becomes dendritic, and a partially molten layer develops. *Fearn et al.* [1981] speculated that the partially molten region would extend to the center of the inner core. However, *Loper* [1983] estimated that the partially molten region would become thinner when convection of the liquid phase is considered.

Sumita et al. [1996] studied the viscous compaction of the solid matrix of a partial melt in a growing inner core and showed that the compaction efficiently expels melt from the inner core. They neglected thermal and compositional effects, and investigated a one-dimensional sedimentary compaction problem in detail. In the Earth's core, the sedimentation velocity (i.e., solidification rate) V_0 is much less than the Darcy velocity defined as

$$V_D = \frac{K}{\eta_f}\Delta\rho g, \tag{23}$$

where K is the permeability, η_f is the viscosity of liquid iron, and $\Delta\rho$ is the density difference between the solid

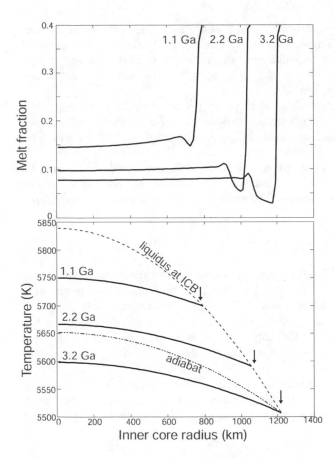

Figure 2. A calculated result of the evolution of (a) melt fraction and (b) temperature structure of the inner core [*Sumita et al.*, 1995]. The core cooling is constant at a rate of $Q_{CMB} = 3 \times 10^{12}$W, and the inner core became the present size after 3.2 Ga for this case. The melt fraction at the ICB is assumed to be 0.4. The dotted line indicates the liquidus curve of iron. Each arrows indicate the inner core radius for corresponding time. The inner core grows as the core cooling proceeds and the liquid is expelled from the inner core by compaction. A "crust" of a small melt fraction forms below the ICB, and a layer of constant melt fraction (low velocity layer) extends to the center.

and liquid iron (i.e. inner and outer cores). In that case, the mushy layer, where melt fraction is large, becomes very thin. Its thickness, given by

$$\delta_{mush} = \sqrt{\frac{\eta_s V_0}{\Delta \rho g}},\qquad(24)$$

where η_s is the viscosity of the solid iron, becomes very small, of the order of 10 m when η_s is 10^{16} Pa s. It shows that the compaction is very effective.

Sumita et al. [1995] included thermal and compositional effects in their calculation, but the essential features were the same as the purely mechanical case.

Fig.2 shows a result of their numerical calculation illustrating how the melt fraction and temperature within the inner core decreases as the inner core grows [*Sumita et al.*, 1995]. Here, coupled equations of mechanical compaction, energy and composition conservation were solved, and the melt fraction was determined from the binary eutectic phase diagram assumed for the inner core. We assumed no internal heat sources, no gravitational energy release, an adiabatic temperature profile in the outer core, and a CMB heat flow of 3×10^{12}W. The radial profile of melt fraction is characterized by three layers. At the top is the mushy layer, which is a thin layer beneath the ICB where the melt fraction decreases sharply. Then there is a transition layer of low melt fraction, followed by a thick layer of constant melt fraction which extends to the center. In the mushy layer, from the combined effect of the deformation of solid matrix and slow inner core growth, the melt is efficiently expelled from the inner core. In the region of constant melt fraction, deformation is limited, and the fluid is expelled by buoyancy driven permeable flow. The melt fraction is depth independent in this region because the gravity is proportional to radius [*Sumita et al.*, 1996]. The region of small melt fraction between these two regions forms because the melt is more efficiently expelled from the inner core for slower growth rates in the recent times. This crust-like region, or a seismic low velocity zone beneath it, may correspond to the seismically detected discontinuity at about 100 km below the ICB [*Souriau and Souriau*, 1989].

The possible existence of the "crust" and a larger melt fraction zone beneath it resembles the structure near the Earth's surface and suggests that plate tectonics might be operating in the inner core. However once the crust subducts, the "plate tectonics" would cease because the crust would not be reproduced. This crust results from gradual decrease in the inner core growth rate, and hence rapid reproduction cannot be expected. It would be interesting to investigate seismologically whether the crust-like region exists or is already disrupted.

2.6. Thermal and Compositional Structure of the Inner Core

Thermal structure of the inner core (Fig.2b), shows that the temperature gradient within the inner core is smaller than the adiabatic temperature gradient. This implies that thermal convection cannot occur in the inner core. Temperature gradients calculated by considering heat conduction alone have also shown that it is less than the adiabat [*Yoshida et al.*, 1996; *Yukutake*, 1998; *Buffett*, 2000]. This result is a consequence of the slow inner core growth rate which allows a long time to cool, and of the large thermal conductivity of metallic iron. Advection of heat by permeable flow has an additional effect in reducing the temperature gradient. However this is estimated to be less important as compared to conduction, and can be evaluated from the Péclet number,

$$Pe = \frac{VL}{\kappa}. \qquad (25)$$

If we take $L \sim 10^6$m and $V \sim 10^{-12}$m/s, a typical velocity of permeable flow, we find $Pe \sim 10^{-1}$.

On the other hand, advection of light elements by permeable flow is generally much more efficient than compositional diffusion through the liquid, because of small diffusion coefficient of light elements such as oxygen and sulfur, in liquid iron of $D \sim 10^{-8}$m^2/s [*Iida and Guthrie*, 1988], resulting in a compositional Péclet number, $Pe_c = VL/D \sim 10^2$. Hydrogen is an exception, however, and has a large diffusion coefficient $D \sim 10^{-4}$ m^2/s [*Iida and Guthrie*, 1988]. For this large diffusion coefficient, compositional super-cooling, would not occur according to the condition of *Fearn et al.* [1981]. In addition, the tendency for compositional cooling is diminished by the partition coefficient of hydrogen close to unity [*Fukai*, 1992; *Okuchi*, 1997]; i.e., considerable amount of hydrogen can enter in the inner core, and not much hydrogen is released upon freezing of the inner core. Hence, if the major light element in the core is hydrogen, the inner core would be completely solid with a sharp boundary.

3. LATERAL VARIATION OF CORE STRUCTURES

In the previous section, we have considered the inner core growth under spherical symmetry. We now proceed to consider how this is modified by the convective pattern of the outer core which is strongly controlled by rotation and mantle structure as we show below.

3.1. Anisotropic Growth of the Inner Core

Convection in the outer core is strongly controlled by rotation and its spherical geometry. *Busse* [1971] proposed from a linear theory that convective pattern in a rapidly rotating spherical shells would form columns aligned parallel to the rotational axis as a result of geostrophic balance (i.e., Taylor-Proudman theorem). Several non-linear numerical calculations [e.g., *Zhang*, 1992] and laboratory experiments [e.g., *Busse and Carrigan*, 1970] have confirmed the two dimensional nature of convective pattern. Because columnar cells are tangential to the inner core at the equator, the heat flow is expected to become anisotropic, becoming largest at the

equatorial regions. This was first shown for weakly non-linear calculations [*Zhang*, 1991, 1992], and was found to be valid even under the presence of a magnetic field [*Olson et al.*, 1999] as well as turbulence [*Sumita and Olson*, 2000].

Yoshida et al. [1996] proposed the consequence of the anisotropic heat flow on the inner core growth, as shown schematically in Fig.3. The inner core would grow predominantly at the equatorial region, but because of the density difference between the outer and inner cores, the inner core would isostatically deform back to its spherical shape. The deformation time scale can be estimated as

$$\tau_{\text{deform}} = \frac{\eta_s}{\Delta\rho g R_{IC}} = 3 \times 10^4\text{y}\left(\frac{\eta_s}{10^{21}\text{Pa} \cdot \text{s}}\right) \qquad (26)$$

The short time scale even for an inner core viscosity as large as 10^{21} Pa·s indicates that isostatic adjustment is instantaneous as compared to the time scale for inner core growth, so the inner core remains close to its hydrostatic shape. Since the anisotropic growth continues throughout the Earth's history, the inner core would always deviate a little away from the hydrostatic state. As a result, deviatoric stress would form, which has a sense of pull in the direction of the axis of rotation, and push in the equatorial regions. This deviatoric stress would

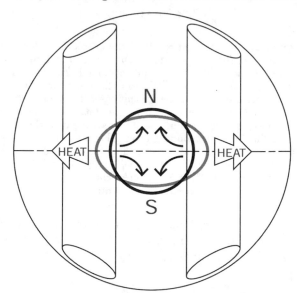

Figure 3.   A model for the formation of inner core anisotropy [*Yoshida et al., J. Geophys. Res., 101*, 28085-29103, 1996, copyright 1996 by the American Geophysical Union]. In the outer core, convection takes the form of columnar rolls parallel to the rotation axis. The resulting anisotropic heat transfer gives rise to the anisotropic growth of the inner core. Viscous relaxation in the inner core toward hydrostatic equilibrium causes the alignment of iron crystals, which is observed as seismic anisotropy.

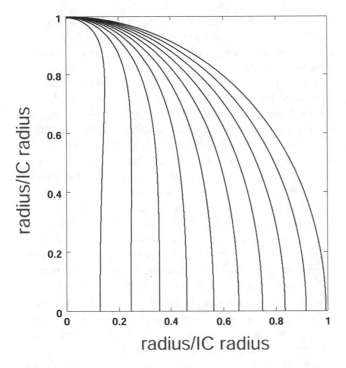

Figure 4. Strata in the inner core, joining the points which have solidified at the same time [*Yoshida et al.*, *J. Geophys. Res.*, *101*, 28085-29103, 1996, copyright 1996 by the American Geophysical Union]. The inner core is assumed to grow proportional to $\sqrt{t}$ and predominantly at low latitudes as shown in Fig.3. Each line corresponds to the strata at units of 1/10 of the inner core age. Inner core strata form approximately parallel to the rotational axis, and the iron which has solidified at an early stage accumulates near the rotational axis.

affect how the crystals align as we shall discuss in section 3.2. One consequence of the anisotropic growth is shown in Fig.4. Here, each lines indicate isochrons joining the points which have solidified at the same time. If the inner core grows isotropically, then the isochrons should form concentric spheres. However, in the case of dominant growth at the equator, the isochrons become nearly parallel to the rotational axis, and the iron which has solidified at the early stage of inner core growth accumulates near the rotational axis.

3.2. Seismic Anisotropy of the Inner Core

Large elastic anisotropy has been observed seismologically in the inner core [for reviews see *Song*, 1997; *Creager*, 2000; *Tromp*, 2001] and various causes have been proposed [for reviews see *Yoshida et al.*, 1998; *Buffett*, 2000]. The anisotropy is transversely isotropic with its symmetry axis being the rotation axis. P-waves traversing through the inner core is fast in the north-south direction, and slow in the equatorial direction.

This type of anisotropy appears as a natural consequence of the anisotropic growth explained above. *Yoshida et al.* [1996] proposed that preferred orientation of iron crystals occurs due to the flow induced by the anisotropic growth. This was calculated using a preferred orientation theory by *Kamb* [1959] and the elastic constants of *Stixtrude and Cohen* [1995]. According to *Kamb* [1959]'s theory, preferred orientation forms by recrystallization under stress, so that it minimizes the elastic strain energy. *Yoshida et al.* [1996] showed that for a deformation caused by anisotropic growth, deformation should be primarily governed by diffusion creep with Newtonian rheology because the stress level is very low, and argued that *Kamb* [1959]'s theory would be applicable. The elastic constants of *Stixrude and Cohen* [1995] was calculated at a core pressure but at 0 K. Recently *Steinle-Neumann et al.* [2001] obtained the elastic constants of hcp iron at core pressures to temperatures as high as 6000 K. They found that the axial ratio c/a increases with temperature, and as a result, the compressional wave becomes faster in the a-axis compared to the c-axis, in contrast to the results of *Stixrude and Cohen* [1995]. Using their elastic constants, we calculate preferred orientation of iron crystals in the inner core following the method described in *Yoshida et al.* [1996]. Fig.5 shows the calculated preferred orientation with the effect of compaction dominating at the uppermost 60 km. As can be seen from this figure, the a-axis, the fast P-wave direction of hcp iron, aligns parallel to the rotation axis in most of the region of the inner core. On the other hand, in the region dominated by compaction, the a-axis aligns in the radial direction because the principal stress axis is in the direction of gravity. As a result, P-waves traversing through the inner core would be fast in the north-south direction in most part of the inner core, and the axial anisotropy would be absent in the shallow parts, which is consistent with some seismic observations [e.g., *Song and Helmberger*, 1995, 1998; *Creager*, 2000; *Ouzounis and Creager*, 2001]. It is also important to note that a perfect alignment would not be required to explain the inner core seismic anisotropy of ~ 3%, because the degree of P-wave anisotropy of single crystal hcp iron obtained by *Steinle-Neumann et al.* [2001] is 13%. We also infer from the isochrons shown in Fig.4 that the anisotropy would be largest at the region close to the rotation axis, because the iron crystals in this region would have been under stress for the longest time, consistent with seismic observations by *Romanowicz et al.* [1996].

There may be conditions where the heat transfer would become larger at the poles, such as the case when a strong toroidal magnetic field inhibits convection out-

pole

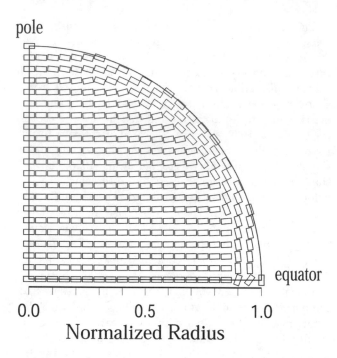

equator

0.0   0.5   1.0
Normalized Radius

Figure 5. An example of the preferred orientation of iron crystals in the inner core calculated using the method of *Yoshida et al.*, [1996] and using the elastic constants of *Steinle-Neumann et al.* [2001]. The mechanism of preferred orientation is assumed to be pressure solution. Grain boundary migration gives a similar result. The longer side of the rectangles represents the *c*-axis of an hcp crystal. We assumed that compaction dominates in the top 150 km beneath the ICB.

side the inner core tangent cylinder. We remark that even under such situations, we obtain the same pattern of preferred orientation, because the principal stress axis is the same. However there may be a stronger anisotropy near the equatorial plane because the crystals would have been under stress for a longer time.

There are other possible mechanisms for the formation of preferred orientation. In usual experimental situations, plastic strain energy is much larger than elastic strain energy, and Kamb's theory is not applicable. The very low stress level and the long time scale envisaged by *Yoshida et al.* [1996] are quite different from experimental conditions and Kamb's theory may still be applicable, but it is of interest to consider other mechanisms as well.

Jeanloz and Wenk [1988] first considered dislocation glide as the dominant deformation mechanism in the inner core. However they assumed thermal convection driven by internal heat generation, and did not provide a reason why the flow should be axisymmetric. The preferred orientation under dislocation glide is determined by the primary slip plane and the flow pattern in the

inner core. *Poirier and Price* [1999] proposed a method of determining the primary slip plane from elastic constants and stacking fault energies, and demonstrated that basal plane should be the primary slip plane for zero temperature ϵ-iron. The finite temperature effect calculated by *Steinle-Neumann et al.* [2001] reinforces the preference for basal slip with larger c/a ratio and smaller c_{66}/c_{44}. *Wenk et al.* [2000a, b] also clarified the preferred orientation of hcp iron under this mechanism. One of their important findings is that basal slip is dominant in determining the overall orientation even if prismatic slip is favored over basal slip. As a consequence, when a simple shear predominates, the a-axis becomes parallel to the shear plane. High pressure deformation experiments show that when the uniaxial strain predominates, the c-axis aligns in the axis of compression, and this alignment can be interpreted to be due to basal slip [*Wenk et al.*, 2000b]. *Wenk et al.* [2000a] assumed a poloidal degree 2 thermal convection in the inner core with a downwelling at the poles and upwellings at the equator, though we doubt its existence because the temperature in the inner core is likely to be subadiabatic as discussed above. They showed that the c-axis aligns in the direction of the rotation axis because of the strong shear near the equatorial plane. On the other hand, for the flow driven by aspherical growth [*Yoshida et al.*, 1996], the shear in the equatorial region is not strong and uniaxial strain is dominant. Therefore if we apply this alignment mechanism to the case of a predominantly equatorial growth, we would obtain the alignment of the a-axis parallel to the rotation axis, and vice versa for the polar growth. Assuming the elastic constants of *Steinle-Neumann et al.* [2001], the equatorial growth is in accordance with the seismic observations.

Maxwell stress due to the magnetic field may be important in driving convection in the inner core. *Karato* [1999] recently argued that the Maxwell stress would be the most important driving force. However, his model does not work as it is because he did not consider the pressure and buoyancy forces, which would balance the Lorentz force. *Buffett and Bloxham* [2000] examined the balance among the magnetic, pressure and buoyancy forces, and showed that complete equilibrium cannot be attained because of the incompatibility between thermodynamic and hydrostatic equilibrium conditions at the ICB. Thus a viscous flow would occur, but they showed that it is weak and confined to the region near the ICB. *Buffett and Wenk* [2001] next used the elastic constants by *Steinle-Neumann et al.* [2001], and considered the azimuthal Lorentz force to be important, and calculated the resulting elastic anisotropy. Because of the azimuthal shear, the a-axis becomes parallel to the rotation axis, and this is consistent with seismic observations if we adopt the elastic constants of *Steinle-Neumann et al.* [2001].

Buffett and Bloxham [2000]'s result also implies that heat transfer in the outer core is important because of thermodynamic equilibrium condition at the ICB. Hence, if we are to modify *Karato* [1999]'s model properly, appropriate heat transfer in the outer core is needed to maintain the inner core away from its hydrostatic shape from solidification and melting. Flow pattern in the outer core determines both the heat transfer and the Maxwell stress pattern at ICB, and their relationship is not yet clarified.

3.3. Effect of Thermally Heterogeneous Mantle

Apart from the rotational control of heat transfer, it is also likely that the core flow is controlled by a thermally heterogeneous mantle, which was first suggested by *Jones* [1977]. *Gubbins and Richards* [1986] noted the correlations between the structure of the lowermost mantle, and persistent features of the CMB. If some features persist for a period longer than the time scales of the fluid motion in the outer core, it implies that they should be produced under the influence of the mantle, because otherwise the geomagnetic features would drift eastward or westward. *Bloxham and Gubbins* [1987] suggested that the stationary flux lobes beneath North America and East Asia is the result of thermal core-mantle interaction. Paleomagnetic data up to 5 My ago have also indicated that these features are persistent (for a review, see *Gubbins*, [1997]; but see also *Kono et al.*[2000], for the problems of previous analyses.) It has also been pointed out that non axial dipole components of the geomagnetic field pattern in the Pacific and Atlantic hemispheres have different characteristics [*Johnson and Constable*, 1998; *Walker and Backus*, 1996]. Thus geomagnetic data up to 5 My ago do suggest that the geomagnetic field pattern is being controlled by the mantle.

We first note that the convective heat flux can be a small fraction of the total heat flux (i.e., the sum of convective and conductive heat flux) across the CMB. This is because the heat flux which is conducted down the adiabat is large. Assuming constant material properties, the adiabatic temperature in the core is expressed as

$$T = T_{ICB} \exp\{-\beta(r^2 - R_{ICB}^2)\} \quad (27)$$

with $\beta = \alpha\gamma/2C_p$, where α is the thermal expansion coefficient, C_p is the specific heat and γ is defined in Eq.(7). Taking $T_{ICB} = 4900$K, the melting point of iron [*Boehler*, 2000], we find that $T_{CMB} \simeq 4000$K and the heat flux at the CMB which is conducted down the

adiabat as $q_{ad}^{CMB} = 18$mW/m^2. A CMB heat flow of $Q_{CMB} = 3 \times 10^{12}$W is equivalent to a heat flux of $q_{CMB} = 20$W/m^2, and this implies that the convective heat flux in the core can be as small as 1/10 of the total heat flux, or even negative (subadiabatic) as pointed out by *Loper* [1978].

The smallness of convective heat flux implies its large lateral variation, which may amount to the order of the mean heat flux ($\delta q_{CMB} \sim q_{CMB}$), because we can expect a lateral temperature variation of the order of 1000 K in the mantle, whereas the temperature drop across the D" layer is also of the order of 1000 K [*Boehler*, 2000]. This means that the lateral variation of convective heat transfer becomes quite large. For example, if the convective transfer is about a tenth of the total heat flux ($q_{conv}^{CMB} \sim 0.1q_{CMB}$), this would lead to a lateral variation of the convective heat transfer of about a factor of 10:

$$\delta q_{conv}^{CMB} \sim 10 q_{conv}^{CMB}. \quad (28)$$

The situation is a little different at the ICB. The heat conducted down the adiabat is not a major part of the total heat flux as we show below. The adiabatic temperature gradient, derived from Eq. (27), is an increasing function of radius, and as a result the adiabatic temperature gradient is smaller at the ICB, giving $q_{ad}^{ICB} \sim 0.4q_{ad}^{CMB}$. On the other hand, the total heat flow at the ICB may be approximated as

$$Q_{ICB} = -C_{latent} \frac{d\Theta}{dt}. \quad (29)$$

Hence the heat flux at the ICB is given by

$$\frac{q_{ICB}}{q_{CMB}} = \frac{Q_{ICB}}{Q_{CMB}} \frac{4\pi R_{CMB}^2}{4\pi R_{ICB}^2} = \frac{C_{latent}}{C_{eff}} \frac{R_{CMB}^2}{R_{ICB}^2} \sim 3. \quad (30)$$

As a result, we have

$$\frac{q_{ad}^{ICB}}{q_{ICB}} \sim 0.1 \frac{q_{ad}^{CMB}}{q_{CMB}}, \quad (31)$$

and in contrast to CMB, the convective heat transfer is a major part of the ICB heat flux. Consequently, if the lateral variation of convection extends deep into the core, we would expect a large lateral variation of heat flux at the ICB, which can result in correspondingly large variation of inner core growth.

Because the core has a flow velocity which is faster than that of the mantle by about 5 orders of magnitude, the core would respond instantaneously to the thermal anomaly of the mantle. If we regard the D" layer as a thermal boundary layer for simplicity, we would expect a high heat flow beneath the cold mantle and vice versa. The resulting flow pattern in the core is, however, com-

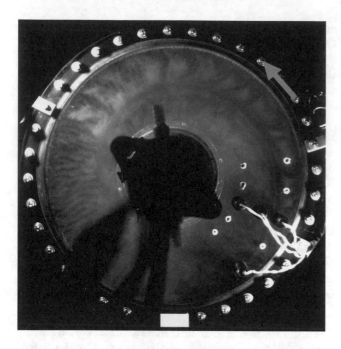

Figure 6. A photographic image of the planform of the convective pattern with a heterogeneous CMB heat flux [Reprinted with permission from *Sumita and Olson*, 1999. Copyright 1999 American Association for the Advancement of Science]. $Ra/Ra_c = 24$ and (peak CMB heat flux) / (mean CMB heat flux) = 95. Ekman number $Ek = 4.7 \times 10^{-6}$. The location of the heater is indicated by a white rectangle. Rotation is counterclockwise. The radially spiraling structure is the front, along which a geostrophic jet flows from the CMB to the ICB.

plicated by sphericity, rotation and the magnetic field. In the following sections, we first describe an experimental model of thermal core-mantle interaction, and then compare it with theoretical (analytical and numerical) models.

3.4. Experimental Model of Thermal Core-Mantle Interaction

There have been a few rotating convection experiments with heterogeneous temperature or heat flow boundaries [*Hart et al.*, 1986a,b; *Bolton and Sayler*, 1991; *Sumita and Olson*, 1999, 2002]. In particular, experiments using rapidly rotating spherical shells have the correct geometrical configuration and are capable of achieving low Ekman number and high Rayleigh number conditions that are similar to the condition in the outer core [*Sumita and Olson*, 2000]. Thermal convection in such experiments are characterized by fine-scale two dimensional plumes (geostrophic turbulence) which are advected westward by the zonal flow driven by the Reynolds stress.

Sumita and Olson [1999] modeled large lateral variations of convective heat flux in laboratory experiments at $Ra/Ra_c < 50$, with a heterogeneous CMB heat flux that has a peak value of up to 100 times its mean. Fig.6 shows an example of how the heterogeneous heat flux affects the convection. Under a homogeneous heat flux boundary condition, the zonal flow is always westward [*Sumita and Olson*, 2000]. However with a heterogeneous CMB heat flux, an eastward flow was generated from the high heat flux region. When the heat flow at the anomaly (i.e., heat flux × size of the anomalous region) exceeded the heat flow of the surrounding region (i.e., heat flux × size of the surrounding region) [*Sumita and Olson*, 2002], we find that these zonal flows flowing in the opposite directions converge, and form a sharp stationary front, along which a geostrophic (2D) jet flows from CMB to ICB (Fig.6). The large-scale flows can be understood by the balance between vortex stretching and buoyancy, with a large eastward phase shift relative to the large heat flux region resulting from advection (see next section for details). Apart from the large-scale flow, there are also fine-scaled radial flows such as plumes and jets (fronts), where non-linear effects become important. These radial flows advect the anomalous heat towards the inner core, thus allowing the thermal interaction to extend across the shell.

Based upon the experiments, *Sumita and Olson* [1999] deduced a core flow model as shown in Fig.7. Here, a high heat flow region is assumed to exist beneath east Asia, corresponding to a significant seismic high velocity region near the CMB common to many seismic tomography models [for a recent review see *Garnero* 2000]. The model of *Sumita and Olson* [1999] shows that there are basically two regions beneath the CMB: a low pressure region with a cold eastward flow originating from east Asia driven by the high heat flux region, and a high pressure surrounding region with a westward mean flow. This zonal flow pattern is similar to that obtained from geomagnetic secular variations under the tangentially geostrophic approximation [e.g., *Bloxham*, 1992], which is consistent with the 2D flows of the experiment. According to this interpretation, the geomagnetic westward drift is the result of the warm mean westward flow, which is inhibited by the cold eastward flow in the Pacific that originates from the high heat flow region in east Asia. The hemispherical dichotomy of the outer core temperature has been previously inferred from geomagnetic secular variation [*Bloxham and Jackson*, 1990] and seismology [*Tanaka and Hamaguchi*, 1993], and is consistent with this model.

The experiments also showed that a high convective heat flux region exists at the region immediately west of the front, which would lead to a hemispherical di-

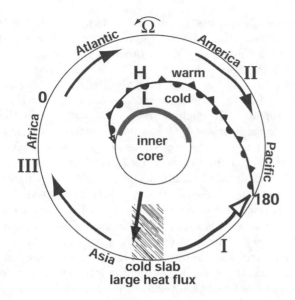

Figure 7. A schematic diagram of the model of core-mantle thermal interaction based on laboratory experiments [Reprinted with permission from *Sumita and Olson*, 1999. Copyright 1999 American Association for the Advancement of Science]. This is an equatorial planform of the Earth's core as seen from the north pole showing a CMB patch with a high heat flux (hatched area). Rotation is counterclockwise. The numbers indicate longitude. Black (white) headed arrows indicate warm (cold) radial flows. H and L indicate the high and low pressure regions, respectively. The shaded region at the ICB is the inferred region of fast solidification.

chotomy of inner core growth. The inferred fast inner core growth region agrees with the region of a large inner core seismic anisotropy [*Tanaka and Hamaguchi*, 1997; *Creager*, 1999, 2000; *Garcia and Souriau*, 2000, 2001]. This can be interpreted using the inner core growth model discussed in the previous section. The region of a larger anisotropy is related to fast inner core growth because of larger strain which drives recrystallization needed for the preferred orientation to form.

The experimental model shows a dichotomy of outer core flow beneath CMB and a dichotomy of inner core growth rate. Because of the eastward spiraling nature of the front, the pattern of inner core dichotomy is shifted east relative to the flow dichotomy beneath the CMB (In Fig.7 this is about 60°). The angle of phase shift is an increasing function of the heat flow at the high heat flux region because of larger advection of heat. This implies that it is possible to constrain the relative magnitude of anomalous total heat flow, provided that the size of the high CMB heat flow region is well constrained by seismic tomography.

If this model is applicable to the Earth, it can have other implications. For the lateral variation of inner core structure to form, the pattern of outer core flow

and the inner core should remain fixed relative to the mantle for the time scale of inner core growth, if *Yoshida et al.* [1996]'s model is valid. This rules out the long-term inner core rotation relative to the mantle. The gravitational torque between the mantle and the inner core [*Buffett*, 1996] would be the mechanism to prevent the relative rotation. It also indicates that there must be a mechanism for the pattern of lateral heterogeneity at the CMB to remain persistent for about ~ 1Ga, the time scale required for inner core growth to produce hemispherical dichotomy.

3.5. Theoretical Model of Thermal Core-Mantle Interaction

There have been a number of numerical studies on thermal core-mantle coupling and how it affects the geomagnetic field pattern [*Zhang and Gubbins*, 1992, 1993, 1996; *Yoshida and Hamano*, 1993; *Sun et al.*, 1994; *Olson and Glatzmaier*, 1996; *Sarson et al.*, 1997; *Glatzmaier and Roberts*, 1997; *Glatzmaier et al*, 1999; *Bloxham*, 2000a, b; *Yoshida and Shudo*, 2000; *Gibbons and Gubbins*, 2000] but a complete understanding including the non-linear and magnetic effects have not been achieved yet. However, the fluid mechanics of linear and weakly non-linear cases are relatively well understood.

Yoshida and Shudo [2000] carried out a detailed study of a linear response of the outer core flow to a thermally heterogeneous mantle. The basic equations they used are for no magnetic field and no basic flow; the velocity is determined by the thermal wind equation

$$2\Omega\frac{\partial v}{\partial z} = -\alpha g \times \nabla T, \qquad (32)$$

and the temperature is governed by the Laplace equation

$$\nabla^2 T = 0, \qquad (33)$$

where T is the temperature, Ω is the rotation vector, v is the velocity and z is the coordinate along the rotation axis. The thermal wind equation is obtained by taking the rotation of the equation of motion, and represents the balance of vorticity generation. They found rigorous analytical solutions for the inviscid problem [*Yoshida*, in preparation]. An example of such solutions is shown in Fig.8. It is for simulating the situation in *Sumita and Olson* [1999]'s experiment (Fig.6). Here a cold region with a Gaussian profile is imposed at the CMB at a longitude centered at 0°, with a half width of 30°. The obtained flow has a large-scale pattern similar to the experiment, with an eastward phase shift relative to the thermal anomaly. It consists of a cyclonic circulation near the cold region and the surrounding anticyclonic circulation. The equatorial section of the flow pattern

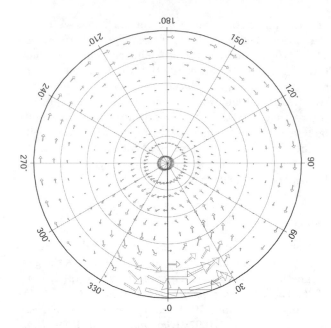

Figure 8. The velocity in the equatorial plane for a linear response of the outer core flow to a thermally heterogeneous mantle. This figure is drawn as seen from the north pole. The rotation is anticlockwise. A cold region with a Gaussian profile was imposed at the longitude 0°, with a half width of 30°. The velocity is constructed from an analytical solution for an inviscid fluid without the inner core [*Yoshida and Shudo*, 2000].

is easily understood by the quasi-geostrophic model of *Yoshida and Hamano* [1993], because the geostrophic component dominates in the equatorial section. In the region where the temperature increases eastward, anti-cyclonic vorticity is generated due to buoyancy. In order to balance the vorticity generation, the geostrophic flow should move inward to produce cyclonic vorticity. That is why downwelling occurs to the east of the cold region.

The similarity of the large-scale flow between the linear theory and the experiment may seem surprising, if one considers the large Péclet number $Pe = VL/\kappa \sim$ 1800 of the experiment, where the shell thickness is taken to be L [*Sumita and Olson*, 2000]. This may be because of the thermal convection occurring in the experiment. Its turbulence would give rise to a large effective eddy diffusion, thereby reducing the effective Péclet number.

Nonlinear effects are of course prominent in the experiment. First, jets and fronts are formed. Second, the eastward phase shift is large and shows spiraling. Also the phase shift increases with the magnitude of the thermal anomaly. These phenomena result from the advective term. Advection is important for fine-scaled features such as plumes whose velocity and length scales are properly understood when non-linear terms are con-

sidered [*Cardin and Olson*, 1994; *Aubert et al.*, 2001; *Sumita and Olson*, 2002]. Improvement in theoretical understanding is necessary in order to understand these nonlinear features well.

The Lorentz force has been omitted in the discussion so far, and the flow was assumed to be nearly geostrophic. Although it seems that the flow in the core is characterized by a columnar pattern, as is supported by the presence of equatorially symmetric flux lobes [*Gubbins and Bloxham*, 1987], Lorentz force is generally considered to be important for the force balance (Elsasser number $\sim O(1)$). In a linear response theory, *Yoshida and Hamano* [1993] showed that the eastward phase shift is suppressed when the Elsasser number $> O(1)$. The apparent consistency of interpreting the core structure by an eastward phase shift may indicate that the advective effects are large enough as to produce a net eastward phase shift.

4. FUTURE PROSPECTS

We describe several future prospects that are particularly important to evaluate the different models and for quantitative comparison with the observation.

A better understanding of the properties of iron at core conditions is a key to an improved quantitative comparison between the models and seismic observations. Evaluation of properties such as diffusion coefficient in the fluid and solid iron would be helpful to understand deformation mechanism in the inner core.

A detailed study on the pattern and variation of heat transfer in the outer core is important to evaluate several candidates of inner core anisotropy. The question of whether heat transfer has a maximum at the equator or at the poles is important to several anisotropy models. If the elastic constants of *Steinle-Neumann et al.* [2001] is adopted, solidification texturing of hcp-alloy [*Bergman*, 1997] requires a polar dominant heat transfer. As already discussed, for a dislocation glide mechanism [*Wenk et al.*, 2000a, 2000b] with a flow driven by anisotropic growth [*Yoshida et al.* 1996], an equatorial dominant heat transfer is compatible. The amplitude of the variation is important for the kinetics of preferred orientation. For example, it affects the strain energy required to drive reorientation by recrystallization under stress [*Yoshida et al.*, 1996] by its square. The weakly non-linear calculations of *Zhang* [1991] indicate that the amplitude of heat flux variation with latitude increases with Rayleigh number. Strong field dynamo calculation by *Olson et al.* [1999] indicates a factor of 7 larger heat transfer at the equator as compared to the poles. There is, however, no systematic parameter study on this problem, which will be needed in the future.

3D seismic tomographies near the CMB with a better spatial resolution in the lateral and radial directions are important to constrain the size of the heterogeneity and the radial temperature gradient distribution, respectively. The amplitude of the heterogeneity is needed to constrain the lateral contrast of heat flux. Since CMB is also known to be a place of chemical heterogeneity, using both P and S waves [e.g., *Wysession et al.*, 1999] would also become important to separate the thermal and compositional effects on seismic anomaly.

In this paper, we showed that thermal interactions between the mantle, outer and inner cores can explain a number of major dynamical structures in the core. The inner core growth is sensitive to how heat is being transferred by convection in the outer core. This implies that the inner core structure can be a good recorder of the heat flow pattern and the thermal history of the outer core and mantle. A continued cross-disciplinary research of the thermally coupled mantle, outer and inner cores should give us a better understanding of how the Earth's interior operates as a system.

APPENDIX: DERIVATION OF EQ.(20)

Here we briefly describe the derivation of Eq.(20). The derivation is based on the theory of *Lister and Buffett* [1998], but modified for the chemical stratification.

Under the hydrostatic approximation, pressure in the core can be expressed as

$$p(r) = p_0 - \Delta p \left(\frac{r}{R_{CMB}} \right)^2, \qquad (A1)$$

where p_0 is the pressure at the center of the Earth, and Δp is defined as

$$\Delta p = \frac{1}{2} \gamma \rho R_{CMB}^2. \qquad (A2)$$

The radial temperature profile in the convecting layer T^- is then given by the adiabatic relationship as

$$T^-(r,t) = \Theta^-(t) - \frac{dT_{\mathrm{ad}}}{dp} \Delta p \left(\frac{r}{R_{CMB}} \right)^2, \qquad (A3)$$

where the adiabatic temperature gradient dT_{ad}/dp is assumed to be constant, and Θ^- is the potential temperature of the convecting lower layer. The potential temperature is defined here as the temperature which the fluid element would have if it were brought adiabatically to the center of the Earth. The temperature distribution of the stratified upper layer T^+ is also given in the form

$$T^+(r,t) = \Theta^+(t) - \frac{dT_{\mathrm{ad}}}{dp} \Delta p \left(\frac{r}{R_{CMB}} \right)^2, \qquad (A4)$$

if the heat flow in the stratified layer is equal to the heat conducting down the adiabat. Here Θ^+ is the potential temperature of the upper layer. By substituting the expression (A4) into the thermal conduction equation, we obtain

$$\Theta^+(t) = T_0 - 6 \frac{dT_{\mathrm{ad}}}{dp} \Delta p \left(\frac{t}{\tau_{\mathrm{therm}}} \right), \qquad (A5)$$

where T_0 is the temperature at the center of the Earth when the inner core began to form, and τ_{therm} is the thermal diffusion time given by

$$\tau_{\mathrm{therm}} = \frac{R_{CMB}^2}{\kappa}, \qquad (A6)$$

where κ is the thermal diffusivity.

We can obtain the expression for the potential temperature of the convecting layer $\Theta^-(t)$ by using the equilibrium condition at the ICB,

$$T^-(R_{IC},t) = T_L(R_{IC},t), \qquad (A7)$$

Where T_L is the liquidus curve given by

$$T_L(R_{IC},t) = T_0 \quad + \quad \frac{\partial T_L}{\partial p} (p_{ICB} - p_0)$$
$$+ \quad \frac{\partial T_L}{\partial C} (C_{ICB} - C_0). \quad (A8)$$

From Eqs. (A1), (A3), (A7), and (A8), we obtain

$$\Theta^-(t) = T_0 \quad + \quad \left(\frac{dT_{\mathrm{ad}}}{dp} - \frac{\partial T_L}{\partial p} \right) \Delta p \left(\frac{R_{IC}}{R_{CMB}} \right)^2$$
$$+ \quad \frac{\partial T_L}{\partial C} (C^- - C_0). \qquad (A9)$$

On the other hand, from Eqs.(11) and (12), we obtain the concentration of light elements in the convecting layer as

$$C^- - C_0 = \Delta C \left(\frac{R_s}{R_{CMB}} \right)^2. \qquad (A10)$$

Substituting Eq. (A10) into Eq.(A9), we have

$$\Theta^-(t) = T_0 \quad + \quad \left(\frac{dT_{\mathrm{ad}}}{dp} - \frac{\partial T_L}{\partial p} \right) \Delta p \left(\frac{R_{IC}}{R_{CMB}} \right)^2$$
$$+ \quad \frac{\partial T_L}{\partial C} \Delta C \left(\frac{R_s}{R_{CMB}} \right)^2. \qquad (A11)$$

At the early stage of the inner core growth ($R_{IC} \ll R_{CMB}$), we can approximate Eq. (A11) as

$$\Theta^-(t) = T_0 + \frac{\partial T_L}{\partial C} \Delta C \left(\frac{R_s}{R_{CMB}} \right)^2. \qquad (A12)$$

Finally, the time evolution of R_s can be obtained from the heat budget of the convective layer, which is written as

$$\frac{4\pi}{3} R_s^3 \rho C_p \frac{d\Theta^-}{dt}$$

$$= 4\pi \left(R_{IC}^2 \frac{dR_{IC}}{dt} \rho L + R_s^2 \rho C_p \kappa \frac{dT^+}{dr}(R_s) \right.$$

$$\left. + R_s^2 \frac{dR_s}{dt} \rho C_p (T^+(R_s) - T^-(R_s)) \right), \quad (A13)$$

where C_p is the specific heat and L is the latent heat of the solidification of the inner core. Here we neglect the conduction from the inner core, and the release of the gravitational energy due to light element because it is small at the early stage of the inner core growth. In addition, we use $R_s \gg R_{IC}$, which is valid at the early stage. Using Eqs. (19), (A3), (A4), (A5), and (A12), we can rewrite Eq. (A13) as

$$\left[\frac{1}{3} \frac{\Delta C}{C_0} \left(\frac{5}{2} \mathcal{C} - \mathcal{L} \right) \left(\frac{R_s}{R_{CMB}} \right)^2 \right.$$

$$\left. + 3 \frac{t}{\tau_{\text{therm}}} \right] \frac{1}{R_s} \frac{dR_s}{dt} = -\frac{1}{\tau_{\text{therm}}}, \quad (A14)$$

where $\mathcal{C}$ and $\mathcal{L}$ are non-dimensional parameters defined as

$$\mathcal{C} = \frac{\partial T_L}{\partial C} \frac{C_0}{\frac{dT_{\text{ad}}}{dp} \Delta p} \quad (A15)$$

and

$$\mathcal{L} = \frac{L}{C_p \frac{dT_{\text{ad}}}{dp} \Delta p}. \quad (A16)$$

The solution of Eq. (A14) is

$$\frac{R_s}{R_{CMB}} = \sqrt{15 \frac{C_0}{\Delta C} \left(\mathcal{L} - \frac{5}{2} \mathcal{C} \right)^{-1} \frac{t}{\tau_{\text{therm}}}}, \quad (A17)$$

which is Eq.(20).

Acknowledgments. We thank Ulrich Christensen and an anonymous referee for carefully reading the manuscript, Kei Kurita for comments about the initial stratification of the outer core, and Ron Cohen for kindly sharing the results prior to publication. This work was supported by Japan Society for the Promotion of Science (I.S.) and by Special Coordination Fund "Earth Simulator Project" from MEXT (Ministry of Education, Culture, Sports, Science and Technology, Japan) (S.Y.). Part of this work was carried out during S.Y.'s stay at the University of Leeds, which was supported by the British Council Fellowship.

REFERENCES

Abbott, D., L. Burgess, J. Longhi, and W. H. F. Smith, An empirical thermal history of the Earth's upper mantle, *J. Geophys. Res., 99,* 13835-13850, 1994.

Abe, Y. Thermal and chemical evolution of the terrestrial magma ocean, *Phys. Earth Planet. Inter., 100,* 27-39, 1997.

Aubert, J., D. Brito, H-C. Nataf, P. Cardin, and J-P. Masson, A systematic experimental study of rapidly rotating spherical convection in water and liquid gallium, *Phys. Earth Planet. Inter., 128* 2001.

Bergman, M. I., Measurements of elastic anisotropy due to solidification texturing and the implications for the Earth's inner core, *Nature, 389,* 60-63, 1997.

Bloxham, J., The Steady part of the secular variation of the Earth's magnetic field, *J. Geophys. Res., 97,* 19565-19579, 1992.

Bloxham, J., The effect of thermal core-mantle interactions on the paleomagnetic secular variation, *Philos. Trans. R. Soc. London, A 358,* 1171-1179, 2000a.

Bloxham, J., Sensitivity of the geomagnetic axial dipole to thermal core-mantle interactions, *Nature, 405,* 63-65, 2000b.

Bloxham, J., and D. Gubbins, Thermal core-mantle interactions, *Nature, 325,* 511-513, 1987.

Bloxham, J., and A. Jackson, Lateral temperature variations at the core-mantle boundary deduced from the magnetic field, *Geophys. Res. Lett., 17,* 1997-2000, 1990.

Boehler, R., Melting temperature of the Earth's mantle and core: Earth's thermal structure, *Annu. Rev. Earth Planet. Sci., 24,* 15-40, 1996.

Boehler, R., High-pressure experiments and the phase diagram of lower mantle and core materials, *Rev. Geophys., 38,* 221-245, 2000.

Bolton, E. W., and B. S. Sayler, The influence of lateral variations of thermal boundary conditions on core convection: Numerical and laboratory experiments, *Geophys. Astrophys. Fluid Dynam., 60,* 369-370, 1991.

Buffett, B. A., Gravitational oscillations in the length of day, *Geophys. Res. Lett., 25,* 2279-2282, 1996.

Buffett, B. A., Dynamics of the Earth's Core, in *Mineral Physics and Seismic Tomography From the Atomic to Global Scale,* edited by S. Karato et al., pp. 37-62, Am. Geophys. Un., Washington DC, 2000.

Buffett, B. A., and J. Bloxham, Deformation of Earth's Inner Core by Electromagnetic Forces, *Geophys. Res. Lett., 27,* 4001-4004, 2000.

Buffett, B. A., H. E. Huppert, J. R. Lister, and A. W. Woods, Analytical model for solidification of the Earth's core, *Nature, 356,* 329-331, 1992.

Buffett, B. A., H. E. Huppert, J. R. Lister, and A. W. Woods, On the thermal evolution of the Earth's core, *J. Geophys. Res., 101,* 7989-8006, 1996.

Buffett, B. A., and H-R. Wenk, Texturing of the Earth's inner core by Maxwell stresses, *Nature, 413,* 60-63, 2001.

Busse, F. H., Thermal instabilities in rapidly rotating systems, *J. Fluid Mech., 44,* 441-46, 1971.

Busse, F. H., and C. R. Carrigan, Laboratory simulation of thermal convection in rotating planets and stars, *Science, 191,* 81-83, 1976.

Campbell, I. H., and R. W. Griffiths, The changing nature of mantle hotspots through time: implications for the chemical evolution of the mantle, *J. Geol., 100,* 497-523, 1992.

Cardin, P., and P. Olson, Chaotic thermal convection in a rapidly rotating spherical shell: consequences for flow in the outer core, *Phys. Earth Planet. Inter., 82*, 235-259, 1994.

Christensen, U., Thermal models for the Earth, *J. Geophys. Res., 90*, 2995-3008, 1985.

Creager, K. C., Large-scale variations in inner core anisotropy, *J. Geophys. Res., 104*, 23127-23139, 1999.

Creager, K. C., Inner core anisotropy and rotation, in *Mineral Physics and Seismic Tomography From the Atomic to Global Scale*, edited by S. Karato et al., pp. 89-114, Am. Geophys. Un., Washington DC 2000.

Davies, G. F., Ocean bathymetry and mantle convection 1. Large-scale flow and hotspots, *J. Geophys. Res., 93*, 10467-10480, 1988.

Fearn, D. R., D. E. Loper, and P. E. Roberts, Structure of the Earth's inner core, *Nature, 292*, 232-233, 1981.

Fukai, Y., The iron-water reaction and the evolution of the Earth, *Nature, 308*, 174-175, 1984.

Fukai, Y., Some properties of Fe-H system at high pressure and temperatures and their implications for the Earth's core, in *High-Pressure research: Application to Earth and Planetary Sciences*, Edited by Y. Syono and M. H., Manghnani, pp. 373-385, Terra Sci. Pub. Comp., Tokyo, 1992.

Garcia, R., and A. Souriau, Inner core anisotropy and heterogeneity level, *Geophys. Res. Lett., 19*, 3121-3124, 2000.

Garcia, R., and A. Souriau, Correction to "Inner core anisotropy and heterogeneity level" by Raphaël Garcia, and Annie Souriau, *Geophys. Res. Lett., 28*, 85-86, 2001.

Garnero, E. J., Heterogeneity of the lowermost mantle, *Annu. Rev. Earth Planet. Sci., 142*, 631-642, 2000.

Gibbons, S. J., and D. Gubbins, Convection in the Earth's core driven by lateral variations in the core-mantle boundary heat flux, *Geophys. J. Int., 142*, 631-642, 2000.

Glatzmaier, G. A., R. C. Coe, L. Hongre, and P. H. Roberts, The role of the Earth's mantle in controlling the frequency of geomagnetic reversals, *Nature, 401*, 885-890, 1999.

Glatzmaier, G. A., and P. H. Roberts, Simulating the geodynamo, *Contemp. Phys., 38*, 269-288, 1997.

Green, D. H., Petrogenesis of Archaean ultramafic magmas and implications for Archaean tectonics, in *Precambrian Plate Tectonics*, edited by A. Kröner, pp. 469-489, Elsevier, Amsterdam, 1981.

Gubbins, D., Energetics of the Earth's core, *Geophys. J., 43*, 453-464, 1977.

Gubbins, D., Interpreting the paleomagnetic field, in *The core-mantle boundary region*, edited by M. Gurnis, M. E. Wysession, E. Knittle, B. A. Buffett, pp. 167-182, Am. Geophys. Un., Washington DC 1997.

Gubbins, D., and J. Bloxham, Morphology of the geomagnetic field and implications for the geodynamo, *Nature, 325*, 509-511, 1987.

Gubbins, D., and T. G. Masters, Driving mechanisms for the Earth's dynamo, *Adv. Geophys., 21*, 1-50, 1979.

Gubbins, D., T. G. Masters, and J. A. Jacobs, Thermal evolution of the Earth's core, *Geophys. J. R. astr. Soc., 59*, 57-99, 1979.

Gubbins, D., and M. Richards, Coupling of the core dynamo and mantle: thermal or topographic? *Geophys. Res. Lett., 13*, 1521-1524, 1986.

Hale, C. J., and D. J. Dunlop, Evidence for an Archean geomagnetic field: a paleomagnetic study of the Komati formation, Barberton greenstone belt, South Africa, *Geophys. Res. Lett., 11*, 97-100, 1984.

Hart, J. E., J. Toomre, A. E. Deane, N. E. Hurlburt, G. A. Glatzmaier, G. H. Fichtl, F. Leslie, W. W. Fowlis, and P. W. Gilman, Laboratory Experiments on Planetary and Stellar Convection Performed on Spacelab 3, *Science, 234*, 61-64, 1986a.

Hart, J. E., G. A. Glatzmaier, and J. Toomre, Space-laboratory and numerical simulations of thermal convection in a rotating hemispherical shell with radial gravity, *J. Fluid Mech., 173*, 519-544, 1986b.

Iida, T., and R. I. L. Guthrie, The physical properties of liquid metals, Clarendon Press, Oxford, 1988.

Jacobs, J. A., The Earth's inner core, *Nature, 172*, 297-298, 1953.

Jeanloz, R., The nature of the Earth's core, *Annu. Rev. Earth Planet. Sci., 18*, 357-386, 1990.

Jeanloz, R., and H. R. Wenk, Convection and anisotropy of the inner core, *Geophys. Res. Lett., , 15*, 72-75, 1988.

Johnson, C. L., and C. G. Constable, Persistently anomalous Pacific geomagnetic fields, *Geophys. Res. Lett., 25*, 1011-1014, 1998.

Jones, G. M., Thermal interaction of the core and the mantle and long-term behaviour of the geomagnetic field, *J. Geophys. Res., 82*, 1703-1709, 1977.

Kamb, W. B., Theory of preferred crystal orientation developed by crystallization under stress, *J. Geol. 67*, 153-170, 1959.

Karato, S., Seismic anisotropy of the Earth's inner core resulting from flow induced by Maxwell stresses, *Nature, 402*, 871-873, 1999.

Kato, T., and A. E. Ringwood, Melting relationships in the system Fe-FeO at high pressures: Implications for the composition and formation of Earth's core, *Phys. Chem Minerals, 16*, 524-538, 1989.

Kono, M., H. Tanaka, and H. Tsunakawa, Spherical harmonic analysis of paleomagnetic data: the case of linear mapping, *J. Geophys. Res., 105*, 5817-5833, 2000.

Kumazawa, M., S. Yoshida, T. Ito, and H. Yoshioka, Archaean-Proterozoic boundary interpreted as a catastrophic collapse of the stable density stratification in the core, *J. Geol. Soc. Japan, 100*, 50-59, 1994.

Kuramoto, K., and T. Matsui, Partitioning of H and C between the mantle and core during the core formation in the Earth: Its implications for the atmospheric evolution and redox state of early mantle, *J. Geophys. Res., 101*, 14909-14932, 1996.

Labrosse, S., J-P. Poirier, and J-L. Le Mouël, On cooling of the Earth's core, *Phys. Earth Planet. Inter., 99*, 1-17, 1997.

Labrosse, S., J-P. Poirier, and J-L. Le Mouël, The age of the inner core, *Phys. Earth Planet. Inter., 190*, 111-123, 2001.

Layer, P. W., A. Kröner, and M. McWilliams, An archean geomagnetic reversal in the Kaap Valley pluton, South Africa, *Science, 273*, 943-946, 1996.

Lister, J. R., and B. A. Buffett, Stratification of the outer core at the core-mantle boundary, *Phys. Earth Planet. Inter., 105*, 5-19, 1998.

Loper, D. E., The gravitationally powered dynamo, *Geophys. J. R. astron. soc., 54*, 389-404, 1978.

Loper, D. E., Structure of the inner core boundary, *Geophys. Astrophys. Fluid. Dyn., 22*, 139-155, 1983.

McElhinny, M. W., and W. E. Senanayake, Paleomagnetic evidence for the existence of the geomagnetic field 3.5 Ga ago, J. Geophys. Res., 85, 3523-3528, 1980.

Mollett, S., Thermal and magnetic constrains on the cooling of the Earth, Geophys. J. R. astr. Soc., 76, 653-666, 1984.

Muromachi, Y., Chemical interaction between the core and the mantle throughout the earth's history, PhD thesis, Univ. of Tokyo, 1991.

Ohta, H., S. Maruyama, E. Takahashi, Y. Watanabe, and Y. Kato, Field occurrence, geochemistry and petrogenesis of the Archean mid-oceanic ridge basalts (AMORBs) of the Cleaverville area, Pilbara craton, Western Australia, Lithos, 37, 199-221, 1996.

Ohtani, E., and A. E. Ringwood, Composition of the core, I. Solubility of oxygen in molten iron at high temperatures, Earth Planet Sci. Lett., 71, 85-93, 1984.

Ohtani, E., A. E. Ringwood, and W. Hibberson, Composition of the core, II. Effect of high pressure on solubility of FeO in molten iron, Earth Planet Sci. Lett., 71, 94-103, 1984.

Okuchi, T., Hydrogen partitioning into molten iron at high pressure: Implications for Earth's core, Science, 278, 1781-1784, 1997.

Okuchi, T., The melting temperature of iron hydride at high pressures and its implications for the temperature of the Earth's core, J. Phys.: Condens. Matter, 10, 11595-11598, 1998.

Olson, P., U. Christensen, and G. A. Glatzmaier, Numerical modeling of the geodynamo: Mechanisms of field generation and equilibration, J. Geophys. Res., 104, 10383-10404, 1999.

Olson, P., and G. A. Glatzmaier, Magnetoconvection and thermal coupling of the Earth's core and mantle, Phil. Trans. R. Soc. Lond., A354, 1413-1424, 1996.

Ouzounis, A., and K. C. Creager, Isotropy overlying anisotropy at the top of the inner core, Geophys. Res. Lett., 28, 4331-4334, 2001.

Poirier, J-P., Light elements in Earth's outer core: A critical review, Phys. Earth Planet. Inter., 18, 319-337, 1994.

Poirier, J. P., and G. D. Price, Primary slip system of ϵ-iron and anisotropy of the Earth's inner core, Phys. Earth Planet. Inter., 110, 147-156, 1999.

Romanowicz, B., X-D. Li, and J. Durek, Anisotropy in the Inner Core: Could it be due to low order convection ?, Science, 260, 1312-1314, 1996.

Sarson, G. R., C. A. Jones, and A. W. Longbottom, The influence of boundary region heterogeneities on the geodynamo, Phys. Earth Planet. Inter., 101, 13-32, 1997.

Sleep, N. H., Hot spots and mantle plumes: some phenomenology, J. Geophys. Res., 95, 6715-6736, 1990.

Sleep, N. H., and R. T. Langan, Thermal evolution of the Earth: some recent developments, Adv. Geophys., 23, 1-23, 1981.

Song, X-D., Anisotropy of the Earth's Inner Core, Rev. Geophys. 35, 277-313, 1997.

Song, X-D., and D. V. Helmberger, Depth dependence of anisotropy of Earth's inner core, J. Geophys. Res., 100, 9805-9816, 1995.

Song, X-D., and D. V. Helmberger, Seismic Evidence for an Inner Core Transition Zone, Science, 282, 924-927, 1998.

Souriau, A., and M. Souriau, Ellipticity and density at the inner core boundary from subcritical PKiKP and PcP data, Geophys. J. Int., 98, 39-54, 1989.

Stacey, F. D., Physics of the Earth, Brookfield Press, Brisbane, Australia, 1992.

Stacey, F. D., and D. E. Loper, Thermal histories of the core and mantle, Phys. Earth Planet. Inter., 36, 99-115, 1984.

Steinle-Neumann, G., L. Stixrude, R. E. Cohen, and O. Gülseren, Elasticity of iron at the temperature of the Earth's inner core, Nature, 413, 57-60, 2001.

Stevenson, D. J., Hydrogen in the Earth's core, Nature, 268, 130-131, 1977.

Stevenson, D. J., Models of the Earth's core, Science, 214, 611-619, 1981.

Stevenson, D. J., Fluid dynamics of core formation, in Origin of the Earth, edited by H. E. Newsom, and J. H. Jones, pp.231-249, Oxford University Press, New York, 1990.

Stevenson, D. J., Formation and evolution of terrestrial planetary cores, in abstracts of the 3rd SEDI Symposium, Mizusawa, 1992.

Stevenson, D. J., T. Spohn, and G. Schubert, Magnetism and thermal evolution of the terrestrial planets, Icarus, 54, 466-489, 1983.

Stixrude, L., and R. E. Cohen, High-pressure elasticity of iron and anisotropy of the Earth's inner core, Science, 267, 1972-1975, 1995.

Sumita, I., and P. Olson, A Laboratory model for convection in Earth's core driven by a thermally heterogeneous mantle, Science, 286, 1547-1549, 1999.

Sumita, I., and P. Olson, Laboratory experiments on High Rayleigh number thermal convection in a Rapidly Rotating Hemispherical shell, Phys. Earth Planet. Inter., 117, 153-170, 2000.

Sumita, I., and P. Olson, Thermal convection experiments in a hemispherical shell with heterogeneous boundary heat flux: Implications for the Earth's core, J. Geophys. Res., 2002 (in press).

Sumita, I., S. Yoshida, Y. Hamano, and M. Kumazawa, A model for the structural evolution of the Earth's core and its relation to the observations, in The Earth's Central Part: Its Structure and Dynamics, edited by T. Yukutake, pp. 231-261, Terra Sci. Tokyo, 1995.

Sumita, I., S. Yoshida, M. Kumazawa, and Y. Hamano, A model for sedimentary compaction of a viscous medium and its application to inner-core growth, Geophys. J. Int., 124, 502-524, 1996.

Sun, Z-P., G. Schubert, and G. A. Glatzmaier, Numerical simulations of thermal convection in a rapidly rotating spherical shell cooled inhomogeneously from above, Geophys. Astrophys. Fluid Dyn., 75, 199-226, 1994.

Tackley, P. J., D. J. Stevenson, G. A. Glatzmaier, and G. Schubert, Effects of multiple phase transitions in a three-dimensional spherical model of convection in Earth's mantle, J. Geophys. Res., 99, 15877-15901, 1994.

Tanaka, S., and H. Hamaguchi, Degree one heterogeneity and hemispherical variation of anisotropy in the inner core from PKP(BC)-PKP(DF) times, J. Geophys. Res., 102, 2925-2938, 1997.

Tromp, J., Inner-core anisotropy and rotation, Annu. Rev. Earth Planet. Sci., 29, 47-69, 2001.

Turner, J. S., Buoyancy effects in fluids, Cambridge Univ. Press, 368pp, 1973.

Walker, A. D., and G. E. Backus, On the difference between the average values of B_r^2 in the Atlantic and Pacific hemispheres, Geophys. Res. Lett., 23, 1965-1968, 1996.

Wenk, H-R., J. R. Baumgardner, R. A. Lebenson, and C.

N. Tome, A convection model to explain anisotropy of the inner core, *J. Geophys. Res., 105*, 5663-5677, 2000a.

Wenk, H-R., S. Matthies, R. J. Hemley, H-K. Mao, and J Shu, The plastic deformation of iron at pressures of the Earth's inner core, *Nature, 405*, 1044-1047, 2000b.

Williams, G. E., Resonances of the fluid core for a tidally decelerating Earth: cause of increased plume activity and tectonothermal reworking events? *Earth Planet. Sci. Lett., 128*, 155-167, 1994.

Wysession, M. E., A. Langenhorst, M. J. Fouch, K. M. Fischer, G. I. Al-Eqabi, P. J. Shore, and T. J. Clarke, Lateral variations in compressional/shear velocities at the base of the mantle, *Science, 284*, 120-125, 1999.

Yagi, T., and T. Hishinuma, Iron hydride formed by the reaction of iron, silicate, and water: Implications for the light element of the Earth's core, *Geophys. Res. Lett., 22*, 1933-1936, 1995.

Yoshida, S., and Y. Hamano, Fluid Motion of the Outer Core in Response to a Temperature Heterogeneity at the Core-Mantle Boundary and Its Dynamo Action, *J. Geomag. Geoelectr., 45*, 1497-1516, 1993.

Yoshida, S., and E. Shudo, Linear response of the outer core fluid to the thermal heterogeneity on the core-mantle boundary, in *abstracts of the 7th SEDI symposium, Exeter*, 66, 2000.

Yoshida, S, I. Sumita, and M. Kumazawa, Growth Model of the Inner Core Coupled with Outer Core Dynamics and the Resulting Elastic Anisotropy, *J. Geophys. Res., 101*, 28085-28103, 1996.

Yoshida, S., I. Sumita, and M. Kumazawa, Models of the anisotropy of the Earth's inner core, *J. Phys. Condens. Matter, 10*, 11215-11226, 1998.

Yoshihara, A., and Y. Hamano, Constraints on paleointensities of the Archean geomagnetic obtained from komatiites of the Barberton and Belingwe greenstone belts, in *abstracts of the 7th SEDI symposium, Exeter*, 117, 2000.

Yukutake, T., Implausibility of convection in the Earth's inner core, *Phys. Earth Planet. Inter., 108*, 1-13, 1998.

Zhang, K., Convection in a rapidly rotating spherical shell at infinite Prandtl number: Steadily drifting rolls, *Phys. Earth Planet. Inter., 68*, 156-169, 1991.

Zhang, K., Spiralling columnar convection in a rapidly rotating spherical fluid shells, *J. Fluid Mech., 236*, 535-556, 1992.

Zhang, K., and D. Gubbins, On convection in the earth's core driven by lateral temperature variations in the lower mantle, *Geophys. J. Int., 108*, 247-255, 1992.

Zhang, K., and D. Gubbins, Convection in a rotating spherical fluid shell with an inhomogeneous temperature boundary condition at infinite Prandtl number, *J. Fluid Mech., 250*, 209-232, 1993.

Zhang, K., and D. Gubbins, Convection in a rotating spherical shell with an inhomogeneous temperature boundary condition at finite Prandtl number, *Phys. Fluids, 8*, 1141-1148, 1996.

_____

I. Sumita, Department of Earth Sciences, Faculty of Science, Kanazawa University, Kanazawa, 920-1192, Japan. (e-mail: sumita@hakusan.s.kanazawa-u.ac.jp)

S. Yoshida, Department of Earth and Planetary Sciences, Nagoya University, Nagoya, 464-8602, Japan. (e-mail: yoshida@eps.nagoya-u.ac.jp)

Decadal oscillations of the Earth's core, angular momentum exchange, and inner core rotation

Stephen Zatman

Washington University in Saint Louis, Missouri

Studies have both suggested and disputed the seismic detection of a differential rotation of the inner core with respect to the mantle. Such differential rotation may have importance for the angular momentum budget of the Earth. Proposed gravitationalcoupling mechanisms imply different relationships between the torque on the mantle and the rate of rotation of the inner core depending on whether the inner core is rigid or viscously deforming. Inner core rotation that is steady or near-steady on long timescales cannot be directly inferred geomagnetically because its releationship to the core flow are the core's surface is unknown. However theory predicts (and observations confirm) that on decadal timescales the flow at the surface of the core reflects that deep within the core. Therefore it is possible to use an observational time-varying model of flow at the core surface in the region that is dynamically linked to the inner core (i.e. the tangent cylinder) to infer decadal variations in inner core rate of rotation. There is a high correlation between estimates of the inner core rate of rotation from the northern and southern hemisphere, although disagreement in amplitude that may result from damping. We use these estimates to test the predicted relationships between mantle torque and variations in inner core rate of rotation: we find that these are inconsistent with our model. If the flow model reliably reproduces the decadal behaviour of the core, this implies that the gravitational couple is not a dominant form of core-mantle coupling on decadal timescales.

1. INTRODUCTION

Models of flow at the surface of the core may be constructed from models of the Earth's magnetic field and its time variation, which are themselves inverted from measurements taken at or above the surface of the Earth. There is a natural ambiguity in the flows calculated in this manner, which may be eliminated or reduced by imposing one or more of several dynamical assumptions when constructing the flow: generally, that the flows are steady in time [*Voorhies and Backus,* 1985], that the flows are toroidal (no upwelling or downwelling at the surface of the core) [*Whaler,* 1980], or that the flows are tangentially geostrophic (that the horizontal part of the force balance in the flow at the surface of the core is primarily between pressure and the Coriolis force) [*Hills,* 1979; *LeMouël,* 1984].

On timescales of centuries and longer, the underlying dynamics of the Earth's core are expected to be primarily magnetostrophic [*Braginsky,* 1964; *Hide,* 1969], i.e. a balance between pressure, buoyancy, Coriolis and Lorentz forces, in which case it is difficult to find a simple relationship between the observed flows at the core's surface and the underlying dynamics of the core. Additionally, it has been suggested that the parts of the

Mountain Building in the Uralides: Pangea to the Present
Geophysical Monograph 132

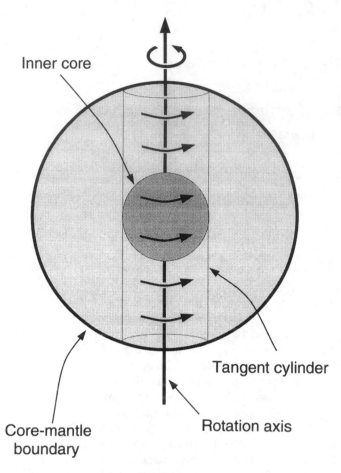

Figure 1. The geometry of the core. The solid inner core lies at the center of the liquid outer core. The tangent cylinder, which is the cylinder coaxial with the Earth's rotation axis that grazes the surface of the inner core, is important in core dynamics on decadal timescales: this cylinder is tightly coupled to the inner core through electromagnetic forces so that it matches variations in the rate of rotation of the inner core. To first order these rotations are invariant with depth so that decadal variations in the rate of rotation observed at the surface of the tangent cylinder (where it meets the core-mantle boundary) reflect those of the inner core.

core flow that are relatively steady cannot be reliably determined due to the presence of magnetic diffusion [*Gubbins and Kelly*, 1996; *Love*, 1999], even if the inversion is technically unique. It is therefore implausible that the part of the rotation of the inner core that is steady or varies on long timescales can be reliably inferred from surface observations of the field, at least without an accurate model of the internal state of the geodynamo. This has been attempted using the results of numerical [*Glatzmaier and Roberts*, 1996b] and theoretical [*Aurnou et al.*, 1996; *Hollerbach*, 1998] models of the core, but there is still considerable uncertainty in the nature of the underlying dynamics due to the dif-

ficulty of reaching the correct parameter regime of the Earth [*Kuang and Bloxham*, 1997a]. However, similarities between structures observed in observational core flow models and flows from numerical models have been used to argue for the presence of particular dynamics within the core [*Pais and Hulot*, 2000], particularly for polar vortices within the portion of the outer core most closely coupled to the inner core, the "tangent cylinder" (see figure 1) [*Olson and Aurnou*, 1999].

Perhaps surprisingly, it is in some ways easier to relate internal core dynamics to surface observations for faster variations. Theory predicts that on decadal timescales the dynamics of the core simplify considerably to those of torsional oscillations [*Braginsky*, 1970], i.e. the differential rotation of coaxial cylinders within the core, with each cylinder rotating as if rigid to first order. This implies that variations in the axisymmetric zonal velocities within the core on decadal timescales should be invariant in the direction parallel to the Earth's rotation axis, and hence can be inferred from those at the core-mantle boundary (CMB). This leads to an observational prediction, that variations in the angular momentum of the core (CAM) calculated using the assumption of torsional oscillations should reflect those predicted by variations in the length of day (LOD, which reflects the rate of rotation and angular momentum in the mantle). This has been confirmed observationally with several core flow models [*Jault et al.*, 1988; *Jackson et al.*, 1993; *Jackson*, 1997]. In addition, the form of the time varying part of the flow at the surface of the core resembles torsional oscillations [*Jault et al.*, 1996; *Zatman and Bloxham*, 1997; *Zatman and Bloxham*, 1998].

The physical significance of the agreement between variations in LOD and models of CAM is complicated by the fact that although this is a fairly common feature of time varying flow models produced by different groups, the variance reduction in fitting geomagnetic field models with time varying flows as opposed to steady flows is very small [*Bloxham*, 1992a]. Presumably this is related to the fact that nonuniqueness can be alleviated in steady flows precisely by fitting the time variation of the secular variation. This makes it very difficult to use the fit of time varying flows to geomagnetic field models to falsify hypotheses of core dynamics if they imply only minor constraints on the flow: the natural ambiguity in the flow will generally allow a good fit while resolving these constraints [*Holme*, 1998a; *Holme*, 1998b].

Differential rotation of the inner core is a feature of numerical simulations of the geodynamo [*Glatzmaier and Roberts*, 1996a; *Kuang and Bloxham*, 1997b; *Sarson et al.*, 1998], although the steadiness, form and magnitude differs between models. Furthermore, these models are poor at predicting decadal variations on account of either excessive viscous damping or exagerrated inertial effects.

2. DECADAL FLOWS IN THE TANGENT CYLINDER

As discussed above, while on the long timescales characteristic of magnetostrophic motion thermal winds in the outer core will decouple motion between the inner and outer core in the tangent cylinder [*Glatzmaier and Roberts*, 1996b], on faster timescales the motion will take the form of a balance between inertia and the Lorentz force leading to the differential motion of geostrophic cylinders, directly relating the surficial motion at the core-mantle boundary to the motion at depth.

On periods of decades the inner core is very tightly coupled to the outer core portion of the tangent cylinder through electromagnetic coupling [*Gubbins*, 1981; *Buffett*, 1996b; *Dumberry and Buffett*, 1999] so decadal variations in inner core rotation rate ($\Delta\Omega_{ic}$) will be reflected in the time variation of core flows at the surface of the core [*Zatman*, 1997]. (In this paper, Ω_{ic} denotes axial rate of rotation of the inner core, and $\Delta\Omega_{ic}$ specifically decadal variations in inner core rate of rotation). If decadal core flows have the form of oscillations $\Delta\Omega e^{i\omega t}$, by "tightly coupled" we mean that $(\Delta\Omega_{\mathbf{tang.cyl.}} - \Delta\Omega_{\mathbf{ic}})/\Delta\Omega_{\mathbf{ic}} << 1$, i.e. that the oscillations in the fluid part of the tangent cylinder and inner core are very close in both phase and amplitude, which means that the lag between motions in the inner core and fluid part of the tangent cylinder is small. From considering the spin-up timescale of the inner core (which will also be similar in magnitude to the spin-up timescale of geostrophic cylinders in the outer core part of the tangent cylinder), this requires:

$$\frac{\rho_i r_i \mu_0}{B'^2} \sqrt{\eta\omega^3} << 1 \qquad (1)$$

where ρ_i is inner core density (assume 10^4 kgm^{-3}), r_i is the inner core radius, μ_0 is the magnetic permeability, η the magnetic diffusivity (assume $\simeq 1$ m^{2}s^{-1}), ω the frequency of oscillations (assume a period of around 70 years), and B' the r.m.s. value of $B_r \sin\theta$ on the inner core boundary where θ is colatitude (assume $B' \simeq 4 \times 10^{-4}$ T, i.e. similar to the magnitude of B_r at the CMB, although it is likely that in fact the field is much stronger at the ICB). With these values, the left hand side of equation 1 is $O(10^{-2})$, so the requirement seems to be satisfied.

It seems likely that torsional oscillations would involve the tangent cylinder (and hence the inner core)

unless the inner core were being held steady with respect to the mantle by some other process. Observationally, torsional oscillations appear to extend into the region of the tangent cylinder [*Zatman*, 1997].

We examine a core flow derived from the ufm1 geomagnetic field model [*Bloxham and Jackson*, 1992b] previously used for investigating internal core processes [*Zatman and Bloxham*, 1997; *Zatman and Bloxham*, 1998]. This flow model is spectral up to spherical harmonic degree 14, is tangentially geostrophic, and is expanded in time by cubic B-splines at a knot interval of 5 years: this model is therefore insensitive to timescales shorter than decadal. In order to constrain the flow within the tangent cylinder (which intersects the CMB at latitudes $\pm 69.5°$) we require data taken relatively close to the pole. Discounting duplicates, 16 of the observatories used in constructing ufm1 lie north of $69.5°$ and 15 south of $-69.5°$. In fact, as the field is downwardly continued from the Earth's surface to the CMB, observations at higher latitudes than $\pm 69.5°$ still depend on the field at the CMB outside the tangent cylinder, and in turn the field on the CMB in the tangent cylinder affects observations at lower latitudes on the Earth's surface. The axisymmetric, equatorially symmetric zonal flows (the part of the velocity which we need for this study) vary smoothly with latitude in the region of the tangent cylinder [*Zatman*, 1997], so we construct a model of variations in inner core rotation rate from the velocities at the latitudes midway between the poles and the edge of the tangent cylinder. This is shown in figure 2. The magnitude of $\Delta\Omega_{ic}$ is of the

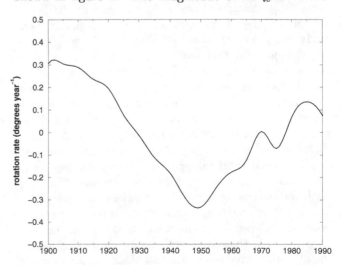

Figure 2. The geomagnetically inferred (short timescale) rate of rotation of the tangent cylinder and inner core, $\Delta\Omega_{ic}$, from 1900-1990 (with the mean over that period subtracted out), from the equatorially and axi-symmetric zonal part of the flow model evaluated at a colatitude of $10°$.

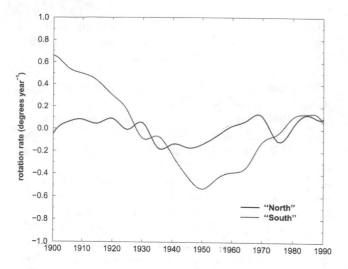

Figure 3. Two different estimates of variations in rotation rate of the tangent cylinder, "North" (using the flow at a colatitude of 10°) and "South" (using the flow at a colatitude of 170°).

same order of magnitude as some seismic estimates of the steady rate of rotation [*Creager*, 1997; *Song and Li*, 1997]. However our results show a clear increase in rotation rate since around 1950. The seismic estimates begin in the 1950s but appear to be most consistent with steady Ω_{ic} (which as noted earlier is more difficulty to constrain geomagnetically).

The full core flow model contains two essentially independent estimates of variations in inner core rotation rate: one inferred from the flow at the North end of the tangent cylinder, and one from the flow at the South end of the tangent cylinder. These flows should individually provide much worse estimates of the variation of the tangent cylinder rotation rate than the axisymmetric case (as each is constrained by fewer data), but the correlations between them are instructive. The rotation estimates are reproduced in figure 3. There is clearly a discrepancy between the amplitudes of the two. However the cross correlations (displayed in figure 4) show a high correlation at low lags (less than around a decade), which suggests that in each case some gross aspects of the form of the flow are recovered and are similar at both ends of the core, even though the amplitude of the flow is not well recovered for both. This is not surprising because the core flow is inverted from a magnetic field model (ufm1) that was itself constructed using damping of both spatial structure and time variations in the field. The flow inversion arises from a relation between the instantaneous magnetic field on the CMB and its time derivative:

$$\frac{\partial B_r}{\partial t} + \nabla_H \cdot (B_r \mathbf{u}) = 0 \qquad (2)$$

from the radial part of the magnetic induction equation with the assumption of frozen flux. If the time variations have been damped to a greater extent than the overall field magnitude, then the flow strength will be underestimated (and vice versa). Therefore modeled flow amplitudes are very sensitive not only to damping in the flow inversion but also to damping in the field inversion. Spatial variations in the importance of damping would then lead to artificial differences in modeled flow in different places. If, however, the rough form of the secular variation (if not its full amplitude) is recovered in the field inversion, then it may still lead to flows of the correct form if not the correct amplitude.

Aside from the difference in amplitude, the major discrepancy between the flow in the North and in the South is the presence in the North of a deceleration of the rotation rate between around 1969 and 1976. Although interpreting subdecadal signal in this flow is risky, it is notable that this interval is thought to be bounded by "geomagnetic jerks" (impulses in the third time derivative of the field).

3. IMPLICATIONS FOR INNER CORE COUPLING

A recent suggestion of Buffett is that core-mantle coupling occurs through the gravitational attraction between mantle inhomogeneities and surface topography of the inner core. In equilibrium, if no torques acted on the inner core, its surface would grow or viscously relax towards the hydrostatic equipotential. Apart from the axisymmetric bulge due to the Earth's rotation, this surface has non-axisymmetric topography due to density variations in the rest of the Earth, primarily the mantle. However the inner core is strongly coupled to the outer core electromagnetically, so flow in the outer core part of the tangent cylinder would tend to move the inner core surface away from the equilibrium position, causing a torque between the inner core and mantle. The axial torque on the mantle Γ_z is related to variations in LOD via:

$$\Gamma_z = I_z \frac{\partial \omega}{\partial t} = -\frac{2\pi I_z}{\text{LOD}^2} \frac{\partial \text{LOD}}{\partial t} \qquad (3)$$

where I_z is the moment of inertia of the mantle about the rotation axis, and ω is the rate of rotation of the mantle. The original suggestion was for a rigid inner core locked to the mantle on long time scales [*Buffett*, 1996b], so that for relatively small rotations the torque on the mantle will be proportional to the angle of displacement of the inner core from its equilibrium position, i.e. (differentiating in time):

$$\Delta \Omega_{ic} \propto \frac{\partial \Gamma_z}{\partial t} \propto -\frac{\partial^2 \text{LOD}}{\partial t^2} \qquad (4)$$

Cross correlations

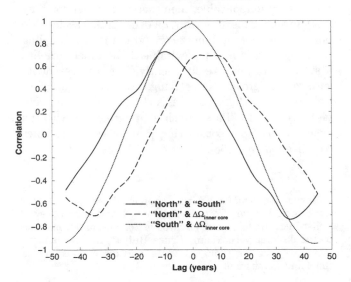

Figure 4. Cross correlations between variations in the flow at the North and South of the tangent cylinder (at colatitudes of 10° and 170°), and the average of the two (which is what we take as our estimate of $\Delta\Omega_{ic}$). As the flow inversion at the North depends on field and secular variation measurements from low colatitudes, and the flow inversion at the South depends on measurements from high colatitudes, these are essentially independent estimates of the rotation rate of the tangent cylinder, so high correlations at small lags suggest that the form of this part of the flow is well recovered.

buffett [1996b] and *Xu et al.* [2000] demonstrated that in this case the amplitude of motion within the tangent cylinder is greatly damped (although our flow model in fact shows that variations in differential rotation are strongest there).

An alternative suggestion was that the inner core is viscous, and is able to relax back to the equipotential on a short timescale [*Buffett*, 1997; *Buffett and Creager*, 1999; *Aurnou et al.*, 2000]. In this case, the displacement between the surface of the inner core and its equilibrium position is approximately proportional to the rate of rotation of the inner core, so that:

$$\Delta\Omega_{ic} \propto \Gamma_z \propto -\frac{\partial\text{LOD}}{\partial t} \qquad (5)$$

This theory is compatible with observations of steady differential rotation of the inner core provided that there is some other form of core-mantle coupling present, as otherwise there would be a large unbalanced torque between the core and the mantle. However, it has been questioned whether the required viscosity of the inner core of around 10^{16} Pa s [*Buffett*, 1997] is too low to be plausible as the implied grain size of less than 5mm [*Frost and Ashby*, 1982] is geophysically implausible

[*Bergman*, 1998]. A comparison of seismic estimates of inner core rotation rate with this theory was inconclusive [*Buffett and Creager*, 1999].

We can test these theories by comparing our estimates of $\Delta\Omega_{ic}$ with predictions from geodetic estimates of LOD using equations 4 and 5. We note that there is a danger of achieving false-positive results in tests like these because even if the inner core is not involved in core-mantle coupling, it is likely to be linked through torsional oscillations to the actual coupling mechanism, which may cause spurious correlations [*Zatman*, 1997]. We use estimates of LOD from *McCarthy and Babcock* [1986] extended to 1990 using more recent data [*Gross*, 2000], with a tidal braking trend of 1.4 milliseconds per century removed and low pass filtered to remove subdecadal signal. These comparisons are shown in figures 5 and 6. The agreement is not good in either case.

A good measure is to compare the correlations for these sets of data with that between CAM for this core flow and LOD (figure 7). Since the net CAM for the core arises from the relatively small difference between parts of the flow whose angular momenta tend to cancel [*Jackson et al.*, 1993], we should expect it to be easier to find a good correlation between $\Delta\Omega_{ic}$ and a derivative of ΔLOD than between CAM and LOD if one of the inner core-mantle coupling theories is correct. On the other hand, if there are particular problems with the

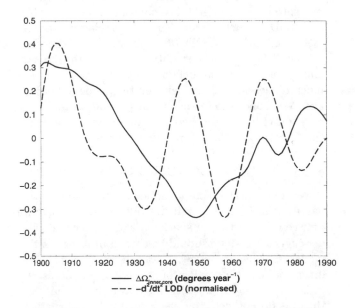

Figure 5. Comparison of variations in $\Delta\Omega_{ic}$ with those implied by observations of changes in the mantle rate of rotation, if the theory of inner core-mantle gravitational coupling with a rigid inner core is valid. The geodetic results are normalised to have the same amplitude as the geomagnetic results. There is agreement between the two over some time periods (eg c. 1905 to c. 1930), but overall there is not a general agreement between the two.

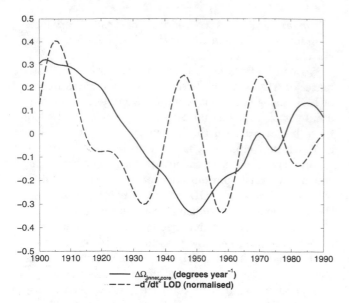

Figure 6. Comparison of variations in $\Delta\Omega_{ic}$ with those implied by observations of changes in the mantle rate of rotation, if the theory of inner core-mantle gravitational coupling with a viscous inner core is valid. The geodetic results are normalised to have the same amplitude as the geomagnetic results. As with figure 5 there is some agreement between the two over limit time intervals but no general agreement between the two.

flow model at high latitudes but not elsewhere then the opposite might be true. The correlation between CAM and LOD is much better than the correlations between $\Delta\Omega_{ic}$ and either $\partial LOD/\partial t$ or $\partial^2 LOD/\partial t^2$, especially for small lags. The correlation between the "North" and "South" tangent cylinder rotation rates was also much higher than those between the geodetic products and $\Delta\Omega_{ic}$, and also (although this is not plotted) those between either of the geodetic products and either of the "North" or "South" rotation estimates.

4. CONCLUSIONS

We conclude that geomagnetic evidence supports decadal variation in Ω_{ic}, although the difference in amplitude between the north and south ends of the tangent cylinder suggests that we should be careful in using the flow model. Our results suggest that neither of the proposed inner core-mantle gravitational coupling mechanisms is consistent with our model. This may indicate that these theories do not describe the real Earth: alternatively this may point to deficiencies in our core flow model or the assumption of torsional oscillations. The study of *Dumberry* [1999] found that although this key assumption breaks down within the tangent cylinder at short time scales (around a year), it should still hold

for decadal oscillations. If the gravitational coupling theories are inapplicable, then there is another coupling mechanism that is important on decadal timescales. We note that with a rigid inner core, gravitational coupling with the mantle is not inevitable if the inner core has a steady rotation rate sufficiently large to prevent its surface from growing or relaxing towards an equipotential, for which Buffett suggests the required inner core viscosity is $> 10^{20}$Pa s, which is in agreement with the expectations from considering grain size in the inner core [*Bergman*, 1998].

Acknowledgments. SZ was supported for part of this work by a fellowship from the Adolphe C. and Mary Sprague Miller Institute for Basic Research in Science, and would like to thank Jeremy Bloxham, Bruce Buffett and two anonymous reviewers for thoughts, ideas, and discussions on this topic and manuscript.

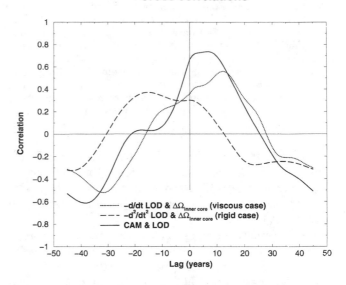

Figure 7. Cross correlations between $\Delta\Omega_{ic}$ and the geodetic predictions for the rigid and viscous inner core theories. Also shown is the cross correlation between core angular momentum and length of day for this flow. A priori we expect $\Delta\Omega_{ic}$ to be better constrained observationally than CAM because the net CAM arises from the relatively minor part of the torsional oscillations in the core that does not cancel out on integration over the whole core [*Jackson et al.*, 1993], although the relative lack of geomagnetic observatories at high latitudes might degrade the estimates of $\Delta\Omega_{ic}$. If one of the inner core-mantle gravitational coupling theories were correct we would hope to see a similar or better correlation at small lags between $\Delta\Omega_{ic}$ and the appropriate geodetic prediction than between CAM and LOD, although as noted in the text there is a danger of a false positive result. In case case, this is not observed.

REFERENCES

Aurnou, J. M., Brito, D., and P. L. Olson, Mechanics of inner core super-rotation, *Geophys. Res. Lett.*, *23*, 3401–3404, 1996.

Aurnou, J. M., and P. L. Olson, Control of inner core rotation by electromagnetic, gravitational and mechanical torques, *Phys. Earth Planet. Int.*, *117*, 111–121, 2000.

Bergman, M. I., Estimates of the Earth's inner core grain size, *Geophys. Res. Lett.*, *25*, 1593–1596, 1998.

Bloxham, J, The steady part of the secular variation of the Earth's magnetic field, *J. Geophys. Res.*, *97*, 19,565-19,579, 1992.

Bloxham, J. and A. Jackson, Time-dependent mapping of the magnetic field at the core-mantle boundary, *J. Geophys. Res.*, *97*, 19537–19563, 1992.

Braginsky, S. I., Magnetohydrodynamics of the Earth's core. *Geomag. Aeron.*, *4*, 698–1712, 1964.

Braginsky, S. I., Torsional magnetohydrodynamic vibrations in the Earth's core and variations in day length, *Geomag. Aeron.*, *10*, 3–12, 1970.

Buffett, B. A., Gravitational oscillations in the length of day, *Geophys. Res. Lett.*, *23*, 2279–2282, 1996a.

Buffett, B. A., A mechanism for decade fluctuations in the length of day, *Geophys. Res. Lett.*, *23*, 3803–3806, 1996b.

Buffett, B. A., Geodynamic estimates of the viscosity of the Earth's inner core, *Nature*, *386*, 571–573, 1997.

Buffett, B. A. and K. C. Creager, A comparison of geodetic and seismic estimates of inner-core rotation, *Geophys. Res. Lett.*, *26*, 1509–1512, 1999.

Creager, K. C., Inner core rotation from small-scale heterogeneity and time varying travel times, *Science*, *278*, 1284–1288, 1997.

Dumberry, M. and B. A. Buffett, On the validity of the geostrophic approximation for calculating the changes in the angular momentum of the core, *Phys. Earth Planet. Int.*, *112*, 81–99, 1999.

Frost, H. J. and M. F. Ashby, *Deformation mechanism maps*, Pergamon Press, Oxford, 1982.

Glatzmaier, G. A. and P. H. Roberts, On the magnetic sounding of planetary interiors, *Phys. Earth Planet. Int*, *98*, 207–220, 1996a.

Glatzmaier, G. A. and P. H. Roberts, Rotation and magnetism of the Earth's inner core, *Science*, *274*, 1887–1890, 1996b.

Gross, R. S., Combinations of Earth orientation measurements: SPACE97, COMB97 and POLE97, *J. Geodesy*, *73*, 627–637, 2000.

Gubbins, D., Rotation of the inner core, *J. Geophys. Res.*, *86*, 11695–11699, 1981.

Gubbins, D. and P. Kelly, A difficulty with using the frozen flux hypothesis to find steady core motions, *Geophys. Res. Lett.*, *23*, 1825–1828, 1996.

Hide, R., Interaction between the Earth's liquid core and solid mantle, *Nature*, *222*, 1055–1056, 1969.

Hills, R. G., *Convection in the Earth's mantle due to viscous shear at the core-mantle interface and due to large-scale buoyancy*, Ph.D. thesis, N. M. State Univ., Las Cruces, 1979.

Hollerbach, R., What can the observed rotation of the earth's inner core reveal about the state of the outer core?, *Geophys. J. Int.*, *135*, 564–572, 1998.

Holme, R., Electromagnetic core-mantle coupling I: Explaining decadal variations in the Earth's length of day, *Geophys. J. Int.*, *132*, 167–180, 1998a.

Holme, R., Electromagnetic core-mantle coupling II: probing deep mantle conductance, In M. Gurnis, M. E. Wysession, E. Knittle, and B. A. Buffett, editors, *The Core-Mantle Boundary Region*, pages 139–151. AGU, 1998b.

Jackson, A., Time dependency of geostrophic core-surface motions, *Phys. Earth Planet. Int.*, *103*, 293–311, 1997.

Jackson, A., Bloxham, J., and D. Gubbins, Time-dependent flow at the core surface and conservation of angular momentum in the coupled core-mantle system, In J.-L. LeMouël, D. E. Smylie, and T. Herring, editors, *Dynamics of Earth's Deep Interior and Earth Rotation*, pages 97–107, Geophysical Monograph 72 IUGG Volume 12, 1993.

ault, D., Gire, C., and J.-L. LeMouël, Westward drift, core motions and exchanges of angular momentum between core and mantle, *Nature*, *333*, 353–356, 1988.

Jault, D., Hulot, G., and J.-L. LeMouël, Mechanical core-mantle coupling and dynamo modelling, *Phys. Earth Plan. Int.*, *98*, 187–191, 1996.

Kuang, W. and J. Bloxham, An Earth-like numerical dynamo model, *Nature*, *389*, 371–374, 1997a.

Kuang, W. and J. Bloxham, On the dynamics of topographical core-mantle coupling, *Phys. Earth Plan. Int..*, *99*, 289–294, 1997b.

LeMouël, J.-L., Outer core geostrophic flow and secular variation of Earth's magnetic field, *Nature*, *311*, 734–735, 1984.

Love, J. J., A critique of frozen-flux inverse modelling of a nearly steady geodynamo, *Geophys. J. Int.*, *138*, 353–365, 1999.

McCarthy, D. D. and A. K. Babcock, The length of the day since 1656, *Phys. Earth Planet. Inter.*, *44*, 281–292, 1986.

Olson, P. and J. Aurnou, A polar vortex in the Earth's core, *Nature*, *402*, 170–173, 1999.

Pais, A. and G. Hulot, Length of day decade variations, torsional oscillations and inner core superrotation: evidence from recovered core surface zonal flows, *Phys. Earth Planet. Int.*, *118*, 291–316, 2000.

Sarson, G. R., Jones, C. A., and A. W. Longbottom, A. W., Convection driven geodynamo models of varying Ekman number, *Geophys. Astrophys. Fluid Dyn.*, *88*, 225, 1998.

Song, X. and A. Li, Rotation of inner core from seismic stations with long recording histories (abstract), *Eos. Trans. AGU*, *78*, F458, 1997.

Song, X. and P. G. Richards, Seismological evidence for differential rotation of the Earth's inner core, *Nature*, *382*, 221–224, 1996.

Souriau, A., New seismological constraints on differential rotation of the inner core from Novaya Zemlyn events recorded at DRV, Antartica, *Geophys. J. Int.*, *134*, F1–F5, 1998.

Su, W. J., Dziewonski, A. M., and R. Jeanloz, Planet within a planet: Rotation of the inner core of the Earth, *Science*, *274*, 1883–1887, 1996.

Voorhies, C. V. and G. E. Backus, Steady flows at the top of the core from geomagnetic field models: The steady motions theorem, *Geophys. Astrophys. Fluid Dynam.*, *32*, 163–173, 1985.

Whaler, K. A., Does the whole of the Earth's core convect?, *Nature*, *287*, 528–530, 1980.

Xu, S., Crossley, D., and A. M. K. Szetob, Variations in length of day and inner core differential rotation from gravitational coupling, *Phys. Earth Planet. Int.*, *117*, 95–110, 2000.

Zatman, S., *Angular Momentum and the Core of the Earth*, Ph.D. thesis, Harvard University, 1997.

Zatman, S. and J. Bloxham, Torsional oscillations and the magnetic field within the Earth's core, *Nature*, *388*, 760–763, 1997.

Zatman, S. and J. Bloxham, J., A one-dimensional map of B_s from torsional oscillations of the Earth's core, In M. Gur-

nis, M. E. Wysession, E. Knittle and B. A. Buffett, editors, *The Core-Mantle Boundary Region*, pages 183–196. AGU/IUGG, 1998.

S. Zatman, Department of Earth and Planetary Sciences, Washington University in St Louis, Campus Box 1169, Wilson Hall, Room 108, One Brookings Drive, St Louis, MO 63130-4899. (e-mail: zatman@levee.wustl.edu)

Can a Stably Stratified Layer in the Core be Detected Using Seismic Normal Modes or Earth Rotation?

David Crossley

Department of Earth and Atmospheric Sciences, Saint Louis University, St. Louis, MO

A chemically homogeneous and adiabatic fluid has been the simplest assumption on which to base seismological starting modes for the outer core. Known otherwise as the Adams-Williamson condition or neutral buoyancy condition, this condition is approximately satisfied in a fluid undergoing thermal convection across the whole of the outer core, which is favored in modeling the geodynamo. The presence of stably-stratified layers follows from geochemical considerations associated with the desegregation of lighter alloying elements that will rise and pool near the core mantle boundary (CMB). Even though there is indirect evidence that radial flow associated with the geomagnetic secular variation might occur close to the CMB, the issue remains somewhat uncertain. Here we re-examine the ability of known observations from seismology and Earth rotation to reveal subtle effects that such a layer might have on free oscillation eigenperiods and the periods of wobble modes, leaving aside the possibility of direct detection of internal gravity waves or torsional magnetic oscillations in such layers. The method is a sensitivity study of the effects of density changes we might make to PREM in layers near the inner core boundary (ICB) and CMB; it is not a full scale inversion of all available data for the core density gradient. The results show that the eigenperiods of the few spheroidal oscillations that have been observed are largely insensitive to such gradients; therefore even a dynamically significant level of stability in layers near the ICB and CMB is allowed under the current seismological resolution. On the other hand, we find that the period of the Chandler Wobble may be sufficiently sensitive to outer core density to discriminate against some of our more extreme models of stability near the CMB.

INTRODUCTION

The possibility of a stably-stratified density profile in the Earth's core seems to have started with the work of Kennedy and Higgins (1973) on the chemical composition of the core and the energetics of the geodynamo. They argued that the core might be stably stratified throughout, a suggestion that would have serious implications for convective models of the geodynamo. The alternative possibility of an entirely small-scale mean field dynamo (e.g. Moffatt 1978; chapters 7-9), operating within a stably stratified fluid, seems no longer to be favored. It is now almost universally agreed that the fluid core maintains itself in a condition very close to a neutral stratification for a model in which the present geodynamo operates by thermal convection (see Stacey, 1992, section 6.7.2).

Restrictions on the stability of the fluid in boundary layers are more difficult to assess, because there are natural

Earth's Core: Dynamics, Structure, Rotation
Geodynamics Series 31
Copyright 2003 by the American Geophysical Union
10.1029/31GD15

processes at both the ICB (inner core boundary) and CMB (core mantle boundary) that tend to form chemical inhomogeneities. Whaler (1980) was one of the first to discuss the possibility of such a stable layer, in which there may departures from the seismic Adams-Williamson equation (see also Gubbins et al;, 1982). However we have no direct evidence on the thickness of such layers or their stability profiles, and we have to rely on indirect modeling arguments, such as whether there is upwelling of fluid near the CMB (e.g. Whaler and Davis, 1997), or whether such a layer can be stable in the presence of liquid convection from below (Lister and Buffett, 1998).

A recent paper by Buffett et al. (2000) has renewed interest in the dynamics and geochemistry of a layer of sediments at the top of the core. Though no mention is made of the pioneering work of Braginsky (e.g. Braginsky, 1984, section 7), Buffett and colleagues strongly argue that a sediment layer must form under the CMB due to the release of light elements from the ICB. They estimate that a compacted layer of several (up to 10) km may form, with possibly a more porous layer forming out to several 100 km (assuming 2×10^9 yr of ICB growth). The presence of such a layer adds significant complexity to the CMB region, and many more degrees of freedom in the interpretation of the geodetic, geomagnetic and seismic data that has traditionally been used to define the properties of the CMB.

Here we re-examine the evidence from seismic normal modes and Earth rotation measurements. These could he used in principle to limit departures of the core boundary layers from PREM, a model that assumes the fluid is close to neutral stratification through the core.

DENSITY PROFILE IN THE OUTER CORE

Our goal is to determine the limits of the density profile in the outer core by examining the sensitivity of this profile to normal mode eigenfrequencies. An important quantity derived from the density profile is N^2, the square of the Brunt-Väisälä frequency:

$$N^2 = -\frac{g}{\rho}\frac{d\rho}{dr} - \frac{g^2}{V_P^2}, \qquad (1)$$

where ρ, g, and V_P are respectively the equilibrium density, gravity, and P wave velocity in a spherical, non-rotating model. If N^2 is positive, the fluid is stably stratified, otherwise it is either neutral ($N^2 = 0$) or negative; in the latter case the fluid will convect. Denis et al. (1997) refer to this as the Schwarzschild stability criterion, and remind us that it is a *local* (pointwise) condition. For the reference model PREM (Dziewonski and Anderson, 1981), we note that it has a nearly neutral core ($N^2 \sim 0$), due to the way the density profile is parameterized by a low degree

polynomial, rather than the usual result of a free layered inversion. We point out that in this paper we use PREM as tabulated in layers, not the version using polynomials, because the latter cannot deal with the modifications to be introduced.

It is well known (e.g. Masters and Shearer, 1990) that seismic models of the deep interior do not constrain the density profile in the core sufficiently well to predict the size or magnitude of N^2. It is therefore reasonable to assume a neutral core as a possible model. It is worth noting here that the seismologically defined quantity (1) applies to a single phase medium that is chemically homogeneous (e.g. Bullen, 1965, section 13.4.2). Stacey (1997) made some useful remarks on this point in reference to the related problem of defining the thermodynamic and seismological parameters for a typical mixture of lower mantle minerals.

Other data sets can of course provide information on core stability. For example, Lay and Young (1990), Souriau and Poupinet (1991), and Garnero et al. (1993) all suggest there might be a small decrease (1-2 %) in P wave velocity in a layer beneath the CMB. Such a change would affect, but not substantially compromise, the type of density modifications envisaged in this paper.

The present study was inspired by Kennett (1998), who examined the density variations throughout the Earth, using normal mode data. He showed that density variations up to ± 40 kg m^{-3} (approximately $\pm 4\%$) are possible, especially in the fluid near the core boundaries. Using this study as a start, we wanted to know explicitly whether the density distribution associated with stably stratified layers was allowed by seismic normal mode observations.

The procedure for changing N^2 requires a re-integration of the density profile in the fluid outer core (OC). The steps are:

(a) integrate the reference model PREM to find the mass M_{OC} and inertia I_{OC} of the OC,

(b) choose a desired value for $N^2(r)$ within the fluid,

(c) choose a trial value of density ρ_{ICB} at the ICB (in the fluid),

(d) integrate, from the ICB to the CMB, the equations:

$$\begin{aligned} \frac{dM_{OC}}{dr} &= 4\pi r^2 \rho \\ \frac{dI_{OC}}{dr} &= \frac{8}{3}\pi r^4 \rho \\ \frac{dg}{dr} &= -\frac{2g}{r} + 4\pi G\rho \\ \frac{d\rho}{dr} &= -\frac{\rho g}{V_P^2} - \frac{\rho N^2}{g} \end{aligned} \qquad (2)$$

(e) at the CMB compare M_{OC} and I_{OC} with those of the reference model,

Table 1. Thickness and location of the stably stratified layers for various models.

Model	Thickness (km)	Location
2a	10	ICB
2b	100	ICB
2c	500	ICB
3a	10	CMB
3b	100	CMB
3c	500	CMB
SOC	80	CMB

(f) adjust ρ_{ICB} and re-integrate equations, and
(g) stop when either M_{OC} or I_{OC} has converged to the reference model.

Evidently with only one free parameter (ρ_{ICB}) we cannot satisfy both mass and inertia, so here we choose to keep M_{OC} the same; later we reconsider this choice. Some practicalities are worth mentioning. First in our PREM reference model, we need a much finer layering near the ICB and CMB in the fluid than given in Table 2 of Dziewonski and Anderson (1981) and we did tests to verify that this enhanced model does not change the normal mode periods. Because the P wave velocity in the OC is not changed from PREM, there is a slight change in bulk modulus in the fluid as the density is perturbed, but this is insignificant in the context of our uncertain knowledge of the core properties.

To explore a number of possible stable layers, we create 6 models (Table 1), of which 3 are ICB models with layers of thickness of 10, 100 and 500 km and a similar 3 at the CMB. The models are based on the desired modifications to N^2 shown in Figure 1, with N scaled to twice the rotation frequency of the Earth (2Ω), as is usual in dynamics of rotating fluids (e.g. Crossley, 1984). We chose a smooth but rapid rise for ($N/2\Omega$) between 0 (neutrality) in the central OC to a constant value of 2.0 in the layers; this is equivalent to a buoyancy period of ~ 6 hr. The actual density perturbations to PREM are shown in Figure 2, indicating that only one model (2c) has a perturbation at the ICB outside the limits indicated by Kennett (1998), otherwise the perturbations are quite modest. Note for future reference that the density is perturbed everywhere in the core in order to constrain the new values of N^2 near the boundaries.

Table 1 also includes a model based on Braginsky's Stratified Ocean of the Core (SOC, Braginsky, 1999). This model has a sharp change in the density excess $C(r)$ at a depth $H = 80$ km (at $r = r_S$) below the CMB:

$$C(r) = -C_S - (C_H / H)(r - r_S), \qquad (3)$$

where $C = (\rho - \rho_a) / \rho_a$ is the density excess over the adiabatic (neutral) density profile ρ_a. In the bulk of the core

$C = C_0 \sim 10^{-8}$, close to that of PREM, whereas in the stable layer near the CMB, $C_S \sim 10^{-2}$ is the value suggested from seismology. The value deduced from analyzing presumed torsional oscillations in the secular variation and length of day is $C_H \sim 10^{-4}$. Braginsky uses the relation $N^2 = - (g/H) C_S$, which is obtained from (1) by ignoring compressibility, and assumes that V_P is also perturbed by about 1% in the layer.

The result of Braginsky's analysis is a stable layer with a normalized buoyancy frequency ($N/2\Omega$) of 7.7, which corresponds to a buoyancy period of about 1.5 hr. This layer, depicted in Figure 3, is therefore much more stable than any of Models 2a-3c (Table 1). Note that in (3) the variation of $C(r)$ with radius is almost negligible because $C_S >> C_H$.

CHOICE OF SPHEROIDAL MODES FOR THIS STUDY

As a starting point, we refer to Kennett who suggested that spheroidal normal modes with periods > 200 s (frequencies f < 5000 µHz) are those most useful for defining the density distribution within the Earth because only at long periods is self-gravitation significant. We therefore extracted from the tables in Masters and Widmer (1995) all observed spheroidal modes with periods > 200 s and added observations from the Reference Earth Model (REM) web site provided by Masters (1999). In all there are 152 modes whose observational errors range from 0.2 µHz for $_2S_0$ to 4.5 µHz (~1 s) for $_1S_{10}$. The complete list is given in Table 2. Although this list is only a small subset of the complete normal mode spectrum, it probably contains most, if not all, of the seismic modes that might be useful for such a study. Note that we use the standard seismological notation for a spheroidal mode $_nS_\ell^m$ with radial number n, degree ℓ, and azimuthal number m.

We also need to compute eigenperiods for these modes as a reference set. The Masters and Widmer tables give

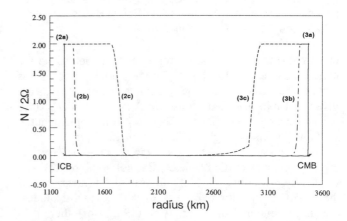

Figure 1. Brunt Väisälä frequency N in the outer core for the 6 models listed in Table 1.

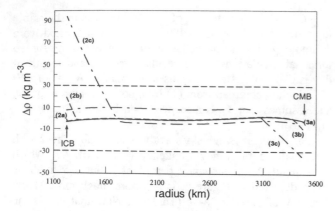

Figure 2. Density departures from PREM for the models shown in Figure 1. The limits suggested by Kennett (1998) are shown as the two horizontal dashed lines at ± 30 kg m^{-3}. According to Kennett, the density departures near the ICB are more likely to be negative (indicating the absence of stable layers near the ICB) than the positive values associated here with models 2a, 2b and 2c.

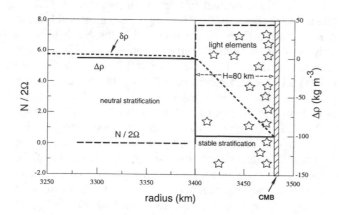

Figure 3. Density departures from PREM and stratification N for the model SOC of Braginsky (1999). Note that the solid curve labeled $\Delta\rho$ is the result of applying Bragnsky's relation $N^2 = gC_S H = g\Delta\rho/H\rho$ so that both N and $\Delta\rho$ are effectively constant within the layer. The dotted curve labeled $\delta\rho$ is the actual density perturbation used in the modified PREM model, obtained by constraining N^2 as described using (2).

eigenfrequencies for an anisotropic Earth model, as do the tables in on the REM web site. For the present purposes it is quite sufficient to use an isotropic model, whose periods are given in Table V of Dziewonski & Anderson (1981), since the anisotropy is confined to layers in the upper mantle, well away from our region of interest in the core.

The program used for the spheroidal mode eigenfrequencies (called SPMODES, not an acronym) was written by the author in 1972 and uses direct integration of the equations of Alterman et al. (1959). For most modes, the starting method of Alsop et al. (1963), for others it is necessary to start at the origin (Crossley, 1975), and for some modes an alternative version is implemented using

Love numbers (e.g. Crossley et al., 1992) in which the IC and mantle integrations are done first and matched to solutions at the ICB and CMB. A comparison of the eigenfrequencies computed with SPMODES with Table V of Dziewonski & Anderson yields differences of 0.04 s or less for all but 3 modes ($_0S_2$, $_1S_2$, and $_{13}S_2$) whose discrepancies of about 0.1 s cannot be reduced by changing any of the integration parameters.

The SPMODES eigenperiods are given in Table 2. Modes are grouped according to whether more than 25% of

Table 2. Classification of modes used in this study[a], based on a region having > 25% of the KE of mode $_nS_\ell$

n	ℓ	Period (s)	O	n	ℓ	Period (s)	O
IC Modes				OC+Mantle Modes (cont.)			
5	1	583.32	-	13	1	222.35	Y
10	1	294.56	-	0	2	3233.57	Y
14	1	202.98	-	3	2	904.04	Y
6	2	414.71	-	7	2	397.20	Y
11	2	246.33	-	8	2	311.06	-
7	3	324.36	-	13	2	206.30	Y
12	3	213.46	-	0	3	2134.46	Y
8	4	270.00	-	2	3	805.18	Y
9	5	231.16	-	6	3	354.35	Y
10	6	203.45	-	9	3	281.23	Y
				11	3	224.02	-
IC+OC Modes				0	4	1545.71	Y
1	1	19514.23	-	2	4	725.32	Y
2	2	1065.94	-	6	4	323.36	-
3	3	705.41	-	9	4	257.79	Y
3	4	545.31	-	11	4	209.70	Y
3	5	447.40	-	2	5	660.51	Y
4	6	380.57	-	8	5	239.95	Y
4	7	331.74	-	1	6	656.99	Y
4	8	294.32	-	9	6	216.37	-
5	9	264.64	-	1	7	604.05	Y
5	10	240.51	-	9	7	205.18	-
6	11	220.46	-	1	8	555.78	Y
6	12	203.54	-	1	9	509.23	Y
				1	10	465.45	Y
IC+OC+Mantle Modes				1	11	425.96	Y
10	2	247.93	Y	1	12	391.36	-
				1	13	361.48	-
OC+Mantle Modes				1	14	336.02	Y
1	0	612.99	Y	2	15	308.60	-
2	0	398.30	Y	2	16	290.42	-
3	0	305.65	Y	2	17	273.42	-
4	0	243.49	Y	2	18	258.10	-
5	0	204.65	Y	2	19	244.35	-
2	1	2475.52	-	2	20	231.98	-
6	1	504.94	Y	2	21	220.80	-
8	1	347.98	Y	2	22	210.66	-
11	1	271.27	-	2	23	201.41	-

[a]O = Y means the mode has been observed

Table 2 (continued)

n	ℓ	Period (s)	O	n	ℓ	Period (s)	O	n	ℓ	Period (s)	O
Mantle Modes				(cont.)				(cont.)			
0	0	1228.03	Y	5	8	283.57	Y	1	17	286.17	Y
3	1	1059.41	Y	6	8	267.50	Y	3	17	242.68	Y
4	1	707.85	Y	7	8	224.55	Y	4	17	204.57	Y
7	1	449.53	-	8	8	203.64	Y	0	18	374.36	Y
9	1	312.65	-	0	9	633.98	Y	1	18	274.29	Y
12	1	232.47	-	2	9	448.96	Y	3	18	233.61	Y
1	2	1471.21	Y	3	9	338.94	Y	0	19	360.37	Y
4	2	580.58	Y	4	9	269.60	Y	1	19	263.50	Y
5	2	478.17	Y	6	9	252.09	Y	3	19	225.14	Y
9	2	309.37	Y	7	9	216.52	Y	0	20	347.66	Y
12	2	230.87	-	0	10	579.61	Y	1	20	253.61	Y
1	3	1064.31	Y	2	10	416.39	Y	3	20	217.23	Y
4	3	488.00	Y	3	10	324.34	Y	0	21	336.04	Y
5	3	460.91	Y	4	10	258.75	Y	1	21	244.49	Y
8	3	304.49	-	6	10	237.36	Y	3	21	209.83	Y
10	3	232.44	-	7	10	209.72	Y	0	22	325.37	Y
1	4	852.81	Y	0	11	537.34	Y	1	22	236.04	Y
4	4	438.71	Y	2	11	388.97	Y	3	22	202.91	Y
5	4	420.19	Y	3	11	310.63	Y	0	23	315.50	Y
7	4	292.93	Y	4	11	249.35	Y	1	23	228.19	Y
10	4	228.19	-	5	11	224.24	Y	0	24	306.34	Y
0	5	1190.16	Y	7	11	203.36	Y	1	24	220.87	Y
1	5	729.81	Y	0	12	502.82	Y	0	25	297.79	Y
4	5	414.75	Y	2	12	365.47	Y	1	25	214.03	Y
5	5	369.84	Y	3	12	297.65	Y	0	26	289.79	Y
6	5	306.08	-	4	12	240.81	Y	1	26	207.63	Y
7	5	273.19	Y	5	12	212.79	Y	0	27	282.27	Y
10	5	223.68	-	0	13	473.66	Y	1	27	201.63	Y
0	6	963.48	Y	2	13	344.95	Y	0	28	275.17	Y
2	6	595.40	Y	3	13	285.34	Y	0	29	268.46	Y
3	6	392.29	Y	4	13	232.84	Y	0	30	262.10	Y
5	6	332.07	Y	5	13	202.91	Y	0	31	256.04	Y
6	6	293.77	-	0	14	448.52	Y	0	32	250.28	Y
7	6	252.55	Y	2	14	326.49	Y	0	33	244.77	Y
8	6	225.38	-	3	14	273.69	Y	0	34	239.51	Y
0	7	812.14	Y	4	14	225.31	Y	0	35	234.47	Y
2	7	536.60	Y	0	15	426.54	Y	0	36	229.64	Y
3	7	372.35	Y	1	15	315.38	Y	0	37	225.00	Y
5	7	303.79	Y	3	15	262.70	Y	0	38	220.55	Y
6	7	281.46	-	4	15	218.11	Y	0	39	216.26	Y
7	7	235.92	Y	0	16	407.13	Y	0	40	212.13	Y
8	7	214.94	Y	1	16	299.50	Y	0	41	208.15	Y
0	8	707.80	Y	3	16	252.37	Y	0	42	204.32	Y
2	8	488.33	Y	4	16	211.20	Y	0	43	200.61	Y
3	8	354.77	Y	0	17	389.85	Y				

their kinetic energy is in a region (IC, OC or mantle). This is not a very sophisticated classification compared to one based on group velocity and compressional energy (e.g. Dziewonski and Gilbert, 1973), but it serves to show which modes are important in our study. No mode in the first two groups of modes (IC, and IC+OC) have yet been observed because they are equivalent to body wave phases trapped within the inner core and they are only marginally excited by surface sources. Several modes in the next group (OC+Mantle) are also not observed, in particular some first and second overtones (e.g. $_2S_{15}$ to $_2S_{23}$) that are Stoneley waves trapped at the CMB. Even the final group, containing modes with most of their energy in the mantle, includes modes that have not yet been observed, e.g. $_7S_1$.

SEISMIC RESULTS FOR MODIFIED MODELS

For each of the 7 new models (Table 1), the eigenfrequencies of the 204 modes in Table 2 are recomputed using SPMODES and compared to the eigenfrequencies of the original PREM model (Table 2). The resulting differences are listed in Table 3a for modes that are not observed and in Table 3b for modes that are observed.

Table 3(a). Frequency differences (in μHz) for modes not currently observed; models as in Table 1.

n	ℓ	2a	2b	2c	3c	3b	3a	SOC
1	1	.00	.00	-1.54	-0.37	.00	.00	-0.11
11	1	.00	.00	.59	.42	.00	.00	.00
2	2	.00	.00	-0.88	-0.17	.00	.00	.00
8	2	.00	.00	.00	-0.23	.00	.00	-0.16
11	2	.00	.00	.24	.23	.00	.00	.16
3	3	.00	.00	-1.52	-0.23	.00	.00	.00
7	3	.00	.00	-0.16	.00	.00	.00	.00
11	3	.00	.00	.00	.21	.00	.00	.27
3	4	.00	-0.13	-2.15	-0.29	.00	.00	.00
6	4	.00	.00	.00	.00	.00	.00	.12
8	4	.00	.00	-0.33	.00	.00	.00	.00
3	5	.00	-0.19	-2.72	-0.33	.00	.00	.00
9	5	.00	.00	-0.57	.00	.00	.00	.00
4	6	.11	-0.25	-3.27	-0.37	.00	.12	.00
6	6	.00	.00	.00	-0.13	.00	.00	.00
8	6	.00	.00	.00	.24	.00	.00	.13
9	6	.00	.00	.00	.38	.00	.00	.12
10	6	.00	.00	-0.81	.00	.00	.00	.00
4	7	.13	-0.32	-3.80	-0.41	.11	.14	.00
6	7	.00	.00	.00	-0.10	.00	.00	.00
9	7	.00	.00	.00	.58	.00	.00	.36
4	8	.15	-0.39	-4.32	-0.45	.13	.15	.00
5	9	.16	-0.47	-4.82	-0.48	.15	.17	.00
5	10	.18	-0.54	-5.31	-0.52	.16	.19	.00
6	11	.20	-0.62	-5.79	-0.55	.18	.21	.00
1	12	.00	.00	.15	1.00	.00	.00	.82
6	12	.21	-0.71	-6.26	-0.59	.20	.23	-0.10
1	13	.00	.00	.15	1.08	.00	.00	.93
2	15	.00	.00	.15	1.11	.00	.00	1.02
2	16	.00	.00	.23	1.67	.15	.00	1.60
2	17	.00	.00	.26	1.96	.18	.00	1.93
2	18	.00	.00	.29	2.16	.21	.00	2.20
2	19	.00	.00	.32	2.34	.24	.00	2.46
2	20	.00	.00	.34	2.53	.27	.00	2.73
2	21	.00	.00	.37	2.73	.30	.00	3.02
2	22	.00	.00	.39	2.92	.33	.00	3.31
2	23	.00	.00	.42	3.10	.36	.00	3.60

Because we are using frequency (μHz), the discrepancies in Table 3a are deceptively small for $_1S_1$, the non-rotating Slichter mode. We adopt the usual seismic nomenclature (n = 1, ℓ = 1) for this mode, even though this classification is not universally accepted (see Denis et al., 1997). In the time domain the period differences are quite large: -25.34, 17.17, 604.77, 142.81, -19.04, and -25.99 s respectively for models 2a, 2b, 2c, 3c, 3b, 3a and SOC. Clearly these discrepancies are much larger than the anticipated errors from superconducting gravimeter data (see e.g. Hinderer et al., 1995) and so the triplet $_1S_1^m$, especially widely split by rotation, will be sensitive to core

Table 3(b). As 3(a), but for observed modes[a]

n	ℓ	Err[b]	2c	3c	3b	SOC
1	0	.05	.46	.00	.00	.00
2	0	.20	.00	.00	.00	.10
3	0	.60	-0.91	-0.24	.00	-0.29
4	0	.10	-0.74	.28	.00	.38
5	0	.30	.32	-0.23	.00	-0.40
6	1	1.50	.00	.31	.00	.00
8	1	.25	.47	-0.33	.00	.00
13	1	.94	-0.40	-0.54	.00	.00
7	2	.40	-0.14	.36	.00	.14
9	2	.80	.00	-0.12	.00	.00
10	2	1.02	.36	.32	.00	.20
13	2	.36	.69	-0.54	.00	-0.38
6	3	.18	-0.12	.33	.00	.20
9	3	.58	-0.20	.00	.00	-0.21
2	4	.28	.00	.11	.00	.00
7	4	.28	.00	.10	.00	.00
9	4	.80	-0.17	.27	.00	.00
11	4	.28	-0.20	-0.27	.00	.13
2	5	.42	.00	.14	.00	.00
7	5	.35	.00	.11	.00	.12
8	5	.32	.00	.43	.00	.11
0	6	.14	.00	.14	.00	.00
1	6	.45	.00	.21	.00	.12
1	7	.25	.00	.33	.00	.21
7	7	.50	.00	-0.11	.00	.00
1	8	.22	.00	.46	.00	.31
3	8	.85	.00	.12	.00	.00
7	8	1.00	.00	-0.17	.00	.00
1	9	.72	.00	.60	.00	.41
3	9	1.80	.00	.14	.00	.00
7	9	2.50	.00	-0.17	.00	.00
1	10	4.53	.12	.75	.00	.56
3	10	1.00	.00	.15	.00	.00
7	10	4.50	.00	-0.12	.00	.00
1	11	1.00	.00	.88	.10	.80
3	11	.70	.00	.14	.00	.00
2	12	.64	.00	.12	.00	.00
3	12	.45	.00	.14	.00	.11
4	12	2.00	.00	.14	.00	.00
2	13	2.02	.00	.19	.00	.16
3	13	.70	.00	.13	.00	.00
4	13	2.00	.00	.14	.00	.00
1	14	1.01	.14	1.01	.00	.90
2	14	1.50	.00	.43	.00	.38
3	14	.60	.00	.11	.00	.00
4	14	1.50	.00	.15	.00	.10
1	15	2.50	.00	.50	.00	.46
4	15	1.50	.00	.14	.00	.11
1	16	1.00	.00	.12	.00	.11
4	16	2.50	.00	.13	.00	.11
4	17	2.00	.00	.11	.00	.00

[a]All values for models 2a, 2b, and 3a are 0.00.
[b]Err = estimated observed error.

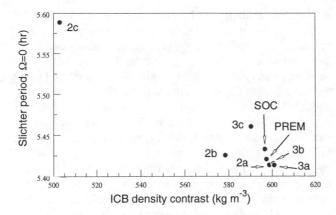

Figure 4. Eigenperiods of the degenerate, non-rotating Slichter translational motion of the inner core for PREM and the various models in Table 1.

structure as well as the density jump at the ICB. Figure 4 shows the non-rotating (degenerate) Slichter mode period for the various models, plotted against the density jump at the ICB. Obviously model 2c, with the largest density perturbation at the ICB (and in the lower fluid core) produces the greatest change in Slichter mode period, whereas models with a stable layer near the CMB have little influence.

As a means of putting all the results in perspective, we plot the RMS misfits and errors from Table 3, separating the modes whose eigenfrequencies have been determined from modes that are not yet observed (Figure 5). It can be seen that models 2c, 3c and SOC all have the potential to be tested if one could use modes that are currently unobserved, because their misfit from PREM are comparable to the current observational errors. The thinner layers (Models 2a, 2b, 3b and 3a) have misfits well below the observational errors, even for modes that have not yet been observed.

On the other hand, none of the currently observed modes have misfits for any model that are close to the observational errors. Hence we are unable to rule out any of our stable-layer models using normal mode data. Without a considerable decrease in observational error estimates, already reduced to a low level by sophisticated processing techniques, and without large moment earthquakes to excite these core-trapped phases, it is unlikely that seismic normal mode data (at periods < 1 hr) alone can decide whether stable layers can exist.

WOBBLE PERIODS FOR MODIFIED MODELS

The eigenperiods of the whole Earth wobble modes are also influenced by the core stratification, as first shown by

Smith (1977). This influence is relatively slight and easily overshadowed by other factors, such as the large discrepancy between the observed and theoretical period of the Nearly Diurnal Free Wobble (NDFW), e.g. Hinderer, 1997. Nevertheless it is worth checking to see whether the 7 models listed in Table 1 have any influence on the theoretical periods of these modes.

To answer this question, we ran the models through the program SPHERCL (again, this is not an acronym, but some readers may have used this program) provided by John Wahr (personal communication, 1999) to first determine the ellipticity profile of the Earth using Clairaut's equation. We then used the program STS (from the same source) to compute the wobble mode eigenperiods in the mantle reference frame. The two modes of interest here are the NDFW and the Chandler Wobble (CW), with associated dependent variables represented as travelling waves by a factor $e^{i(\omega t + m\varphi)}$, where ω is the angular frequency and φ is east longitude. Each of these wobble modes can be represented by the truncated harmonic sum $T_1^m + S_2^m + T_3^m$; for positive ω, the NDFW is retrograde ($m = 1$) and the CW is prograde ($m = -1$).

For PREM the period of the CW is 401.413 sidereal days (sd), slightly changed from the period of 401.581 sd obtained with the PREM version supplied by Wahr because we use a different solid surface layer. The STS program supplied by Wahr did not work with a fluid surface layer, but this is not a significant problem in this study. The NDFW period for PREM is 0.9978216 sd, which is equivalent to a FCN with a period in the space frame of 458.053 sd. In addition STS was able to find an

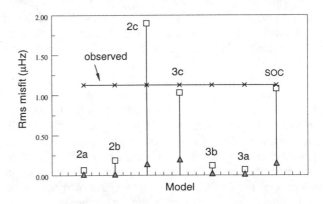

Figure 5. Comparison of the computed RMS deviation of normal mode eigenfrequencies for various models. Squares refer to the modes that have not yet been observed, but would be potentially useful for determining core structure; triangles are for observed modes (Table 2). Also shown as a horizontal line is the RMS error of currently observed modes.

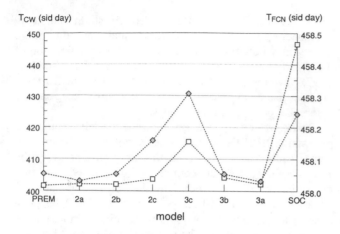

Figure 6. Eigenperiods of two wobble modes of the Earth for different models. T_{CW} is the Chandler Wobble (open squares) and T_{FCN} is the eigenperiod of the Free Core Nutation, which is the motion in space of the NDFW of the outer core (filled diamonds).

inner core wobble (m = -1) at about 675 sd, but as Smith (1977) comments, the solution is unreliable due to the heavy truncation. We were unable to clearly find either a diurnal tilt-over mode, or a nearly diurnal free inner core wobble, for reasons discussed by Rogister (2001).

The results for the CW and FCN periods are shown in Figure 6 where the left hand axis is the eigenperiod of the CW and the right hand axis is the eigenperiod change of the NDFW, both in sidereal days (sd). The results generally confirm those of Smith (1977), who reported (his Table 2) on two models, both with uniform core stability profiles. The differences between one model with $N^2 = 0$, and the other with $N^2 = 3.38 \times 10^{-7}$ (giving $N/2\Omega \sim 8$), were less than 2 sd for the CW and < 0.0001 sd for the NDFW. Smith did not discuss the procedure for deriving these models, so it is not clear what constraints (i. e. mass, inertia) were used.

The results for models 3c and SOC, however, seem unusual and they represent significant departures from PREM. It was thought that this might be due to the method for modifying the density profile outline in Section 2, where we kept the mass of the core fixed and we let the inertia be unconstrained. This naturally will also lead to changes in the inertia factor I/MR^2, where I is the moment of inertia of the equivalent spherical Earth of mean radius R. To check this we recomputed the density profiles necessary to produce our 7 models in Table 1, but this time requiring that I/MR^2 for the whole Earth be preserved. Note that we keep the inertia coefficient as for PREM and do not delve into the complications of re-defining it as in Denis et al. (1999).

Figure 7 shows density departures in the outer part of the liquid core for 2 models (3c) and SOC, for both the originally modified model (M_{OC} constant) and inertia

(I/MR^2) constant. It can be seen that the density profiles differ only slightly, and of course both are negative because, see equation (1), we require a positive value of N^2 in the stable layer. The eigenperiods of CW and NDFW are very little changed by this new modeling procedure (Table 4). Nevertheless C/MR^2 is almost perfectly preserved in the new models, shown in Figure 8 as a relationship between I/MR^2 and the period of the CW. In this plot a cluster of models with the smaller stable layers lies close to PREM, but models 3c and SOC in particular are strongly affected by the density changes just under the core-mantle boundary. One of the reviewers commented on the large effect of the density perturbations on the CW, compared to the FCN eigenperiod, as both are governed by the same ratio of mantle/whole Earth inertia. This is a good point that applies not only to this study, but also Smith's results (above). Work is currently underway to refine the wobble mode calculations to answer this and other similar questions.

CONCLUSIONS

We therefore arrive at two different conclusions, one for the normal modes and one for the wobble modes. On the one hand we find that the sensitivity of the seismic normal modes to stable layers ranges from marginal, including modes that are very difficult to observe, to insignificant for the modes that have actually been observed. Note that these results place the detection of stable layers in the best possible light. When Earth models are constructed based on travel time and normal mode data, all sorts of compromises and trade-offs come into play that lessen the effect of the core density profile, because this is not the only region of the Earth that can be adjusted to fit for the eigenfrequencies of the modes.

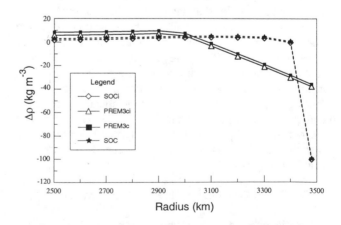

Figure 7. Density departures from PREM for 4 selected models. Model PREM3c and SOC are the same as in Table 1, whereas model PREM3ci and SOCi are similar models where the inertia coefficient I/MR^2 is kept constant instead of the total mass.

Table 4. Wobble eigenperiods (sd) for Table 1 models.

Model	CW	CWi[a]	NDFW	NDFWi[b]
PREM	401.41		0.9978216	
2a	402.08	402.08	0.9978215	0.9978216
2b	402.09	402.09	0.9978216	0.9978216
2c	403.73	403.70	0.9978221	0.9978219
3c	415.48	415.42	0.9978228	0.9978223
3b	404.17	404.24	0.9978216	0.9978216
3a	402.17	402.21	0.9978225	0.9978223
SOC	446.51	448.68	0.9978225	0.9978223

[a]CWi is CW with the inertia coefficient conserved
[b]NDFWi is NDFW with the inertia coefficient conserved

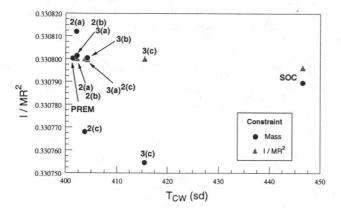

Figure 8. Variation of the inertia coefficient I/MR^2 with the eigenperiods of the CW for the models in Table 1 when mass is conserved (filled circles). The eigenperiods obtained for similar models obtained when the inertia coefficient is kept constant are shown as filled triangles.

On the other hand, we are surprised to find that two of our models, (3c) and SOC, both with significant stable layers near the CMB, are apparently ruled out by the known eigenperiod of the CW. Allowance needs to be made for the modification due to mantle anelasticity and the oceans in comparing these eigenperiods to the observed CW. This result clearly needs further investigation, preferably with a computer algorithm that might be more accurate than that first developed by Smith, e.g. the one developed by Rogister (2001). Again we note that the approach here is to assume N^2 is a constraint, as is a choice of whether M_{OC}, I_{OC}, or I/MR^2 is preserved in the core, and we assume V_P is unchanged. This does not give the same flexibility as used by seismologists, who construct Earth models based on total mass and inertia as constraints, but adjust V_P, V_S and ρ globally to all available seismic data.

Acknowledgments. I thank M. Ritzwoller for informing me of the existence of the Reference Earth Model Web Site and to G. Masters and colleagues for making the observations available in a convenient form. I appreciate the help of John Wahr for supplying the programs SPHERCL and STS for computing the wobble modes of realistic Earth models; Veronique Dehant also helped me to understand the programs. The 3 referees (M. Rochester, J. Hinderer, and C. Denis) gave comments that helped improve the manuscript; Y. Rogister also supplied useful input. Funding through the NSF EAR and International Programs supported this study.

REFERENCES

Alterman, Z., H. Jarosch, and C. L. Pekeris, Oscillations of the Earth, *Proc. Roy. Soc. Lond. A, 252*, 80-95, 1959.

Alsop et al., Free spheroidal vibrations of the Earth at very long periods, Part I - Calculation of periods for several Earth models, *Bull. seism. Soc. Am., 53*, (3), 483-501, 1963.

Braginsky, S. I., Short-period geomagnetic secular variation, *Geophys. Astrophys. Fluid Dynam., 30*, 1-78, 1984.

Braginsky, S., Dynamics of the stably stratified ocean at the top of the core, *Phys. Earth Planet. Int., 111*, 21-34, 1999.

Buffett, B. A., E. J. Garnero, and R. Jeanloz, Sediments at the top of the Earth's core, *Science, 290*, 1338-1342, 2000.

Bullen, K. E., *An Introduction to the Theory of Seismology*, 3rd. edition, pp. 227-228, Cambridge University Press, 1965.

Crossley, D. J., M. G. Rochester and Z. R. Peng, Slichter modes and Love numbers, *Geophys. Res. Lett., 19*, 1679-1682, 1992.

Crossley, D. J., Oscillatory flow in the liquid core, *Phys. Earth Planet. Int., 36*, 1-16, 1984.

Crossley, D. J., The free oscillation equations at the center of the Earth, *Geophys. J. R. astr. Soc., 41*, 153-163, 1975.

Denis, C., Y. Rogister, M. Amalvict, C. Delire, A İbrahim Denis, and G. Munhoven, Hydrostatic flattening, core structure, and translational mode of the inner core, *Phys. Earth Planet. Int., 99*, 195-206, 1997.

Denis, C., M. Amalvict, Y. Rogister, M. Amalvict, and S. Tomecka-Suchon, Methods for computing internal flattening, with applications to the Earth's structure and dynamics, *Geophys. J. Int, 132*, 603-642, 1999.

Dziewonski, A. M. and D. L. Anderson, Preliminary reference Earth model, *Phys. Earth Planet. Int., 25*, 297-356, 1981.

Dziewonski, A. M. and F. Gilbert, 1973. Observations of normal modes from 84 recordings of the Alaskan earthquake of 1964 March 28 - II. Further remarks based on new spheroidal overtone data, *Geophys. J. Roy. astr. Soc., 35*, 401-437, 1973.

Garnero, E. J., D. V. Helmberger, and S. P. Grand, Constrain-ing outermost core velocity with SmKS waves, *Geophys. Res. Lett., 20*, 22, 2463-2466, 1993.

Gubbins, D., C. J. Thompson, and K. A. Whaler, Stable regions in the Earth's liquid core, *Geophys. J. R. astr. Soc., 68*, 241-251, 1982.

Hinderer, J., Constraints on Earth's deep structure and dynamics from superconducting gravimetry, in *Earth's Deep Interior*, edited by D. Crossley, pp 167-195, Gordon & Breach, Amsterdam, 1997.

Hinderer, J., D. Crossley, and O. Jensen, A search for the Slichter triplet in superconducting gravimeter data, *Phys. Earth Planet Int., 90*, 183-195, 1995.

Kennedy, G. C. and G. H. Higgins, The core paradox, *J. Geophys. Res. 78*. 900-904, 1973.

Kennett, B. L., On the density distribution within the Earth, *Geophys. J. Int., 132*, 374-382, 1998.

Lay, T. and C. J. Young, The stably-stratified outermost core revisited, *Geophys. Res. Lett., 17,* 11, 2001-2004, 1990.

Masters, T. G., The Reference Earth Model Internet Site: http://mahi.ucsd.edu/Gabi/rem.html, 1999.

Masters, T. G. and R. Widmer, Free oscillations: frequencies and attenuation, in *Global Earth Physics: a Handbook of Physical Constants, 1,* edited by T. J. Ahrens, pp. 104-125, AGU, Washington, D.C., 1995.

Masters, T. G. and R. M. Shearer, Summary of seismological constraints on the structure of the Earth's core, *J. Geophys. Res., 95,* (B13), 21,691-21,695, 1990.

Moffatt, H. K., *Magnetic field generation in electrically conducting fluids,* Cambridge U. P., Cambridge, 343 pp., 1978.

Smith, M., Wobble and nutation of the Earth, *Geophys. J. R. astr. Soc., 50,* 103-140, 1977.

Rogister, Y., On the diurnal and nearly-diurnal free modes of the Earth, *Geophys. J. Int., 144,* 459-470, 2001.

Souriau, A. and G. Poupinet, A study of the outermost liquid core using different travel times of the SKS, SKKS and S3KS phases, *Phys. Earth Planet. Int., 68,* 183-199, 1991.

Stacey, F., *Physics of the Earth,* 3rd edition, Brookfield Press, Brisbane, 513 pp., 1992.

Stacey, F., Bullen's seismological homogeneity parameter, η, applied to a mixture of minerals; the case of the lower mantle, *Phys. Earth Planet. Int., 99,* 189-193, 1997.

Whaler, K. A. and R. G. Davis, Probing the Earth's core with geomagnetism, in *Earth's Deep Interior,* edited by D. Crossley, pp. 115-166, Gordon & Breach, Amsterdam, 1997.

Whaler, K. A., Does the whole of the Earth's core convect?, *Nature, 287,* 528-530, 1980.

David Crossley, Department of Earth and Atmospheric Sciences, Saint Louis University, 3507 Laclede Ave., St. Louis, MO 63103.

The Core and Fluctuations in the Earth's Rotation

S. Ponsar, [1] V. Dehant, [1] R. Holme, [2] D. Jault, [3] A. Pais, [4] T. Van Hoolst [1]

On decadal time scales, length of day (LOD) and geomagnetic field variations appear to be correlated, reflecting most probably angular momentum transfer between the core and the mantle. Different mechanisms have been suggested for this transfer: gravitational coupling (torque due to the density anomalies in the fluid core), topographic coupling (torque due to the fluid pressure acting on bumps at the core-mantle boundary), and electromagnetic coupling (torque related to the Lorentz force at the core-mantle boundary). The viscous torque is usually considered as a minor contributor to the core-mantle coupling. Length of day variations could also result from angular momentum transfer between the inner core and the mantle by gravitational coupling (torque due to the mass heterogeneities produced by the misalignment between the inner core and the mantle equipotential surfaces). The aim of this paper is to review these mechanisms and the information that can be deduced from analyses of geodesy observations.

1. INTRODUCTION

Variations in the Earth's rotation appear at different time scales. Depending on the time scale, different physical mechanisms can account for these variations.

On very long time scale, tidal friction is responsible for a secular decrease in the Earth's rotation rate and thus for an increase in the length of day (LOD) (for more details, see for instance *Brosche and Sündermann* [1978]). Post glacial rebound also contributes to the long-term trend in Earth rotation variations [*Dicke*, 1966, *Peltier*, 1986].

The variation of the length of day can be measured very precisely using Very Long Baseline Interferometry (VLBI) measurements. These measurements provide Earth's orientation parameters at the sub-centimeter level. On short time scale (annual and subannual), these variations are caused

principally by the atmosphere [*Rosen*, 1993], and to a lesser extent, by the oceans [*Marcus et al.*, 1998] and the hydrosphere [*Chen et al.*, 2000]. On subannual time scales down to periods of about 30 days, *Zatman and Bloxham* [1997a] have shown that the phase of the LOD variations leads that of the atmospheric angular momentum (AAM). They showed that if changes in the LOD variations are driven mainly by the atmosphere, this discrepancy could arise from either core-mantle coupling or from the action of the oceans. *Zatman and Bloxham* [1997a] argue that the moment of inertia of the ocean is much smaller than that of the core and that therefore the core will be dominant. With a simple three-layer model of the Earth (atmosphere, core and mantle) and treating the core as a rotating body coupled to the mantle, they show that the core can cause a phase lead of the LOD variations of the correct magnitude. On the other hand, *Dickey et al.* [2000], adding the ocean angular momentum in the angular momentum budget equation, find for periods between 30 and 60 days a phase lag close to zero between the LOD and the atmosphere angular momentum plus the ocean angular momentum (see also *Johnson et al.* [1999] and *Ponte and Rosen* [2001]).

On decadal time scales, the length of day varies by a few milliseconds (these variations are 3 to 5 times larger than those on annual and subannual scales). *Ball et al.* [1969] have

---

[1] Royal Observatory of Belgium, Brussels, Belgium
[2] GeoForschungsZentrum Potsdam, Germany
[3] LGIT, University of Grenoble, France
[4] Department of Physics, University of Coimbra, Portugal

Earth's Core: Dynamics, Structure, Rotation
Geodynamics Series 31
Copyright 2003 by the American Geophysical Union
10.1029/31GD16

shown that these variations are correlated with the variations of the geomagnetic field, suggesting that the fluid iron core is important at this time scale. In this paper, we review the role of the core in the Earth's rotation variations in detail. Other properties of the Earth's core, such as its dynamics, thermal evolution, chemical properties and the geodynamo are treated elsewhere in this volume (see *Hollerbach, Kuang and Chao*).

First, we describe how geomagnetic information can be used to constrain the flow in the Earth's core. Using this information, the link between LOD variations and core dynamics can be studied either by considering the transfer of angular momentum from the core to the mantle (see section "Core angular momentum") or by calculating the torques at the core-mantle boundary (see section "Core mantle coupling mechanisms"). A recent determination from high precision observation of nutations by Very Long Baseline Interferometry (VLBI) has allowed constraints to be obtained on the magnetic field at the outer and inner core boundaries. We refer to these results in the frame of the analysis of the core-mantle coupling mechanisms. These constraints are further discussed by *Dehant and Mathews* [this volume].

The data needed to study the influence of the core on LOD variations are gathered by the "Special Bureau for the Core". This data center is part of the Product Center for Global Geophysical Fluids (CGGF) of the IERS (International Earth Rotation Service) [*Chao et al.*, 1999].

2. GEOMAGNETISM AND FLOW MODELING

If the flow were known throughout the core, the core angular momentum could be calculated and from that, the core contribution to the length of day could be determined. Obtaining the core flow in the entire core seems impossible, but fortunately the core angular momentum variation can be estimated using only the variations of the axisymmetric toroidal part of the horizontal velocity field at the CMB (see section "Core angular momentum computation"). The horizontal flow at the CMB can be determined from the secular variation of the geomagnetic field. In this section, we review the practicalities and difficulties in core flow modeling.

2.1. Frozen Flux Hypothesis

Inside the fluid outer core, the magnetic field plays an important role and cannot be neglected. The time evolution of the magnetic field inside the core is governed by the induction equation:

$$\frac{\partial \mathbf{B}}{\partial t} = \nabla \times (\mathbf{u} \times \mathbf{B}) + \eta \nabla^2 \mathbf{B} \qquad (1)$$

where $\mathbf{u}$ is the velocity field in the fluid core, $\mathbf{B}$ is the magnetic field inside the core, η is the magnetic diffusivity. The first

term in the right-hand side of this equation is the advection term and the last term is the diffusion term.

In the limit of perfect electrical conductivity of the fluid core ($\eta = 0$), the lines of the magnetic field become frozen in the fluid, so changes of magnetic field are only due to the advection of the field by the flow. This assumption (due to Alfvén) is called the *frozen flux approximation*. The diffusion time scale in the core is generally thought to be much larger (of the order of 10^4 years) than the advection time scale, which is of the order of the decadal time scale. Therefore diffusion is usually neglected when considering decadal time scale variations, and, as suggested by *Roberts and Scott* [1965], the frozen flux approximation is adopted.

See also *Bloxham and Jackson* [1991] for a discussion of the use of this approximation in core flow studies.

2.2. Toroidal and Poloidal Components of the Magnetic Field

The Earth's magnetic field varies on time scales ranging from seconds to millions of years [*Courtillot and Le Mouël*, 1988]. The variations on time scales shorter than 5 years are dominated by sources of external origin located in the ionosphere and in the magnetosphere. Variations over longer time scales, collectively known as the secular variation (SV), are predominantly of internal origin and are linked to the geodynamo. They provide an extremely useful tool for investigating the dynamics of the Earth's fluid outer core.

In general, a magnetic field in a conducting medium such as the liquid outer core, has both a toroidal component and a poloidal component :

$$\mathbf{B} = \mathbf{B}_T + \mathbf{B}_P = \nabla \times (T\mathbf{r}) + \nabla \times \nabla \times (P\mathbf{r}), \qquad (2)$$

where $\mathbf{r}$ is the position vector, and T and P are the toroidal and poloidal scalars dependent on the radius, the latitude and the longitude. Any solenoidal vector field (vector field with zero divergence) admits such a decomposition.

In an insulator such as the mantle, the magnetic field has only a poloidal component and can be expressed as

$$\mathbf{B} = -\nabla V, \qquad (3)$$

where V is the magnetic potential satisfying Laplace's equation. The potential V can further be expressed in terms of spherical harmonics as :

$$V = a\sum_{n=1}^{\infty} \sum_{m=0}^{n} \left(\frac{a}{r}\right)^{n+1} [g_n^m \cos m\phi + h_n^m \sin m\phi] P_n^m(\cos\theta)$$

$$(4)$$

where (r, θ, ϕ) are spherical coordinates, n and m are respectively the degree and the azimuthal order of the associated Legendre functions P_n^m, and a is the mean surface radius of the Earth.

The knowledge of the Gauss coefficients $\{g_n^m, h_n^m\}$ is sufficient to represent the geomagnetic field at a given epoch. Their derivatives $\{\dot{g}_n^m, \dot{h}_n^m\}$ represent the secular variation of the magnetic field. At the Earth's surface, the low degree terms in the expression of V are generally attributed to the CMB field, and the higher degree terms $n \geq 14$ to sources in the Earth's crust.

2.3. Frozen Flux Radial Magnetic Induction Equation

Assuming that the CMB is spherical, and that there is no flow through it, the velocity field there must satisfy the constraint $\mathbf{u} \cdot \hat{\mathbf{r}} = 0$, so the flow can be written $\mathbf{u} = \mathbf{u}_h$, where $\mathbf{u}_h$ is the horizontal velocity in the liquid core. Thus, following *Bullard and Gellman* [1954], the frozen flux radial magnetic induction equation is:

$$\frac{\partial B_r}{\partial t} = -\nabla_h \cdot (B_r \mathbf{u}_h) \tag{5}$$

2.4. Statement of the Inverse Problem

If the flow is assumed incompressible (so that the velocity field is solenoidal) then the fluid velocity at the top of the core can be split into toroidal and poloidal components (similar to the magnetic field). Because there is no flow through the CMB, at the top of the core, we may write:

$$\mathbf{u} = \mathbf{u}_h = \nabla \times (\mathcal{T}\mathbf{r}) + r \, \nabla_h \mathcal{S} \tag{6}$$

where $\nabla_h = \nabla - \hat{\mathbf{r}}(\hat{\mathbf{r}} \cdot \nabla)$ is the horizontal gradient. The potentials $\mathcal{T}$ and $\mathcal{S}$ are further expanded in spherical harmonics:

$$\mathcal{T} = \sum_{n=0}^{\infty} \sum_{m=0}^{n} t_n^m Y_n^m(\theta, \phi) \tag{7}$$

$$\mathcal{S} = \sum_{n=0}^{\infty} \sum_{m=0}^{n} s_n^m Y_n^m(\theta, \phi) \tag{8}$$

Define vectors $\mathbf{t}$ with elements t_n^m, $\mathbf{s}$ with elements s_n^m, and $\mathbf{b}$ with elements $\{g_n^m, h_n^m\}$. Substituting in the radial induction equation yields the matrix equations

$$\dot{\mathbf{b}} = E\mathbf{t} + G\mathbf{s} \tag{9}$$

where E and G are interaction matrices whose coefficients are linear combinations of the Gaunt-Elsasser integrals [*Whaler*, 1986] depending on the value of the magnetic field. The radial component of the magnetic field just above the CMB – which is customarily deduced from downward continuation of the observed surface field – is equal to the radial component of the field immediately below the CMB.

The inverse problem is to compute the elements of the vectors $\mathbf{t}$ and $\mathbf{s}$ from the known vector $\dot{\mathbf{b}}$ and matrices E and G.

2.5. Non-Uniqueness Problem

The determination of the two components of the horizontal velocity field $\mathbf{u}_h$ with this single equation is non-unique – many different flows can generate the same secular variation. To resolve this non-uniqueness, it is necessary to make assumptions as to the behavior of the flow at the CMB (essentially, to include some assumptions as to the flow dynamics). The main hypotheses discussed in the literature are:

• toroidal flow: the flow at the CMB has only an horizontal component which is assumed to be purely toroidal. If there is no up- or downwelling near the CMB, which can be the case if the upper part of the core is stably stratified, then the flow will be purely toroidal. See *Jault and Le Mouël* [1991] and *Lister and Buffett* [1998] for a discussion of the stratification of the fluid core at the CMB.

• (piecewise) steady flow: by assuming the flow does not change with time over a finite time interval, it can be determined uniquely. This approximation is often used because of its simplicity and is dynamically motivated by the long time scales (≥ 300 years) expected for phenomena in the core [*Davis and Whaler*, 1996]. When, however, the decadal LOD variations are attributed to the core, the core flow can evidently not be steady at this time scale. Moreover, steady flow is not adequate to explain fine scale features in the secular variation.

• steady flow in a drifting frame: a generalisation of the steady flow hypothesis. The core flow is considered steady in a frame which is allowed to drift with respect to the mantle [*Davis and Whaler*, 1996], [*Holme and Whaler*, 2001].

• (tangentially) geostrophic flow: the flow is governed by the balance between the (tangential components of the) Coriolis force and the pressure gradient at the CMB. It is the most commonly used assumption in core flow inversion modeling. Because the mantle is almost an insulator, the current near the core-mantle boundary is expected to be small. As a result, the Lorentz force is likely to be small. Of the remaining forces, the viscous force is thought to be small, and the buoyancy force is large only in the radial direction. The Coriolis force remains large, and can only be balanced in the horizontal direction by a pressure gradient force: this is precisely the required condition for tangential geostrophy to apply [*Le Mouël*, 1984] (indeed, without the Lorentz force, the force balance is very similar to that for the atmosphere). For more discussion as to whether neglect of the Lorentz force is appropriate, see *Bloxham and Jackson* [1991].

2.6. Restrictions to the Knowledge of the Magnetic Field

Due to the precision and spatial distribution of observations and to the use of truncated spherical harmonics expansions, there are limitations to the precision of the Gauss coefficients [*Whaler*, 1986]. The secular variation for the last century is

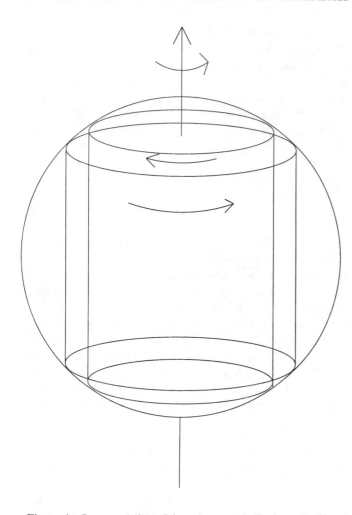

Figure 1. Representation of the axisymmetric Taylor cylinders

well known only up to degree six [*Hulot et al.*, 1992]. New satellite measurements, such as those provided by the Ørsted mission [*Olsen et al.*, 2000], will give insight into higher degrees of both the field and its secular variation.

Following *Roberts and Scott* [1965], the toroidal and poloidal scalars as well as B_r are expanded in a truncated series of spherical harmonics, see also *Voorhies* [1986] and *Bloxham and Jackson* [1991]. Let N_B be the truncation degree of the series for B_r. N_B is mainly defined by observational constraints. It is usually considered that $N_B = 13$ or 14 is an adequate choice to avoid significant crustal field contamination of the main field [*Langel*, 1987]. As for the flow, it can be truncated at degree 12 to 14, provided the spatial spectra of the remaining terms is ensured to converge (see, e.g., *Hulot et al.*, [1992], *Jackson et al.*, [1993]). This convergence is achieved, in practice, by applying a regularization to the flow solution, which can be the requirement that the spectrum of the flow falls as n^{-2} or faster [*Celaya and Wahr*, 1996].

The core flows we consider are time-dependent, and most are derived from the field model ufm1 of *Bloxham and Jackson* [1992], spanning the period 1840-1990. Steady flows over the same period do not provide an adequate fit to the fine-scale secular variation and so must be discounted.

2.7. Limits on the Knowledge of the Flow

There are several other difficulties in estimating the core-surface flow by inversion of the radial magnetic induction equation. The non-uniqueness of the inversion is not completely resolved by the assumption of tangential geostrophy [*Backus and Le Mouël*, 1986] and a further assumption, that the flow is large scale, must be used. The frozen-flux hypothesis can also be questioned as well as the fidelity of a CMB poloidal magnetic field model. Furthermore, the presence of noise in the secular variation models blurs the velocity field resolution [*Hulot et al.*, 1992, *Celaya and Wahr*, 1996]. This remains an obstacle independently of the way inversions are parametrized in space and time.

3. CORE ANGULAR MOMENTUM

The preceding section was dedicated to the difficulties occurring when modeling the core flow. Here, we explain how the core angular momentum can be estimated from the core flow and how its variations only depend on the variations of the axisymmetric toroidal part of the horizontal velocity flow at the CMB.

From an analysis of the magnitude of the various terms in the equation of motion, it is believed that the core flow is essentially governed by a balance between the Lorentz force, the Coriolis force, the pressure gradient and buoyancy, called magnetostrophic balance [*Gubbins and Roberts*, 1987, *Bloxham*, 1998]. Central in the discussion of core flows are the columns of fluid (called Taylor cylinders) coaxial with the rotation axis, as represented in Figure 1.

For a magnetostrophic solution to exist, the axial Lorentz torque on the Taylor cylinders must be zero. This state is referred to as the Taylor state. When, however, such a Taylor state is disturbed, so-called torsional oscillations about the Taylor state result, with periods of a few tens of years [*Braginski*, 1970] (c.f., the arrows showing the angular velocities of the cylinders in Figure 1). In that case the cylinders are rotating rigidly relative to each other and their motions are characterized by a zonal toroidal flow at the CMB. The rotation of the liquid core at the CMB is then latitude dependent, in agreement with the observation that the westward drift of the magnetic field is also latitude dependent [*Jault et al.*, 1988].

The torsional oscillations cause variations in the core angular momentum. The surface expression of the oscillations is a purely zonal ($m = 0$) toroidal flow; hence, the angular momentum variations can be calculated using only this part of the surface flow. Using the poloidal-toroidal decomposition of equation (6), the zonal component of tangentially geostrophic motions is a function of the t_n^0 coefficients:

$$\mathbf{t}_{ax}(\theta) = -\mathbf{r} \times \nabla(\sum_{n=1}^{\infty} t_n^0 P_n^0(cos\theta)) = t_{ax}(\theta)\hat{\phi} \quad (10)$$

where $\hat{\phi}$ the unit azimutal vector.

For the equatorial symmetry required by the Taylor cylinders, even n harmonics are excluded. Were the flow spherically symmetric, only the t_1^0 component would contribute to the angular momentum. However, the cylindrical symmetry and invariance in z of the Taylor cylinders leads to an additional contribution from other odd n terms. If the core is assumed of uniform density, and the inner core ignored, then this contribution is restricted to the t_3^0 term. The core angular momentum is then transformed into an equivalent change in the length of day by conservation of angular momentum, which is given by

$$dT = \frac{T_0^2}{2\pi} \left(\frac{I_c}{I_c + I_m}\right) \left(dt_1^0 + \frac{12}{7}dt_3^0\right) \quad (11)$$

with I_c the core moment of inertia, I_m the mantle moment of inertia, and $T_0 = 86400s$ the reference value for the length of day, $dT = T - T_0$ the predicted length of day variation. The notations dt_1^0 and dt_3^0 indicate that the values used for the prediction are taken relative to a uniform reference flow. In practice, this means that t_1^0 and t_3^0 can be used if an arbitrary value dT_{arb} is subtracted. So one has to compute

$$dT = 3.98 \left(t_1^0 + \frac{12}{7}t_3^0\right) - dT_{arb}. \quad (12)$$

The predictions give only relative changes in the length of day, therefore an arbitrary offset has to be chosen so that the predictions agree reasonably well with the length of day series given from astronomical observations.

In Figure 2, we compare the observed LOD variations with the predicted values for three geostrophic flows [*Jault et al.*, 1988, *Jackson et al.*, 1993, *Pais and Hulot*, 2000] and a drifting flow [*Holme and Whaler*, 2001]. The annual means of the excess LOD have been corrected for a linear trend of 1.8 ms/cy to account for tidal friction and post-glacial rebound effects (see, e.g., *Stephenson and Morrison*, [1995]). (The effect due to recession of the Moon has the same order of magnitude, see *Christodoulidis et al.*, [1988]). The offset between the observed and predicted curves for the excess LOD has already been eliminated. The agreement between these curves is good starting from around 1900, but less good before this date (although the agreement for the drifting flow is perhaps slightly better than for the geostrophic flows). It has been suggested that this could be due to the fact that part of the signal in t_1^0 and t_3^0 is not immediately related to the bulk geostrophic Taylor cylinders as was invoked by *Jault et al.* [1988], or otherwise that the observed disagreement could simply be due to the limited accuracy of the geomagnetically inferred predictions [*Pais and Hulot*, 2000].

These studies have considered only the overall angular momentum budget of the core. Extending this, assuming tangentially geostrophic flow, *Hide et al.* [2000] studied the latitude dependence of the core angular momentum (CAM) fluctuations. The CAM fluctuations are most pronounced in the mid-latitudes, where they are generally out of phase with those occurring in the equatorial regions. They are roughly in phase with decadal LOD fluctuations, especially after about 1870, with a dominant variability period of approximately 65 years. Both the latitude dependence of CAM and its latitudinal correlation with length of day fluctuations suggest that the CAM anomalies propagate from the equatorial to polar regions. *Zatman and Bloxham* [1997b] used similar ideas (considering the torsional oscillations as torsional waves) to probe the interior of the core.

4. CORE-MANTLE COUPLING MECHANISMS

In the previous section, the core angular momentum variations computed using different hypothesis have been compared with the LOD fluctuations. The results show that core angular momentum transfer to the mantle is a possible explanation of LOD variations. In this section, four coupling mechanisms for transferring core angular momentum to the mantle are presented: viscous, gravitational, topographic and electromagnetic coupling. We examine the relative importance of these four mechanisms in accounting for LOD variations.

4.1. Viscous Torque

Rochester [1976, 1984] has investigated the possible influence of a viscous torque. Adopting an Earth model which is a simple extension of Poincaré's model of a perfect (inviscid, incompressible, homogeneous) liquid core, he treats the combined effects of inertial (pressure) and frictional (viscous and electromagnetic) couplings across the CMB, assuming a laminar flow in the hydromagnetic boundary layer. Because this layer is thin, the dissipative part of the core-mantle interaction is treated as a small perturbation of the non-dissipative inertial coupling. For commonly used estimates of core viscosity (see, for example, *de Wijs et al.*, [1998]), the effect of viscosity is small.

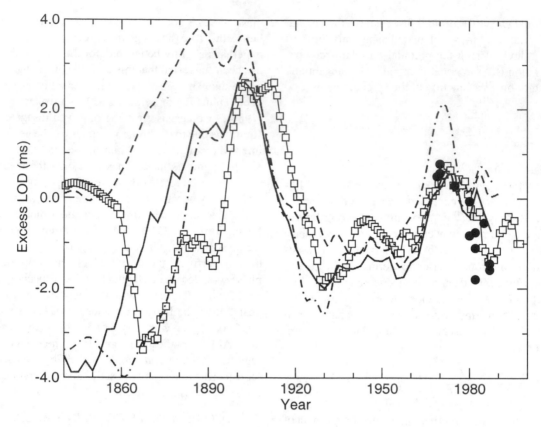

Figure 2. Comparison of LOD observations (squares) with predictions from the tangential geostrophic flow obtained when using the inversion parameters of Pais and Hulot (2000) (dashed line), Jackson, Bloxham and Gubbins (1993) (solid line), Holme and Whaler (2001) (dot dashed line), Jault et al. (1988) (circles).

4.2. Gravitational Torque

Length of day variations can also result from a gravitational torque acting on the mantle. First, the fluid outer core exerts on the mantle a torque due to the interaction of core density heterogeneities which may drive core convective motions with non-spherical equipotential surfaces in the core (resulting from density anomalies in the mantle and topography at the core-mantle boundary) [*Jault and Le Mouël*, 1989]. *Buffett* [1996a] argues that this effect is likely to be too small to explain LOD variations, as the density heterogeneities in the core are too small. Instead, he proposes a second form of gravitational coupling, between the mantle and inner core. The inner core is gradually solidifying from the fluid outer core. When it solidifies, its boundary will follow the gravitational equipotential produced by the mantle. Seismic tomography suggests that this will have a sizeable departure from zonal symmetry, with a particularly strong degree 2, order 2 spherical harmonic component. If the inner core is rotated out of alignment with the mantle, a strong gravitational torque will result to return it to its aligned position. The result would be a locking of the inner core and mantle, but allowing the

possibility of small oscillations of decadal period about the equilibrium position which through magnetic coupling across the inner core boundary can excite torsional oscillations in the fluid core [*Buffett*, 1996b].

Such a gravitational locking would seem to be contradicted by observations of secular inner core rotation with respect to the mantle [*Song and Richards*, 1996]. (For further discussion of inner core superrotation, see the papers by *Laske and Masters, Li and Richards, Song, Souriau and Poupinet*, [this volume]). However, *Buffett* [1998] argues that viscous relaxation of the inner core can allow both gravitational coupling and inner core rotation. Including viscous inner core deformation and gravitational coupling between inner core and mantle in geodynamo simulations, *Buffett and Glatzmaier* [2000] found that, on average, the inner core rotates faster than the mantle by a small amount. So, gravitational coupling is not ruled out by inner core rotation.

4.3. Topographic Torque

The idea that the hydrodynamical interaction between core motions and shallow, irregular undulations of the core-mantle interface might be an effective coupling mechanism was first suggested by *Hide* [1969]. The topography at the CMB is produced by convective motions in the mantle [*Defraigne et al.*, 1996]. The so-called topographic torque is produced by the effect of the pressure of the fluid in the core on the non-elliptical contributions to the boundary topography. The strength of the topographic torque, and whether it is possible to calculate the torque from observations, has been a matter of controversy. Early estimates of the torque from models of the flow and topography were made by *Jault and Le Mouël* [1989]. *Hide* [1993, 1995a] argued that a topography of no more than 500 m would suffice to produce such a torque. This view was challenged by *Bloxham and Kuang* [1995] and by *Kuang and Bloxham* [1997], who argued that the dominant part of the topographic torque, associated with the Coriolis force, should be identically equal to zero. Their analysis has been strongly contested by *Hide* [1995b] and *Jault and Le Mouël* [1999].

The question as to whether the topographic torque is strong or weak can be restated as: Does the topography significantly affect the pattern of the flow? Define a small parameter δ, equal to the ratio of the scale of the topography to the core radius. If the flow is not strongly influenced by the topography, as advocated by *Jault and Le Mouël* [1999], the torque scales with δ, and can be large. If the flow pattern is affected, the torque scales with δ^2, and is unlikely to be important. To try to resolve this controversy, *Kuang and Chao* [2001] have simulated the effects of topographic coupling in a numerical dynamo model. The effect of pressure at the boundary is calculated by considering the perturbation to an ambient flow from boundary topography. Their calculation leads to a torque proportional to δ^2, suggesting that topographic torque in the Earth will be small. With an amplitude of the topography of 3 km, the computed torque is only 20% of that required to explain the Earth's length of day variations. The most recent seismological study of core-mantle boundary topography [*Garcia and Souriau*, 2000] suggests that the amplitude of CMB topography is less than about 2 km. Thus, in the frame of the results of *Kuang and Chao* [2001] a torque from such a topography, while not insignificant, would be insufficient to explain the Earth's length of day variations.

However, the parameters used in the dynamo model are still far from Earth values [*Dormy et al.*, 2000]. Further, it would certainly be possible to build a self-consistent model by assuming that the torque is proportional to δ as required by the analysis of *Jault and Le Mouël* [1999]. There is still no consensus on the importance of the topographic torque.

4.4. Electromagnetic Torque

The magnetic field at the CMB induces electric currents in the electrically conducting lower mantle; this induces a poloidal torque (related to the poloidal part of the magnetic field), and an advective torque (related to the transport of the poloidal magnetic field). A leakage torque arises from the "leakage" of electrical currents diffusing from the core to the mantle. This last torque cannot be calculated from surface magnetic field observations [*Stix and Roberts*, 1984]. The advective torque is a possible candidate for explaining LOD variations [*Holme*, 1998a].

A recent determination from high precision nutation observations by Very Long Baseline Interferometry (VLBI) has allowed to obtain constraints on the magnetic torque at the outer and inner core boundaries [*Mathews et al.*, 2001]. Note that this magnetic torque is perpendicular to the poloidal magnetic torque explaining LOD variations. The analytical formulation of the theory of nutations reveals the combinations of Earth's interior parameters and global parameters that govern the nutational response of the Earth to gravitational (tidal) forcing by celestial bodies. Forced nutations are almost (except for a small part related to oceanic and atmospheric forcing) entirely due to periodic spectral components of the torques resulting from the gravitational action of celestial bodies (the Sun, the Moon, and to a minor extent, the planets) on the equatorial bulge of the Earth. The Earth's response to this gravitational forcing involves resonances due to normal modes inside the Earth and in particular to the excitation of the Free Core Nutation (FCN) and the free inner core nutation (FICN).

The dissipation necessary to explain the observed damping of the free core nutation and of the free inner core nutation can be used to constrain the electromagnetic torques acting on both core boundaries. To that aim, and in particular for explaining the large residuals in the out-of-phase retrograde annual nutation and in the out-of-phase prograde semi annual nutation, a large magnetic field is necessary at the ICB and a conductive layer at the bottom of the mantle is considered [*Buffett*, 1992]. *Mathews et al.* [2001] use a 200m layer with a conductivity of 5×10^5 Sm^{-1} and a magnetic field of 0.7mT at the CMB and of 6.8mT at the ICB to explain the large out-of-phase nutation residuals and the period of the observed nutations with an electromagnetic torque. This layer conductance of 10^8S was also the value determined by *Holme* [1988b] as the lower limit for effective influence of electromagnetic coupling on LOD variations.

Is such a highly conductive thin layer consistent with other geophysical observables? The conductivity of the mantle can be estimated in three ways. First, geomagnetic depth sounding can provide a conductivity depth profile from modeling of the induction effects of external field variations [e.g., *Olsen* 1999]. To penetrate the deep mantle, long period variations (such as the 11 years solar cycle) are required, but even these do not provide sufficient resolution for a detailed picture of the lower mantle, and certainly do not provide fine resolution at the core mantle boundary. Secondly, the observation of so-called geomagnetic jerks at the Earth surface provides a constraint on mantle conductivity. *Alexandrescu et al.* [1999] argue that a strongly conducting deep mantle would make observation of these jerks most unlikely. However, they show that for a conductance of $10^8 S$, this effect becomes significant for a uniform layer thickness of the order of 250km (their Figure 9). The layer postulated by *Buffett et al.*, [1992] and *Mathews et al.*, [2001] is three orders of magnitude thinner than this. Hence, the conductivity structure necessary to explain the damping of the FCN, and to influence LOD by EM coupling, is in no way precluded by the observation of geomagnetic jerks.

Thirdly, mantle conductivity can be estimated experimentally. Experiments by *Poirier et al.* [1992] suggest that deep mantle silicate conductance will be no higher than $10 \, Sm^{-1}$, far below the required value. However, various mechanisms exist which may significantly increase the conductivity of the very deepest mantle. A particularly attractive mechanism was recently provided by *Buffett et al.* [2000] - they propose a combined mechanism of core-mantle reaction and sedimentation of light material from the core at the core-mantle boundary to enlarge the thickness of the conductive layer, which would be sufficient to achieve the required conductance value.

5. CONCLUSION

In this paper, we have reviewed the links between core dynamics and variations in the Earth's rotation. We have detailed the different possibilities of modeling the core flow at the core-mantle boundary (CMB) from a downward extrapolation of the magnetic field and its secular variation observed at the surface of the Earth. The value of LOD variations can be known either by computing the torques acting on the mantle at the CMB or by computing the core angular momentum transfer to the mantle. As explained in this paper, the core angular momentum variations can be calculated from the CMB velocity field. For determining this velocity field, physical assumptions are necessary. Using the frozen flux hypothesis, the magnetic induction equation can be simplified and allows to relate the two components of the CMB velocity field to the known radial magnetic field. However, additional physical assumptions on the flow are necessary to reduce the non uniqueness of this latter equation. The computations lead to LOD variations close to the observations whatever the physical assumptions taken for the flow at the CMB.

We have also reviewed the different mechanisms of transfer and have detailed the possible torques at the core-mantle boundary. While the viscous torque is believed to be small, the amplitudes of the other torques are still subject to debate. In particular, the controversy surrounding the description of the topographic torque does not allow statement of a definitive value. Inner core-mantle graviational coupling could also be important. The observed LOD fluctuations can be explained by the electromagnetic torque if a layer of rather high conductance exists a the bottom of the mantle. The conductance needed also allows explanation of the VLBI observed nutations, assuming that the only torque at the core mantle boundary and inner core boundary is electromagnetic. The presence of the Free Core Nutation, an Earth normal mode related to the existence of a fluid core rotating in an ellipsoidal deformable rotating mantle, in the frequency band where the nutations can be observed (long periods in an inertial frame, diurnal in a frame tied to the Earth) implies a resonance which provides us with information about the Earth's core. In particular, the dissipation observed for the retrograde annual nutation and the prograde semi annual nutation gives constraints on the electromagnetic torques at the CMB and the ICB [see *Dehant and Mathews*, this volume].

Acknowledgments. One of us (S. Ponsar) acknowledges the Belgian National Fund for Scientific Research for financial support. We also gratefully acknowledge A. Jackson for providing us with his data concerning the core angular momentum computation. Two anonymous reviewers are also acknowledged for constructive reviews.

REFERENCES

Alexandrescu, M. M., Gilbert, D., Le Mouël, J. L., Hulot, G., and Saracco, G., An estimate of average lower mantle conductivity by wavelet analysis of geomagnetic jerks, *J. Geophys. Res.*, *104*, 17735-17745, 1999.

Backus, G. E., and Le Mouël, J. L., The region on the core-mantle boundary where geostrophic velocity field can be determined from frozen-flux magnetic data, *Geophys. J. R. astron. Soc.*, *85*, 617-628, 1986.

Ball, R. H., Kahle, A. B., and Vestine, E. H., Determination of surface motions of the Earth's core, *J. Geophys. Res*, *74*, 3659+, 1969.

Bloxham, J., Dynamics of Angular Momentum in the Earth's Core, *Annu. Rev. Earth Planet. Sci.*, *26*, 501-517, 1998.

Bloxham, J., and Kuang, W., Comment on "The topographic torque on a bounding surface of a rotating gravitating fluid and the excitation by core motions of decadal fluctuations in the Earth's rotation" by R. Hide, *Geophys. Res. Lett.*, *22*, 3561-3562, 1995.

Bloxham, J., and Jackson, A., Fluid Flow near the Surface of Earth's Outer Core, *Reviews of Geophysics, AGU*, *29*, 97-120, 1991.

Bloxham, J., and Jackson, A., Time-dependent mapping of the magnetic field at the core-mantle boundary, *J. Geophys. Res.*, *97*, 19537-19563, 1992.

Braginsky, S. I., Torsional magnetohydrodynamic vibrations in the Earth's core and variations in day length, *Geomag. Aeron.*, English Transl., *10*, 1-8, 1970.

Brosche, P., and Sündermann, J., Tidal Friction and the Earth's rotation, Springer Verlag, 1978.

Buffett, B. A., Constraints on Magnetic Energy and Mantle Conductivity From the Forced Nutations of the Earth, *J. Geophys. Res.*, *97*, 19581-19597, 1992.

Buffett, B. A., Gravitational Oscillations in the Length of Day, *Geophys. Res. Lett.*, *23*, 2279-2282, 1996a.

Buffett, B. A., A mechanism for decade fluctuations in the length of day, *Geophys. Res. Lett.*, *23*, 3803-3806, 1996b.

Buffett, B. A., Free Oscillations in the Length of Day : Inferences on Physical Properties near the Core-Mantle Boundary, in: *The Core-Mantle boundary Region, eds. M. Gurnis, M. E. Wysession, E. Knittle, B. A. Buffett, AGU Monograph, Geodynamics, 28*, 1998.

Buffett, B. A., and Glatzmaier, G. A., Gravitational Braking of Inner Core Rotation in Geodynamo Simulations, *Geophys. Res. Lett.*, *27(19)*, 3125-3128, 2000.

Buffett, B. A., Garnero, E. J., and Jeanloz, R., Sediments at the Top of Earth's Core, *Science*, *290*, 1338-1342, 2000.

Bullard, E. C., and Gellman, H., Homogeneous dynamos and terrestrial magnetism, *Philos. Trans. R. Soc. London, Ser. A*, *247*, 213-278, 1954.

Celaya, M., and Wahr, J., Aliasing and noise in core-surface flow inversions, *Geophys. J. Int.*, *126*, 447-469, 1996.

Chao, B. F., Dehant, V., Gross, R. S., Ray, R. D., Salstein, D. A., Watkins, M. M., and Wilson, C. R., Space Geodesy Monitors Mass Transports in Global Geophysical Fluids, *EOS, AGU Publication*, *81(22)*, 247-250, 2000.

Chen, J. L., Wilson, C. R., Chao, B. F., Shum, C. K. and Tapley, B. D., Hydrological and oceanic excitations to polar motion and length of day variation, *Geophys. J. Int.*, *141*, 149-156, 2000.

Christodoulidis, D. C., Smith, D. E., Williamson, R. G. and Klosko, S. M., Observed tidal braking in the Earth/Moon/Sun system, *J. Geophys. Res.*, *93*, 6216-6236, 1988.

Courtillot, V., and Le Mouël, J. L., Time variations of the Earth's magnetic field : from daily to secular, *Ann. Rev. Earth Planet. Sci.*, *16*, 389-476, 1988.

Davis, R. G., and Whaler, K. A., Determination of a steady velocity in a rotating frame of reference at the surface of the Earth's core, *Geophys. J. Int.*, *126*, 92-100, 1996.

Defraigne, P., Dehant, V., and Wahr, J. M., Internal loading of an inhomogeneous compressible Earth with phase boundaries, *Geophys. J. Int.*, *125*, 173-192, 1996.

Dehant, V., and Mathews, P. M., Information about the core from Earth's nutation, this volume.

de Wijs, G. A., Kresse, G., Vocadlo, L., Dobson, D., Alfe, D., Gillan, M. J., and Price, G. D., The viscosity of liquid iron at the physical conditions of the Earth's core, *Nature*, *392*, 805-807, 1998.

Dicke, R.H., The secular acceleration of the Earth's rotation and cosmology, *The Earth Moon System*, eds Mardsen, B. G., and Cameron, A. G. W., Plenum Press, New York, 1966.

Dickey, J. O., Marcus, S. L., and de Viron, O., The Earth angular momentum budget on subseasonal time scales: exchange among the solid Earth, atmosphere and ocean subsystems. *EOS Trans. AGU, 81(48)*, G61B-25, Fall meet. suppl., 2000.

Dormy, E., Valet, J. P., and V. Courtillot, Numerical models of the geodynamo and observational constraints, *Geochemistry, Geophysics, Geosystems*, *1*, 2000.

Garcia, R., and Souriau, A., Amplitude of the core-mantle boundary topography estimated by stochastic analysis of core phases, *Phys. Earth Planet. Int.*, *117*, 345-359, 2000.

Gubbins, D., and Roberts, P., Magnetohydrodynamics of Earth's core, in *Geomagnetism 2*, ed. J.A. Jacobs, Academic Press, London, 1-183, 1987.

Hide, R., Interaction between the Earth's Liquid Core and Solid Mantle, *Nature*, *222*, 1055-1056, 1969.

Hide, R., Angular Momentum Transfer between the Earth's Core and Mantle, in: *Dynamics of Earth's Deep Interior and Earth Rotation, AGU Geophysical Monograph*, *72*, 109-112, 1993.

Hide, R., The topographic torque on the rigid bounding surface of a rotating gravitating fluid and the excitation by core motions of decadal fluctuations in the Earth's rotation, *Geophys. Res. Lett.*, *22*, 961-964, 1995a.

Hide, R., "Reply to the comment" by J. Bloxham and W. Kuang on a paper entitled "The topographic torque on the rigid bounding surface of a rotating gravitating fluid and the excitation by core motions of decadal fluctuations in the Earth's rotation" by R. Hide, *Geophys. Res. Lett.*, *22*, 3563-3565, 1995b.

Hide, R., Boggs, D. H., and Dickey, J. O., Angular momentum fluctuations within the Earth's liquid core and torsional oscillations of the core-mantle system, *Geophys. J. Int.*, *143*, 777-786, 2000.

Hollerbach, R., The range of time scales on which the geodynamo operates, *this volume*.

Holme, R., Electromagnetic Core-Mantle Couping I : Explaining decadal changes in the length of day, *Geophys. J. Int.*, *132*, 167-180, 1998a.

Holme, R., Electromagnetic Core-Mantle Couping II : Probing Deep Mantle Conductance, in: *The Core-Mantle Boundary Region*, eds. M. Gurnis, M. E. Wysession, E. Knittle, B. A. Buffett, *AGU Monograph, Geodynamics*, 28, 139-151, 1998b.

Holme, R., and Whaler, K. A., Steady core flow in an azimuthally drifting reference frame, *Geophys. J. Int.*, *145*, 560-569, 2001.

Hulot, G., Le Mouël, J. L., and Wahr, J., Taking into account truncation problems and geomagnetic model accuracy in assessing computed flows at the core-mantle boundary, *Geophys. J. Int.*, *108*, 224-246, 1992.

Jackson, A., Bloxham, J., and Gubbins, D., Time-dependent flow at the core surface and conservation of angular momentum in the coupled core-mantle system, in: *Dynamics of Earth's Deep Interior and Earth Rotation*, eds. J. L. Le Mouël, D. E. Smylie, T. Herring, *AGU Geophysical Monograph*, 72, 97-107, 1993.

Jault, D., Gire, C., and Le Mouël, J. L., Westward drift, core motions and exchanges of angular momentum between core and mantle, *Nature*, *333*, 353-356, 1988.

Jault, D., and Le Mouël J. L., The topographic torque associated with a tangentially geostrophic motion at the core surface and inferences on the flow inside the core, *Geophys. Astrophys. Fluid Dynamics*, *48*, 273-296, 1989.

Jault, D., and Le Mouël, J. L., Physical properties at the top of the core and core surface motions, *Phys. Earth Planet. Inter.*, *68*, 76-84, 1991.

Jault, D., and Le Mouël, J. L., Comment on 'On the dynamics of topographical core-mantle coupling' by Weijia Kuang and Jeremy Bloxham, *Phys. Earth Planet. Inter.*, *114*, 211-215, 1999.

Johnson, T. J., Wilson, C. R., and Chao, B. J., Oceanic angular momentum variability estimated from the Parallel Ocean Climate Model, 1988-1998, *J. Geophys. Res.*, *104*, 25183-25195, 1999.

Kuang, W., and Bloxham, J., On the dynamics of topographical core-mantle coupling, *Phys. Earth Planet. Inter.*, *99*, 289-294, 1997.

Kuang, W., and Chao, B. F., Topographic Core-Mantle coupling in Geodynamo modeling, *Geophys. Res. Lett.*, *28(9)*, 1871-1874, 2001.

Kuang, W., and Chao, B. F., Geodynamo Modeling and Core-Mantle Interactions, *this volume*.

Langel, R. A., The main field, in *Geomagnetism 1*, J.A. Jacobs, ed., Academic Press, NewYork, 249-512, 1987.

Laske, G., and Masters, G., The Earth's free oscillations and the differential rotation of the inner core, *this volume*.

Le Mouël, J. L., Outer-core geostrophic flow and secular variation of Earth's geomagnetic field, *Nature*, *311*, 734-735, 1984.

Li, A., and Richards, P. G., Study of Inner Core Structure and Rotation Using Novaya Zemlya Nuclear Explosions, *this volume*.

Lister, J. R., and Buffett, B. A., Stratification of the outer core at the core-mantle boundary, *Phys. Earth. Planet. Inter.*, *105*, 5-19, 1998.

Marcus, S. L., Chao, Y., Dickey, J. O., and Gegout, P., Detection and Modeling of Nontidal Oceanic Effects on Earth's Rotation Rate, *Science*, *281*, 1656-1659, 1998.

Mathews, P. M., Herring, T. A., and Buffett, B. A., Modeling of Nutation-Precession : Insights into the Earth's Interior and New Nonrigid Earth Nutation Series, *J. Geophys. Res.*, in press, 2001.

Olsen, N., Long period (30 days-1year) electromagnetic sounding and the electrical conductivity of the lower mantle beneath Europe, *Geophys. J. Int.*, *138*, 179-187, 1999.

Olsen, N., Holme, R., Hulot, G., Sabaka, T., Neubert, T., Tøffner-Clausen, L., Primdahl, F., Jørgensen, J., Léger, J.-M., Barraclough, D., Bloxham, J., Cain, J., Constable, C., Golovkov, V., Jackson, A., Kotze, P., Langlais, B., Macmillan, S., Mandea, M., Merayo, J., Newitt, L., Purucker, M., Risbo, T., Stampe, M., Thomson, A. and Voorhies, C., Ørsted initial field model, *Geophys. Res. Lett.*, 27, 3607-3610, 2000.

Pais, A., and Hulot, G., Length of day decade variations, torsional oscillations and inner core superrotation: evidence from recovered core surface zonal flows, *Phys. Earth Planet. Inter.*, *118*, 291-316, 2000.

Peltier, W. R., Drummond, R. A., and Tushingham, A. M, Post-glacial rebound and transient lower mantle rheology, *Geophys. J. R. astron. Soc.*, *87*, 79-116, 1986.

Poirier, J.-P., and Peyronneau, J., Experimental determination of the electrical conductivity of the material of the Earth's lower mantle, in: *High-Pressure Research : Applications to Earth and Planetary Sciences, AGU Geophysical Monograph Series*, *67*, edited by Y.Syono and Manghnani, 77-87, 1992.

Ponte, R. M., and Rosen, R. D., Atmospheric torques on land and ocean and implications for Earth's angular momentum budget, *J. Geophys. Res*, *106*, 11793-11799, 2001.

Roberts, P. H., and Scott, S., On Analysis of the Secular Variation. 1. A Hydrodynamic Constraint: Theory, *Journal of Geomagnetism and Geoelectricity*, *17(2)*, 137-151, 1965.

Rochester, M. G., The Secular Decrease of Obliquity due to Dissipative Core-Mantle Coupling, *Geophys. J. R. astron. Soc.*, *46*, 109-126, 1976.

Rochester, M. G., Causes of fluctuations in the rotation of the Earth, *Phil. Trans. R. Soc. Lond* A *313*, 95-105, 1984.

Rosen, R. D., The axial momentum balance of Earth and its fluid envelope, *Surveys Geophys.*, *14*, 1-29, 1993.

Song, X., Three-dimensional inner core structure and differential rotation of inner core, *this volume*.

Song, X. D. and Richards P. G., Seismological evidence for differential rotation of the Earth's inner core, *Nature, 382*, 221-224, 1996.

Souriau, A., and Poupinet, G., Inner core rotation: a critical approach, *this volume*.

Stephenson, F. R., and Morrison, L. V., Long–term fluctuations in the Earth's rotation: 700 BC to AD 1990, *Phil. Trans. R. Soc. Lond. A, 351*, 165–202, 1995.

Stix, M., and Roberts, P. H., Time-dependent electromagnetic core-mantle coupling, *Phys. Earth Planet. Int., 36*, 49-60, 1984.

Voorhies, C. V., Steady flows at the top of Earth's core derived from geomagnetic field models, *J. Geophys. Res., 91*, 12444-12466, 1986.

Whaler, K. A., Geomagnetic evidence for fluid upwelling at the core-mantle boundary, *Geophys. J. Res., 86*, 563-588, 1986.

Zatman, S., and Bloxham, J., The phase difference between length of day and atmospheric angular momentum at subannual frequencies and the possible role of core-mantle coupling, *Geophys. Res. Lett., 24*, 1799-1802, 1997a.

Zatman, S. and Bloxham, J., Torsional oscillations and the magnetic field within the Earth's core, *Nature, 388*, 760-763, 1997b.

S. Ponsar, V. Dehant, T. Van Hoolst, Royal Observatory of Belgium, Avenue Circulaire 3, B-1180 Brussels, Belgium. (email: stephap@oma.be, veroniq@oma.be, timvh@oma.be)

R. Holme, GeoForschungsZentrum Potsdam, Telegrafenberg, Potsdam, Germany. (email: holme@gfz-potsdam.de)

D. Jault, LGIT, University of Grenoble, Grenoble, France. (email: Dominique.Jault@.obs.ujf-grenoble.fr)

A. Pais, Department of Physics, University of Coimbra, Coimbra, Portugal. (email: pais@teor.fis.uc.pt)

Information About the Core From Earth Nutation

V. Dehant

Royal Observatory of Belgium, 3avenue Circulaire, B1180 Brussels, Belgium

P. M. Mathews

Department of Theoretical Physics, University of Madras (Guindy Campus), Chennai 600025, India

The nutations are periodic variations of the Earth's orientation in space (orientation of the Earth's figure axis or rotation axis or angular momentum axis). They are induced by the gravitational forcing due to the external bodies (the Sun, the Moon, and, to a minor extent, the other planets), which are moving relative to the Earth. As the Earth is not a rigid planet, its nutational response depends on the physics of the Earth's interior. Resonances in the nutational motion, arising from the presence of fluid outer and solid inner core regions, are sensitive to the parameters and properties of these regions, such as the flattening of the outer core boundaries and the magnetic fields present at the core-mantle boundary (CMB) and at the inner core boundary (ICB). The Earth orientation data extracted from the high precision Very Long Baseline Interferometry (VLBI) observations available now provide very robust constraints on the parameters of the Earth's interior. For example, the root mean square values of the radial component of the magnetic fields over these boundaries are estimated to be 6.9 gauss (or 0.69 mT) at the CMB, and 72 gauss (or 0.72 mT) at the ICB. This paper is a review on the retrieval of information regarding properties of the core from VLBI data, elucidating the hypotheses behind the computations involved.

INTRODUCTION

For the description of the motion of the Earth in space under the gravitational attraction of the Moon, the Sun, and the planets, the Earth may be (and is) considered as a point mass. In reality, the Earth is of finite size, and the gravitational attraction is not the same at all the points of the planet. The spatially varying part, over the volume of the Earth, of the gravitational potential due to external bodies is called the tidal potential (also referred to as the tidal perturbation). The Earth undergoes tidal deforma-tions under the action of this potential. Additionally, as the Earth is ellipsoidal (axisymmetric in the first approximation), the gravitational attraction is not symmetric and there exists a torque that tends to reorient the equator into the direction of the attracting body, i.e. towards the ecliptic. This torque is proportional to the Earth's dynamical flattening, which is the fractional difference between the polar moment of inertia and the mean equatorial moment of inertia. It mainly represents the existence of an equatorial bulge on the moment of inertia. Because of its rotation, the Earth responds to that torque as a gyroscope and its rotation axis describes a motion in space. The secular part of the motion is called precession. The remaining motion, consisting almost wholly of a superposition of periodic components with a discrete set of periods from several hours and up, is called nutation.

The periods mentioned are relative to an inertial reference system (in space). They correspond to quasi-diurnal

Earth's Core: Dynamics, Structure, Rotation
Geodynamics Series 31
Copyright 2003 by the American Geophysical Union
10.1029/31GD16

periods in the terrestrial rotating reference system. The link between them is given by

$$\sigma = \Omega \left(-1 + \frac{1}{T}\right)$$

where σ is the angular frequency in the terrestrial reference system, T is the period (in sidereal days) in an inertial reference system, and Ω is the mean angular velocity of Earth rotation. T is negative for retrograde nutations, and positive for prograde nutations.

It has become possible, from the study of the Earth's response to periodic gravitational forcing, to infer certain types of information about the Earth's interior with a precision never reached by other kinds of observation. The nutational response of the Earth has been studied theoretically in a very elaborate way, and has been observed at a very high precision (millimeter level at the Earth surface) by using Very Long Baseline Interferometry (VLBI). VLBI makes it possible to determine the time dependent Earth orientation from the observed arrival times of extragalactic radio-source signals at a number of stations at the Earth's surface, the radio-sources being practically stationary in an inertial reference system. One determines precession and nutations (e.g., *Ma et al.* [1997]) from the Earth orientation variations. These motions in space are provided as functions of time in two directions: in longitude (parallel to the ecliptic) and in obliquity (in a plane perpendicular to the ecliptic). Nutation is a linear response of the Earth to gravitational (tidal) forcing, apart from very small contributions from nonlinear effects and from non-tidal forcing. It is made up of elliptical motions at a large number of frequencies. (The forcing being a sum of periodic components, the nutational motions are a sum of periodic motions.) Every elliptical motion is decomposable into prograde and retrograde motions (prograde meaning in the sense of the Earth rotation, and retrograde, in the opposite sense). The two (prograde/retrograde) parts are due to forcing at two separate frequencies present in the tidal potential. The amplitudes of all the prograde and retrograde nutations can be determined from the frequency domain version of the time series of nutations in longitude and obliquity. The largest of the nutations are the prograde and retrograde 18.6-year nutations of which the period is related to the motion of the node of the Moon. The torques producing them, which are of very low frequency in an inertial frame (and have amplitudes that are not too small), act over long periods of time and produce a large response, proportional to the period in space. In mathematical terms, such a response results from the appearance of the very low frequency in denominator in consequence of an integration of the angular velocity variable over time. It is thus not surprising that the 18.6-year nutations are the most important ones.

VLBI observations provide a very precise knowledge of nutational motion. Because the numerical values of specific physical parameters inferred from geophysical data are not precise enough, close agreement between theoretical predictions and such precise observations cannot be achieved without optimizing the parameter values that are not obtainable with high precision from the geophysics of the Earth's interior, through a least squares fit to the observed nutation and precession. Thanks to the high precision of the observations (currently at the 40 to 5 microarcsesond (μas) level depending on frequency), we have strong constraints on the fitted parameters thus enabling precise determination of the corresponding properties of the Earth's interior. For example, the presence of a liquid core causes a normal mode, called the Free Core Nutation (FCN), and an associated resonance in the forced nutation. Enhancement of the amplitudes of the nutations at frequencies close to the frequency of the FCN is clearly observable, and information about the core can be inferred from the characteristics of the resonance. *Gwinn et al.* [1986] and *Herring et al.* [1986] have indeed shown, from VLBI data analysis, that the flattening of the core is higher than in a hydrostatic equilibrium Earth, and that the excess flattening corresponds to an increase of about 500 meters in the difference between the equatorial radius and the polar radius. This result has been revisited on the basis of the more precise estimates of nutation amplitudes (see *Herring et al.* [2001]) derived from the currently available VLBI data set. Using these estimates, *Mathews et al.* [2001] have been able to infer additional properties of the core. In particular, estimates were obtained for the electromagnetic coupling constants representing the torques due to magnetic fields at the Inner Core Boundary (ICB) and at the Core Mantle Boundary (CMB). Thus, the study of the nutations of the non-rigid Earth gives complementary and independent information to the scientific community involved in the study of the core. This paper aims at presenting the implications of these new estimates related to the core, and at reviewing the methods and assumptions in the determination of core properties from nutation study. We will first review nutation modeling with particular emphasis on the role of the core regions. Then, we will focus on the information related to the core-mantle boundary topography and to the inner core boundary topography. We begin with a discussion on the form of the displacement field within the Earth (introducing rotational displacements induced by tidal forcing). Then we proceed to review how the core electromagnetic coupling is computed and what this information obtained from nutation tells us about magnetic fields at the core boundaries. We also present the possible consequences of the existence of viscosity of the inner core for the nutations.

NUTATION MODEL

The nutation is computed as a linear response to external gravitational (tidal) forcing, supplemented by a calculation of small nonlinear effects. A small part of the nutation is due to forcing not directly associated with the gravitational interaction. In particular, atmospheric forcing by pressure changes arising mainly from solar heating gives rise to small incremental nutations at some of the frequencies present in the tidal potential. Ocean tides raised by the tidal potential make their contributions too, which may be computed separately and added to the results of the linear formalism, as is often done, or may be integrated into the linear formalism. In the following paragraphs, we shall focus on the linear response to the tidal potential, taking it for granted that all effects that are not taken into account in the linear equations are accounted for separately.

Nutation models are built in two steps: (1) from the ephemerides of the Sun, the Moon, and the other planets, one computes the gravitational torque induced by these celestial bodies, and the nutational response of a rigid Earth to this torque, taking account of the Earth's global density structure (such as the global dynamical flattening and the triaxiality of the Earth); (2) the "transfer function" representing the nutational response of the non-rigid Earth relative to that of the rigid Earth is computed, and a convolution is performed between rigid-Earth nutations and the transfer function which accounts for the non-rigidity of the Earth and displays resonances resulting from the presence of the core regions. In the frequency domain, the convolution is simply a product. The nonlinear effects not included in the above general scheme make small contributions to both nutations and precession, and these are computed through a separate perturbative calculation [Mathews et al., 2001]. Such effects include the action of the tidal potential on the deformations produced by potential, and the coupling between nutational motion and length-of-day variations (rotation rate).

1. Rigid-Earth Precession and Nutations

The rigid-Earth nutations are computed from high precision celestial ephemerides of the Sun, the Moon and the other planets, and are known to better than a few tenths of a microarcsecond (μas). These models are provided by Souchay and Kinoshita [1996, 1997], Roosbeek and Dehant [1998], and Bretagnon et al. [1998]. The differences between them (at a level of a few tens of μas) arise from differences in the approaches used (the torque approach or the Hamiltonian approach) and in the value used for the dynamic flattening of the Earth, in turn deduced by each author from the observed precession.

2. Non-rigid Earth Transfer Function

There are presently different approaches developed to determine the non-rigid Earth contributions that lead to different numerical values. All these approaches include resonance effects associated with normal modes that are important for the nutations, namely the Free Core Nutation (FCN, also called the Nearly Diurnal Free Wobble, NDFW), the Free Inner Core Nutation (FICN, also called the Prograde Free Core Nutation, PFCN), and the Chandler Wobble (CW). The FCN is a mode owing its existence to the fact that misalignment of the axis of rotation of the fluid core relative to the symmetry axis of the core mantle boundary (CMB) generates a pressure torque on this boundary (which is itself deformable); this torque produces a motion in space of the axis of core rotation, which is distinct from that of the rotation axis of the mantle. The restoring forces when there exists a differential rotation between the core and the mantle are the pressure force, the friction force, and the Lorentz force. This differential rotation is the FCN. The larger the flattening of the CMB (α_f), the shorter the period of the FCN mode in a space-fixed reference frame. The FICN is a mode of similar origin, except that the differential rotation is between the rotation axes of the solid inner core and the mantle. In this case, the relevant pressure torque is that which acts on the flattening (α_s) of the ellipsoidal ICB; the gravitational torque between the mantle and the inner core also comes into play because the differential rotation is necessarily accompanied by a tilt of the frame axis of the inner core rotation relative to the mantle. Once again, the larger the flattening is, the shorter is the period in space. The motions associated with both these free modes are retrograde and quasi-diurnal in a frame tied to the Earth. The FCN is retrograde in space, and the FICN, prograde. Their frequencies in a frame tied to the Earth are given by (e.g. Dehant et al. [1993]):

$$\lambda_{FCN} = -\Omega\left(1 + \frac{A}{A^m}(\alpha_f - \frac{q_0}{2}\bar{h}_1^f) + \text{termsin}\frac{A^s}{A^f}\right) \quad (1)$$

$$\lambda_{FICN} = -\Omega\left(1 - \left[\frac{\rho_f}{\rho_s}(\alpha_s - \frac{q_0}{2}\bar{h}_2^s) - \frac{\Delta\rho}{\rho_s}(\alpha_s\gamma_s - \frac{q_0}{2}\xi_s)\right]\right.$$
$$\left. + \text{termsin}\frac{A^s}{A^f}\right) \quad (2)$$

where A, A^m, A^f, and A^s are the principal equatorial moments of inertia of the whole Earth, the mantle, the outer core, and the inner core (A^s being less than a thousandth of A). The h-, γ-, ξ-types of parameters are related to deformations of the boundaries, to mass redistribution potential contributions, or to the tilt of the inner core (case of γ_s). ρ_f is the outer core density, ρ_s is the inner core density, $\Delta\rho$ is the density jump at the ICB (and positive), and q_0 is $\dfrac{\Omega^2 a_{surf}}{g_0}$, the ratio of the centrifugal acceleration and the gravity (g_0) at the Earth's surface. In these expressions, the viscosity of the inner core and the electromagnetic coupling at the core boundaries are not considered.

Computations of non-rigid Earth nutation transfer function have been made from different approaches:

1. Models using numerical integration inside the Earth such as the one developed by Wahr (see *Wahr* [1981]) and adopted by the IAU (International Astronomical Union) in 1980. The newer generation of models comprises those of *Dehant and Defraigne* [1997], *Schastock* [1997], *Rogister* [2000], *Huang et al.* [2000], and *Huang* [2001]. These are purely theoretical models that are constructed starting from internal profiles of the rheological properties (density, shear and bulk moduli) such as those given by *Dziewonski and Anderson* [1981]. These profiles are based on the equations of static equilibrium inside the Earth (assumed to be in uniform rotation) and their numerical integration from the center up to the surface. They take the Earth to be ellipsoidal, with elastic or inelastic deformable inner core, a liquid outer core, and an elastic or inelastic deformable mantle. The nutations of an Earth with such a structure are computed from the momentum balance equations for the displacement field produced by the tidal forcing, employing spherical harmonic expansions for obtaining the solution.

2. Models using a resonance formula of which the parameters are fitted on the observations. In the work of *Getino and Ferrandiz* [2000] based on a Hamiltonian approach, and in their other works cited therein, the frequency of the FCN is used as an adjustable parameter, along with other geophysical and astronomical parameters, to optimize the fit between their theory and nutation observations. The ocean effect on nutation, computed separately, is accounted for through corrections to the nutation amplitudes before parameter fitting. Their parameter estimates include one that implies a value for the ratio of the Earth and Moon masses; its deviation from the otherwise accurately determined value (see the IERS Conventions [*McCarthy*, 1996]) is a few orders of magnitude larger than the uncertainty in the known value. Their estimates for other parameters are then unlikely to be useful for understanding the Earth's interior.

3. The model built by Herring for the IERS in 1996 (see the IERS Conventions [*McCarthy*, 1996]) is a purely empirical model. It was constructed just to provide users with a compact formula, which most closely agrees with observations, and was not intended for geophysical interpretation. Similarly, *Shirai and Fukushima* [2001] have built a new empirical model based on recent observations.

4. The last generation of models is the MHB2000 model of *Mathews et al.* ([2001], see also *Herring et al.* [2001], and *Buffett et al.* [2001]) adopted by the IAU in 2000. It considers the amplitudes of the "wobbles" of the inner core, the outer core, and the Earth's mantle, from which nutation amplitudes can be obtained directly. (The wobble of any of these regions of the Earth consists of a motion of the axis of global rotation of that region relative to an Earth-fixed reference frame.) The model is based on a geophysical theory of nutation, with the selected basic geophysical parameters of which the values are not known with a sufficient precision optimized for best fit of the theoretical predictions to nutation data. (Resonance frequencies and strengths are not the fitted parameters.) A feature of the procedure employed is that in using all the nutation angle estimates besides the estimated precession rate to constrain the dynamical ellipticity (as well as other parameters), full advantage is taken of the fact that the dynamical ellipticity is the parameter to which all the nutations are most sensitive. The estimates obtained for the fitted parameters may be interpreted in terms of the physics of the Earth's interior. The model MHB2000 incorporates a number of physical processes not considered in the other models in points 1 to 3. The ocean tidal contributions to nutations, computable with aid of a four-parameter empirical formula, are taken into account for all the nutations; the parameters are estimated by making use of observational data on the ocean response to tidal forcing, available for just a few large tides, and employing the frequency dependence of the solid Earth deformation due to ocean tidal loading and currents. This frequency dependence is derived from the nutation-wobble theory fitted to nutation and precession data. This procedure minimizes the chances of oceanic effects contaminating estimates of parameters of the Earth's interior. Inelasticity effects and electromagnetic couplings also are accounted for. Details of these computations can be found in *Mathews et al.* [2001]. At this point, it is important to note that it is impossible to represent exactly the nutation amplitudes using a simple resonance formula because of the ocean tide effects for instance. However, one can use an approximate resonance formula; then some nutations need additional corrections. It must be noted that a time-variable amplitude of the 'free' FCN is given in addition to the model for forced nutations. The amplitude of this free mode is estimated from the data because the mechanism of excitation of this mode is not well understood.

Some of the numerical integration models allow departure from hydrostatic equilibrium within the Earth and may reproduce the data quite well. This is the case for *the Dehant and Defraigne* [1997] model based on lateral heterogeneities within the mantle observed by seismic tomography (see next section). But, although mantle inelasticity is considered, it does not reproduce perfectly the data, especially the out-of-phase components of the nutations. The main problem is the lack of dissipative couplings at the core boundaries. Nevertheless those models do yield information on the physics of the Earth's core that will be developed in the next section. The semi-analytical model MHB2000 contains electromagnetic coupling at the CMB and at the ICB. The model of *Getino and Ferrandiz* [2000] also allows for these out-of-phase contributions. But, as already mentioned, it does not allow the physical interpretation of the parameters obtained from the fit on the VLBI observations. We refer the reader to *Dehant et al.* [1999] for details on all these models, which are not really the focus of this paper.

Figure 1 shows the non-rigid Earth nutation residuals with respect to the hydrostatic model as computed by *Wahr* [1981], for a non-hydrostatic model as computed by *Dehant and Defraigne* [1997], and for MHB2000 as computed by *Mathews et al.* [2001], compared with the residual VLBI nutation observations. Figure 2 shows the transfer functions for the theoretical models used in Figure 1, i.e. *Wahr*'s [1980] transfer function for an hydrostatic equilibrium model, *Dehant and Defraigne*'s [1997] model for a non-hydrostatic Earth without electromagnetic torques at the core boundaries, and *Mathews et al.*'s [2001] model including electromagnetic torques. Because the main differences between these models concern the core modeling (shape and electromagnetic coupling), it is to be expected that this kind of confrontation with observations will provide us with information about the core. In the following sections, we shall point out what we can learn about the core from nutation. In a first section, we will focus on the topography of the core boundaries. We will then show that the "wobble" approach is adequate for treating the effects of the core on the nutations. The subsequent sections will be devoted to the MHB2000 model and will highlight what we can learn about the core from VLBI nutation data using this semi-analytical approach.

NUTATIONS AND TOPOGRAPHY AT THE CORE-MANTLE BOUNDARY AND AT THE INNER CORE BOUNDARY

The topographies at the core-mantle boundary and at the inner core boundary have an influence on the nutation. This is particularly true for the degree two-order zero spherical harmonic of the topography corresponding to the flattenings of the boundaries. The influence of the flattenings appears in the FCN and FICN eigenfrequencies and in the strengths of the resonances affecting the nutation amplitudes. Therefore a change in the flattening causes a change in the nutation amplitudes.

On the nutation time scale, mantle convection can be considered as a steady state. The internal boundary topographies, such as the γ-spinel/ provskite boundary between

the lower mantle and the upper mantle (at 670km depth) and the inner core and outer core boundaries, are shaped by this convection. This steady state, characterized by an equilibrium between the buoyancy force (Archimedian force) due to mantle mass heterogeneities, the pressure gradient, and the viscous drag, accounts for the induced mass redistribution; the convective flow within the mantle induces deformations of the Earth's surface and of internal boundaries. In particular the flattenings of the Earth's surface and of the CMB become different from their respective hydrostatic values (Clairaut's equation).

Using mantle tomography data i.e. mantle density heterogeneities and seismic anomalies, it is possible to compute the CMB and ICB deformations due to internal loading. *Defraigne et al.* [1996] have computed the boundary deformation and in particular, the changes in the flattening of the boundaries induced by different mantle tomography models. They found that the CMB flattening corresponds to an increase, relative to hydrostatic equilibrium, of a few hundred meters to a couple of kilometers in the difference between the equatorial and polar

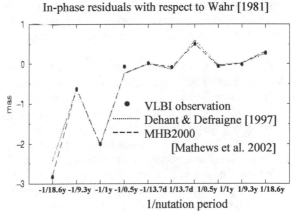

In-phase residuals with respect to Wahr [1981]

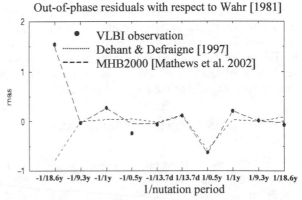

Out-of-phase residuals with respect to Wahr [1981]

Figure 1. Nutation amplitudes from VLBI observations and from different theoretical model, presented as corrections to the IAU 1980 adopted model of Wahr [1981].

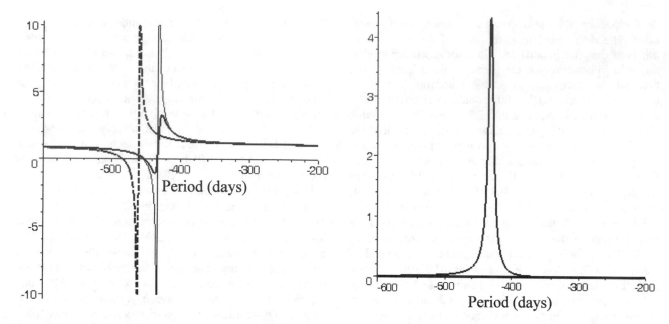

Figure 2. Transfer functions (no unit) for different hypotheses concerning the core flattening and concerning the coupling at the core interfaces, as a function of the nutation period in an inertial frame (in days). The dashed line corresponds to the hydrostatic equilibrium case, the thin solid line, to the non-hydrostatic CMB (increase of the equatorial radius of about 500 m) without electromagnetic coupling, and the thick solid line, to the non-hydrostatic CMB (increase of the equatorial radius of about 500 m) with electromagnetic coupling.

radii. The non-hydrostatic flattening of the ICB is much smaller than that of the CMB [*Defraigne*, 1995], because the density anomalies in the inner core are smaller than those in the mantle, and also because the response of the Earth to loading inside the inner core decreases rapidly as the point at which the loading takes place moves closer to the Earth's center. In any case, a small change in the ICB flattening has only a negligible effect on nutations. *Defraigne* [1997] built a model corresponding to an increase of the core equatorial radius of about 500 meters and a global Earth dynamical flattening corresponding to the observed value, as inferred from nutations. One could then obtain from mass heterogeneities given by tomography, a model of boundary displacements; the geoid associated with this model is in agreement with the observed one. This is an example of a constraint from VLBI nutation data on the model of the Earth's interior.

CONSTRAINT TO THE POINCARE MOTION

The angular momentum budget equations are used for computing the nutations, considering coupling at the interfaces of the outer core with the inner core and the mantle. In this kind of formulation, only the global rotations of the inner core, the outer core, and the mantle are considered and solved for. In the liquid core, in particular, this is accomplished by choosing the angular velocity

vector of the "rigid" global rotation to be such that the angular momentum of the global rotation equals the total angular momentum of the flow within the core exactly. This is the so-called Poincaré motion, where the flow is related to the pressure and the gravitational force by a simple equation relating the total acceleration, in the mantle reference frame, arising from the gradient of the pressure, and of the centrifugal and gravitational potentials, to the angular acceleration. This acceleration can then in turn be expressed in terms of a gradient of a potential field directly proportional to the global differential rotation of the core with respect to the global rotation of the mantle.

Another method to compute nutation is based on numerical integration within the Earth as mentioned earlier. The latter approach involves computation of the velocity field or displacement field throughout the Earth, while, for the "wobble" approach, only the global rotations of the core regions and of the mantle are considered, apart from the incremental moment of inertia induced by the deformation. The results of the numerical integration, presented in Figure 3, show that the toroidal displacements are essentially global rotations. This is valid at the first order. In addition to these global rotations, which play the central role in the "wobble" approach, only the deformations of the boundaries and global mass redistributions are important, and they are accounted for in the wobble approach.

ELECTROMAGNETIC COUPLING IN MHB2000

The electromagnetic torques at the core boundaries are dealt with in detail in the paper of *Buffett et al.* [2001]. In this section, we will review the way they are incorporated in MHB2000 [*Mathews et al.*, 2001], and report on the information obtained about the core through a fit to the VLBI data [*Herring et al.*, 2001].

First of all, we present the equations used to solve the nutation problem in a semi-analytical approach; these are the angular momentum budget equations for the whole Earth, the core, and the inner core. In a frame tied to the Earth's mantle, they can be written:

$$A\dot{\omega} - i\Omega\alpha A\omega + \Omega\dot{c} + i\Omega^2 c + A^f \dot{\omega}_f + i\Omega A^f \omega_f + A^s \dot{\omega}_s + i\Omega A^s \omega_s = \Gamma^{ext} \quad (3)$$

$$A^f \dot{\omega} + A^f \dot{\omega}_f - i\Omega\alpha_f A^f \omega + \Omega\dot{c}^f + i\Omega^2 c^f + i\Omega A^f \omega_f = \Gamma^f \quad (4)$$

$$A^s \dot{\omega} + A^s \dot{\omega}_s - i\Omega\alpha_s A^s \omega + \Omega\dot{c}^s + i\Omega^2 c^s + i\Omega A^s \omega_s = \Gamma^s \quad (5)$$

where Ω is the mean Earth rotation; ω, ω_f, ω_s represent the incremental rotation vectors of the mantle with respect to the mean Earth rotation, and of the fluid outer core and the solid inner core relative to the mantle, respectively (these quantities are usually called the Earth wobble or polar motion, the outer core wobble, and the inner core wobble, respectively); Γ^{ext} represents the external torque acting on the whole Earth; and Γ^f and Γ^s stand for the respective torques on the fluid core and the inner core, arising from external gravitational forcing on the respective regions, gravitational coupling with other regions, and pressure torques at the relevant boundaries. Each of these symbols stands for a complex combination of the first two (equatorial) components of the relevant vector, e.g., $\omega = \omega_1 + i\,\omega_2$. c, c^f, c^s are complex sums of components of the incremental moments of inertia of the whole Earth, the fluid outer core and the solid inner core. All the other quantities appearing in these equations have been defined earlier. Γ^f and Γ^s are sums of the torques on the respective regions due to external gravitational forcing and gravitational coupling with other regions, and the pressure torques at the relevant boundaries. The electromagnetic couplings at the CMB (Γ_{CMB}) and the ICB (Γ_{ICB}) can be accounted for by considering an additional contribution to the total torques Γ^f and Γ^s.

It can be shown (as summarized below) from theoretical magneto-hydrodynamic considerations that, under certain assumptions, the electromagnetic torques important for nutations and acting on the core are directly proportional to the equatorial components of the differential rotations of the fluid core with respect to the solid parts of the Earth. These differential rotations are ω_f for the core

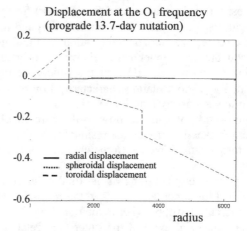

Displacement at the O_1 frequency (prograde 13.7-day nutation)

— radial displacement
...... spheroidal displacement
– – toroidal displacement

radius

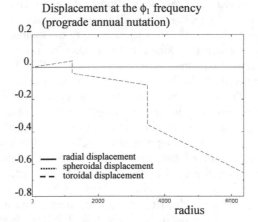

Displacement at the ϕ_1 frequency (prograde annual nutation)

— radial displacement
...... spheroidal displacement
– – toroidal displacement

radius

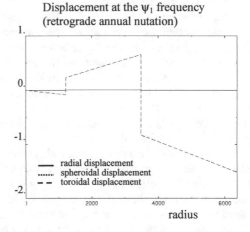

Displacement at the ψ_1 frequency (retrograde annual nutation)

— radial displacement
...... spheroidal displacement
– – toroidal displacement

radius

Figure 3. Transfer functions (no unit) for displacement field within the Earth (as a function of the radius in km) computed as a response of the Earth to a gravitational tidal forcing corresponding to the prograde 13.7-day nutation, to the prograde annual nutation, and to the retrograde annual nutation. The units are cm for a 'unit' tidal potential forcing corresponding to g_0 at the surface in cm^2/s^2 (the tidal potential is considered to be 978 cm^2/s^2).

with respect to the mantle, and to $(\omega_s - \omega_f)$ for the inner core with respect to the outer core, where ω_f and ω_s represent the respective incremental rotations of the fluid outer core and the solid inner core relative to the mantle, as mentioned before. The flow in the fluid core does not constitute a "rigid" rotation characterized by ω_f, but ω_f is chosen in such a way that the contribution from the residual "non-rigid" part of the flow to the core angular momentum is exactly zero; so the residual flow plays no role in the torque equation (see *Mathews et al.* [1991], Appendix B).

Electromagnetic coupling of the fluid outer core to the mantle and the inner core arises in the following manner. The differential wobble (i.e. a rotational motion about a moving equatorial axis) of the outer core relative to the mantle involves a relative motion of the conducting material on the two sides of the CMB, which causes a motion of the material on both sides relative to the magnetic field lines crossing the boundary. The electric current that is generated as a consequence on both sides of the boundary interacts with the magnetic field to produce a Lorentz force. The associated torque tends to oppose the relative wobble motion between the mantle and the outer core. This is what is referred to as the electromagnetic coupling at the CMB. The coupling at the ICB arises in a similar manner. Both of the torques are oscillatory, having a nearly diurnal period equal to that of the tidal component that gives rise to the wobbles.

The induced electric currents of nearly diurnal frequency, and the oscillatory magnetic fields generated by these currents, are confined to layers of a few hundred meters thickness on the two sides of the boundary, the actual thickness being determined by the conductivity of the layer. On the core side, the Lorentz force perturbs the flow of the fluid within this layer, while the flow in the rest of the core remains essentially a rigid rotation. The perturbation of the fluid flow, in turn, induces an additional magnetic field at the tidal frequency. The torque can be written in terms of this induced field and the main magnetic field. Therefore, in order to compute the torque, one must solve two coupled vectorial equations: the equation of fluid motion (the Navier-Stokes equation with the Lorentz and Coriolis forces included) and the induction equation relating the induced magnetic field to the perturbed flow velocity in the core relative to the solid side. The induced magnetic field and the perturbation of the fluid velocity are the two unknowns to be solved for from the coupled equations. For details of the solution, see *Buffett et al.* [2001].

The following justifiable assumptions are appropriate for solving this system: (1) the radial derivative of the induced magnetic field in the layer considered is large compared to the tangential derivatives which are deemed negligible; (2) the radial variation of the main field over this thin layer is negligible; (3) the Lorentz force is primarily balanced by the pressure gradient, the Coriolis force, and the centrifugal force (this is the classical magnetogeostrophic approximation).

The electromagnetic torques at the CMB and the ICB can be deduced from the solution of the vectorial equations (with the mentioned simplifications) and are written in the form

$$\Gamma_{CMB} = i\widetilde{K}_{CMB}\Omega A^f \omega_f \qquad (6)$$

where

$$\widetilde{K}_{CMB} = K_{CMB} + iK'_{CMB} = \frac{\pi a_{CMB}^4}{\mu \Omega A^f}\sqrt{\frac{1}{2|\lambda|}}\sqrt{\frac{1}{\eta}} \times$$
$$\left[\begin{array}{c}\text{complex function of}\\ \eta, \langle B_{r\,dipole}^2(a_{CMB})\rangle \text{ and } \langle B_{r\,uniform}^2(a_{CMB})\rangle\end{array}\right] \qquad (7)$$

and

$$\Gamma_{ICB} = i\widetilde{K}_{ICB}\Omega A^s (\omega_s - \omega_f) \qquad (8)$$

where

$$\widetilde{K}_{ICB} = K_{ICB} + iK'_{ICB} = \frac{\pi a_{ICB}^4}{\mu \Omega A^s}\sqrt{\frac{1}{2|\lambda|}}\sqrt{\frac{1}{\eta}} \times$$
$$\left[\begin{array}{c}\text{complex function of}\\ \eta, \langle B_{r\,dipole}^2(a_{ICB})\rangle \text{ and } \langle B_{r\,uniform}^2(a_{ICB})\rangle\end{array}\right] \qquad (9).$$

Here a_{CMB} and a_{ICB} are the outer core and inner core mean radii, μ is the magnetic permeability, and λ is the frequency of the tidal forcing. Note that the torques are proportional to the differential rotations. The coupling constants appearing in these equations are complex ($\widetilde{K}_{CMB} = K_{CMB} + i K'_{CMB}$, for the core-mantle coupling, and $\widetilde{K}_{ICB} = K_{ICB} + i K'_{ICB}$, for the inner core-outer core coupling), leading to a phase lag between the torque and the response. They are inversely proportional to the square root of the magnetic diffusivity η (itself inversely proportional to the electrical conductivity) near the boundaries; they are also proportional to $<B_r^2>$, the mean square, over the boundary, of the radial component of the magnetic field, if the field is weak. Expressions for the coupling constants may be obtained by solving the pair of coupled equations referred to above.

The main magnetic fields at the two boundaries have been modeled by *Buffett et al.* [2001] as a composite of a dipole field ($B_{r\ dipole}$) and a "uniform" field ($B_{r\ uniform}$) or more precisely, one that has a constant value for its radial component all over the boundary in question. (The latter is meant to represent the effective contributions from all the shorter wavelength parts of the magnetic field.) A mix of the two types of field configurations was found to be essential in order to simultaneously account for the values estimated from the fit for both the real and the imaginary parts of either coupling constant, in terms of the expression derived from electromagnetic coupling theory.

If the field at the CMB were weak, the coupling constant for this boundary would have the relatively simple form

$$\widetilde{K}_{CMB} = K_{CMB} + iK'_{CMB} = (1-i)\frac{\pi a_{CMB}^4}{\mu \Omega A^f}\sqrt{\frac{1}{2|\lambda|}}\sqrt{\frac{1}{\eta}} \times$$

$$\left[\frac{16}{5}\langle B_{r\,dipole}^2(a_{CMB})\rangle + \frac{8}{3}\langle B_{r\,uniform}^2(a_{CMB})\rangle\right] \quad (10),$$

which leads to equal magnitudes for the real and imaginary parts.

Actually, the weak field approximation is not good enough at the CMB, as seen for instance from the inequality of the magnitudes of the imaginary and real parts of the coupling constant derived from VLBI data, and is quite inapplicable for the ICB where the field as inferred by *Mathews et al.* [2001] is very strong. The solution of the coupled equations (the equation for the perturbation in the fluid flow velocity and the induction equation) then becomes complicated, but the coupling constant can still be expressed in terms of parameters pertaining to the inner core as presented before.

For the purpose of the theoretical computations of the electromagnetic coupling constants [*Buffett et al.*, 2001] the material in a conducting layer of thickness 210 m at the bottom of the mantle is taken to have the conductivity of pure iron ($5 \cdot 10^5$ Sm^{-1}), the same as for the fluid core and the inner core. The conductance of the mantle layer is then about 10^8 S, which is the same as the conductance needed for explaining decadal variations in length-of-day in terms of CMB electromagnetic coupling as done by *Holme* [1998] (see also this volume, *Ponsar et al.* [2001]). With this value for the conductance, the *Mathews et al.* [2001] estimate for the CMB coupling constant, $2.24 \times 10^{-5} - i\,1.85 \times 10^{-5}$, calls for an RMS (root mean square) radial field of 6.9 gauss (or 0.69 mT) at the CMB, assuming the dipole part to be 2.64 gauss (or 0.26 mT) as estimated from magnetic satellite observations [*Langel and Estes*, 1982]. A lower conductance would necessitate an even stronger magnetic field. On the other hand, laboratory experiments on diffusion of molten iron into silicates [*Poirier and Peyronneau*, 1992] do not favour a conductivity at

the CMB as large as that indicated above, which requires then highly conducting material embedded within the lower-most mantle as proposed by *Buffett et al.* [2001].

The presence of the magnetic field at the CMB is expected to strengthen the coupling between the mantle and the core; it provides a way of 'rigidifying' the core-mantle set. Since increased flattening also increases the (inertial) coupling between the mantle and the core, the magnetic field acts in the same way as the flattening of the core boundaries i.e. its presence decreases the FCN period in space, and consequently, moves its period further away from the one-sidereal-day cycle in the terrestrial frame (a nutation period in space corresponds to a quasi-diurnal retrograde period in a frame tied to the Earth). The FICN is prograde because its frequency in space, namely $\lambda_{FICN} + \Omega$ is positive as may be seen from equation (2). Indeed, the sign of the main term $\Omega\dfrac{\rho_f}{\rho_s}\alpha_s$ is positive; this term is due to the inner core–outer core coupling involving the fluid pressure acting on the ICB of which the normal is in the opposite direction with respect to the CMB (the equivalent term for the FCN is $-\Omega\dfrac{A}{A^m}\alpha_f$).

There is also an influence of the density jump at the ICB on the FICN period; the physical insight related to this is that the mode is related to a gravitational coupling involving the density of the liquid core ρ_f replaced by the density of the inner core ρ_s due to the inner core tilt, $\rho_f - \rho_s = -\Delta\rho$, which is negative. See *Dehant et al.* [1993] for further details. For the FICN, the effect of the magnetic coupling is in the opposite direction than for the FCN, as may be seen from *Mathews et al.* [2001]; the magnetic coupling makes the period closer to the retrograde one-sidereal-day period in the terrestrial frame. Thus, there is a decrease of the FCN period and an increase of the FICN period in an inertial reference frame, and a larger imaginary part for both (because of the imaginary part of the electromagnetic coupling constants). The out-of-phase contributions from the imaginary parts of the FCN and the FICN frequencies are important for the fit to the retrograde annual, prograde semi-annual, and prograde and retrograde 18.6 year nutations. Intuitively, a shorter FCN period in the inertial frame, bringing it closer to the annual period, increases the influence of the resonance on the annual nutation; similarly, a larger FICN period in space, which is closer to the 18.6 year period, would affect especially the VLBI fit to the 18.6 year nutations. It will also decrease the influence of the FICN on the prograde annual nutation. The importance of the electromagnetic couplings then becomes evident. The difference between the FICN period needed by the data fit and the hydrostatic value is very large, which indicates that a large ICB magnetic field will be deduced from the fitting. But, because the magnetic constant enters into the least squared fit in a combination with other parameters, one needs to consider possible deviations of the core density, the core boundary flattenings, inner core viscosity, etc., from their default values while assessing the strength of the ICB magnetic field, as will be discussed below.

After optimizing the geophysical parameters (such as the rheology properties of the mantle, the electromagnetic coupling constants, the core flattenings, and the global Earth dynamical flattening) for best fit to the amplitudes and phases of the nutations, we can interpret the values obtained for these parameters in terms of the physics of the Earth's interior and, in particular, get information about the core. We will not discuss the results concerning parameters not related to the core; their values can be found in *Mathews et al.* [2001].

From the geophysical parameters so-obtained, the FCN and FICN frequencies and resonance strengths can be computed:

$$\widetilde{\lambda}_{FCN} = -\Omega\left(1 + \frac{A}{A^m}(\alpha_f - \frac{q_0}{2}\overline{h}_1^f) + \text{termsin}\frac{A^s}{A^f} + \widetilde{K}_{CMB}\right) \quad (11)$$

$$\widetilde{\lambda}_{FICN} = -\Omega\left(\begin{array}{c}1 - (\frac{\rho_f}{\rho_s}(\alpha_s - \frac{q_0}{2}\overline{h}_2^s) - \frac{\Delta\rho}{\rho_s}(\alpha_s\gamma_s - \frac{q_0}{2}\xi_s) \\ + \text{termsin}\frac{A^s}{A^f} - \widetilde{K}_{ICB}\end{array}\right) \quad (12)$$

$$f(\lambda) = R + R'\lambda + \frac{R_{CW}}{\lambda - \widetilde{\lambda}_{CW}} + \frac{R_{FCN}}{\lambda - \widetilde{\lambda}_{FCN}} + \frac{R_{FICN}}{\lambda - \widetilde{\lambda}_{FICN}} \quad (13).$$

In the expression (13) of the nutation transfer function $f(\lambda)$ there are different physical parameters which may themselves be combinations of other physical parameters. This is for instance the case for the real part of the FCN period: it appears not only in the denominator, but also in the numerator of one of the terms of the resonance formula in the transfer function (in R_{FCN}), and it is itself a linear combination of the flattening of the core, a Love number expressing the deformation of the CMB boundary due to the pressure (called 'compliance' hereafter), and the real part of the electromagnetic coupling constant (see equation (11)); the Earth's inner core being small, the contributions in A^s are small. The nutation amplitudes deduced from VLBI data are sensitive to the resonance but are not able to separate the different parameters involved in the frequency of the resonance (except if these parameters appear in another combination resolved in the data). The imaginary parts of the electromagnetic constants K'_{CMB} and K'_{ICB} appear directly as parameters in the fit, thus enabling us to interpret the estimates obtained for them in terms of the magnetic field strength at the core boundaries. For the CMB, the fit gives $K'_{CMB} = -1.85 \ 10^{-5}$.

The radial magnetic field at the CMB is known from downward extrapolation of the field observed by a dedicated satellite; the RMS value of the radial component of the dipole part of the field over the CMB is found to be 2.64 gauss (or 0.264 mT). On requiring that the theoretical value for K'_{CMB} be equal to the above estimate when the magnetic field has a dipole part agreeing with the RMS 2.64 gauss (or 0.26 mT), it turns out that the uniform field has to be 6.41 gauss (or 0.64 mT) (leading to a combined RMS value of 6.93 gauss (or 0.69 mT)), and that $K_{CMB} = 2.24. \ 10^{-5}$. For the ICB, the fit gives $K'_{ICB} = -77.5 \ 10^{-5}$ and $K_{ICB} = 111.0 \ 10^{-5}$ (ignoring for the present any deviation from the assumed value for other parameters entering in combination with K_{ICB}). The RMS value needed for the radial component of the magnetic field in order to account for both of the above estimates simultaneously is 71.7 gauss (or 0.72 mT), with the RMS of the dipole part being 76% of the uniform part.

For the real part of the electromagnetic constants K_{CMB} and K_{ICB}, we have to take account of the fact that they appear in combination with other parameters: the relevant ellipticities and compliances, and in the case of the ICB, also the density jump $\Delta\rho$ across the ICB. Compliances are non-dimensional parameters characterizing the deformability of the relevant regions (e.g., outer core, inner core, or whole Earth) as manifested through the magnitudes of off-diagonal elements of the inertia tensor produced by forcing such as by tidal and centrifugal potentials (the tidal force is derived from a tide generating potential). At the CMB, the combination is $(\alpha_f + K_{CMB} - compliance)$; it is estimated from the nutation fit. The compliance here characterizes the deformability of the outer core boundaries in response to the centrifugal force acting on the outer core alone, and is obtained accurately enough from the Earth model. K_{CMB} is already found as described in the last paragraph. The value of the flattening α_f of the core then follows from the above. It is found to be about 0.002646, in the interval [0.0026426, 0.0026485]. Its difference from the hydrostatic value may be viewed as a change in the difference between the equatorial radius of the core and the polar radius. The value obtained from the VLBI data fit corresponds to a mean excess of about 3.8%, i.e. about 390 meters excess for the difference between the equatorial and polar radii of the core, instead of the 5% (500 meters) obtained if no electromagnetic torque is considered. The excess flattening found is in agreement with CMB topography determined recently from seismic data [*Sze et al.*, 2001].

INNER CORE VISCOSITY EFFECTS

The estimate quoted in the last section for the strength of the ICB magnetic field is well above what is obtained by dynamo simulations (about a factor two). Is there a

Table 1. Contributions to nutations (in µas) of the electromagnetic coupling and the effects of different viscosities of the inner core. The results from two papers are considered in that table: MHB refers to the electromagnetic coupling at the different boundaries of the core of *Mathews et al.* [2001]; 'Gr' refers to *Greff et al.* [2000 and 2001]. 'elast + coupling' refers to the elastic inner core case (high viscosity) with electromagnetic coupling, 'viscous + coupling' refers to a very viscous inner core case (low viscosity) with electromagnetic coupling, and 'range' to the range of all the values obtained for the possible inner core viscosity (from 10^{12} Pa s to 10^{17} Pa s) and for an electromagnetic coupling going from 0 to high coupling at the ICB (high magnetic field of 60 gauss).

	MHB at CMB	MHB at ICB	Total MHB	Gr elast + coupling	Gr range	Gr viscous + coupling
retro 18.6y in-ph	-275	-100	-375	-10	(-800,100)	-800
retro 18.6y out-of-ph	270	280	550	60	(-60,60)	-60
pro 18.6y in-ph	30	10	40	10	(-15,110)	110
pro 18.6y out-of-ph	-30	-50	-80	500	(-60,500)	-60
retro 1y in-ph	-395	-15	-410	-60	(-150,10)	-150
retro 1y out-of-ph	440	15	455	-40	(-50,40)	-50

possibility of bridging the gap? One response to that question was thought to be found in the failures related to the numerical simulations of geodynamo. Indeed, in a review paper, *Dormy et al.* [2000] have drawn attention to the large differences between the values of parameters used in the simulations and those estimated for the Earth and to failures of geodynamo models to actually reproduce a number of fundamental characteristic features.

In considering this question, we begin with the fact that for the ICB, it is not the K_{ICB} by itself but the combination $(\alpha_s \alpha_2 - K_{ICB} + compliance)$ that is really estimated (where α_s is the flattening of the inner core, α_2 depends on a parameter measuring the strength of the gravitational torque between the tilted inner core and the rest of the Earth (γ_s in equation (2)) and on a parameter directly related to the density contrast $\Delta\rho$ between the inner core and the fluid outer core at the CMB (see *Mathews et al.* [1991]). If the inner core is viscous rather than strictly elastic, the value of the compliance becomes higher. This point is initially ignored in the work of *Mathews et al.* [2001], except for accounting for small inelasticity effects. The parameter α_s was also taken as the hydrostatic value, and the density jump $\Delta\rho$ at the ICB was taken from seismic data [*Dziewonski and Anderson*, 1981]. Since the above mentioned combination involves the difference between the compliance and the electromagnetic coupling constant, and since the compliance is higher for a viscous inner core, the estimates for K_{ICB} and K'_{ICB} in the viscous case will have to be higher than the values stated in the previous section. In other words the ICB magnetic field would need to be higher if the inner core were considered to be viscous.

Greff et al. [2000] have studied theoretically the effect of the inner core viscosity on the normal modes (FCN and FICN). In particular, they have studied how those effects would affect the amplitudes of the main nutations, considering various viscosity values and a range of inner core electromagnetic coupling constants. The viscosity effects on the modes and on the nutations can be quite large

(even at the level of a few hundreds of µas). For example, they have drawn, for each important nutation component (18.6 year, annual, semi-annual) graphs of in-phase and out-of-phase amplitudes for a range of inner core viscosities (from the fluid behavior to the elastic behavior) and for a range of electromagnetic coupling constants at the ICB (strong and weak couplings). From the comparison of these theoretical results with the fitted value of *Mathews et al.* [2001], some new insights can be obtained, such as the danger of using the weak field approximation in the computation of the electromagnetic torque.

INNER CORE VISCOSITY AND ELECTROMAGNETIC COUPLING; RESULTS AND DISCUSSION

We have compared the results of *Mathews et al.* [2001] and those of *Greff et al.* [2001]. Table 1 presents the partial contributions of the CMB and the ICB electromagnetic couplings to three main nutations, as well as the total electromagnetic coupling effects of MHB2000. The table presents also the ranges of the nutation contributions obtained by *Greff et al.* [2001] for different viscosities of the inner core, ranging from 10^{12} to 10^{17} Pa s, and different electromagnetic field strengths, ranging from 2 gauss (the same value as their CMB field) to a maximum of 60 Gauss at the ICB. The table gives also the contributions for a highly viscous inner core and an elastic inner core, for the highest electromagnetic coupling considered in *Greff et al.* [2001]. The results of these authors for an elastic inner core without viscosity and with a high electromagnetic coupling at the ICB, can be compared with the contribution of the inner core electromagnetic torque to the three nutations as computed by *Mathews et al.* [2001]. The values do not correspond entirely, even after allowing for a difference in the sign convention on the out-of-phase components (as we did for Table 1).

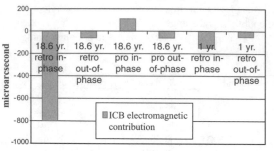

contributions of core electromagnetic coupling at ICB, by Greff et al. (2001)

Figure 4. Amplitudes of the contributions to the principal nutation from electromagnetic coupling at the ICB from electromagnetic coupling at the ICB from *Greff et al.* [2000 and 2001].

In the graphics presented here (Figures 4 and 5), we have reproduced the electromagnetic coupling contributions found by *Greff et al.* [2000 and 2001] from theoretical considerations with a relatively large electromagnetic torque (but not as large as the one of *Mathews et al.* [2001]), and the contributions found by *Mathews et al.* [2001] from the VLBI fitting. The figure shows the importance of these effects on nutations. We have presented here the nutations on which the effect is maximum and for a maximum coupling constant. The main differences between the results of *Greff et al.* [2000 and 2001] and *Mathews et al.* [2001] are believed to be related to the fact that *Greff et al.* do not take values for the coupling constant and for the magnetic field as high as those of *Mathews et al.* In *Greff et al.*, no additional uniform field is considered, and they do consider the weak field approximation.

The model presented in *Greff et al.* [2000 and 2001], built to measure the relative influence of the coupling parameters on the nutation, is certainly not competitive with MHB2000 as a model for practical use by astronomers. This model did not take advantage of VLBI estimations, unlike MHB2000, and is based on the approximation that the magnetic field is weak. However, the magnetic field values from geodynamo simulations [*Glatzmaier and Roberts*, 2000; *Kuang and Bloxham*, 2000]), which are less than half the value estimated from fits to VLBI data, are still strong enough that the weak-field approximation does not hold at the ICB. Nevertheless, *Dormy et al.* [2000] have shown the limitations of geodynamo numerical simulations; in particular they found that the simulations provide even too large fields when taking into account parameters that correspond to the "real" Earth; and so, the results from present-day geodynamo simulations, while very helpful to understand the phenomena (see *Kuang* [2001], this issue), must be taken with caution.

The model MHB2000 includes two main a priori considerations: (1) a highly conducting lower-most mantle and (2) a strong small-scale magnetic field. It is not obvious that both exist, but they allow to explain the observed nutations. As mentioned in *Jault* [2001] we are entering a satellite era that will allow to clarify that problem.

Another important remark is that the results of *Greff et al.* [2001] concerning the relative influence of the viscosity and the electromagnetic coupling show that the interpretation of VLBI data using strong electromagnetic coupling and ignoring viscous coupling might not be totally correct because of the possibility of a contribution from inner core viscosity effects. But it turns out, as we have noted above, that if the inner core is not strictly elastic but viscous, the ICB magnetic field called for would be even higher than the 72 gauss (or 0.72 mT) required otherwise. There is one possibility, however, to reduce the magnetic field strength needed: by increasing the density contrast $\Delta\rho$ across the ICB. If it could be increased by as much as 560 kg/m^3, the field strength required could be brought down to about 46 gauss (or 0.46 mT) if the inner core were elastic (see *Mathews et al.* [2001]). Such a high density-contrast seems very unlikely, since results derived from seismic travel times and normal mode studies lead independently to the same value for the density jump [*Shearer and Masters*, 1990]. Additionally, an even larger increase would be needed for a viscous inner core to get the same reduction in the field strength.

Residuals do remain after the fit on the VLBI data of the MHB2000 model. They amount to 22 μas on the out-of-phase of the retrograde 18.6 year nutation, 72 μas on the out-of-phase of the prograde 18.6 year nutation, -38 μas on the in-phase of the prograde 9.3 year nutation and -50 μas on the retrograde one. The only other residuals exceeding 20 μas are -28 μas in the prograde out-of-

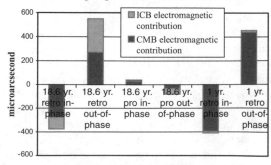

contributions of core electromagnetic coupling to nutations for MHB2000

Figure 5. Amplitudes of the contributions to the principal nutation from electromagnetic coupling at the CMB and the ICB from *Mathews et al.* [2001].

phase and retrograde in-phase 9.3-year nutations. From the work of *Greff et al.* [2001], it may be noted that a low viscosity inner core induces "amplification" in the 18.6-year prograde out-of-phase amplitude, which could provide the explanation for the large residuals at that nutation frequency. On the other hand, it is not demonstrated that considering such a viscosity would keep all the nutation residuals with respect to VLBI below the observation uncertainties. It might nevertheless be the case when considering the atmospheric effects on all the nutations. *Yseboodt et al.* [2001] have shown that the amplitudes and phases of the atmospheric corrections at the main nutation frequencies are at the level of magnitude of the inner core viscosity effects. In particular, for the retrograde annual nutation, a low viscosity would induce a high contribution (−150 μas) on the in-phase amplitude, which is possible to reconcile with VLBI data if the atmospheric contribution is considered.

CONCLUSIONS

In this paper, we have reviewed the information that one can obtain about the core from the modeling and observation of the Earth's nutation.

First, the nutation modeling based on numerical integration enables one to show, through internal loading and mantle convection computations based on mantle density heterogeneities, that the difference between the equatorial and polar radii of the core should be larger than in hydrostatic equilibrium. Moreover, the core flow associated with the response to a tidal potential has been shown to be mainly a global rotation around an axis in the equator (nutational notion).

The Earth has rotational normal modes such as the Free Core Nutation (FCN) and the Free Inner Core Nutation (FICN). These modes induce resonances in the nutational response of the non-rigid Earth to the tidal torque. The characteristics of the resonances are sensitive to the properties of the core. Consequently, it is possible to infer properties of the core from the resonances seen in the nutation. We have examined the effects of the electromagnetic coupling acting at the core boundaries, and of the viscosity of the inner core, on the nutations. We have discussed the values of the parameters used in the expression for the coupling, as estimated from a fit to high precision VLBI data. The viscosity of the inner core has been shown to have a significant effect, and may need to be considered in future models of nutation, though the higher magnetic fields needed with a viscous inner core may pose a problem.

It must be noted that the ocean and atmospheric effects on the nutation have been taken into account in MHB2000, as explained above, in the process of fitting to data. Any departure from the models used for these, or any problem with rigid-Earth nutation values, would introduce small changes in the other parameters determined by the fit. It is likely that these changes would not affect nutation amplitudes by more than a few tens of μas, except for the atmospheric effects on the nutations. So, apart from that and from the possibilities examined when considering the inner core viscosity, the model MHB2000 and the values determined for the Earth parameters can be considered to be very close to reality. *Dehant et al.* [2001] have made some tests in order to see how far the theory could be 'cheated'. They have examined how one particular nutation amplitude or phase and their uncertainties influence the determination of the parameters involved in the fit. Their conclusion is that there are not many possibilities to change the parameters determined in MHB2000, even if other atmospheric corrections are introduced, or if we consider departures from the International Celestial Reference Frame stability. It is thus possible to interpret the fit on VLBI data, with some confidence, in terms of physics of the Earth's core.

Dehant et al. [2001] have shown that, unless the retrograde annual, the prograde semi-annual, and the prograde and retrograde 18.6 year nutations are, for some unsuspected reason, completely out of the range determined by the MHB2000 model, the parameters obtained for the interior of the Earth should be very close to reality. For instance, changing their amplitudes by accounting for additional atmospheric effects such as those derived from atmospheric angular momentum series by *Yseboodt et al.* [2001], moves the FCN period as well as the FICN period determined by the fit; but leaving the periods unchanged is still a possible option in view of the large uncertainties in these corrections, considering the differences among the atmospheric global circulation models from which these corrections are evaluated. Unfortunately, our knowledge of atmospheric effects cannot be considered precise enough at present to be useful for the estimation of corrections to nutation data.

Knowing that the accuracy of the model may be nevertheless limited by the influences of external geophysical fluids (atmosphere and ocean) that may be not well corrected for, and knowing that the inner core viscosity could have a non-negligible influence, it is difficult to pinpoint a definitive value for the ICB electromagnetic coupling. However we cannot change much the values obtained for the parameters because, as explained above, these parameters are estimated jointly and the effects of changes in individual parameters on different nutations have quite different signatures.

To illustrate this again, let us mention that a reasonable change in the inner core viscosity does not change the out-of-phase annual retrograde nutation when a high electromagnetic coupling is considered, while it does for the out-of-phase 18.6-year retrograde nutation. For low coupling, the signs of the corrections on the out-of-phase 18.6-year retrograde nutation do depend very much on the inner core viscosity. These observations could be of relevance to the differences discussed above and seen on the graphics shown.

The theory behind MHB2000 is based on several assumptions, e.g., the existence of a thin and highly conducting layer at the base of the mantle. Subject to these assumptions, MHB2000 provides with a geophysics-based nutation model close to the observations. The values found for the electromagnetic coupling parameters are derived from high precision VLBI data, and provide the scientific community with constraints on the magnetic field at the two core boundaries. Such constraints, especially at the ICB, have not been available from other sources. The coupling mechanisms involved at the core boundaries are based on the gravitational, pressure and electromagnetic interactions between the inner core, the outer core, and the mantle (see *Jault* [2001], and *Ponsar et al.* [2001], this issue). The ellipsoidal shape of the core-mantle boundary is found to have a flattening well in excess of the hydrostatic equilibrium value. But the shorter scale structure of the topography is not considered in this computation. *Wu and Wahr* [1997] have shown that particular degrees of the topography could produce important contributions. This aspect of the topographic coupling could be a candidate when considering future improvements to nutation theory. In the review of *Jault* [2001], it is shown that taking into account uncertainties such as the height of the CMB topography or the electrical conductivity in the lower-most mantle, it is impossible to select one or more coupling mechanisms to reproduce the observed length-of-day variations. These coupling mechanisms are the pressure, the gravity and electromagnetic torques. The question remains thus open.

Acknowledgments. We are thankful to P. Defraigne, O. de Viron, M. Greff-Lefftz, T. Herring, H. Legros, S. Ponsar and T. Van Hoolst, for reading our manuscript and for helpful comments.

REFERENCES

Bretagnon, P., Francou G., Rocher P., and Simon J.-L., SMART97: a new solution for the rotation of the rigid Earth, *Astron. Astrophys.*, *329, 329-338, 1998.*

Buffett, B.A., Mathews P.M., and Herring T.A., Modeling of nutation-precession: effects of electromagnetic coupling, *J. Geophys. Res.*, in press, *2001.*

Chao, B.F., Ray R.D., Gipson J.M., Egbert G.D., and Ma C., Diurnal/semidiurnal polar motion excited by oceanic tidal angular momentum, *J. Geophys. Res.*, *101, 20,151-20,163, 1996.*

Defraigne, P., *Modèle de la convection actuelle dans le manteau terrestre*, Ph.D. Thesis, Université Catholique de Louvain, 243 p., *1995.*

Defraigne, P., Geophysical model of the Earth dynamical flattening in agreement with the precession constant, *Geophys. J. Int.*, *130, 47-56, 1997.*

Defraigne, P., Dehant V., and Wahr J.M., Internal loading of a homogeneous compressible Earth with phase boundaries, *Geophys. J. Int.*, *125, 173-192, 1996.*

Dehant, V., Dehant V., Hinderer J., Legros H., and Lefftz M., Analytical approach to the computation of the Earth, the outer core and the inner core rotational motions, *Phys. Earth planet. Inter.*, *76, 259-282, 1993.*

Dehant, V., and Defraigne P., New transfer functions for nutations of a non-rigid Earth, *J. Geophys. Res.*, *102, 27,659-27,688, 1997.*

Dehant, V., et al., Considerations concerning the non-rigid Earth nutation theory, *Celest. Mech. Dynamical Astron.*, *72 (4), 245-310, 1999.*

Dehant, V., Feissel M., de Viron O., Ma C., and Yseboodt M., Nutation at the sub-milliarcsecond level, *J. Geophys. Res.*, in preparation, *2001.*

Dormy, E., Valet J-P., and Courtillot V., Numerical models of the geodynamo and observational constraints, *Geochemistry Geophysics Geodynamics*, *1, 2000GC000062, 2000.*

Dziewonski, A.M., and Anderson D.L., Preliminary reference Earth model, *Phys. Earth planet. Inter.*, *25, 297-356, 1981.*

Getino, J., and Ferrandiz J.M., Advances in the unified theory of the rotation of the nonrigid Earth, in Proceedings of the IAU Colloquium 180 on `Towards Models and Constants for Sub-Microarcsecond Astrometry', eds. K. Johnston, D. McCarthy, B. Luzum and G. Kaplan, Washington, USA, pp. 236-241, *2000.*

Glatzmaier, G., and Roberts P., Rotation and magnetism of Earth's inner core, *Science*, *274, 1887-1891, 1996.*

Gwinn, C.R., Herring, T.A., and Shapiro I.I., Geodesy by radio interferometry: studies of the forced nutations of the Earth. 2. Interpretation, *J. Geophys. Res.*, *91, B5, 4755-4765, 1986.*

Greff-Lefftz, M., Legros H., and Dehant V., Influence of the inner core viscosity on the rotational eigenmodes of the Earth, *Phys. Earth planet. Inter.*, *122, 187-203, 2000.*

Greff-Lefftz, M., Dehant V., and Legros H., Effects of inner core viscosity on the gravity changes and spatial nutations induced by luni-solar tides, *Phys. Earth planet. Inter.*, in press, *2001.*

Herring, T.A., Gwinn C.R., and Shapiro I.I., Geodesy by radio interferometry: studies of the forced nutations of the Earth. 1. Data analysis, *J. Geophys. Res.*, *91, B5, 4745-4754, 1986.*

Herring, T.A., Mathews P.M., and Buffett B., Modeling of precession-nutation: Very Long Baseline Interferometry results, *J. Geophys. Res.*, in press, *2001.*

Holme, R., Electromagnetic Core-Mantle Couping. 1. Explaining decadal changes in the length of day, *Geophys. J. Int.*, *132, 167-180, 1998.*

Huang, C.L., Jin W.J., and Liao X.H., A new nutation model of non-rigid Earth with ocean and atmosphere, *Geophys. J. Int.*, *146 (1), 126-133, 2000.*

Huang, C.L., The scalar boundary conditions for the motion of the elastic Earth to second order in ellipticity, *Earth Moon Planets*, *84, 125-141, 2001.*

Jault, D., Electromagnetic and topographic coupling, and LOD variations, in 'Earth's core and lower mantle', eds. C.A.

Jones, A.M. Soward, and K. Zhang, *The Fluid Mechanics for Astrophysics and Geophysics Series*, eds. A. Soward and M. Ghil, in press, *2001*.

Kuang, W., and Bloxham J., On the dynamics of topographical core-mantle coupling, *Phys. Earth planet. Inter.*, *99, 289-294, 1999*.

Langel, R.A., and Estes R.H., A geomagnetic field spectrum, *Geophys. Res. Lett.*, *9, 250-253, 1982*.

Ma, C., Arias E.F., Eubanks T.M., Fey A.L., Gontier A.-M., Jacobs C.S., Sovers O.J., Archinal B.A., and Charlot P., The International Celestial Reference Frame as realized by Very Long Baseline Interferometry, *Astron. J.*, *116, 516-546, 1998*.

Mathews, P.M., Buffett B.A., Herring T.A. and Shapiro I.I., Forced nutations of the Earth: influence of inner core dynamics. 1. Theory, *J. Geophys. Res.*, *96, B5, 8219-8242, 1991*.

Mathews, P.M., Herring T.A., and Buffett B.A., Modeling of nutation-precession: new nutation series for nonrigid Earth, and insights into the Earth's interior, *J. Geophys. Res.*, in press, *2001*.

McCarthy, D.D., IERS Conventions, *IERS Technical Note*, *21*, 95 p., 1996.

Poirier, J.-P., and Peyronneau J., Experimental determination of the electrical conductivity of the material of the Earth's lower mantle, in '*High-Pressure Research: Applications to Earth and Planetary Sciences*', *AGU Geophysical Monograph Series* 67, eds. Y.Syono and Manghnani, pp. 77-87, 1992.

Ponsar, S., Dehant V., Holme R., Jault D., Pais A., and Van Hoolst T., The Core and the Earth's Rotation, '*Core dynamics, structure and rotation*', *AGU Monograph Series*, this volume, 2001.

Rogister, Y., *Etude théorique des déformations globales de la Terre*, Ph.D. Thesis, Université de Liège, in French, 215 p., 2000.

Roosbeek, F., and Dehant V., RDAN97: an analytical development of rigid Earth nutation series using the torque approach, *Celest. Mech. Dynamical Astron.*, *70, 215-253, 1998*.

Sasao, T., and Wahr J.M., An excitation mechanism for the free core nutation, *Geophys. J.R. astron. Soc.*, *64, 729-746, 1981*.

Schastok, J., A new nutation series for a more realistic model Earth, *Geophys. J. Int.*, *130, 137-150, 1997*.

Shearer, P., and Master G., The density and shear velocity contrast at the inner core boundary, *Geophys. J. Int.*, *102, 491-498, 1990*.

Shirai, T., and Fukushima T., Contribution of a new forced nutation theory of the non-rigid Earth, *Astron. J.*, *121, 3270-3283, 2001*.

Souchay, J., and Kinoshita H., Corrections and new developments in rigid Earth nutation theory: 1. Lunisolar influence including indirect planetary effects, *Astron. Astrophys.*, *312, 1017-1030, 1996*.

Souchay, J., and Kinoshita H., Corrections and new developments in rigid Earth nutation theory: 2. Influence of second-order geopotential and direct planetary effect, *Astron. Astrophys.*, *318, 639-652, 1997*.

Sze, E., and Van der Hilst R.D., Core mantle boundary topography from short period PcP, PKP, and PKKP data, *Phys. Earth planet. Inter.*, in press, *2001*.

Wahr, J.M., The forced nutations of an elliptical, rotating, elastic and oceanless Earth, *Geophys. J.R. astron. Soc.*, *64, 705-727, 1981*.

Wu, P., and Wahr J.M., Effects of non-hydrostatic core-mantle boundary topography and core dynamics on Earth rotation, *Geophys. J. Int.*, *128, 18-42, 1996*.

Yseboodt, M., de Viron O., Chin T.M., and Dehant V., Atmospheric excitation of the Earth nutation: Comparison of different atmospheric models, *J. Geophys. Res. (Solid Earth)*, in press, *2001*.